Manufacturing Facilities

Manufacturing Facilities

Location, Planning, and Design

D. R. SULE
Louisiana Tech University

PWS–KENT Publishing Company
Boston

PWS–KENT
Publishing Company

20 Park Plaza
Boston, Massachusetts 02116

THE PWS–KENT SERIES IN INDUSTRIAL ENGINEERING

Engineering Economy: Capital Allocation Theory, Fleischer

Manufacturing Facilities: Location, Planning, and Design, Sule

Probability and Statistics for Engineers, Second Edition, Scheaffer and McClave

Probability and Statistics for Modern Engineering, Lapin

PWS–KENT Publishing Company is a division of Wadsworth, Inc.

Library of Congress Cataloging-in-Publication Data

Sule, D. R. (Dileep R.)
  Manufacturing facilities.

  Includes index.
  1. Factories—Design and construction. 2. Factories—Location. I. Title.
TS177.S84 1988   725′.4   87-15680
ISBN 0-534-91971-5

Printed in the United States of America
91 92 — 10 9 8 7 6 5 4 3

Sponsoring Editor: Robert Prior
Production Coordinator: Robine Andrau
Production: Technical Texts, Inc./Jean T. Peck
Interior Design: Sylvia Dovner
Cover Design: Robine Andrau
Interior Illustration: Boyd & Fraser Publishing Company
Typesetting: University Graphics, Inc.
Cover Printing: New England Book Components, Inc.
Printing and Binding: Arcata Graphics/Halliday

Preface

Many graduates from industrial engineering and business schools start their careers in manufacturing facilities. Regardless of which facet of industry they enter, these graduates should be well versed in the planning and design involved in the overall development of a manufacturing plant. *Manufacturing Facilities: Location, Planning, and Design* is intended as a textbook for facility planning and design courses in industrial engineering and business curricula. It provides the information and analytical tools that are necessary to convert a product design into production plans and then describes in sufficient detail all of the planning techniques needed to build an efficient manufacturing facility that will make the production feasible. With access to computers now quite common, it places some emphasis on the use of computers in facility planning analysis.

The book is divided into two major parts: (1) facility location and (2) planning and design for manufacturing. Chapter 1 provides an introduction to the topics. Chapters 2, 3, and 4 are devoted to facility location analysis. Some heuristic methods are illustrated, thus making location analysis readily usable.

Chapter 5, on product development, shows the importance of market research and forecasting in estimating demand as well as the basic steps in converting a product design into manufacturing steps. Chapter 6 gives a brief introduction to manufacturing processes and topics in automation such as sensors, numerically controlled machines, and robots. Chapter 7 illustrates production charts and systems and shows how the volume of production, available initial capital, and desired flexibility determine the production system to use.

Chapter 8 describes how to select machines and labor for the project and how to write project specifications. Chapter 9 provides information on the building structure, organization, and other requirements necessary to develop a manufacturing plant. Chapters 10 and 11 discuss the principles and equipment in material handling and the methods used in equipment selection, in development of flow lines within a plant, and in packaging.

Chapter 12 concentrates on the workings of storage/warehouse facilities, including automated storage and retrieval systems. Chapters 13 and 14 focus on plant layouts. Chapter 13 describes a conventional method for obtaining a layout and suggests the ways of verifying and presenting it to a group. Chapter 14 discusses computerized methods for plant layout and evaluates their strong points and shortcomings.

Chapter 15 describes policymaking and financial considerations for in-plant services such as utilities, insurance, and safety, each of which could substantially influence production cost. The financial statements included in the chapter show how all the costs and incomes must be integrated. Chapter 16 describes computer-integrated manufacturing systems, including their components, functions, and vendors.

Manufacturing Facilities places emphasis on design by including a few problems requiring independent analysis. These problems provide opportunities for readers to challenge their creativity and analytical skills.

A disk listing the computer programs mentioned in the text is available from the publisher. The programs are written in BASIC for IBM-PC or compatible machines. The listing consists of a number of the more interesting problems in location analysis, assembly line balancing, forming groups for group technology, M/M/1 and M/M/C/K queuing models, and generating financial analysis reports. The programs are instrumental in providing an insight into the analytical methods presented in the text. The disk also lists the data guide describing the necessary formats for running some of the well-known programs in computer-aided plant layout such as CRAFT, CORELAP, ALDEP, and PLANET.

Depending on the available time and the backgrounds of their students, instructors may develop a facility planning and design course around different chapters. If desired, location theory (Chapters 2, 3, and 4) may be omitted; students who have had a course in manufacturing processes may skim through the first part of Chapter 6 concerning that subject; or an instructor in a technology program may omit Chapter 14 on computerized plant layout. Other chapters are developed in a logical sequence, and it is suggested that they be covered in that sequence; but again, instructors may apply their judgment in selecting the depth of coverage.

Acknowledgments

Many people have helped me tremendously during the preparation of this book. My special thanks to John Cochran for spending days in editing the manuscript. His constant advice was extremely valuable in the overall formulation of the book. Gene Marshall from AT&T provided a number of photographs useful in showing the major principles of facilities planning. Mohsen M. D. Hassan contributed the chapter on computer-aided plant layout (Chapter 14) and sections in the chapter on material handling (Chapter 10). Randall Barron and C. Ray Wimberly's encouragement and support are appreciated.

Special thanks also go to my students Donna Williams, Patrick Seamands, Bryan Dodson, and Reza Ziai for their contributions.

Typing of the manuscript was described by four student workers, Rhonda Yates, Patty Patrick, Alayne Roques, and Brenda Clawson as a "no-fun" job (especially at minimum wages). I thank them for their dedication and service beyond the call of duty.

And, of course, no book can properly achieve its goals without the evaluations and suggestions of extremely competent reviewers. I was fortunate to have as reviewers Kailash M. Bafna, Western Michigan University; John R. Canada, North Carolina State University; Thomas P. Cullinane, Northeastern University; James M. Moore, Virginia Polytechnic Institute; and Vinay Vasudev, Louisiana State University. To them I express my profound gratitude.

Finally, to my wife, Ulka, and children, Sangeeta and Sandeep (my wife wonders why family always comes last), my thanks for their understanding during this project.

Contents

CHAPTER

1

Introduction

Planning is vital for the efficient utilization of available resources. It is necessary throughout the economy, in both the public and private sectors, manufacturing and service organizations, business and education.

Manufacturing organizations are especially challenging, since they involve the performance of multiple activities; all must be planned, designed, coordinated, and executed by a team effort. Even though each member (designer, production worker, material handler) of the team may be an expert in his or her own area, each person's work must achieve the overall objective of the plant. Only an efficient and productive organization can survive in today's competitive market.

Because of this, managers and supervisors must have broad backgrounds in all phases of manufacturing. They should be able to communicate with each other and understand problems and their proposed solutions as well as their effects on the performance of the entire plant. In essence, managers should have an overview of all the functions that are involved in successfully designing and operating a plant.

In this book we will approach the subject of manufacturing planning in two phases. Part I covers basic location/allocation analysis, which determines the most economic location for a plant or plants within a country (or machines within a plant) and determines which customers should be served by which plant (or machine) if there is more than one manufacturing facility. Part II covers all phases of manufacturing within the plant such as designing a product, choosing manufacturing processes, developing production systems and associated plant layout, designing material-handling and storage systems, selecting the essential labor resources to operate the plant, and considering the cost factors that go into the design and operation of a plant.

1.1 NATURE AND SCOPE OF PART I

One problem area in which planning can improve efficiency and save considerable effort and resources is that of facility location and allocation. The basic questions here are how to choose from all available locations those in which to install the facilities and then how to assign customers to each facility so as to minimize the overall cost of operation. What a facility is and what a customer is are defined by the nature of the problem. For example, in determining suitable locations for industrial plants so that they can best serve demands from various regions in the country, the plants are the facilities and the product users are the customers. In determining the market territories to assign to sales personnel, the territories are the customers and the salespeople are the facilities.

Examples of facility location/allocation problems are numerous. Although the few methods that we will illustrate in this book might not address all the applications listed below, it is still useful to note the fields in which the problems of facility location may arise. They are:

- Selecting sites for emergency service facilities such as hospitals and fire stations.
- Determining the best locations for toolrooms, machines, water fountains, wash areas, concession areas, and first aid stations in a manufacturing plant.
- Choosing sites for warehouses and distribution centers.
- Selecting subcontractors and assigning the appropriate work to each.
- Picking vendors and determining items to purchase from each, with or without quantity discounts.
- Choosing sites for maintenance departments in plants or, on a larger scale, sites for state agencies such as a highway department or garage.

- Selecting sites and capacities of machines to meet expected demands from customers distributed throughout a given area.
- Developing a layout for a machine shop or an instrument panel.

In this book we will study only the basic models in facility location. Other methods and extensions are illustrated in the vast literature of industrial engineering and management science. Interested readers may refer to Francis and White (1974) for further bibliography.

1.2 NATURE AND SCOPE OF PART II

Achieving efficiency in operations does not, of course, end with the location analysis. Designing a manufacturing plant is a bold undertaking that necessitates many complex decisions, each of which significantly affects the profitability of the venture. In broad generalities, we must select a product, design it, produce it, and sell it. Besides the financial consideration of raising capital, which we will not discuss, numerous decisions must be made in each step that affect the time and cost of manufacturing the product. For example:

1. Improper design or layout of a plant can add considerably to the cost of manufacturing. It may increase the throughput time, setup times, and in-process inventories and may contribute to overall inefficiency of operations.

2. For repeated high-volume and tedious jobs, use of robots might reduce production cost, but the same robots could actually add to the cost if the facility must handle many different jobs, each with different characteristics and having a low volume of production.

3. Being overinsured or underinsured can add to costs and potential liabilities.

4. A careful design, appropriate processing methods, and a well-laid-out plant with well-trained personnel might produce a product that is functionally superior but still not competitive. This could be because the plant was located at a site far from its market, thus increasing the distribution cost; or because not enough attention was paid to long-term performance (reliability) of the product; or perhaps because the packaging was not attractive enough.

5. The choice of the material-handling system and equipment can define the efficiency of operation, and, once selected, the system is generally expensive to change.

6. The use of computers, numerically controlled machines, and robots can add to integration and automation in manufacturing, but it also adds to the initial (installation) cost.

7. Storage and warehouse operations are critical in maintaining smooth production and low inventory costs.

These are just some of the numerous problems that a decision maker must face. The purpose of Part II of the book is to provide the necessary background information so that such decisions are based on an understanding of the alternatives. Successful introduction of products or services in the market must follow consideration of some or all of the following:

- Market research and forecasting of demand.
- Designing and drawing of the product.
- Manufacturing processes and automation.
- Operations analysis and make-or-buy decisions.
- Machine and labor selection.
- Production processes.
- Building size and type of construction.
- Development of organization and determination of support personnel.

- Material-handling requirements.
- Development of material-handling systems.
- Selection of material-handling equipment.
- Packaging.
- Storage and warehousing operations.
- Dock design and development.
- Plant layout.
- Plant site selection.
- Utilities.
- Safety.
- Insurance and taxes.
- Computer integration in manufacturing.

We will discuss each of these topics in some detail in Chapters 5–16. Those chapters also present descriptions of computer programs on selected topics of interest for which the analysis could prove to be tedious.

In this book we also plan to show how a computer can be effectively used in facility planning analysis. It is, however, hoped that the reader will first become familiar with the illustrated techniques and then use a computer to solve large-scale problems and to do sensitivity analysis with the input/output parameters. Like any other prepackaged computer programs, the programs listed in Appendix D should not be regarded as instant providers of answers. Each one is built on certain assumptions and logic, and only by understanding the methods shown in various chapters can one appreciate the results presented by the programs.

The listings in Appendix D concern both location analysis and the planning and design segment of the book. The arrangement for the introduction to a program follows a general pattern within a given chapter. First, a specific procedure is explained; then a numerical example is solved. At the end of the chapter, input requirements for the computer program associated with the procedure are listed. The text of the program itself is displayed in Appendix D.

The programs are written in either BASIC

or FORTRAN (WATFIV) languages; both are readily available in almost any computer facility. The plant-layout program is, however, written in Better-BASIC. A disk containing all programs for an IBM-PC or compatible machine is available from the publisher and includes the compiled version of the plant-layout program.

SUGGESTED READINGS

Francis, R.L., and White, J.A., *Facility Layout and Location—An Analytical Approach,* Prentice-Hall, Englewood Cliffs, N.J., 1974.

Phillips, D.T.; Ravindran, A.; and Solberg, J.J., *Operations Research—Principles and Practice,* John Wiley and Sons, New York, 1976.

Zanakis, S.H., and Evans, J.R., "Heuristic Optimization: Why, When, and How to Use It," *Interfaces,* No. 5, Oct. 1981, pp. 84–86.

PART

I

Facility
Location

CHAPTER
2

Basic Facility Location Problems

Facility location problems deal with the question of where to place facilities and how to assign customers. We will initiate our analysis for choosing locations by discussing very basic models where there is only one type of cost to consider, that of transportation. We will assume that each facility is sufficiently large that it can serve all the customers if necessary, that is, that there is no limitation on a facility's capacity. Sample problems are presented in the following discussion, which is divided into two parts: the single-facility placement problem and the multiple-facility placement problem.

2.1 SINGLE-FACILITY PLACEMENT PROBLEM

Suppose that a new copying machine is to be installed in an office building complex. The machine can be set up in any one of five possible locations, and six offices will be assigned to use it. The demand from each office (that is, the expected number of trips made daily) and the cost (measured in terms of the time that a person spends in walking to and from each location) are given in Table 2.1.

The table shows that it takes three minutes for Customer A to go to and return from Location 1. We wish to determine where the machine should be installed to minimize the total cost, that is, the total time spent by all

Table 2.1   Time and demand

Customer (Office)	Location					Demand
	1	*2*	*3*	*4*	*5*	
A	3	5	4	1	6	15
B	5	2	4	4	2	20
C	3	6	3	4	5	10
D	5	3	8	5	6	12
E	3	9	5	2	8	10
F	9	10	7	9	2	7

users walking to and from the copying machine.

Solution Procedure

When there is only one facility to place, the solution procedure consists of three very basic steps. They are:

    1. Calculate the total transportation cost

Portions of this chapter have been adapted from the *Journal of Operations Management,* Vol. 1, No. 4, May 1981, with permission of the American Production and Inventory Control Society, Falls Church, VA.

matrix (also called the demand cost matrix). An element of the matrix is obtained by multiplying the cost of assigning a unit of demand, from a customer to a location, by the total demand from the customer. For example, the cost of assigning Customer A to Location 1 is $3 \times 15 = 45$ units of time.

2. Take the sum of each column. These values represent the total cost of assigning all customers to each location.

3. Select the location with the minimum total cost. This is the location in which the machine should be placed.

Application to the Sample Problem

The total transportation cost matrix for this problem, generated in Step 1, is given in Table 2.2. The last row, the total, is the sum of each column. Since the total cost for Location 4 is the least, this site is selected, and all the customers are assigned to this machine. The procedure is quick and requires a minimum of effort.

2.2 MULTIPLE-FACILITY PLACEMENT PROBLEM

Let us now turn to an example of a more complex type of problem. A manager of automobile

Table 2.2 Demand cost

| Customer | Location | | | | |
	1	*2*	*3*	*4*	*5*
A	45	75	60	15	90
B	100	40	80	80	40
C	30	60	30	40	50
D	60	36	96	60	72
E	30	90	50	20	80
F	63	70	49	63	14
Total	328	371	365	278	346

Table 2.3 Transportation time and demand data

| Customer | Time for Location | | | | | Demand |
	1	*2*	*3*	*4*	*5*	
A	5	3	2	8	5	100
B	3	5	2	6	7	50
C	5	2	0	1	0	150
D	2	1	8	2	3	200
E	3	2	4	0	4	300

service stations has sufficient funds to buy two identical diagnostic machines. He currently operates five service stations within the city, and he has identified five areas from which demands will occur. The transportation time from the center of gravity of each area cluster to each location is given in Table 2.3. Also shown is the demand from each customer (area) for such a machine. The requirement is to place the two new machines in locations that will minimize the total transportation time, since the manager considers this to be the measure of efficiency.

A unique feature of this type of problem is that since the capacity of each facility is unlimited, there is no need to divide a customer's demand and assign parts of it to the two different facilities. Indeed, a customer should be assigned to the facility for which the cost of transportation is at a minimum. The problem is then twofold. First, choose from the available locations those in which the facilities should be placed, and then assign the customers to these facilities so that the cost of these assignments is a minimum.

We will present two approaches to solving this problem: first the brute force approach and then a heuristic approach. As the name suggests, the brute force approach tries out all possibilities. However, even here some reduction in the required calculations is possible by using the basic idea that the customer should be assigned to a facility for which the assignment cost is a minimum.

2.3 BRUTE FORCE APPROACH

Suppose we wish to place K facilities in M available locations. In how many ways can this be done? A theorem in statistics gives the answer.

There are $\binom{M}{K}$ ways of selecting K distinct objects from a set of M objects where

$$\binom{M}{K} = \frac{M!}{K!(M-K)!}$$

In our problem, since there are five locations and two facilities to place, we have $\binom{M}{K}$ = 5!/2!3! = 10 possible ways of placing these facilities. If all combinations are examined and demand is assigned to each such that we obtain the minimum cost for the combination, then the overall minimum can be easily found. The following procedure achieves this objective.

First, calculate the total cost of assigning all the demand from a customer to a location. The procedure is similar to one illustrated in Section 2.1. In Table 2.4, for example, the cost of assigning Customer A to Location 3 would be the product of the cost of assigning a unit of demand from Customer A to Location 3 (2 units of time) multiplied by that customer's demand (100 units), that is, 200. All other like products are calculated in a similar fashion and entered in Table 2.4.

Table 2.4   Demand cost

Customer	\multicolumn{5}{c}{Location}

Customer	1	2	3	4	5
A	500	300	200	800	500
B	150	250	100	300	350
C	750	300	0	150	0
D	400	200	1600	400	600
E	900	600	1200	0	1200

Let us consider one possible combination, say, Locations 1 and 2. This means that one of the two available machines is placed in Location 1 and the other in Location 2. If that is the case, then each customer should be assigned to the location for which the cost of transportation is a minimum. Customer A is assigned to Location 2, Customer B to Location 1, Customer C to Location 2, Customer D to Location 2, and Customer E to Location 2. This solution has the cost of 300 + 150 + 300 + 200 + 600 = 1550. Similar analyses for the other ten combinations lead to the data in Table 2.5.

The minimum cost is that associated with the combination of Locations 3 and 4; hence these locations are selected as the best in which to place the facilities, and the customers are assigned in the manner that led to the minimum cost. Formally, then, the steps of the procedure are as follows:

1. Construct a demand cost table. An entry ij in the demand cost table represents the cost of allocating all the demand from Source (Customer) i to Location j.

2. Formulate all $\binom{M}{K}$ possible combinations.

3. In the first combination, determine the demand assignment by assigning demand to a facility where the cost of the assignment is the minimum. This is done in the following manner:

   a. Consider only the columns of the demand cost matrix associated with the particular combination.
   b. Assign each demand to the facility having the smallest demand cost entry in its row for those columns selected in Step a.

4. Calculate the total cost for the combination by taking the sum of the costs of the assignments in Step 3.

5. Repeat Steps 3 and 4 for all remaining combinations.

Table 2.5   Minimum cost

Customer	\multicolumn{10}{c}{Location Pairs (Combinations)}
	1 & 2	*1 & 3*	*1 & 4*	*1 & 5*	*2 & 3*	*2 & 4*	*2 & 5*	*3 & 4*	*3 & 5*	*4 & 5*
A	300	200	500	500	200	300	300	200	200	500
B	150	100	150	150	100	250	250	100	100	300
C	300	0	150	0	0	150	0	0	0	0
D	200	400	400	400	200	200	200	400	600	400
E	600	900	0	900	600	0	600	0	1200	0
Total	1550	1600	1200	1950	1100	900	1350	700	2100	1200

6. Select the combination with the overall minimum cost as the optimum combination; the associated demand assignments are the optimum assignments.

Computationally, the method is fast and simple, and relatively small problems can be quickly worked out by hand. For a given number of combinations the computer (CPU) time is linearly proportional to the number of customers. The time increases very rapidly, almost exponentially, with the number of locations that need to be examined, and therefore the method is not recommended for problems involving a large number of combinations. Table 2.6 shows that for $K > 1$ the number of possible combinations increases dramatically for M

$> K + 1$. The decision of whether to use the brute force method should take into consideration the number of combinations that must be examined. To solve larger problems, we recommend using the heuristic method presented in the next section.

2.4 HEURISTIC METHOD FOR PROBLEMS INVOLVING FACILITIES WITH UNLIMITED CAPACITIES

Most heuristic methods follow certain logical steps that generally lead to optimum or near-optimum results. For the problem at hand the heuristic method suggested in this section works well. The procedure is based on the following logic.

If we are to place the facilities one at a time, then we should place each in the location where the maximum benefit is derived from it. To begin, if we have only one facility, it should be located where the cost of assigning all customers to that location is the minimum. If additional facilities are available, they should be placed one at a time in positions that contribute to the maximum savings. Thus each subsequent facility that is placed in an additional location should successively improve upon each

Table 2.6   Possible combinations

M, Available Locations	\multicolumn{6}{c}{K, Facilities to Be Placed}
	1	2	3	4	5	6
1	1	NA	NA	NA	NA	NA
2	2	1	NA	NA	NA	NA
3	3	3	1	NA	NA	NA
4	4	6	4	1	NA	NA
5	5	10	10	5	1	NA
6	6	15	20	15	6	1

previous solution. If the solution does not improve, then we do not need the additional facility. Therefore not only does the method indicate where the facilities should be located and how the customers should be assigned, but it also points out the optimum number of facilities we should have. The procedure continues until either all the facilities are located or there is no further improvement in the solution.

As illustrated in the flowchart in Figure 2.1, the steps involved in the procedure are as follows:

1. Formulate the total cost table. An entry *ij* in the total cost table represents the cost of allocating all demand from Customer *i* to Location *j*.
2. Total each column. The sum represents the total cost if demands from all customers are assigned to that location.
3. Assign the first facility to the location with the minimum total cost. We will denote a location with a facility as an assigned location and denote one without a facility as an unassigned location.
4. If no additional facility is available for assignment, go to Step 8; otherwise continue.
5. Determine the savings to be derived by moving each customer from the assigned location(s) to a nonassigned location(s). The savings may occur because of a change in transportation cost, but if there are no savings, mark "—" in the appropriate column.
6. Total each unassigned column. The sum represents the savings that could be achieved if an assignment is made in that location.
7. Make an assignment in the location that indicates the maximum savings. Transfer the customers that contributed to these savings to the new location, which now becomes an assigned location. Return to Step 4.

Figure 2.1 Basic facility relocation problem flowchart for heuristic procedure in Section 2.4

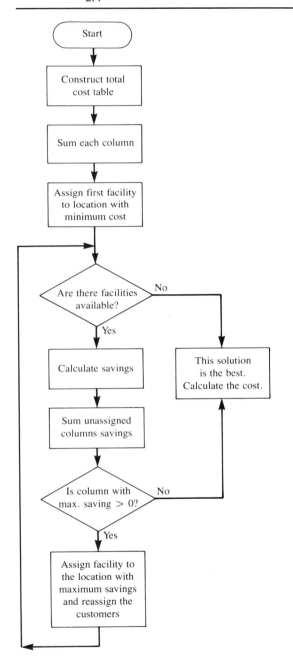

Table 2.7   Demand cost (time)

Customer	Location 1	2	3	4	5	Demand
A	500	300	200	800	500	100
B	150	250	100	300	350	50
C	750	300	0	150	0	150
D	400	200	1600	400	600	200
E	900	600	1200	0	1200	300
Total	2700	1650	3100	1650	2650	800

8. All the assignments are made. Calculate the minimum cost and schedule.

Application

To demonstrate the heuristic procedure, we will apply it to the auto shop manager's problem in Section 2.2. The cost data, in terms of time, are the same as in Table 2.3; however, suppose now that the manager has three machines to place. Table 2.4, a demand cost table, is reproduced for convenience as Table 2.7 and is expanded to include the customers' demands and the totals of the columns.

Since both columns 2 and 4 have the minimum total cost of 1650, the first machine could be assigned to either Location 2 or Location 4. We arbitrarily select Location 4. Now, to decide where to place the second machine, we must calculate the savings (if any) to be gained by moving customers that are currently assigned to Location 4 to any other unassigned Location j. For example, consider Customer A. If Customer A is moved from Location 4 to Location 1, it will save $800 - 500 = 300$. Similarly, for Location 2 the savings are $800 - 300 = 500$. Other savings are calculated in the same manner, and Table 2.8 provides the results. It should be noted that if moving to a different location results in negative savings, then that location is marked by "—". For example, there are negative savings when Customer B is moved from Location 4 to Location 5 ($300 - 350 = -50$), and accordingly, the "—" notation is entered in the appropriate position. As shown in Table 2.8, the savings are calculated for all the customers and for each of the unassigned locations.

The total of the savings for each location is obtained by adding the column entries. For example, the total of the savings for Location 1 is only $300 + 150 = 450$. The second machine should be placed in a location that has the maximum savings, in this case, Location 3. The customers that contributed to these savings, Customers A, B, and C, are now reassigned to Location 3. The others, Customers D and E, are retained at Location 4. The assignments at this stage are shown by circles in Table 2.8.

Table 2.8   Savings table I for moving from Location 4

Customer	Location 1	2	3	4	5
A	300	500	(600)		300
B	150	50	(200)		—
C	—	—	(150)		150
D	—	200	—	○	—
E	—	—	—	○	—
Total	450	750	950		450

Table 2.9   Savings table II for moving from Location 3 or 4

Customer	Location				
	1	2	3	4	5
A	—	—	○		—
B	—	—	○		—
C	—	—	○		0
D	0	(200)	—		
E	—	—	—	○	—
Total	—	200	—		—

To decide where to place the third machine, we must recalculate the savings to be realized by moving the customers from their currently assigned locations to the unassigned locations. Again, if the amount saved is negative, we will mark the unassigned location by "—". Customer A is presently attached to Location 3, and if we move this customer to Location 1, the savings are $200 - 500 = -300$. This -300 is a cost and is therefore marked by "—". Similarly, for Location 2 the savings are $100 - 150 = -50$, and for Location 5 they are $200 - 500 = -300$. For Customer D, currently assigned to Location 4, the corresponding calculations are as follows: for Location 1 the amount saved is $400 - 400 = 0$; for Location 2 the figure is $400 - 200 = 200$; and for Location 5 the savings are $400 - 600 = -200$, which is marked by "—". All of the results are shown in Table 2.9.

Since Location 2 has the only savings, a total of $200, we will place the third machine in this location. The customer contributing to these savings, Customer D, is now assigned to Location 2. All the assignments are shown by circles in Table 2.9.

Thus our final assignment is as follows:

Customer	A	B	C	D	E
Location	3	3	3	2	4
Cost	200	100	0	200	0

The total cost for this solution is 500. Incidentally, we would have reached the identical solution had we chosen Location 2 instead of Location 4 for assigning the first machine.

Now suppose that a fourth machine is available. Could we profitably use it? We decide by determining whether any additional savings are possible; the results of our calculations are displayed in Table 2.10.

The only unassigned locations are 1 and 5. We find no savings in either of these two locations, so there is no need for the fourth machine. Our heuristic approach has identified the optimum number of machines as three.

Note that Table 2.10 indicates that Customer C could be assigned to a fourth machine at Location 5 at no increase in transportation cost; but since there would likewise be no savings, any additional installation, operating, and maintenance costs for a fourth machine could never be recovered.

Validity

How good is the method? We have found by experimentation that this procedure provides the exact result about 90 percent of the time. The remaining 10 percent of the time, when it deviated from the theoretical minimum, it differed by an average of 5 percent of the optimum solution. (In one example it did deviate by as

Table 2.10   Savings table III for moving from Locations 2, 3, or 4

Customer	Location				
	1	2	3	4	5
A	—				—
B	—				—
C	—				0
D	—				—
E	—				—

much as 18 percent.) In most applications the data are estimated to start with, and the 5 percent variation is negligible. This is especially important when one observes the simplicity of the procedure and the speed with which it can be applied.

The input to the computer program for the method is outlined in Section 2.6; the program itself is given in Appendix D. To give some idea of how quickly the program can be run, a 30-customer, 30-location, 5-machine problem took 4.57 seconds on an IBM 370/165 computer; and a 50-customer, 50-location, 6-machine problem took 16.3 seconds.

2.5 OTHER METHODS

We will briefly review the mathematical formulation of the model for readers who are interested in the exact mathematical solution. Let us define:

X_{ij} = 1 if the demand from Source i is assigned to Location j

   = 0 otherwise

C_{ij} = cost of assigning a unit of demand from Source i to Location j where i = 1, 2, . . . , n and j = 1, 2, . . . , m

d_i = demand from Source i

K = number of facilities available for placement

Then we wish to minimize TC (total cost), where

$$TC = \sum_{j=1}^{m} \sum_{i=1}^{n} C_{ij} \times d_i \times X_{ij}$$

subject to the constraints that are developed as follows.

Since the demand for each source must be assigned to at least one location, we have

$$\sum_{j=1}^{m} X_{ij} > 1 \qquad \text{for each } i$$

Similarly, since only K locations are to be selected and only X_{ij} variables assigning demands to these locations can take positive values, we have

$$\sum_{i=1}^{n} X_{ij} < nI_j \qquad \text{for each } j$$

$$\sum_{j=1}^{m} I_j = K$$

I_j = 1 if the facility is assigned to Location j

   = 0 if the facility is not assigned to Location j

The above formulations can be solved by integer programming.

The other methods suggested are branch and bound, implicit enumerations, and decomposition. We will not dwell on any of these methods, since the purpose of this chapter is to present a method that requires hardly more than basic algebraic skills.

2.6 COMPUTER PROGRAM DESCRIPTION

Heuristic Method

1. *Description of parameters*

   IROW = Number of customers
   ICOL = Number of possible locations
   IMC = Number of facilities to be located
   COST(I,J) = Total cost of assigning Customer i to Location j (that is, product or demand from Customer i and unit cost of assigning Customer i to Location j)

2. *Input lines*

   Line type 1:   IROW, ICOL, IMC
   Line type 2:   COST(I,J)   (in row)

3. *Printed output:*

   a. Listings of input data
   b. Final assignment for each customer (that is, each customer number assigned to a location number)

SUMMARY

This chapter has covered two types of facility location problems in which the only factor considered is transportation cost: the single-facility placement problem and the multiple-facility placement problem. In the former case the minimum cost is determined simply by multiplying each customer's demand by each location's unit cost. The facility is then placed in the location that is found to have the lowest overall cost when all of the customers' demand costs are totaled by location.

The multiple-facility problem is slightly more complicated, and we discussed two methods for its solution: the brute force method and the heuristic approach. Brute force attacks the problem directly by setting up all possible combinations of locations and calculating the minimum transportation cost for each customer to be served by each location combination. The best combination of locations is that which has the overall minimum cost when the totals for all customers at each of the combinations are compared. This procedure becomes unwieldly when the number of possible combinations becomes large.

The heuristic method assigns the facilities to one location at a time. The first is assigned to service all customers and is placed in the location having the lowest total demand cost for all customers. Then by working with one location at a time, the savings that would be gained if each customer were to be served from that location are determined. Only positive savings are of interest, and the location is selected if there is a net gain. Redistribution of the customers follows.

PROBLEMS

2.1 A warehouse is to be built in one of two possible locations to store merchandise to be shipped to three stores.

| | Location | | |
Store	A	B	Demand
1	15	24	50
2	21	13	35
3	17	8	60

a. Given the distances from the stores to the possible locations and the demands per week, as shown in the accompanying table, use the heuristic method to determine the best location for the warehouse.

    b. Solve the problem by mathematical formulation and compare the results with the solution to part a.

2.2 A water fountain can be located in one of three locations to accommodate four offices. The company would like to minimize the time spent away from the offices.

		Location		
Office	*1*	*2*	*3*	*Employees*
1	15	45	30	2
2	38	27	25	4
3	70	63	58	5
4	40	90	50	3

    a. Given the distances from the offices to the possible locations and the number of employees in each location, as shown in the accompanying table, use the brute force method to determine the best location.

    b. Solve by heuristic methods and compare the result with that of part a.

    c. Determine the best locations if two water fountains can be installed.

2.3 Four locations in a department are being considered for the installation of a new drill to be operated by five workers who will have to travel to and from their stations in order to use it. The time required for each worker to travel to each location and the number of trips to be made by each are shown in the accompanying table.

		Location			
Worker	*A*	*B*	*C*	*D*	*Demand*
1	0.50	1.00	1.20	0.80	8
2	0.70	0.30	2.00	0.75	15
3	1.60	2.20	1.30	0.90	22
4	0.40	0.60	1.40	2.10	17
5	1.30	0.30	0.50	0.80	11

Determine the best location for the drill. If another drill machine could be purchased, what would be the two best locations for these machines?

2.4 Three microcomputers are purchased for use by six departments in a small appliance manufacturing factory. The transportation time from the center of each department to each of the possible locations for the computers is shown in the accompanying table, along with each department's daily demands.

Department	Location 1	2	3	4	5	6	Demand
A	7.0	4.0	2.0	6.0	3.5	1.0	8
B	4.0	2.5	3.0	0.0	5.0	2.0	16
C	2.0	3.0	1.0	4.0	4.0	3.0	24
D	1.0	6.0	7.0	5.0	2.0	0.0	10
E	3.0	8.0	2.0	7.5	6.0	8.0	4
F	5.0	0.0	9.0	4.0	4.0	5.0	15

a. Use the brute force method to determine the best locations for the computers.

b. Solve by using the heuristic method.

c. Determine whether more microcomputers can be profitably added.

Note: Problem 2.4 can be solved by using the computer.

2.5 Discuss possible instances in which it would become difficult or impractical to use the brute force method.

2.6 a. The diagram in Figure P2.6 shows a department with five stations (1, 2, 3, 4, and 5) and three possible locations for a new piece of machinery (A, B, and C). Using rectilinear distances from the points in the stations to the possible locations, determine the best location. Assume that the machines and demands are located at the centers of their respective squares, as marked. The demands are as shown in the accompanying table.

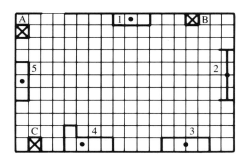

☐ = 5 × 5 feet

Figure P2.6

Station	1	2	3	4	5
Demand	10	13	9	21	15

b. Repeat part a if two new machines are to be located.

2.7 Conduct a 10×10 problem (ten customers and ten possible locations) and solve it by the brute force and heuristic methods. Compare the results when placing
a. Two facilities
b. Three facilities

Note: Problem 2.7 can be solved by using the computer.

CHAPTER
3

Location Analysis with Fixed Costs

Under most circumstances, the site where a facility is to be located is not in a suitable condition to accept the facility on an "as-is" basis, and site preparation or modification work is required. This involves an additional expenditure; however, it is a one-time, initial cost, known as a fixed cost. The numerical values of the fixed costs for installing a particular facility may vary from location to location. For example, in choosing a site for a plant, locations might differ in available conveniences. In some locations, land might have to be purchased and a building constructed; in others, suitable existing buildings might be available that could be converted to the desired plant. In yet other locations, modifications to transportation facilities might be needed to use the existing buildings as production facilities. Thus the one-time costs to build the warehouses (or other facilities) in each location are generally different.

In some other instances, location (or relocation) of a piece of equipment may be planned. Examples include relocation of a lathe in a machine shop, placement of a new X-ray machine in a hospital, and installation of a new copy machine in an office building. In each case a number of suitable locations might be available; however, the cost of placing a machine might be different in each location. It could depend upon factors such as the present condition of the site and/or the modifications required to the foundation as well as the surroundings. In some areas, preparation of the site would demand no additional expenditure, but in other areas it might require extensive work such as the installation of plumbing fixtures, sewage systems, and electric outlets; redesigning the area; and even perhaps the installation of air-conditioning systems.

Environmental regulations might also dictate an additional fixed cost for a site. For example, in locating a chemical or power plant, regulations governing exhaust emissions may be important. Such regulations within a city might require a special piece of equip-

Portions of this chapter have been adapted from the *Journal of Operations Management,* Vol. 1, No. 4, May 1981, with permission of the American Production and Inventory Control Society, Falls Church, VA.

ment, such as, a scrubber for a coal-burning power plant, while in outlying areas these restrictions might not exist.

The objective for the problems to be studied in this chapter is to select the location(s) and assign the demand(s) in a manner that minimizes the total cost, that is, the sum of transportation and fixed costs. Before proceeding, however, it is important to understand the time units for which these costs are defined. The fixed cost is a one-time cost and is therefore to be recovered over the entire life of the facility. On the other hand, a transportation cost resulting from assigning a customer's demand(s) to a location depends on the period of time for which the demands are defined. A demand may be expressed for any convenient unit of time—a week, a month, or a year—provided that all the demands and all the fixed costs are for the same time period. If this is not already the case, they must be converted to a convenient unit. For fixed costs this can be easily done by estimating the life of each facility and then using an annuity factor to spread the fixed cost (depreciation and interest cost) for each facility over a unit time period. (See Appendix A.) The effect, if any, of income taxes is ignored in all illustrative examples throughout this chapter. If the tax rate is known, we suggest developing after-tax figures and then using the procedures, which are applicable in either case without loss of generality.

3.1 HEURISTIC METHOD FOR SOLVING PROBLEMS WITH FIXED COSTS

The following heuristic method has been found to be very effective in solving problems with fixed costs. As illustrated in the flowchart in Figure 3.1, the steps involved in the procedure are as follows:

1. Construct the demand cost table as before (Chapter 2). If the number of facilities available for placement is two or more, go to Step 2. Otherwise, take the sum of each column, which will represent the total variable (transportation) cost if the demands from all sources are assigned to that location. Then add the total variable cost for each column to the corresponding fixed cost. Select the location with the minimum overall cost and assign all customers to that location. This is the optimum solution if only one facility is available, and so the solution process terminates.

2. A good initial solution when there is more than one facility is obtained by constructing a "minimum savings table" as follows. For each customer, calculate the difference between the cost of assigning the customer to the minimum cost location and the next higher cost location. Let us call this difference the "minimum increment." This represents the minimum amount saved if the customer is assigned to the location with the minimum cost. All other locations to which this customer could be assigned would only increase the cost of assignment. In constructing the minimum

Figure 3.1

Location analysis with fixed
cost schematic diagram for the
procedure in Section 3.1

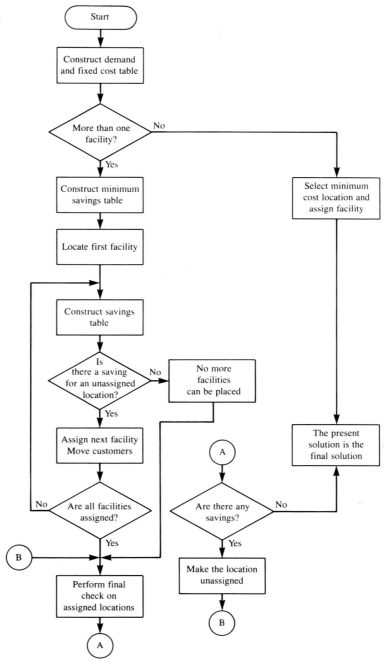

savings table, assign each customer's minimum increment to its minimum cost location.

3. Total the savings for each column in the minimum savings table, and subtract the fixed cost for each location from the corresponding column total. This gives the net savings for each location. Since we must have at least one facility to satisfy the demands, the first facility should be located where the cost is the least. Assign all of the customers to this facility. This now becomes an assigned location.

4. Determine the savings in moving each customer from the assigned location(s) to the nonassigned location(s). If there is no saving for a particular location, mark it by "—" in the appropriate column.

5. Total each unassigned column. Each sum represents the savings that could be achieved if the customers were moved to that column. From each sum, subtract the fixed cost associated with that location and make the next assignments in the location that has the maximum net savings. This is because we would use the additional facility only if it were going to reduce the total cost. The location just chosen now becomes an assigned location and a part of the "assigned group." Move the customers that contributed to the savings to this location. If all the facilities have been assigned, or if no column with savings can be found, proceed to Step 6. Otherwise return to Step 4.

6. This step is the final check analysis. For each assigned location, determine the cost that would be incurred if each of the currently assigned customers for that location were moved to the assigned location with the next higher transportation cost. (Note that this may result in the customers' being moved to different assigned locations.) Total these costs. If this sum is less than the fixed cost for the assigned location under consideration, then additional savings could be achieved by not selecting the present location for placing a facility, that is, by

making this location an unassigned location. Perform this check on all the assigned locations, and select the assigned location with the maximum savings if it exists. Make this location an unassigned location by moving each of the customers currently assigned to this location to the assigned location with the next higher transportation cost. Repeat Step 6 until no further savings can be obtained. We then have the optimum solution.

3.2 APPLICATION

Consider the auto shop problem analyzed in Chapter 2. Recall that we had five customers and five possible locations in which the facilities could be placed. In Chapter 2 we assumed that the costs were in units of time; in this chapter we will measure them in terms of dollars.

Table 3.1 provides the transportation costs and demands and leads directly to Table 3.2, the demand cost table for the problem.

Table 3.1   Transportation cost and demand

Customer	Location					Demand
	1	2	3	4	5	
A	5	3	2	8	5	100
B	3	5	2	6	7	50
C	5	2	0	1	0	150
D	2	1	8	2	3	200
E	3	2	4	0	4	300

Table 3.2   Demand cost

Customer	Location				
	1	2	3	4	5
A	500	300	200	800	500
B	150	250	100	300	350
C	750	300	0	150	0
D	400	200	1600	400	600
E	900	600	1200	0	1200

Suppose now that the fixed cost associated with each location, if a facility is placed there, is as follows:

Location	1	2	3	4	5
Fixed Cost	5000	4000	4500	6000	4200

In addition, assume that the life of the equipment is affected by the location in which it is placed. This may be due to such things as climate, available maintenance, or operating procedures. The estimated life expectancy for a facility in each location is as follows:

Location	1	2	3	4	5
Expected Life (years)	8	7	10	10	9

We wish to place the facilities and assign the customers in a manner that will minimize the total cost of operation per year.

Solution

Since the economic life of the equipment is location-dependent, the first step is to convert all costs to the same basic time units. The demand cost matrix is based on annual demand; therefore it is convenient to convert the fixed costs to an annual basis. If the interest rate is known, we can use the capital recovery factor (Appendix A) to obtain the annual equivalent fixed cost. If the interest rate is not known or is considered insignificant, then an approximate estimate of the annual cost is obtained simply by dividing the fixed cost by the expected useful life.

Assume that the interest rate is 10 percent. Then by using the capital recovery factor, A_i, the fixed cost per year for Location i is as follows:

$$A_1 = \$5000 \, (A/P, 10\%, 8) = 5000(.1874)$$
$$= 937$$
$$A_2 = \$4000 \, (A/P, 10\%, 7) = 4000(.2046)$$
$$= 818$$
$$A_3 = \$4500 \, (A/P, 10\%, 10) = 4500(.1627)$$
$$= 732$$
$$A_4 = \$6000 \, (A/P, 10\%, 10) = 6000(.1627)$$
$$= 976$$
$$A_5 = \$4200 \, (A/P, 10\%, 9) = 4200(.1736)$$
$$= 729$$

Since we were given the total cost table in the data and there is more than one facility to place, we will proceed directly to Step 2 of the procedure. For Customer A, Location 3 has the minimum transportation cost of $200, and the next larger amount is $300 for Location 2. The difference of $100 is the minimum savings if this customer were serviced by Location 3. This amount for Customer A is entered in Location 3 of the minimum savings table (Table 3.3).

Table 3.3   Initial assignment minimum savings table

		Location				
Customer	1	2	3	4	5
A			100		
B			50		
C			0		
D		200			
E				600	
Fixed Cost	−937	−818	−732	−976	−729
Net Savings	−937	−618	−582	−376	−729

Similar analyses are performed for all other customers, and the results are entered in the table. Finally, the fixed cost for each location is entered as a negative saving.

The net savings for each location are determined by taking the sum of each column. In this case, each one has a net negative saving that actually indicates additional cost rather than a saving. Since the equipment must be placed and customers assigned, the first machine is placed where the cost is minimum, that is, at Location 4. All customers are assigned to this location, as shown in Table 3.4.

The next step is to determine whether a second machine is needed and, if so, where it should be placed. To do so, we calculate the savings resulting from moving a customer from its currently assigned location to an unassigned location. Let us do this for Customer A. The saving that results from moving Customer A from Location 4 to Location 1 is 800 − 500 = $300, which is entered in Table 3.5 for Customer A in Location 1. Similar calculations are performed for all customers for each unassigned location. Table 3.5 shows the results.

Again, a negative saving is indicated by a dash. Taking the total of the savings and the fixed cost for each location makes it clear that the maximum net savings are at Location 3. The facility is thus placed in Location 3, and the customers that contributed to these savings—A, B, and C—are assigned there. The present demand assignments are shown in Table 3.6.

Table 3.4   Demand assignments I

Customer	Location				
	1	2	3	4	5
A				X	
B				X	
C				X	
D				X	
E				X	

Table 3.5   Savings table I for moving from Location 4

Customer	Location				
	1	2	3	4	5
A	300	500	600		300
B	150	50	200		—
C	—	—	150		150
D	—	200	—		—
E	—	—	—		—
Total Savings	450	750	950		450
Fixed Cost	−973	−818	−732		−729
Net Savings	−487	−68	218		−279

Table 3.6   Demand assignments II

Customer	Location				
	1	2	3	4	5
A			X		
B			X		
C			X		
D				X	
E				X	

Table 3.7 details the process of selecting a location for the third machine (facility). The transportation cost saving is calculated on the basis of moving a customer from the present assignment. For example, moving Customer A from Location 3 to Location 2 will result in 200 − 300 = −$100 savings (see Table 3.2) and is marked by a dash in Location 2; whereas moving Customer D (presently assigned to Location 4) to Location 2 will result in 400 − 200 = $200 savings as indicated for Customer D in the second column. The results of similar calculations for all remaining combinations of customers and currently unassigned locations are shown.

After including the fixed cost, the net savings are found to be negative for all unassigned locations; therefore the third machine is not

Table 3.7   Savings table II

Customer	Location				
	1	2	3	4	5
A	—	—			—
B	—	—			—
C	—	—			—
D	—	200			—
E	—	—			—
Total	0	200			0
Fixed Cost	−937	−818			−729
Net Savings	−937	−618			−729

Table 3.8   Total cost for the final solution

Customer	Location					Total
	1	2	3	4	5	
A			200			200
B			100			100
C			0			0
D				400		400
E				0		0
Demand Cost			300	400		700
Fixed Cost			732	976		1668
Total Cost			1032	1376		2368

profitable, and the process is stopped here. Thus the previous solution is still the best at this stage. Table 3.8 presents the costs for those assignments.

The next step is to perform a final check analysis. Since there are only two assigned locations, the final check is rather simple. Customers A, B, and C are currently assigned to Location 3. If all of them are moved to Location 4, since the next minimum transportation cost assigned location for each customer happens to be the same in this example, the transportation cost is increased by 600 + 200 + 150 = $950. The fixed cost saving associated with making Location 3 unassigned is $732 (Table 3.3). Since the saving is smaller than the cost, this move is not made.

Similarly, if customers D and E are moved from Location 4 to Location 3, the transportation cost is increased by 1200 + 1200 = $2400. The fixed cost saving for Location 4 is $976. Consequently, this move is also rejected. Thus the present solution is indeed the best solution.

Incidentally, one may also start the solution procedure by locating the first facility in the minimum cost location obtained in Step 1 and ignoring Steps 2 and 3. At times this starting point has given a better result. The reader should examine each problem using both starting assignments and then select a solution that gives the best results.

3.3 UNASSIGNABLE LOCATION

Quite often we come across a problem in which one or more customers cannot be assigned to certain locations. For example, consider the placement of a fire station in a community. Fire codes might dictate the maximum allowable response time of the fire department to an alarm, and travel time is obviously a major component of that time. This restriction on the maximum value for response time might not permit the selection of certain locations as suitable sites for placement of fire stations to serve the outlying areas. To solve this type of problem, we modify the fixed cost procedure by converting the "*" wherever it appears in the demand cost table (which indicates that designated customers cannot be assigned to that location) to some large cost. However, assigning an infinite cost (∞) will make it difficult to determine the net savings (they will be indistinguishable from one location to another) as we move the demands from assigned locations to unassigned locations. As a general rule, this cost should be more than the largest fixed cost plus the two largest transportation costs. With this modification to the demand cost table, we can then apply the method described in Section 4.1 to solve the resulting problem.

Table 3.9   Demand cost and fixed cost

Customer	Location				
	1	2	3	4	5
A	120	210	180	210	170
B	180	*	190	190	150
C	100	150	110	150	110
D	*	240	195	180	150
E	60	55	50	65	70
F	*	210	*	120	195
G	180	110	*	160	200
H	*	165	195	120	*
Fixed Cost	200	200	200	400	300

Table 3.10   Modified demand cost and fixed cost

Customer	Location				
	1	2	3	4	5
A	120	210	180	210	170
B	180	900	190	190	150
C	100	150	110	150	110
D	900	240	195	180	150
E	60	55	50	65	70
F	900	210	900	120	195
G	180	110	900	160	200
H	900	165	195	120	900
Fixed Cost	200	200	200	400	300

Example: Fire Station Problem

Consider the case of eight customers that are to be assigned to facilities placed in five locations. The demand cost data (that is, demand multiplied by cost per unit demand) and the fixed cost data are given in Table 3.9.

Remember, an asterisk (*) for a cost entry means that the indicated customer cannot be assigned to the associated location. The first task is to determine the appropriate value for the "*" based on the data in Table 3.9. The largest fixed cost is 400; the largest transportation cost is 240, and the next largest is 210. With these data the "*" could be replaced by $400 + 240 + 210 = 850$ or, say, $900. This is shown in Table 3.10.

The next step is to determine the values for the minimum savings table, Table 3.11. We construct it by taking the difference between the minimum and the next higher cost of assignment for each customer and entering that amount for the location having the minimum cost for that customer.

Because Location 1 has the greatest net savings (lowest net cost), the first facility is placed there, and all the customers are assigned to that location. To determine the placement of the second facility, we continue the procedure

Table 3.11   Minimum savings for the initial placement

Customer	Location				
	1	2	3	4	5
A	50				
B					30
C	10				
D					30
E			5		
F				75	
G		50			
H				45	
Total	60	50	5	120	60
Fixed Cost	−200	−200	−200	−400	−300
Net Savings	−140	−150	−195	−280	−240

by calculating the values for the first savings table, Table 3.12.

Since Location 2 has the maximum savings, place the second machine there, and assign Customers D, E, F, G, and H to it. The revised assignment is shown in Table 3.13.

To check for placement of a third machine with additional savings, let us calculate the second savings table, Table 3.14.

Since there are no net savings in any unassigned location, the demand assignment in Table 3.13 provides the present best solution as shown in the total cost table, Table 3.15.

The next step is to perform the final check. If we move the customers assigned to Location 1—A, B, and C—to Location 2, we will increase the transportation cost by 90 + 720 + 50 = $860 (see Table 3.10). The corresponding saving by making Location 1 an unassigned location is its fixed cost, $200. Since the saving is less than the cost, this move is not made. Similarly, if the customers currently assigned to Location 2 are moved to Location 1,

Table 3.12 Savings table I for moving from Location 1

Customer	Location				
	1	2	3	4	5
A	—	—	—	—	—
B	—	—	—	—	30
C	—	—	—	—	—
D		660	705	720	750
E		5	10	—	—
F		690	—	780	705
G		70	—	20	—
H		735	705	780	—
Total		2160	1420	2300	1485
Fixed Cost		−200	−200	−400	−300
Net Savings		1960	1220	1900	1185

Table 3.13 Demand assignment

Customer	Location				
	1	2	3	4	5
A	X				
B	X				
C	X				
D		X			
E		X			
F		X			
G		X			
H		X			

Table 3.14 Savings table II for moving from Location 1 or 2

Customer	Location				
	1	2	3	4	5
A			—	—	—
B			—	—	30
C			—	—	—
D			45	60	90
E			5	—	—
F				90	15
G			—	—	—
H			—	45	—
Total			50	195	135
Fixed Cost			−200	−400	−300
Net Savings			−150	−205	−165

Table 3.15 Total cost for the final solution

Customer	Location					Total
	1	2	3	4	5	
A	120					120
B	180	*				180
C	100					100
D	*	240				240
E		55				55
F	*	210	*			210
G		110	*			110
H	*	165			*	165
Demand Cost	400	780				1180
Fixed Cost	200	200				400
Total Cost	600	980				1580

* indicates unpermitted customer/location combination.

it will increase the transportation cost by 660 + 5 + 690 + 70 + 735 = \$2160. The fixed cost saving for Location 2 is \$200. Again, the move is rejected. The solution in Table 3.15 is the optimum solution.

It may seem fortuitous that we have customers assigned only to locations where the assignments are possible; however, this is not a matter of luck. As long as we are free to choose the number of facilities, and as long as there is a location to which a customer can be assigned, then a valid assignment for each customer is always possible.

The problem may become infeasible, that is, one without solution, if there is only a limited number of facilities. For instance, if we had only one facility in the previous example, all of the customers would have been assigned to Location 1 because of its lowest total cost. But Customers D, E, and H cannot be assigned to this location, and so the solution is infeasible. One way to resolve this is to drop from the data the location where a customer with an artificially large cost is assigned in the final solution and solve the problem again with the remaining data. If a solution exists, continued application of this modification process will obtain the minimum cost solution. If there is no solution, the process will so indicate.

Let us return to the example. It is obvious that if there is only one facility to locate, it will not be in Location 1. By using the modification process we find that it will be placed in Location 4 because this is the only location where it is possible to assign all customers.

3.4 COMPUTER PROGRAM DESCRIPTION

Purpose

The computer program for the fixed cost facility location problem using heuristics finds the optimum facility location/allocation for both single-facility and multiple-facility problems. It will also determine the optimum number of facilities, provided that the number of facilities is not restricted.

Program Description

1. *Usage:* The program consists of a main program and a subroutine. All array sizes are specified in DIMENSION or COMMON statements, which may be altered for the particular problem. The program is currently set to handle up to 50 locations and 60 customers. If the number of locations and/or the number of customers is to be increased, the DIMENSION and COMMON statements must be modified. The program will calculate the total variable cost per unit of demand, convert fixed cost values to the same period as the variable cost values, and handle unassignable locations.

2. *Description of parameters, including dimensions of arrays*

ASIGN(60) = Assignment matrix,
        customer to location,
        used for the final
        output

ASIGNB(60) = Assignment matrix,
        customer to location,
        used in the final check
        table calculation

COUNT = Counter that holds the
        number of locations
        that have been
        assigned; used in the
        final check table
        calculation

FINFC(50) = Final check fixed cost
        table

FINMAT(60,50) = Final check variable
        cost table

FLAG = Flag used to control

the NET subroutine
operation

FXCM(50) = Fixed cost table

FXCMB(50) = Fixed cost table B used
to maintain the
original fixed cost
values

IDMND = Customer demand

K = Used for location
assignment from
subroutine NET; also
used as a DO loop
counter

LOCST = Cost of assigning all
demand at the location
(demand × unit cost)

MINSAV = Used to hold
minimum saving for
(customer, location)
assignment in
minimum savings table

MSAV = Savings table counter

NOCUST = Number of customers

NOFAC = Maximum number of
facilities available for
placement in locations

NOLOC = Number of locations

EXPLIF = Expected life of the
facility

PC = Print code, where 0 is
print input tables and
the solution and 1 is
print input tables,
intermediate tables,
and the solution

PIDU(50) = Period in demand
units (demand/period)

PRNTR = Set to installation
device code for printer
(i.e., 6)

RATE = Interest rate

RDR = Set to installation
device code for reader
(i.e., 5)

SAVMAT(60,50) = Savings table

SAVNET(50) = Net savings table

STK(60) = List of assigned
locations, used in final
check table calculation

VARCM(60,50) = Variable cost table

3. *Input formats* (Figure 3.2)

Line type 1: NOCUST, NOLOC,
NOFAC, PC

Line type 2: IDMND, one per customer

Line type 3: LOCST, one per location

Line type 4: FXCMB, fixed cost per
location

Line type 5: EXPLIF, expected life for
each facility

Line type 6: RATE

If the rate is greater than 0, a type 7 line is required.

Line type 7: PIDU, per location

4. *Output* (Figure 3.3): This program prints, as appropriate, a single-facility table, a minimum savings table, savings tables, final check tables, a customer allocation table, and the cost of the solution.

5. *Summary of user requirements*

a. Specify values of NOCUST, NOLOC, NOFAC, and PC. If the program is to optimize the number of facilities to be allocated, then enter NOFAC = NOLOC.

b. Enter values for FXCMB.

c. Enter values of EXPLIF.

d. If fixed costs are for different periods, enter the current rate (e.g., 10% = .10) and PIDU for each location.

Figure 3.2

Sample example input for facility location with fixed cost program

```
                                                        INPUT

                                                        COLUMN

     12345678901234567890123456789012345678901234567890123456789012345 6

                                                        1ST LINE
                                                        --------

           5       5       0       3       1

                                                        2ND LINE
                                                        --------

        100    50 150 200 300

                                                        3RD LINE
                                                        --------

                 5       3       2       8       5
                 3       5       2       6       7
                 5       2       0       1       0
                 2       1       8       2       3
                 3       2       4       0       4

                                                        4TH LINE
                                                        --------

           5000    4000    4500    6000    4200

                                                        5TH LINE
                                                        --------

        0807101009

                                                        6TH LINE
                                                        --------

        0000010
```

Figure 3.3

Output of the program from the sample example

```
                                                OUTPUT
                                                ------

                                INPUT DATA REPRESENTED AS DATA
                                ------------------------------

                         5       5       3       1
                      100    50 150 200 300
                      000050000003000000200000080000005 00
                      000030000005000000200000060000007 00
                      000050000002000000000000010000000 00
                      000020000001000008000000200000003 00
                      000030000002000004000000000000004 00
                      050000004000000600000042000 0
                      0807101009
                      0000010

                                        FINAL ASSIGNMENT
                                        ----------------

              CUSTOMER            1       2       3       4       5
              LOCATION            3       3       3       4       4
        COST OF ASSIGNMENT 200.00 100.00    0.00   400.00    0.00
              TOTAL COST OF ASSIGNMENT = 700.00
```

SUMMARY

New concepts in this chapter are the fixed cost, the savings table, and the minimum savings table. Also, we saw that problems with unassignable locations could be readily solved by modifying the demand cost table to show a high cost for unassignable locations.

Both sample problems that were presented assumed that a product (or service) was already complete (production costs were not considered) and required only being transported to the customer. The total cost table showed the solution as the total of the fixed cost (site preparation, construction, etc.) and the transportation cost. Articles that expand the cost structure are published frequently in management science and industrial engineering journals; however, such analysis is beyond the scope of this book.

PROBLEMS

3.1 Define fixed cost and give three examples in which such cost is significant and may not be ignored.

3.2 A farmer wishes to locate a tobacco-drying barn in one of three locations to receive tobacco from his three fields. His fixed cost is associated with building the facilities to accommodate his trucks and house the equipment. These costs vary from location to location and are $1200, $1050, and $1500 per year for locations A, B, and C, respectively. The cost of truck travel is estimated to be $0.70/mile. The distance from fields to locations and truckloads of tobacco that are expected from each field are given in the accompanying table. Determine the best location for the barn.

| Field | Location | | | Loads |
	A	B	C	
1	20	2	11	200
2	7	14	6	175
3	1	28	21	160

3.3 In Problem 2.2 the cost of placement of the water fountain may vary from location to location because of the cost of installation and plumbing.

| Office | Location | | | Employees |
	1	2	3	
1	15	45	30	2
2	38	27	25	4
3	70	63	58	5
4	40	90	50	3
Cost	110	125	102	

a. Given the cost per trip between each office and a location, the number of employees making trips from each office, and the cost per location for installing the water fountain (in the same units as trips), as shown in the accompanying table, determine by the heuristic method the best location(s) for the fountain.

b. Solve by any other method and compare the results.

3.4 A plant wishes to install two pieces of machinery for use by three departments. Using the travel cost from each department to each possible new location (in dollars per trip), the demand per month for the new machinery by each department, the building cost for each location (in dollars), and the life expectancy of the machinery at each location, given in the accompanying table, choose the two best of the five locations being considered. Assume 10 percent interest. *Note:* Problem 3.4 can be solved by using the computer.

Cost per Trip

Department	A	B	C	D	E	Demand
1	15	8	2	5	0.5	2000
2	4	3	7	2	1	5000
3	5	1	2	2.5	3	7500

Location	A	B	C	D	E
Building Cost	100,000	150,000	136,000	95,000	125,000
Life Expectancy	15	12	17	10	8

3.5 Three locations are available for two machines to be operated by five people. Each of the five people is located at a different operating station. The accompanying table gives the fixed cost per week for each of the three possible locations, the travel cost per trip from each operating station to each possible location, and the number of trips per day each operator will make to the new machine. Determine the best possible locations for the two machines and identify the customers each machine should serve.

Location Cost

Customer	1	2	3	Demand
A	2.0	0.5	0.75	10
B	1.0	2.1	1.4	17
C	0.8	1.2	1.9	25
D	0.6	0.4	1.1	8
E	1.75	2.2	2.4	14
Fixed Cost	200	140	175	

3.6 In the diagram in Figure P3.6, 1, 2, 3, 4, and 5 are possible locations, and A, B, C, D, and E are the departments that will be using the machines to be located. Using rectilinear distances, determine the optimum number of machines to be placed and the best locations. Assume that machines and demands are located at the centers of the squares. The operating cost for the machine at each possible site and the demand for use of the machine by each department are given in the accompanying table. The travel cost is $0.01 per foot. *Note:* Problem 3.6 can be solved by using the computer.

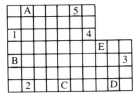

☐ = 10 × 10 feet

Figure P3.6

Location	1	2	3	4	5
Operating Cost/Month	400	350	375	420	315
Department	A	B	C	D	E
Demand/Month	40	25	15	36	20

3.7 Discuss when it would be better or worse to use a computer program to solve a location problem instead of doing the computations by hand.

3.8 Four mailboxes are to be erected in a certain city, which is divided into twelve sections. (See Figure P3.8.) Seven possible locations are shown. Dark lines indicate sections of the city. Use the demand for the use of mailboxes for each section, the fixed cost for using each location, and the life expectancy of each of the mailboxes at each of the locations provided in the accompanying table. To determine the best locations, assume that the customers are evenly dispersed throughout the city. Use rectilinear distances from the center of an area to the location as the average distance; but if the location is in the middle of the section in question, use 2.5 miles as the average distance. *Note:* Problem 3.8 can be solved by using the computer.

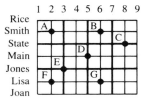

Figure P3.8

	A	*B*	*C*	*D*	*E*	*F*	*G*
Fixed Cost	500	600	750	640	475	550	580
Life	10	8	12	14	9	11	13

Section	*Demand (in thousands)*
State to 3rd	10
State, 3rd, 5th	17
State, 5th, 7th	12
State, 7th, 9th	15
State, Jones, 1st, 3rd	9
State, Jones, 3rd, 5th	8
State, Jones, 5th, 7th	11
State, Jones, 7th, 9th	13
Jones, Joan, 1st, 3rd	16
Jones, Joan, 3rd, 5th	14
Jones, Joan, 5th, 7th	11
Jones, Joan, 7th, 9th	7

SUGGESTED READINGS

Atkins, R.J., and Shriver, R.H., "New Approach to Facilities Location," *Harvard Business Review,* May–June 1985.

Barcelo, J., and Casanovas, J., "Heuristic Lagrangean Algorithm for Capacitated Plant Location Problem," *European Journal of Operations Research,* Vol. 15, No. 2, Feb. 1984, pp. 212–226.

Davis, P.S., and Ray, T.L., "A Branch-Bound Algorithm for the Capacitated Facilities Location Problem," *Naval Logistic Quarterly Review,* Vol. 16, Oct. 1969.

Francis, R.L., and White, J.A., *Facility Layout and Location—An Analytical Approach,* Prentice-Hall, Englewood Cliffs, N.J., 1974.

Khumawala, B.M., "An Efficient Branch and Bound Algorithm for the Warehouse Location Problem," Krannert School of Industrial Administration Institute Paper Series No. 294, Purdue University, Dec. 1980.

Phillips, D.T.; Ravindran, A.; and Solberg, J.J., *Operations Research—Principles and Practice,* John Wiley and Sons, New York, 1976.

Plastria, F., "Localization in Single Facility Location," *European Journal of Operations Research,* Vol. 18, No. 2, Nov. 1984, pp. 208–214.

Revelle, C.; Marks, D.; and Liebman, J.C., "An Analysis of Private and Public Sector Location Models," *Management Science,* Vol. 16, July 1970.

Sa, G., "Branch and Bound and Approximate Solutions to the Capacitated Plant Location Problem," *Operations Research,* Nov.–Dec. 1969.

CHAPTER
4

Continuous Facility Location

The facility location models that we have studied thus far are based on the assumption that there are a few preselected locations that are suitable for placing the new facilities and that the problem is to select from among these locations one or more that would minimize a particular cost function. Preselection of sites may follow some qualitative considerations; for example, in selecting a site for a plant, one might evaluate factors such as availability of labor, land, transportation facilities, and market proximity. (Details can be found in Chapter 15.) Many locations may satisfy the basic requirements, but we then must select from these the one that would minimize a cost function. The solution of most industrial problems follows the above-mentioned two-step procedure.

There is another class of problems, however, for which one is free to choose any site and place the new facility there; it is called the continuous or universal facility location problem. This method can be used to determine the ideal location only when the sole consideration involved is transportation cost. The term "ideal" is used to stress the fact that on such a site, construction or placement of a facility may or may not be physically possible. It does, however, guide the analyst to the location(s) that would result in the minimum transportation costs; and if it should prove infeasible to utilize such a location, the "penalty" incurred by selecting an alternative location is better understood.

In this chapter we shall study the following variations of the problem: (1) single-facility location, (2) one-at-a-time facility location, and (3) multiple-facility location of the same type. In each case the demand points or the customers and their $x-y$ coordinates and the necessary number of trips from each customer to a facility are known. The cost is measured in terms of the number of trips multiplied by the distance a customer has to travel to reach the facility. In manufacturing plants the travel is in aisles, and so the distance is measured in rectilinear fashion. When direct travel by the shortest path is possible, the cost can be measured as either quadratically proportional to the distance or linearly proportional, called Euclidean distance. (See Figure 4.1 for examples of rectilinear and Euclidean measurements distance.)

Figure 4.1

(a) Rectilinear distance and (b)
Euclidean distance

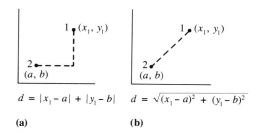

$$d = |x_1 - a| + |y_1 - b|$$ $$d = \sqrt{(x_1 - a)^2 + (y_1 - b)^2}$$

(a)   **(b)**

Quadratic cost implicitly states that customers that are farther away will contribute much more to the cost than those nearby. This is appropriate when one also wants to implicitly include time as another parameter in the cost. For example, in placing a public facility such as a hospital within a community, it might be important that the hospital not be very far from any center of population that it is expected to serve, timely treatment in case of emergency being a critical criterion.

Euclidean cost, on the other hand, maintains the direct proportionality of the cost with distance and is appropriate when only the travel distance, and not time, is worthy of attention.

4.1 SINGLE-FACILITY LOCATIONS

By means of examples we will explore the various possible conditions that might be encountered in single-facility location problems considering only the cost of transportation.

Rectilinear Distance Cost

Suppose the introduction of a new product will require parts from a new molding machine to be transported to six existing machines in the plant. The coordinates of the existing machines (x, y) and the number of trips that would be made each day from the molding machine to these machines (t) are given in Table 4.1. Figure 4.2 shows the plotted coordinates. The

problem is to determine the appropriate location for the new molding machine.

The cost is assumed to be directly proportional to the distance traveled. Therefore if the location of the molding machine is given by (a, b), the objective is to minimize

$$d = \sum_{i=1}^{n} t_i \left[|(x_i - a)| + |(y_i - b)| \right]$$

where n is the number of existing machines.

The function is linear, however; hence it is not possible to obtain the optimum by the usual method of taking the first derivative, setting it equal to zero, and solving the resulting expression.

To demonstrate a basic concept of how such a function can be optimized, suppose we

Table 4.1   Coordinates of existing machines

Machine	Coordinates (x, y)	Trips/Day (t)
1	20, 46	20
2	15, 28	15
3	26, 35	30
4	50, 20	18
5	45, 15	20
6	1, 6	15

Figure 4.2   Coordinates of existing machines

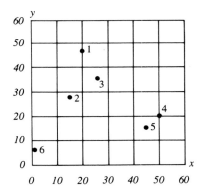

Figure 4.3   Cost analysis for the *x* axis

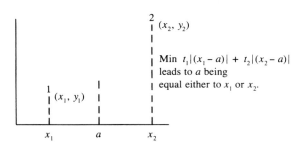

Min $t_1|(x_1 - a)| + t_2|(x_2 - a)|$ leads to a being equal either to x_1 or x_2.

Table 4.2   Data for machine location problem

Customer	Abscissa	Number of Trips	Cumulative Trips
1	13	10	10
3	21	15	25
2	24	35	60

take another example in which there are only two customers, who are located at (x_1, y_1) and (x_2, y_2), respectively, and who are making t_1 and t_2 trips to a facility. The cost of travel is given by

$$t_1|(x_1 - a)| + t_1|(y_1 - b)| \\ + t_2|(x_2 - a)| + t_2|(y_2 - b)|$$

the value of which is to be minimized. The expression can be separated into two parts, min: $t_1|(x_1 - a)| + t_2(x_2 - a)|$ and min: $t_1|(y_1 - b)| + t_2|(y_2 - b)|$.

In minimizing on the *x* axis, since the function is linear, the optimum is on the boundary; that is, *a* is equal to either x_1 or x_2. Similarly, for *y* the minimum cost is obtained when *b* is either y_1 or y_2. In general, the new facility should be located at the *x* and *y* coordinates of one of the existing customers. (See Figure 4.3.)

The details of the procedure are as follows. To determine the best abscissa *x* for the new machine, list the existing machines (facilities) in their order of increasing values of abscissae and the corresponding trips. The optimum *x* coordinate is one associated with the 50th percentile, or the median value of the trips. It should be understood that the trips originate from discrete points, and therefore the median is associated with one of those points. For example, if the cumulative trips are as shown in Table 4.2, the median is the 30th trip, which is made by Customer 2. Repeat the same process for the *y* ordinates of the existing machines to obtain the optimum *y* coordinate of the new facility.

To illustrate the procedure, Table 4.3 lists the machine arrangement based on the abscissa values for the data in Table 4.1.

Table 4.3 The *x* axis (abscissa) determination

Machine	Abscissa	Number of Trips	Cumulative Trips
6	1	15	15
2	15	15	30
1	20	20	50
3	26	30	80
5	45	20	100
4	50	18	118

Table 4.4 The *y* axis (ordinate) determination

Machine	Ordinate	Number of Trips	Cumulative Trips
6	6	15	15
5	15	20	35
4	20	18	53
2	28	15	68
3	35	30	98
1	46	20	118

The 50th percentile would be the 59th trip (118/2). That trip occurs at Machine 3, with an abscissa of 26, which thus is the optimum *x* axis value for the new molding machine.

To obtain the optimum ordinate, arrange the machines in ascending order of ordinate values as shown in Table 4.4. The 50th percentile trip, that is, the 59th trip, occurs with Machine 2, whose ordinate is 28. This becomes the *y* axis value of the new facility. Thus the location for the new molding machine is (26, 28).

Single-Facility Location— Quadratic Cost

Now suppose that the cost is proportional to the square of the distance traveled rather than being linearly proportional as in the previous case. The procedure would be to minimize the objective function

$$d = \sum_{i=1}^{n} t_i((x_i - a)^2 + (y_i - b)^2)$$

Taking partial derivatives with respect to a and b, equating them to zero, and solving the resultant expressions, we obtain the following solution:

$$a = \frac{\sum_{i=1}^{n} t_i x_i}{\sum_{i=1}^{n} t_i}$$

and

$$b = \frac{\sum_{i=1}^{n} t_i y_i}{\sum_{i=1}^{n} t_i}$$

The solution is referred to as a centroid or the center-of-gravity solution. Referring to the example with data in Table 4.1, the centroid solution is

$$a = [(20 \times 20) + (15 \times 15) + (26 \times 30) \\ + (50 \times 18) + (45 \times 20) + (1 \times 15)] \\ \div (20 + 15 + 30 + 18 + 20 + 15) \\ = 27.3$$

$$b = [(20 \times 46) + (15 \times 28) + (30 \times 35) \\ + (18 \times 20) + (20 \times 15) + (15 \times 6)] \\ \div (20 + 15 + 30 + 18 + 20 + 15) \\ = 26.6$$

Thus the location for the new molding machine is (27.3, 26.6).

Euclidean Distance Cost

The Euclidean distance between two points *T* and *S* with coordinates (*x*, *y*) and (*a*, *b*), respectively, is defined as

$$d = [(x - a)^2 + (y - b)^2]^{1/2}$$

When the cost is proportional to Euclidean distance, the objective is to minimize the resulting cost expression:

$$d = \sum_{i=1}^{n} t_i[(x_i - a)^2 + (y_i - b)^2]^{1/2}$$

Following a procedure similar to that in the previous subsection—that is, taking partial derivatives with respect to a and b and setting them equal to zero—leads to

$$\frac{\partial}{\partial a} = 0 = \sum_{i=1}^{n} \frac{t_i(x_i - a)}{[(x_i - a)^2 + (y_i - b)^2]^{1/2}}$$
(1)

$$\frac{\partial}{\partial b} = 0 = \sum_{i=1}^{n} \frac{t_i(y_i - b)}{[(x_i - a)^2 + (y_i - b)^2]^{1/2}}$$
(2)

The optimum location should be obvious if we could solve the resulting expressions. However, if the new machine could be placed, at least mathematically, at the same coordinates as one of the old machines—that is, for some i, $x = a$ and $y = b$— both expressions (1) and (2) are undefined, and there is great difficulty in trying to obtain a unique solution.

An alternative is to engage in an iterative solution procedure. It can be proved mathematically that the objective function is convex as shown in Figure 4.4.

Figure 4.4 (a) Cost as a function of x dimension and (b) cost as a function of y dimension

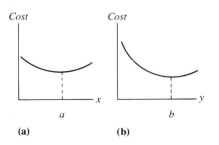

(a)   **(b)**

One can begin with a centroid solution and then perform a search with respect to each dimension until the optimum solution is obtained.

The procedure is easy to follow, and for the small problems most often encountered in practice it provides sufficient accuracy and speed. There are a few alternative rules that permit a reduction in the computational effort (see Feldman et al., 1966; Scriabin and Vergin, 1985). If a problem is very large or computational time is expensive, the reader is advised to review these rules. Because of the level of mathematics involved, however, they will not be discussed here.

4.2 MULTIPLE FACILITIES

The two prominent variations in multiple-facility problems are the following:

1. *Multiple facilities of different types.* Each type is independently demanded by the existing customers. Furthermore, there may also be demand between the new facilities.

2. *Multiple facilities of the same type.* Customers are divided among them so that any one customer is served by only one new facility.

Multiple Facilities of Different Types

Even though alternative methods have been developed for obtaining the solution to each type of cost function, the simplest procedure remains the iterative technique in which the new facilities are located one at a time. The procedure for the rectilinear type of cost calculation is presented in the following subsection, and quadratic and Euclidean types are discussed in the subsection after that.

Rectilinear Cost

Consider a manufacturing plant within which four machines (customers) will require the services of two new facilities that are to be installed in the plant. The present location of each machine and the number of trips between them and each of the new facilities are given in Table 4.5. There are also nine trips expected between the two new facilities. The problem is to determine the optimum locations for the new facilities.

We will begin the process, called "one-at-a-time," by determining the abscissae for the new facilities. As a first step, ignore the trips between the new facilities and determine the x axis value for Facility 1 (F_1) with respect to the present machines (customers). Recall from Section 4.1 that the procedure calls for listing the present machines in order of increasing abscissae and selecting the x value associated with the 50th percentile of the cumulative number of trips. (See Table 4.6.)

The 50th percentile is ten trips (20/2), and the corresponding abscissa is 8. Therefore as a first try the abscissa for the first facility is fixed at 8.

We now repeat the process for the second facility (F_2) except that we will also consider the trips between the facilities, since the first facility is at least temporarily located on the x axis. Table 4.7 lists the data in proper sequence to make the determination.

The 50th percentile is 32.5, and the associated x value, which must be the same as that of one of the existing customers, is 10. Therefore the temporary abscissa for the second facility is 10.

Now go back and determine the abscissa for Facility 1 knowing that the temporary x location for Facility 2 is 10. Table 4.8 shows the necessary update of Table 4.6 to find the new F_1 abscissa.

The 50th percentile, 14.5, is associated with $x = 10$, and the new temporary abscissa for F_1 is different from the previous value for

Table 4.5   Multiple facilities of different types

Machine	Coordinates (x, y)	Number of Trips	
		Facility 1	Facility 2
1	10, 15	10	8
2	8, 20	4	6
3	4, 7	6	12
4	15, 9	0	30
F_1	a_1, b_1	—	9
F_2	a_2, b_2	9	—

Table 4.6   The x axis (abscissa) data for Facility 1 (first iteration)

Machine	Abscissa	Number of Trips	Cumulative Trips
3	4	6	6
2	8	4	10
1	10	10	20
4	15	0	20

Table 4.7   The x axis (abscissa) data for Facility 2 (first iteration)

Machine	Abscissa	Number of Trips	Cumulative Trips
3	4	12	12
2	8	6	18
F_1	8	9	27
1	10	8	35
4	15	30	65

Table 4.8   The x axis (abscissa) data for Facility 1 (second iteration)

Machine	Abscissa	Number of Trips	Cumulative Trips
3	4	6	6
2	8	4	10
1	10	10	20
F_2	10	9	29
4	15	0	29

the same facility. The process must be repeated with $x = 10$ as the new temporary abscissa for Facility 1. The details are shown in Table 4.9.

The abscissa associated with the 50th percentile of 32.5 trips is 10; that is, $x = 10$, which is the same as before for F_2. Therefore the search for the abscissa is complete, and the latest temporary value becomes permanent.

Similarly, the search process for the F_1 and F_2 ordinates leads to Tables 4.10 and 4.11. Keep in mind that the nine trips between the two new facilities are ignored during the first iteration.

The median cumulative trip value is 20/2; therefore the temporary ordinate is 15 (Machine 1). Table 4.11 presents the data in the appropriate order to determine the first temporary ordinate for F_2.

The median cumulative trip value is 32.5, and this corresponds with an ordinate of 9, determined by Machine 4. Table 4.12 continues the procedure, providing the second iteration data for F_1.

The median cumulative trip value is 14.5, which equates to an ordinate of 9, determined by Facility 2. Since this is a different value from that found during the first iteration (Table 4.10), we must perform at least one more iteration as shown in Table 4.13.

The median is 32.5, and this leads to an ordinate of 9, determined by Machine 4. This y value remains unchanged from that of the first iteration; hence the procedure terminates. The optimum locations for the new facilities are

Facility 1: (10, 9)

Facility 2: (10, 9)

The solution yields the same location for both facilities, which, of course, could present a problem. It does, however, point out a drawback of the procedure: it might try to place multiple facilities at the same point. In practice, the facilities would be set side by side within existing practical limitations, including the shape

Table 4.9 The x axis (abscissa) data for Facility 2 (second iteration)

Machine	Abscissa	Number of Trips	Cumulative Trips
3	4	12	12
2	8	6	18
1	10	8	26
F_1	10	9	35
4	15	30	65

Table 4.10 The y axis (ordinate) data for Facility 1 (first iteration)

Machine	Ordinate	Number of Trips	Cumulative Trips
3	7	6	6
4	9	0	6
1	15	10	16
2	20	4	20

Table 4.11 The y axis (ordinate) data for Facility 2 (first iteration)

Machine	Ordinate	Number of Trips	Cumulative Trips
3	7	12	12
4	9	30	42
1	9	8	50
F_1	15	9	59
2	20	6	65

Table 4.12 The y axis (ordinate) data for Facility 1 (second iteration)

Machine	Ordinate	Number of Trips	Cumulative Trips
3	7	6	6
4	9	0	6
F_2	9	9	15
1	15	10	25
2	20	4	29

Table 4.13   The y axis (ordinate) data for Facility 2 (second iteration)

Machine	Ordinate	Number of Trips	Cumulative Trips
3	7	12	12
4	9	30	42
F_1	9	9	51
1	15	8	59
2	20	6	65

and size of the facilities, space required for operating and maintenance personnel, and the area necessary for material flow and inventory storage. The other techniques might provide closer-to-optimum solutions, but the required mathematics is beyond the scope of this book. For most practical purposes the "one-at-a-time method" works quite satisfactorily.

Quadratic and Euclidean Distance Cost

The one-at-a-time method is also applicable when the cost function is either quadratic or Euclidean. The only modification needed is using the appropriate placement method as described in Section 4.1.

4.3 MULTIPLE FACILITIES OF THE SAME TYPE

With multiple facilities that are of the same type the problem of facility placement becomes that of determining the demand points (customers) that should be assigned to each facility and then deciding the locations of the facilities to minimize the cost of travel from those assigned customers. The solution procedure is based on recognizing such a cluster of customers that may be assigned to a single facility and then applying the procedures of single-facility placement to obtain the optimum location for each facility. We will illustrate the procedure, called the combinational approach, by means of an example.

Suppose there are four customers with the data listed in Table 4.14, and two facilities are to be located.

We begin by scaling the demand from each customer using the smallest demand (50) as the scaling factor. For Customer 1, for example, the scaled demand is 150/50, or 3.0. The complete results are shown in Table 4.15.

As the next step, using each customer's x and y coordinates from Table 4.14, we plot the locations of the customers with the same scale on both x and y axes on graph paper, as shown in Figure 4.5(a), to visualize the present layout.

Now we are ready to begin the combinational process. The cost of travel for a customer is given by the Euclidean distance between the facility and the customer, multiplied by the demand from the customer. If we fix a cost value and divide it by the individual demands, we will obtain the distances that each customer may move and still have the same costs for travel. The value of the constant can be kept small by using scaled demands rather than the actual demands. Using the distance calculated for each customer, we can draw circles centered at each customer to represent the identical cost of travel between the customer and a facility

Table 4.14   Multiple facilities of the same type

Customer	Demand	Coordinates (x, y)
1	150	60, 20
2	100	20, 100
3	50	10, 20
4	60	30, 60

Table 4.15   Customer demand (scale factor of 50)

Customer	Scaled Demand
1	3.0
2	2.0
3	1.0
4	1.20

Figure 4.5   (a) Determination of cost constant and (b) the initial layout

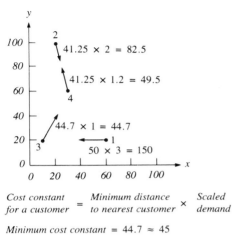

$$\begin{matrix} \text{Cost constant} \\ \text{for a customer} \end{matrix} = \begin{matrix} \text{Minimum distance} \\ \text{to nearest customer} \end{matrix} \times \begin{matrix} \text{Scaled} \\ \text{demand} \end{matrix}$$

Minimum cost constant = 44.7 ≈ 45

(a)

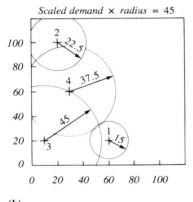

(b)

function. In this case we will use the centroid rule (Section 4.1).

To determine the cost constant, use the following procedure: Calculate the distance from each customer to the nearest neighboring customer, and multiply by the scaled demand for the customer. Choose the minimum value as the cost constant for this setup.

Once two customers are combined, the procedure is repeated, the new layout being considered as an initial layout. The procedure continues until we have exactly as many remaining customers as the number of facilities needing placement. The facilities are placed in the positions identified by the remaining customers in the final graph.

In our example, as shown in Figure 4.5(b), with a cost constant of 45 from Figure 4.5(a), Customers 3 and 4 are first combined. Their equivalent facility coordinates are

$$x = \frac{(1) \times 10 + (1.2) \times 30}{2.2} = 20.9$$

$$y = \frac{(1) \times 20 + (1.2) \times 60}{2.2} = 41.8$$

with a total equivalent scaled demand of 2.2.

With a cost constant of 96.8, obtained by the method explained earlier, as shown in Figure 4.6, the radius from the equivalent facility

Figure 4.6   The second layout

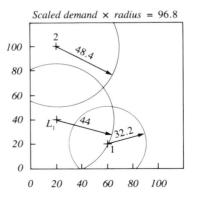

anywhere on the periphery of that circle. The customers are combined on the basis of the principle of encompassment; that is, when the circle from one customer encompasses or encloses the location of another customer, those two customers should be combined and represented by a single equivalent customer with its total demand equal to the sum of the demands of the two customers. Its equivalent location is calculated on the basis of the appropriate cost encompasses Customer 1. Combining these two leads to a subsequent equivalent customer with a scaled demand of 5.2 (2.2 + 3) and coordinates of

$$x = \frac{(2.2) \times 20.9 + (3) \times 60}{5.2} = 43.45$$

$$y = \frac{(2.2) \times (41.8) + (3) \times 20}{5.2} = 29.22$$

Now only two customers remain, and since there are two facilities to place, the procedure terminates. The final solution is given in Table 4.16 and includes the assignment of specific customers to the individual facilities. (See Figure 4.7.)

It should be noted that the combinational procedure is heuristic in nature and provides a nearly optimum solution. Furthermore, be-

Table 4.16   Combinational approach (final solution)

Facility	Coordinates	Assigned Customers
1	(43.45, 29.22)	1, 3, 4
2	(20, 100)	2

Figure 4.7   The final solution

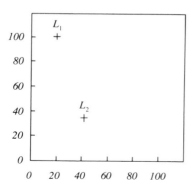

cause it is assumed that the facility capacities are unlimited, this procedure results in the customer demands never being split between any two or more facilities.

4.4 COMPUTER PROGRAM DESCRIPTION

Three programs are listed in Appendix D that represent computerization of some of the methods from this chapter. Two can be used for solving single-facility location problems; one, for multiple facilities of the same type problems.

Single-Facility Problems

Rectilinear Distance Cost Method
The program is in interactive mode with the information to be supplied based on the following questions.

1. *How many machines are there?* (Enter number of facilities.)
2. *What is the x coordinate for the machine; the y coordinate; the number of trips?* (Enter x and y coordinates and number of trips for each machine.)

This program outputs the x-y coordinates of the location for the new machine (facility).

Quadratic Cost Model
The program is in interactive mode with information supplied with the same set of questions as in the rectilinear distance cost method. This program outputs the x-y coordinates of the location for each new machine (facility).

Multiple Facilities of the Same Type Problems

The program is in interactive mode with data supplied as answers to the following questions.

1. *How   many   facilities   need   placement?* (Enter number of new facilities to
   be located.)
2. *How many customers are there?* (Enter
   the number of customers.)
3. *What is the x coordinate of the customer;
   the y coordinate; the demand?* (Enter *x*
   and *y* coordinates and demand for each
   customer.)

This program outputs (1) the coordinates
of the new facilities and (2) customer assignments to the new facilities.

SUMMARY

Three models for placing facilities are presented when one is free to choose any location
within the planning area for their placement. A
major component of the cost, which is minimized, is the distance between facilities. This
distance is measured either in a rectilinear or
Euclidean manner.

In a single-facility placement problem, the
unit processed on the existing facilities also requires services of the new facility so that the
total cost of demand times distance traveled is
minimized.

Two prominent variations for multiple-
facility location problems exist. In the first,
each new facility is independently demanded
by the unit processed on the existing facilities.
Furthermore, there may also be some flow between the new facilities. In the second variation, all new facilities are of the same type. The
problem is to determine the best location for
each of the new facilities and also to assign existing customers to them, so that one customer
is served by one new facility only.

The procedures illustrated here are simple
and efficient. Some of the procedures are also
computerized, listings for which are presented
in Appendix D.

PROBLEMS

4.1 Define the continuous facility location problem.

4.2 Distinguish between rectilinear and Euclidean distances.

4.3 Using Figure P4.3,

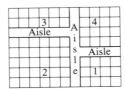

Figure P4.3

   a. Calculate the distance from 1 to the center of each of the other locations
   assuming you are traveling by forklift truck.
   b. Calculate these distances assuming you are traveling by overhead
   monorail.

4.4 a. Using Figure P4.3, determine the best location for a new piece of equipment. Use the rectilinear distance from the center of each location. As-

sume demand from each section is proportional to the area of the section.

b. Solve again using Euclidean distance.

4.5 On the 15th floor of an office building, a coffee maker is to be placed. There is no limit to the number of trips that may be made as long as work is done. A study was done to see how many trips per day on the average each person makes. Given the data in the accompanying table, determine where the coffee maker should be placed.

Customer	Coordinates	Trips/Day
1	5, 5	3
2	10, 40	5
3	70, 45	2
4	65, 20	1
5	25, 25	3

4.6 a. In the Gulf of Mexico, Tell Oil has four rigs. The location of each is given in Figure P4.6. Food for all the rigs is brought out once a month by ships to a central warehouse. From here the food is distributed according to the data in the accompanying table. Where should the warehouse be located?

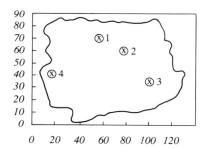

Figure P4.6

Rig	1	2	3	4
Trips/Month	7	10	8	12

b. Rework part a assuming there are two warehouses to place.

4.7 The island of Margarita is famous for its delicious lobsters. They are caught at sea and brought to three main harbors around the island. A refrigerated warehouse is needed to which the lobsters will be shipped by airplane from the harbors, as shown in the accompanying table. Since the product's freshness is so important, where should the warehouse be located to minimize the cost of losing market due to lack of freshness?

Harbors	Coordinates	Trips/Day
1	4, 2	5
2	4, 7	7
3	10, 5	11

4.8 A manufacturing company is to be located along an east-to-west route in the southern section of the United States. The company will serve the three states of Louisiana, Mississippi, and Arkansas. The accompanying table lists the coordinates and the number of trips to be made to each city where the products will be sold. Determine the optimum location for the plant site.

City	Coordinates	Trips/Month
1	100, 440	6
2	20, 300	8
3	200, 300	9
4	400, 400	11

4.9 After being cut at saws 1 and 2, wood is sent either to the sander at F_1 or the lathe at F_2. Determine the best location for the sander and the lathe so as to minimize the distance traveled. The number of trips per hour between each existing machine and the new machine, along with the coordinates of the existing machine, are given in the accompanying table.

Machine	Coordinates (x, y)	Number of Trips	
		Facility 1	Facility 2
1	3, 5	5	2
2	10, 8	4	3
F_1		—	—
F_2		3	—

4.10 a. A corporation wishes to add on a new wing for its president and vice-president. Secretaries are located at 1, engineering at 2, accounting at 3, and sales at 4. Locate these two offices so as to minimize the distance traveled. The average number of trips per day that the president and vice-president make to these offices is given in the accompanying table.

Office	Coordinates (x, y)	Number of Trips	
		Facility 1	Facility 2
1	1, 1	10	15
2	1, 3	8	7
3	3, 1	12	10
4	5, 1	7	6
F_1		—	6
F_2		6	—

      b. Rework part a using the quadratic cost function.

4.11 Describe the variations of the multiple-facility problem.

4.12 Solve Problem 4.5 for the placement of two coffee dispensers.

4.13 Solve Problem 4.6b for the placement of two identical warehouses.

SUGGESTED READINGS

Blair, E.L., and Miller, S., "Interactive Approach to Facilities Design Using Microcomputers," *Computers in Industrial Engineering,* Vol. 9, No. 1, 1985, pp. 91–102.

Brady, S.D.; Rosenthal, R.E.; and Young, D., "Interactive Graphical Minimax Location of Multiple Facilities with General Constraints," *IIE Transactions,* Vol. 15, No. 3, Sept. 1983, pp. 242–254.

Feldman, E.; Lehrer, F.A.; Ray, T.L., "Warehouse Locations Under Continuous Economies of Scale," *Management Science,* Vol. 2, May 1966.

Francis, R.L., and White, J.A., *Facility Layout and Location—An Analytical Approach,* Prentice-Hall, Englewood Cliffs, N.J., 1974.

Khumawala, B.M., "An Efficient Branch and Bound Algorithm for the Warehouse Location Problem," Krannert School of Industrial Administration Institute Paper Series No. 294, Purdue University, Dec. 1980.

Phillips, D.T.; Ravindran, A.; and Solberg, J.J., *Operations Research—Principle and Practice,* John Wiley and Sons, New York, 1976.

Scriabin, M, and Vergin, R.C., "A Cluster-Analytic Approach to Facility Layout," *Management Science,* No. 1, Jan. 1985, pp. 33–49.

PART II

Planning and Design

CHAPTER
5

Product Development

This chapter introduces three topics: market research, forecasting, and product design. Market research is performed to analyze the sales potential for a new or revised product that is being proposed. Forecasting estimates the expected demand for the product in each time period. Design is used to transfer an idea into engineering drawings that form the basis for developing the necessary manufacturing facilities.

5.1 MARKET RESEARCH

In the business world, companies very seldom remain constant in terms of the products they manufacture. New products are often introduced, and existing items are modified to accommodate the changing needs and preferences of consumers. For old and newly emerging corporations alike, information about what people desire, the price they are willing to pay, and the extent of the market potential is of tremendous value.

Market research involves the use of scientific methods, mainly statistical techniques, in collecting and analyzing data to discover consumer needs and desires in relation to the product a company might wish to manufacture and/or introduce in the market. This research could determine consumers' attitudes (for example, the need for the product and the quality and packaging requirements) and their buying habits (for example, frequency of purchase, brand loyalty, and acceptable price) and might help to develop sales and lower marketing costs when the product is finally introduced.

Product planning plays an important part in sales potential, since customers are influenced by factors such as design, price, performance, ergonomics, and the aesthetic appeal of the product. These factors, however, are interrelated. The customer must have some reason to purchase the product; perhaps the item performs a useful task, or the purchaser has a desire to possess the object. The design itself must meet the visual requirements of the consumer; it often influences the purchasing decision and the price the consumer is willing to pay. While failure to provide a satisfactory ergonomic interface for frequently used items may not affect initial sales, it will definitely have an impact on future sales to the same customer, as well as to the customer's associates.

Market surveys can also be used to determine customer-desired characteristics that might influence the design—for example, aesthetic appeal. The public might have preferences in color, shape, handling, and other factors that appeal to their senses. The survey can further indicate an approximate selling price and how this price might change on the basis of additional design features.

Introducing a new product is especially challenging. The company must be sure that the item will not just meet today's competition but will include all of the features that will be available at its introduction and soon after by a competing product selling at approximately the same price. To be assured of sales, the product has to cost less than that of the competitors or must offer features that they do not include. Unless the product is better or cheaper, the customer will not choose it; the objective should not be to match the competitors, but to beat them.

Market Potential

If similar products are already on the market, it is vital to determine the volume produced and sold by the other manufacturers in order to estimate the sales potential. This information can often be obtained from the manufacturers themselves with surveys by letter and telephone and at times by research at the library in government documents, stockholders' reports, and trade journals. Analysis of past and present data on sales, along with the factors that affect those sales, such as population, income level, frequency of purchase, and changes in attitudes and fashions may indicate the market potential for the foreseeable future.

Government documents provide vast amounts of data that otherwise would be almost impossible or very expensive to gather. For instance, the U.S. Department of Commerce, Bureau of the Census, publishes annual reports about the various industries in the United States. These reports include information such as production, sales, imports, and exports. The government also publishes reviews on the "Revised Monthly Retail Sales and Inventories," which record production and sales figures for each industry. These monthly reports divide the retail industries into two segments: durable and nondurable. Durable goods include commodities such as furniture and automobiles, items that will be used at least for a short duration; nondurable goods include food, gasoline, and other things that are suitable for immediate consumption.

Some other sources of data are:

- The Conference Board, *Guide to Consumer Market,* New York. Data on population, employment, income, expenditures, and related items.
- *Business Week.* The first issue each year presents data on expected trends for industries during that year.
- Dun's *Census of American Business* and *The Future Directory* published by *Time.* Both provide financial data about American businesses.
- Market research firms. Many sell their findings; prominent among these are:

  a. A.C. Nielsen Company
  b. Market Research Corporation of America
  c. Market Facts, Inc.
  d. National Family Opinion

Many markets are changing rapidly. Not only is sales history not necessarily a good indicator of the future market, but as a result of the introduction of new technology, often no information is available on which to base a forecast. How do you accurately predict the future of videocassette recorders when you are the first one to enter the market? The use of microchips in appliances, automobiles, and communications has created many new opportunities for companies introducing unique products in the market. In such cases, estimation of the market potential is especially difficult and is mainly based on factors that the company thinks might influence the sales. The data for some of these factors may be obtained either from sources mentioned earlier or by the methods described below.

Collection of Data

Although data can be gathered in many different ways, there are three basic types of surveys that obtain personal views and values, each with its own advantages and disadvantages. The methods are mail survey, telephone survey, and personal interview.

The mail survey is relatively easy to perform and requires a minimum of manpower. A survey form is developed and distributed to the people on a mailing list. However, it can be difficult to obtain responses from the individuals. Maintenance of mailing lists of customers who might be interested in the product can also be a difficult task. It may be possible to buy an appropriate mailing list from any of several organizations.

The telephone interview requires more manpower than does the mail survey. As the name implies, questions are asked over the phone, and responses are tabulated. Again, it is important to converse with the "right type" of customer—people who would be interested in buying the product. The mail survey (with a low response rate) and the telephone survey (with a high response rate) can both be conducted at minimal cost but at the expense of personal contacts. To convey all the features of the product may at times necessitate displaying the product or describing it with visual aids; both alternatives are impossible in mail and telephone surveys.

The personal interview technique requires the polling personnel to contact the potential customers at their homes or places of business. This is a slower method in which the interviewer and the consumer may discuss the questions on the survey form, and the observer can note such things as the consumer's attitude and enthusiasm about the product. However, this form of data collection can be very expensive if the group to be surveyed is spread over a large area.

Survey Form

As an example of a survey form, a questionnaire from a prospective manufacturer of a coffee maker is illustrated in Figure 5.1. The questions should be asked for both home and office use of coffee makers, and the form should display the information the company wishes to acquire. Questions 1, 2, and 3 are asked to determine whether there is a cross section of population with a special interest in the product; if such a group could be identified, advertising and sales campaigns might be geared to appeal to these people. Questions 4, 5, and 6 are included to learn what percentage of the people might be interested in the company's product. Knowledge of what features to include in the product is important; thus questions 7, 10, 11, 14, 15, and 16. Questions 8 and 9 are designed to determine the extent of the competition. To aid in setting prices, questions 12 and 13 are asked. Comments might suggest new design features to be considered for future inclusion.

The questionnaire is generally developed on the basis of the information that is needed and can be used. Only very infrequently should questions that have no information value be included in this form. Collection and analysis of the data are expensive and time consuming, and there is hardly any room for collection of trivial information. Occasionally, a simple question might be used at the beginning of the survey to help put the subject at ease.

Basic Steps in Market Research

We will now formally define the steps required to perform market research. The time and effort required in each step depend on how de-

Figure 5.1

Market research questionnaire

In order to better serve our customers, the ABC Company is conducting a survey, the results of which will be used in developing our latest coffee maker. We value your opinion and would appreciate your assistance in our research. Please take a few minutes to complete the following questionnaire and return it in the enclosed envelope. Thank you very much for your time and cooperation.

(1) Circle appropriate age:
20 or under   21 to 30   31 to 40   41 to 50   Over 50

(2) Sex: Male   Female

(3) Occupation:

(4) Are you a coffee drinker?   Yes   No
How many cups per day?   1   2   3   4   5   Other

(5) Present method of making coffee:
Instant   Percolator   Drip   Other

(6) Preferred method of making coffee:
Instant   Percolator   Drip   Other

(7) What is your preference in color for coffee makers?
White   Yellow   Brown   Black
Other (Please specify)

(8) If you own a drip type coffee maker, what brand is it?

(9) How long have you owned your drip coffee maker?
Less than 6 months   6–12 months   1–2 years
More than 2 years

(10) How often do you make coffee?
Per day: 0   1   2   3   More than 3
Per week: 0   1–4   5–8   9–15   More than 15

(11) How many cups do you make at a time?
2   4   6   8   10

(12) How much did you pay for your present coffee maker?
Below $30 $30–$50   More than $50

(13) How much would you be willing to pay for your coffee maker today?
Below $30 $30–$50   More than $50

(14) Which do you prefer?
Porcelain   Stainless   Glass   Plastic   Other

(15) Would you like a timer on your coffee maker?   Yes   No

(16) Do you use filters?   Yes   No   Cost?

Comments:

tailed the analysis must be for each of the factors involved.

1. *Analyze the situation.* Review company records, trade and professional publications, library material, and available past market research reports to understand the markets. Be aware of the company's potentialities and limitations. Know why the firm wishes to enter into this venture and what it stands to lose in the process.

2. *Plan the research.* Determine what information is needed. What is the purpose of this research, and what are the methods for gathering this information? Decide how the study is to be organized.

3. *Collect data.* Which survey, if any, may be used to collect the information from the consumer? Can government reports, reports released by private corporations, and/or trade magazines be used to discover present sales trends? What facts should be gathered concerning conditions affecting the sales?

4. *Analyze the data.* Which statistical method should be used to analyze the collected data? It is important to review the information and determine its implications. An attempt should be made to find out whether the market size would increase if the public were (more) interested in the product, what price could be charged, and the answers to other related questions about design and sales.

5. *Report the findings.* All results and interpretations should be properly documented and reported to the appropriate levels of management.

Market Decision

The above type of analysis should enable management to decide whether it will be profitable for the company to enter the present market. By using the expected life of the product and the anticipated future consumer needs, the market potential for the company can be estimated. Initially, rough calculations should be made of the production cost to determine the estimated profitability of the product. The cost may depend upon what the company feels its share of the market will be or on how many units it must produce and sell to make the project worthwhile. If the venture seems profitable, arrangements can be made to develop manufacturing facilities to produce the targeted amount.

An Example of Market Research

An equipment-manufacturing company wishes to begin production and sales of steel tables. The plant is located in a large city that has several colleges and two universities in addition to the usual public school system. The company's equipment sales are distributed among many customers: about 30 percent are to the government and offices; and the major portion, about 70 percent, are to the schools. Management expects that 90 to 100 percent of the new steel tables will be sold to the schools. New school buildings are seldom constructed, so this will not be a major source of sales; therefore the company will depend heavily on its present customers.

There are currently four major suppliers of school tables. One of these has decided to stop making shipments to the area. That company's plant is located quite a distance from the city, and when the cost for shipment was added to that for production, the price to the consumer no longer made the company competitive. This has opened the market to someone already in the area who can supply the tables.

The company plans to produce three sizes of tables. The smallest table seats two students on one side or four if both sides are used. The next larger table seats three on one side or up to eight using both sides and the ends. The largest table will be a conference table seating up to

Table 5.1   Physical characteristics of the proposed tables

Table Capacity in Persons		Dimensions (length × width × height) (inches)
One Side	Total	
2	4	54 × 36 × 30
3	8	84 × 36 × 30
6	14	180 × 45 × 40

fourteen people. Table 5.1 shows the dimensions of the three proposed tables.

The company would first like to survey its present customers to determine the potential demand for the product. A large percentage of these customers are located in the vicinity of the plant; and because sales personnel often visit the schools and institutions, the market research teams decide to use the personal interview survey.

Of all the people questioned in the survey, 92 percent responded; all of those were current customers of the company. Seventy percent of the customers are schools that are required by law to take bids. About 9 percent of the customers were under contract with their present supplier. All of the other customers, being private schools, are under no particular restrictions for buying.

Virtually all of the respondents expressed at least some interest in the product. Forty percent of the schools planned to change over from desks to tables as soon as their desks needed replacing. All of the other schools would consider the changeover; and even without a general replacement, they would still have use for some tables. Most expected their desks to have a life of from five to ten years, with an average of eight years.

The respondents also replied concerning the sizes of tables they think are, or would be, in demand; their relative demand ratios; and the average price they are presently paying for

these tables. The statistical analysis provided the distribution shown in Table 5.2.

The company plans to distribute its products through both direct sales and sales to independent distributors. All major sales of 25 units or more will be handled directly by the company. Smaller sales will go through independent distributors.

Production plans call for start-up in January of next year. It is estimated that to make the venture profitable, the company will have to obtain 17 percent of the area market by the end of five years. The sales personnel will strive for 8 percent at the end of the first year, 11 percent the second, and a 2 percent increase in the market share each year for the following three years. These figures were developed after studying past market reports for similar products that the company had manufactured and sold. Demand for school tables in the area has been steadily increasing for the past five years. The desks currently in use have an expected life of eight years, which means that each year approximately 12.5 percent of the desks should be replaced. More and more schools are replacing their desks with tables because the latter are found to be more comfortable and lend flexibility to the classroom setting. The forecast demand for the next five years is shown in the graphs presented in Figure 5.2. Because only a short period of time was to be considered, a linear forecast was performed for each type of table.

The company that is planning to cease business currently has 25 percent of the market.

Table 5.2   Estimated relative demand and price for each type of table

Table Capacity in Persons/Side	Percentage Sales	Price
2	63	$ 63.00
3	30	88.00
6	7	210.00

Figure 5.2   Five-year demand forecast

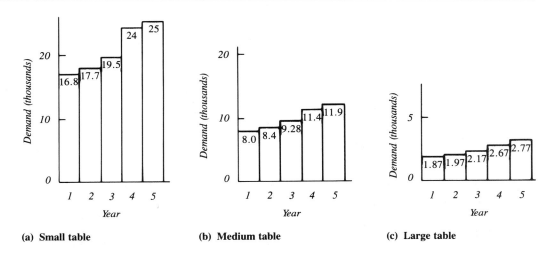

(a) **Small table**   (b) **Medium table**   (c) **Large table**

Using this information and all the other statistics gathered, the team concludes that it is realistic to strive for 17 percent of the market. The number of units to be produced and sold each year is calculated by multiplying the percentage of the market the company plans to get each year by the annual demands. The results are shown in Table 5.3.

At these production rates is it feasible to produce the product? The answer depends on its cost. A cost estimate of direct material, equipment, and labor to produce the parts must now be made. To this, overhead costs to account for such items as material handling, utilities, and supervision must be added. The overhead cost could be considerable; 150 to 200 percent of the direct labor is not uncommon. A planning sheet, as shown in Figure 5.3, describing all operations and the equipment and tools to be used could help in developing the cost estimate for the product. Chapter 7 describes production charts that are helpful in planning the detailed steps necessary in developing cost estimates.

With an estimated efficiency of 90 percent the direct cost would be $32.57 = 29.32/0.9. The indirect cost for the plant is 160 percent of the direct labor, or $20.27 (11.40/0.90 $\times$ 1.6), giving a cost for manufacturing of $52.84. If we add $1.20 for administration and packaging/shipping, the net cost to the company is $54.04. Since the unit is expected to sell for $63.00, it will generate a profit of $8.96.

Similar analysis on other types of tables results in profit of $10.20 for medium tables and $18.15 for large tables. Based on projected sales from the previous chart, the yearly profit is as follows:

Table 5.3   Projected sales

Year	Percent of Total Market	Units Produced and Sold		
		Small	Medium	Large
1	8%	1300	640	150
2	11%	1947	924	217
3	13%	2535	1204	282
4	15%	3160	1710	401
5	16%	4000	1904	444

Figure 5.3   Sample planning sheet

Small Table

| OPERATION | | MACHINE | TOOLS/EQUIP. | DIRECT COST | | |
NO.	DESCRIP.	NAME	REQUIREMENTS	LABOR	MATERIAL	EQUIP.
10	Cut to size (legs)	Radial arm saw		5 min. at $5/hr = $.42	Steel tubing 4 × 29" × 1¹/₄" dia. × ¹/₈" = $2.90	5 min. at $6/hr = $.50
20	Weld bracket	Welding M/C	Clamps	10 min. at $6.60/ hr = $1.10	Rods $.10 Brackets $2.20	10 min. at $4/hr = $.67
			Total	$11.40	$12.20	$5.72

The direct cost of production is $29.32 (11.40 + 12.20 + 5.72).

Year	Profit
1	$21,292.74
2	$30,808.47
3	$40,143.30
4	$56,976.15
5	$63,319.40

This gives a rough estimate of the average yearly profit of $42,508.01. The management expects the need for an additional investment of $125,000 to engage in the planned level of activity. Assuming a five-year life for the project, the internal rate of return for this investment is 17 percent, and, since the company's minimum attractive rate of return before taxes is 15 percent, the project is accepted.

5.2 FORECASTING

Most people like to have sufficient lead time to plan for future activities. This is especially true of engineers and managers who are responsible for plant installation and operations. They must decide, for example, how large the plant should be, which machines to buy, how many workers to employ and train, and how to arrange for financing. The answers might depend on projected or forecast sales of the products.

Forecasting techniques can be divided into two major categories, quantitative and qualitative. Quantitative techniques are applicable when some quantifiable data are available about the past performance and it is assumed that the same pattern will continue in the foreseeable future. Qualitative forecasting, on the other hand, requires no specific data but is mainly based on intuition, judgment, and the opinions of people who are knowledgeable and have experience in that specific activity or product. When past quantitative data are not available, as may be the case in introducing a new product, qualitative methods might be the only way of forecasting. Such techniques could also supplement quantitative methods when the pattern for future activities might change by factors that were not prominent in the past. We will study some of the basic techniques of quantitative forecasting, but before that we must characterize the means of identifying what a good forecast is.

Error Measuring

The difference between the forecast value, F, and the actual demand, D, is the error in forecasting, e. The objective is to choose a forecasting method and/or associated parameters such as weight that will minimize this error.

Mathematically, an error in period t can be expressed as $e_t = F_t - D_t$. If there are n periods for which both forecast and actual demands are known, then we have several ways of recording the total error:

$$ME = \sum_{t=1}^{n} \frac{e_t}{n}$$

$$MAE = \sum_{t=1}^{n} \frac{|e_t|}{n}$$

$$MSE = \sum_{t=1}^{n} \frac{e_t^2}{n}$$

$$SDE = \sqrt{\sum_{t=1}^{n} \frac{e_t^2}{n-1}}$$

where

ME = Mean error

MAE = Mean absolute error

MSE = Mean squared error

SDE = Standard deviation of error

There are also other error measures that are more suitable under certain conditions [interested readers may refer to Makidakis et al. (1984),] but MSE is one of the more popular procedures to follow in evaluating the forecasting methods described herein.

Forecasting Methods

Moving Average

A simple forecasting method for averaging the past relevant data, the moving average method requires us to define the number of observations that will be included in the calculation of the average. As new data become available, the oldest observation is dropped and the latest observation is included in calculating the moving average, the prediction for the next period. If we are predicting demand, for example, mathematically we have

$$F_t = \frac{\sum_{i=1}^{n} D_{t-i}}{n}$$

where F_t is the forecast demand during the period t, D_{t-i} is the actual demand in period $t - i$, with $i = 1, 2, \ldots, n$, and n is the number of observations to be included in the moving average.

For example, given that $n = 3$ and the actual demands for the past six months are as follows:

Month	Observed Demand
January	500
February	515
March	600
April	620
May	595
June	635

the predicted demand for July is

$$F_{\text{July}} = \frac{(620 + 595 + 635)}{3} = 616.66$$

or 617 units

If the actual demand for July were found

to be 625, then the predicted demand for August would be

$$F_{\text{August}} = \frac{(595 + 635 + 625)}{3} = 618.33$$

$$\text{or } 618 \text{ units}$$

Weighted Moving Average

In the preceding method, each observation within the moving average calculation made an equal contribution, or had the same weight, in predicting the forecast value. It may be desirable to give more importance to the latter observations and less to the more distant past. This is achieved by "weighting" each data point. In the previous example, suppose we assign the weights shown in Table 5.4.

Then the forecast demand in July would be

$$\frac{1}{8}(620) + \frac{1}{4}(595) + \frac{5}{8}(635) = 623$$

or, in general,

$$F_t = \frac{\sum_{t=1}^{n}(W_{t-i} \times D_{t-i})}{n}$$

where W_{t-i} is the weight placed on observation $t - i$.

Exponential Smoothing

Though it is generally preferred that recent values be given more weight in forecasting

than older observations, determining the appropriate weight for each as is required in the weighted moving average method is normally a complex task. Exponential smoothing is another alternative in which the weight assigned to observations decreases with age, specifically in an exponential manner. Furthermore, it is accomplished by knowing only the actual and forecast demands for the last period and the smoothing constant. In practice this method is mainly used for short-term forecasting.

The method is developed with the following thoughts in mind. The variation between the forecast demand and the actual demand may be caused by two factors: (1) a trend and (2) random fluctuations called noise. Ideally, the next forecast should follow the trend and not react to the noise. We must decide what percent (called the smoothing constant) of the difference between the actual and forecast demands is due to the trend. The forecast for the next time period is adjusted for the trend.

Mathematically, it can be summarized as

$$F_{t+1} = F_t + \alpha(D_t - F_t)$$

where

F_{t+1} = Demand forecast in period $t + 1$
D_t = Actual demand in period t
α = Smoothing constant where $0 < \alpha < 1$

By rearranging the terms we can write

$$F_{t+1} = \alpha D_t + (1 - \alpha)F_t$$

To show how the influence of each observation decreases as it becomes older, substitute for F_t in the above expression:

$$F_{t+1} = \alpha D_t + (1 - \alpha)(\alpha D_{t-1} + (1 - \alpha)F_{t-1})$$
$$= \alpha D_t + \alpha(1 - \alpha)D_{t-1} + (1 - \alpha)^2 F_{t-1}$$

Continuing the process, one obtains

Table 5.4   Relative weights on the demands

Observation	*Weight*	*Relative Weight*
One month old	5	5/8
Two months old	2	1/4
Three months old	1	1/8
Total	8	

$$F_{t+1} = \alpha D_t + \alpha(1 - \alpha)D_{t-1} + \alpha(1 - \alpha)^2 D_{t-2}$$
$$+ \alpha(1 - \alpha)^{n-1}D_{t-(n-1)} + (1 - \alpha)^n F_{t-(n-1)}$$

as $\alpha < 1$, and the weight for each successive observation is decreasing at the rate of $1 - \alpha$; therefore the weight of the older observation is decreasing exponentially.

Initially, we must first fix the value of α. Suppose in the prior example that $\alpha = 0.2$; then the forecast demands for each period would have been as shown in Table 5.5.

To begin the procedure, the forecast value is assumed to be the same as the actual demand for the first period, January. The forecast for February is the same as that for January, since there is no forecasting error in January. The forecast for March is $500 + 0.2(515 - 500) = 503$, and so on. The forecast values for the last six periods do not compare very well with the actual values. Try another value for α, for instance 0.6, chosen at random for illustration. (See Table 5.6.)

Table 5.5 Demand forecast for $\alpha = 0.2$

Month	Forecast	Actual Demand
January	500.0	500.0
February	500.0	515.0
March	503.0	600.0
April	522.4	620.0
May	541.9	595.0
June	552.5	635.0
July	569.0	

Table 5.6 Demand forecast for $\alpha = 0.6$

Month	Forecast	Actual Demand
January	500.0	500.0
February	500.0	515.0
March	509.0	600.0
April	563.6	620.0
May	597.4	595.0
June	594.9	635.0

Now the forecast is improved, but it is desirable to determine the optimum value of α. One method is to try different values of α, calculate the mean square error associated with each, and select the one that gives the minimum *MSE*. In this example the best value for α, to one-digit accuracy, is 0.9.

Curve Fitting

One way to forecast the demand is simply to plot the curve or calculate the value of an equation from available data and extend it to make predictions. The method allows a number of variables to be independent, the values for which can be either preset, controlled, or measured. The estimate of the expected demand can then be made when these values are known. The coefficients of the equations are evaluated following the least squares fit method—that is, minimizing the sum of squares of the errors.

Suppose the equation to be fitted is

$$Y = b_0 + b_1 X_{1i} + b_2 X_{2i} + \cdots + b_m X_{mi}$$

where there are m independent variables and n observations, $\hat{Y}_i$ is the estimated value of the demand, and Y_i is the actual observed demand in period i. The error in measurement is $e = (Y_i - \hat{Y}_i)$, and the least square fit requires

$$\min e^2 = \min \sum_{i=1}^{n} (Y_i - \hat{Y}_i)^2$$
$$= \min \sum_{i=1}^{n} (Y_i - (b_0 + b_1 X_{1i} + b_2 X_{2i} + \cdots + b_m X_{mi}))^2$$

The minimization is obtained by taking the first derivative with respect to each coefficient, that is, $b_0, b_1, \ldots, b_m$, equating them to zero, and solving the resulting equations simultaneously to obtain the values of each coefficient. In its simplest form it is known as simple linear regression, which uses only one independent variable, for example, the time period, in the equation $Y = b_0 + b_1 X_i$. The curve is a straight line, and the corresponding coefficients are obtained by

$$b_1 = \frac{n\Sigma X_i Y_i - (\Sigma X_i)(\Sigma Y_i)}{n\Sigma X_i^2 - (\Sigma X_i)^2}$$

$$b_0 = \bar{Y} - b_1 \bar{X}$$

where

$$\bar{Y} = \frac{\Sigma Y_i}{n} \quad \text{and} \quad \bar{X} = \frac{\Sigma X_i}{n}$$

Example: Demand Prediction by Curve Fitting

Consider a lumber manufacturer that has sawdust as its by-product and is contemplating developing a fire log producing facility to utilize the sawdust. After contacting the stores within a hundred mile radius where it believes it can sell its product, the manufacturer has received the following information on the demand for the fire logs in this area, over the last six seasons:

Season (X)	Demand (in thousands) (Y)
1980	102
1981	105
1982	110
1983	108
1984	115
1985	113

The manufacturer expects that it can capture about 10 percent of the market in the first year and can increase its share by 3 percent in each of the next three years. Determine the expected demand in each of the next four years.

Solution: The plot of the data points is shown in Figure 5.4. It seems that the relationship between years and demand can be expressed satisfactorily by a linear fit, that is,

$$\hat{Y}_i = b_0 + b_1 X_i .$$

Figure 5.4   Relationship between the years and demand

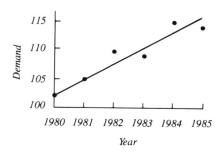

When $\hat{Y}_i$ is the predicted demand in year X_i, note that the calculations can be made with X_i as perhaps 1980 or 1981 or by coding the years by subtracting 1980 to result in 0, 1, 2, 3, etc. Using X_i as 1980, 1981, ..., we get the following results:

$$b_1 = \frac{n\Sigma X_i Y_i - (\Sigma X_i)(\Sigma Y_i)}{n\Sigma X_i^2 - (\Sigma X_i)^2}$$

and

$$b_0 = \bar{Y} - b_1 \bar{X}$$

$\Sigma X_i = 11,895, \qquad \Sigma X_i^2 = 23,581,855,$

$\bar{X} = 1982.5$

$\Sigma Y_i = 653, \quad (\Sigma X_i \, Y_i) = 1,294,614,$

$\bar{Y} = 108.83 \quad (\Sigma X_i)(\Sigma Y_i) = 7,767,435$

Subtracting the appropriate values in the above expressions, we have

$$b_1 = 2.3714 \qquad \text{and} \qquad b_0 = -4592.47$$

giving the following equation for predicting demand:

$$\hat{Y} = -4592.47 + 2.3741 X_i$$

Now we can forecast the total demand for the next four years by extending this graph and by knowing the percent of the market the manufacturer has predicted for itself. The manufacturer estimates demand as shown in Table 5.7.

Table 5.7 Forecasted demand

X_i	$\hat{Y}_i$ (in thousands)	Market Share	Predicted Demand (in thousands)
1986	117.13	10%	11.71
1987	119.50	13%	15.54
1988	121.87	16%	19.50
1989	124.24	19%	23.61

Table 5.8 Domestic consumption

Year	Chrome Faucets	Chrome Showerheads
1981	579,337	566,170
1982	524,958	552,970
1983	514,989	458,936
1984	519,239	551,481
1985	632,444	595,291

Incidentally, had we used the coded values for the years (1980 = 0, 1981 = 1, . . .), the calculations would have resulted in the following equation:

$$\hat{Y}_i = 102.90 + 2.3714X_i$$

Using 1986 as 6 would have resulted in the same market predictions as we have observed before.

This solution assumes a continuous linear trend. However, predicting so far into the future is very subjective. For example, the demand might saturate and remain at some constant value. If the manufacturer thinks that this might happen in the planning period, it must change the equation of the curve that will take into consideration this belief. For instance, the equation might be of the form

$$\hat{Y}_i = b_0 X_i^{b_1}$$

Demand Forecasting for a New Company

The following two cases indicate how forecasting could be accomplished by using government documents and surveys. The first case is for chrome faucets and showerheads. A U.S. government document entitled "Current Industrial Reports of Plumbing Fixtures" shows domestic sales of chrome faucets in the United States between 1980 and 1984. (See Table 5.8.) The data were obtained by adding U.S. manufacturers' production to imports and subtracting exports from the total.

Simple linear regression analysis results in the following forecasting equations.

For chrome faucets:

Demand in current year = 524,004.9
+ 10,049.5 (Current year − 1980)

For showerheads:

Demand in current year = 527,943.7
+ 5,675.3 (Current year − 1980)

The results from the regression equations could be improved by including factors that affect major sales of the faucets and showerheads, such as new building construction and retail replacement sales. But as an initial estimate, the above equations are fairly good predictors.

The second case consists of a company trying to forecast the demand for a product called "CompuTable." This is a set consisting of a chair and a table to accommodate a desk computer, its monitor, disk drives, printers, and floppy disks. The product is to be used mainly by owners of home computers, although it can also be used in small businesses.

The government document does not list such an item; therefore the future demands could not be predicted on the basis of the past known sales. Surveys were conducted by telephone and by personal interviews, and the data associated with home computer users were analyzed further. The results indicated that 61.8 percent of home computer users would like to own a product similar to CompuTable. A search in *Facts on File*, April 1985 (published by Facts on File Inc., 460 Park Ave., New York, NY 10016), showed that 10 percent of U.S. households owned a home computer as of 1985. Using statistics from the U.S. Bureau of Commerce as a guide to the number of households in 1984, the number of home computer owners is estimated to be approximately 8.5 million. Further research in *Business Week* magazine (June 24, 1985) supported these findings. *Business Week* predicted a leveling off of home computer demand in the near future to about 2 million units per year and a slight rise in the sale of portable computers, which can also use CompuTable.

Neglecting minor losses due to death and the "fad factor," a conservative estimate of the potential buying population as of February 1985 would be 61.8 percent of 8.5 million computer owners, or 5.25 million people. This population would increase annually by 61.8 percent of 2 million new computer buyers, or by 1.23 million per year. The surveys indicate that approximately 10.9 percent of the above market would purchase a table in any given year. Hence the expected demand for the next five years, starting in 1987, which is the first production year for the company, are as follows, where demand is in thousands:

1987: $0.109((5.25) + 3(1.23)) = 974.46$ (projected from 1985 to 1987)

1988: $0.109(0.70(8.94) + 1.23) = 1002.23$

1989: $0.109(0.891(7.49) + 1.23) = 861.491$

1990: $0.109(0.891(6.473) + 1.23) = 762.72$

1991: $0.109(0.891(5.76) + 1.23) = 693.475$

where 8.94 was the total prospective population in 1987, 10.9 percent of which bought a table, leaving 89.1 percent as prospective buyers.

If the company targets 2 percent, 4 per-

cent, 5.5 percent, 6.5 percent, and 7.5 percent of the total market in each consecutive year, its potential demand would be 19,489, 40,089, 47,382, 49,577, and 52,011 for each year, respectively.

5.3 DESIGN AND DRAWINGS

Once the market analysis is performed and it has been determined that the product has a sufficient probable market, the next step is to develop the detailed design that is suitable for production. Resource requirements and complexity in design are almost directly proportional to the extent to which the prospective product requires an original analysis. If a similar product is already on the market, one can take advantage of the situation by synthesizing its design features. However, if a new product is being introduced, the development and testing phases can be very expensive and time consuming.

Before one initiates the design of a product, it is helpful to keep a few points in mind.

Close tolerances are important because they define the quality of the product, but expensive machine tools and machine operations are required to produce parts with great accuracy. Figure 5.5 shows the cost-tolerance relationship for a machining operation; the cost of

Figure 5.5   Cost-tolerance relationship for machining

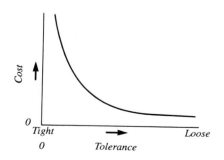

production increases very rapidly as the tolerance is reduced.

A product should not be overdesigned. There always is a need to allow for some safety factor, but overdesigning is expensive. Most items need not be the best on the market; they must, however, be competitive.

A product is often judged by its appearance as well as how well it performs its designated function. As is the case with tolerance, however, the cost for the appearance (surface finish, coating, trim) increases rapidly.

A designer must be aware of the estimated volume of production of the finished product. The product designed for mass sales must be adaptable for manufacture on mass production machines with a minimum of different setups.

The product designed for processing with automatic equipment such as a numerically controlled machine and/or automated material-handling unit such as a robot should include some or all of the following characteristics.

It should be able to self-align and provide a reference surface to eliminate adjustments during setups to the extent possible. The number of screws in the part should be minimized by providing snap-on fasteners; and where screws are required, the ratio of the length to the diameter should be at least 1½. A rectangular rather than square part shape is easier to orient in handling operations. A design that will allow parts to become entangled with each other should be avoided. For example, closed end springs are preferable to open end springs, and parts with cables tend to tangle more often than parts without such cables.

Design

Development of design follows basically the same steps as any engineering analysis, namely:

1. Identify the problem and develop preliminary ideas.

2. Refine the ideas.
3. Analyze and select the design. Test the suitability of that design.
4. Implement the decision.

Each step can be further subdivided to analyze its specific properties. For example, refinement may include modification of

· shapes and forms,
· weights and volumes,
· physical properties such as strength, elasticity, and/or impact resistance, and/or
· scale drawings.

Or implementation may include development of

· working drawings and specifications,
· models and details, and/or
· testing and modification of a prototype.

Drawings

The final solution (design) must have a sufficient degree of detail and a complete set of working drawings (or models) that may form the legal basis for an outside contractor to bid on the job, if so desired.

These drawings should include an assembly drawing, detail drawings, a bill of materials or parts list, and perhaps a special, exploded pictorial drawing.

The assembly drawing is necessary to demonstrate how the various parts of the finished product are finally assembled or fit together and to reveal, as far as is practical, how the finished product might function. Usually, only overall dimensions and those necessary for assembly are given on the drawings, which themselves can take many different forms. For example, they may consist of one or more of the following, depending on the designer's desire for clarity and economy:

· orthographic sectional views (cut-aways) (see Figure 5.6),
· axonometric or oblique pictorials (see Figure 5.7),
· perspective drawings (artist sketches), and/or
· exploded axonometric views and perspective sketches.

The detail drawings, such as that shown in Figure 5.8, are inevitably orthographic projections of each individual piece (part) of the product and provide a detailed description of the shape and size (dimensions) of the finished piece. Detailed drawings for standard parts

such as standard nuts, bolts, and keys need not be made, but such parts are shown in the assembly drawing and are included in the bill of materials or parts list.

The specifications for each piece (in addition to the shape and size as shown on the detail drawing) are given either in the bill of materials, described in the next section, or on the parts list, which is usually included on the face of the assembly drawing. (Exceptions are permitted when clarity is thus enhanced, and in-house standards might vary from company to company.) The parts list includes the following information for each piece required for the final assembly: piece number, piece name,

Figure 5.6 Assembly drawing of teakettle

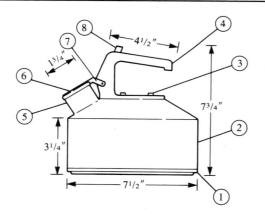

NO.	PART NAME	REQ'D	MAT'L
1	BASE	1	ALUM
2	BODY	1	ALUM
3	SCREW	2	STEEL
4	HANDLE	1	BAKELITE
5	SPOUT	1	ALUM
6	UPPER LID PART	1	BAKELITE
7	ROLL PIN	1	STEEL
8	LID LEVER	1	BAKELITE

Original drawing scale: $1'' = 2''$

Figure 5.7 Isometric drawing of teakettle

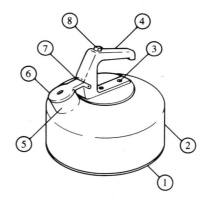

NO.	PART NAME	REQ'D	MAT'L
1	BASE	1	ALUM
2	BODY	1	ALUM
3	SCREW	2	STEEL
4	HANDLE	1	BAKELITE
5	SPOUT	1	ALUM
6	UPPER LID PART	1	BAKELITE
7	ROLL PIN	1	STEEL
8	LID LEVER	1	BAKELITE

Original drawing scale: $1'' = 2''$

Figure 5.8 Detail drawing of teakettle

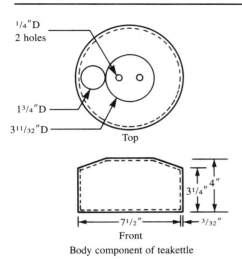

Body component of teakettle

Original drawing scale: $1'' = 2''$

quantity required for the final assembly, material, sometimes the stock size of the raw material, detail drawing numbers, weight, and any other pertinent details. A space is also provided beside each piece for any remarks, such as heat treatment.

Bill of Materials

A bill of materials displays a list of parts that are directly required to make a complete assembly. At a minimum it should indicate, for each part, its number (or drawing number), its description, the quantity necessary in the assembly, and whether the part is to be made within the plant or to be purchased from a supplier. The listing may include additional information such as material description, weight, and unit price. The name and part number or the model number of the assembly are also shown at the top of the chart. Since a bill of materials should list only the parts and subassemblies that go directly into that assembly, a separate bill of ma-

terials is required for each assembly and then for each subassembly that is a part of the assembly, and so on. Figure 5.9 shows a bill of materials for the teakettle illustrated earlier.

Additional Drawings

In addition to the assembly and detail drawings, there may be auxiliary drawings that include special needs of the user. These might include, for example, in the case of a machine, plans for the foundation to accommodate the machine, oiling diagrams for use with the preventive maintenance program, and wiring diagrams showing electrical connectors.

The next step is to develop route sheets, one for each part that is to be produced in the plant. They show, step by step, how the part is to be made, which machines to use, the operations to perform, and standard time allocations. A detailed routing sheet is illustrated in Chapter 7, but at this point it should be noted that machines and standard times are required to develop such a sheet. Chapter 6 briefly describes various manufacturing processes that could be used in making a part. The choice of processes influences the selection of machines and thus affects the entries in the routing sheets.

5.4 COMPUTER-AIDED DESIGN

Computers are being used in design development with increasing frequency. Besides performing engineering calculations, computer-aided design (CAD) can be used to improve speed and accuracy in designing. CAD is extremely useful in modifying an existing design or in developing a new design in a family of parts, such as printed circuit boards. The computer permits pretesting of the proposed product by simulating its use under various conditions. It is also possible to use the computer to

Figure 5.9   Bill of materials for teakettle

PRODUCT: TEAKETTLE (FINAL ASSEMBLY)
STOCK NUMBER: 100

STOCK # DRAW. #	PART NAME	UNITS REQ.	MATERIAL DESCRIP.	WT (LB)	UNIT COST ($)	SOURCE FOR MAT./PART
1	Base	1	3/64" sheet aluminum 3003 alloy (AL3003)	.1678	.150	Reynolds Aluminum Supply (RAS) Make
2	Body	1	3/64" (AL3003)	.201	.18	RAS Make
3	Machine screw	2	Low-carbon plated steel	.02	.008	Lone Star Screw Co. (LSS) Buy
4	Handle	1	Thermo-setting phenolic compound mineral (Bakelite)	.1830	.10	Plastic Engr. Co. (PEC) Buy
5	Spout	1	3/64" (AL3003)	.0368	.040	RAS Make
6	Upper lid part	1	Bakelite	.0166	.010	PEC Buy
7	Roll pin	1	Stainless steel spring	.010	.006	Schinder's Machine Works Buy
8	Lid lever	1	Bakelite	.020	.008	LSS Buy

Total weight = 0.6407 lb
Total material cost = $0.518

Computer-aided design. A designer can test different product features on a computer that is programmed to display data in the form of design drawings. When the designer modifies the data to change features like shape and size, the effect shows up instantly in the drawing. The computer can also compare the effects on factors such as durability, aerodynamics, and buoyancy of using various materials. Revisions to one of a set of drawings automatically modify all related drawings, and changes can be made—and unmade—quickly and easily.

test many different design features without having to actually manufacture the model.

The computer-aided design system consists of three major components: the designer/draftsman, hardware (a computer of any reasonable size from a large mainframe down to a standard personal computer, a graphics display terminal, keyboard(s), a light pen, printer, disk drives, and so forth), and software. In the interactive mode the computer is used to display data in the form of pictures and symbols. The designer can modify the data and immediately see the effects on the design. The software calculates the results of the changes and displays the corresponding figures on the cathode-ray tube (CRT) screen. The designer may similarly change shapes and forms, and the computer

will show the effects of those changes on individual parts and on the assembly. The designer may alter design tolerances and observe the effects on mating parts. The designer can also enlarge a design on the screen and look for any irregularities such as interference between moving parts.

CAD systems have also been used in drafting. A draftsman may use a light pen, a keyboard, or a mouse to input pictures and symbols into a computer. Because of the ability of the computer to repeat symbols that have been drawn, enlarge or reduce the size, transfer the picture from one position to another on the screen, store and return information and drawings, and ultimately print a hard copy, a threefold increase in productivity in drafting is quite common when computers are used. Some companies have reported as much as sevenfold productivity gains. Revisions to existing drawings, a task that is often put off, becomes very easy to accomplish because of the system's ability to store and retrieve drawings. Software that is capable of producing three-dimensional views has proved to be quite effective in automatic interference drawings of components. CAD systems have also been developed to produce parts lists and bills of materials from drawing specifications and to check for the validity of specifications.

The advantages of a CAD system are as follows:

1. It allows visualization of an item being designed. The computer analyzes and displays the effects of any design changes on the part and/or the subassembly and the entire assembly.

2. It can be used to produce drawings, specifications, and bills of materials.

If an organization has more than ten fulltime draftsmen, it can probably afford some type of a CAD system. Which type is most suit-

Figure 5.10   Functional tasks

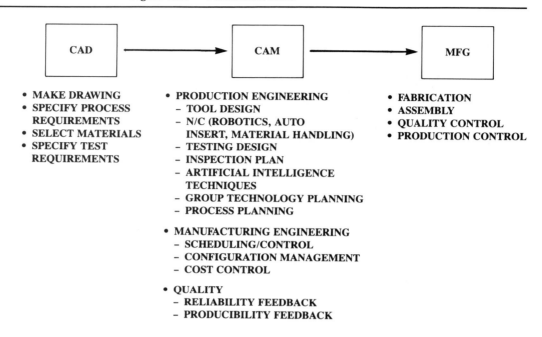

- **MAKE DRAWING**
- **SPECIFY PROCESS REQUIREMENTS**
- **SELECT MATERIALS**
- **SPECIFY TEST REQUIREMENTS**

- **PRODUCTION ENGINEERING**
  - **TOOL DESIGN**
  - **N/C (ROBOTICS, AUTO INSERT, MATERIAL HANDLING)**
  - **TESTING DESIGN**
  - **INSPECTION PLAN**
  - **ARTIFICIAL INTELLIGENCE TECHNIQUES**
  - **GROUP TECHNOLOGY PLANNING**
  - **PROCESS PLANNING**

- **MANUFACTURING ENGINEERING**
  - **SCHEDULING/CONTROL**
  - **CONFIGURATION MANAGEMENT**
  - **COST CONTROL**

- **QUALITY**
  - **RELIABILITY FEEDBACK**
  - **PRODUCIBILITY FEEDBACK**

- **FABRICATION**
- **ASSEMBLY**
- **QUALITY CONTROL**
- **PRODUCTION CONTROL**

able depends on the work being done and the expected growth in the use of the machine. Careful thought during the initial installation phase is necessary, since each manufacturer's system is different and most are not compatible with those of other manufacturers.

There are three basic types of CAD systems: those based on a microcomputer that can be a stand-alone system, a minicomputer-based system, or a mainframe-based system. The introduction of work stations in microprocessors and the networks of modems that are available are making the distinctions between types of systems less well defined.

Characteristics of the basic types of CAD systems are as follows:

1. A stand-alone system has a small memory and only one work station. Its main advantage is its low cost.

2. A system based on a minicomputer, such as a DEC Vax 11/780, can support per-

haps 20 computer terminals. This type of system has a high initial cost, but it is expandable.

3. Both the microcomputer-based and the minicomputer-based systems require a minimum of data processing or management information system support. The CAD system user can usually purchase or develop the necessary operating systems and the software support.

4. A mainframe-based system uses a large computer, such as an IBM 3033, and can support a large number of work stations, but it requires a large staff to operate and maintain. The cost for this type of system is obviously much greater than that for any other, but it can be justified if a large volume of work is required.

Even though CAD systems can be developed by the user (an expensive undertaking), most are purchased from software vendors. Indeed, it is generally more convenient to have a vendor provide the hardware and the software but to retain in-house responsibility for all maintenance. In most cases, some modifications to the purchased programs will be necessary for a system to fit the user's specific needs.

The use of the CAD system may be extended further if its data base can be directly connected to computer-aided manufacturing (CAM) machines, such as numerically controlled machines. Research activity is currently moving in that direction. Figure 5.10 shows functional relationships between CAD-CAM and manufacturing. More discussion on this integration is saved until the last chapter (Chapter 16), by which point the reader will be familiar with more aspects of automation.

5.5 COMPUTER PROGRAM DESCRIPTION

The program calculates the rate of return given the initial investment and yearly income. It is in interactive mode with the information to be supplied based on the following questions.

1. *What is the initial investment?* (Enter the dollars invested at year 0.)
2. *How many years will the project operate?* (Enter the estimated life of the project.)
3. *What is the income for year j?* (Enter the estimated earnings for year *j*. One entry is required for each year of operation.)

This program outputs the internal rate of return.

SUMMARY

To decide whether the introduction of a new product is potentially profitable, market research must be conducted. This can be accomplished by gathering information from well-established marketing firms or by performing market research of our own. Mail surveys, telephone surveys, and personal interviews are three techniques for conducting such research, and each has advantages and disadvantages. An analysis with a planning sheet, along with the data from market research, may provide clues to the profitability of a venture.

Forecasting future markets based on available information plays an important part in estimating customer demands. If the data are insufficient or inconclusive, one must resort to a qualitative prediction based mainly on judgment and experience. When data are available, quantitative methods, such as a moving average, exponential smoothing, and trend analysis, may be utilized. The question of which method to use is to a large extent answered by determining which forecasting method provides the minimum error.

Close attention must be paid to the details of product design during the initial planning phases. A product should be functional and reliable without being overdesigned; however, the method of manufacturing also influences some of the design features. Mechanical drawings are the link between design engineers and manufacturing personnel. These drawings should include assembly drawings, detail drawings, a bill of materials, parts lists, and even special and exploded views for clarity.

Computer-aided design (CAD) can significantly improve speed and accuracy in design. It is especially valuable if the new product can be built using or modifying existing components. CAD allows one to visualize the items, test them for design changes, simulate their use to understand potential problems, and produce detail drawings and bills of materials for plant use—all without going to the expense of actual production.

In examining market research, forecasting, and product design, the chapter covers the initial planning phases for a product. Whether the product is successful in the market or not is to a large extent dependent on the thoroughness of the analysis.

PROBLEMS

5.1 What is market research?

5.2 Where can information about market potential be found?

5.3 Review the steps for performing market research.

5.4 Contrast quantitative and qualitative forecasting techniques. Is any one technique preferable to another?

5.5 What is meant by error in forecasting? What are the best means of measuring such error? How do you know if you have a fairly good forecasting model?

5.6 Known demand for a product (in hundreds) for the last eight periods is listed below.

Period	1	2	3	4	5	6	7	8
Demand	5.0	8.3	13.9	16.2	15.4	18.6	16.4	17.5

   a. Using the three-months moving average, forecast the demand for Period 9. If the actual demand for Period 9 is 18.3, forecast the demand for Period 10.

   b. Predict the demand for Periods 9 and 10 using the exponential smoothing technique. Select the best value of α.

   c. Use linear regression to forecast the demand for these two periods.

5.7 A fiberglass insulation manufacturer is trying to decide how much insulation to produce in August. The manufacturer knows that it will sell about twice as much insulation in the first months of summer and winter (June, July and December, January) than in any other month of the year. The eight previous months of data that are available are given in the accompanying table. Predict demand for July and August.

Month	Number of Rolls
September	820
October	790
November	835
December	1575
January	1724
February	783
March	811
April	827
May	845
June	1710

5.8 A furniture manufacturer is thinking of adding a new line of waterbeds to its existing line of products. The manufacturer believes it must sell at least 750 beds per month to make any profit. Being a new product, waterbed sales will probably not amount to more than 10 percent of the market in the first year. Market research showed the sales of waterbeds for the last five years given in the accompanying table. Apply the least squares method to predict the demand and to determine whether or not the manufacturer should add the new line.

Year	Total Sales/Month of Waterbeds
1981	1000
1982	3000
1983	4430
1984	5129
1985	6538

5.9 BP Manufacturing is trying to enter the fishing equipment market by introducing a sound-emitting device called the Dial-A-Fish, which is scientifically proven to attract fish in laboratories. This new device should gain 10 percent of the present market according to market research. The facilities needed are 5000 square feet for production, 1350 square feet for packaging, and 1500 square feet for shipping and receiving. The plant will be located in a region where the labor rate is $5 an hour for assembly line workers, and overhead costs are 150 percent of direct labor cost. Land cost is $7000 per acre, and the construction cost is $200 a square foot.

   The present market for similar devices is 250,000 per year, and it is estimated that on the average a total of 0.5 hour of labor is required per piece. The unit will sell for $15.00. Over a five-year period, will BP Manufacturing make a profit if the present interest rate is 10 percent?

5.10 Why is it beneficial for a designer to consult with a manufacturing engineer in the plant during the design phase of a product?

5.11 Review the steps in developing a product design.

5.12 As a manufacturer, list six considerations, in order of importance, that should be kept in mind in designing an electric toaster and a lawn mower. How would the rankings change when viewed by a consumer?

5.13 What information should be included in the product drawing? In the bill of materials?

5.14 Using Figure P5.14, develop a bill of materials for an eight-inch aluminum frying pan including weight of parts (aluminum weighs 0.098 pound per cubic inch, and Bakelite 0.162 pound per cubic inch), number of each part needed, part name, material, material description and supplier, and unit cost. The cost per 1000 pounds of aluminum is $1350.00. The cost per 1000 pounds of Bakelite is $700.00.

BP Manufacturing

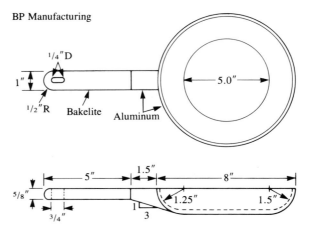

Original drawing scale: $1'' = 3''$

Figure P5.14

5.15 Using the bill of materials developed for Problem 5.14, calculate the yearly material cost for 200,000 of these frying pans.

5.16 Develop a planning sheet for manufacturing finished picture frames. The costs are as follows: labor, $4.80 an hour; wood, $0.50 a foot; nails, $0.02 apiece; and varnish, $1.30 a quart. Estimate equipment usage and cost.

5.17 How can a computer assist in product design? What are the components of such a system? Does CAD make communication between a designer and a manufacturer easier?

5.18 List important functional tasks associated with CAD, CAM, and manufacturing.

SUGGESTED READINGS

Market Research

Cox, W.E., Jr., *Industrial Marketing Research,* John Wiley and Sons, New York, 1979.

Ferber, R., Ed., *Handbook of Market Research,* McGraw-Hill Book Company, New York, 1974.

Freiman, D.J., *The Marketing Path to Global Profits,* AMACOM, New York, 1979.

Williams, R., Jr., *Technical Market Research,* Roger Williams Technical and Economic Services, Inc., Switzerland, 1962.

Forecasting

Abramson, A.G., *Operations Forecasting,* American Marketing Association, 1967.

Makidakis, S.; Wheelwright, S.C.; and McGee, V.E., *Forecasting Methods and Applications,* 2nd edition, John Wiley and Sons, New York, 1984.

Wenzel, C.D., "Look at the Foreseeable Trends in Parts Handling Offers Strategies for System Planners," *Industrial Engineering,* Vol. 16, No. 3, March 1984, pp. 46–54.

Design and Drawing

Eide, A.; Jenison, R.; Mashaw, L.; Northup, L.; and Sanders, C.G., *Engineering Graphics Fundamentals,* McGraw-Hill, New York, 1981.

Engineering Design Graphics Journal, American Association for Engineering Education, Ohio State University, Columbus, Ohio, 1985.

Giesecke, F.; Mitchell, A.; Spencer, H.; Hill, I.; and Loving, R., *Engineering Graphics,* 3rd edition, Macmillan, New York, 1981.

Computer-Aided Design

Groover, M.P., and Zimmers, E.W., *CAD/CAM Computer-Aided Design and Manufacturing,* Prentice-Hall, Englewood Cliffs, N.J., 1984.

Industrial Design Magazine, Design Publications, Inc., New York, 1985.

Pao, Y.C., *Elements of Computer-Aided Design and Manufacturing,* John Wiley and Sons, New York, 1984.

CHAPTER
6

Manufacturing Processes and Topics of Automation

In developing a detailed plan for production, selection of the proper manufacturing process is essential, a choice that is often governed by whether the product is new or a modification of an existing product. An older firm introducing a modified product will try to use, to the degree possible, equipment and processes that are already operational in the plant. The new firm, however, unencumbered by existing investments, generally has more latitude in making its selection. This is mainly because there will probably be several ways in which the necessary operations can be performed. For example, if a part requires a flat surface, the plant might use any one of several machines including planers, shapers, milling machines, and surface grinders. In general, metal products must pass through three steps in their manufacture: the rough shape, machining to fixed dimensions, and obtaining a surface finish. A detailed analysis of product specifications should indicate all the operations that are needed in manufacturing. Starting with the main product, each subassembly and its parts are identified by size, shape, material, tolerance, and operations required. One could design a product that would be impossible to manufacture; therefore a basic knowledge of the available processes is important in developing a sequence of operations that will translate the design into the final assembly.

Economical production depends to a large extent on the proper selection of the machine that will deliver a satisfactory finished product. The choice of machine is influenced by the quantity of items to be produced. For small-lot manufacturing, general-purpose machines such as lathes and drill presses may prove to be best. A special-purpose machine would be more suitable for large quantities of a standard product. Selecting the best machine requires a knowledge of possible production methods, the volume of production, desired quality of the finished product, and the advantages and disadvantages of the various types of equipment used in each method.

In this chapter we will briefly introduce several different manufacturing processes, our purpose being simply to familiarize the reader with these methods and their applicable terms. Most processes are explained in terms of metal working, since metal continues to be the main component in the majority of products being manufactured. With minor modifications, however, many of these processes can be applied to most nonmetallic materials. The following are the broad classifications of activities involved in manufacturing:

1. Changing the shape of material
2. Machining parts to a fixed dimension

    a. Traditional machining
    b. High-technology machining

3. Obtaining a surface finish
4. Joining parts or materials
5. Changing physical properties

6.1 PROCESSES USED TO CHANGE THE SHAPE OF MATERIAL

Most metal products originate as an ingot when the metal, obtained by one of the many ore-reducing or ore-refining processes, is poured into molds to form solid blocks of convenient size. The next step is to transfer the ingot into a preliminary shape needed in the product.

Table 6.1 is a listing of the more common processes for changing the shape of materials. It also includes typical machines and equipment that are used in each process. Each of the methods is briefly described in the text. The information provided is helpful in planning and in dealing with vendors.

Casting

Casting is pouring or injecting molten metal into molds, where it cools and hardens in the desired form. The various casting processes differ principally in the materials of the molds and the methods of introducing the metal, such as pouring or forcing under pressure. The principal casting processes are sand casting, shell mold casting, plaster mold casting, permanent mold casting, investment casting, and die or pressure casting.

Forging (Hot and Cold)

Hot forging is working metal heated up to 60 percent of melting point (absolute temperature) into shape by means of pressure, usually with a hammering action. Years ago, forging was done manually with the help of a forge (furnace) and an anvil. Now, however, most forging is done by mechanically operated presses. Forging is still the best method for developing great strength and toughness in steel, bronze, brass, copper, aluminum, and magnesium. Open-die forging, drop forging, upset or machine forging, and press forging are various forging methods. In addition to forging hot metals by these methods, much is now being done cold, in which solid metal is caused to flow under tremendous pressure into the desired shape.

Table 6.1   Machines or equipment used to change the shape of materials

Process	*Machines/Equipment*
Casting	Dies, molds, shells, die cast machine, continuous cast machine
Forging (hot and cold)	Hammer, steam hammer, gravity-drop hammer, forging press, forging machine
Extruding	Presses, dies
Rolling	Two-high reversing mill, three-high continuous rolling mill
Roll forming	Similar to the above, but smaller in size
Drawing	Draw bench and dies
Swaging	Swaging machine, cold header machine
Bending	Crank, eccentric, and cam-operated presses
Punching	Punch press, dies
Shearing	Shear press
Spinning	Speed lathe
Stretch forming	Stretch-draw forming machine
Torch cutting	Oxyacetylene torch, cutting machine
Explosive forming	Explosives, dies
Electrohydraulic forming	Electrodes, capacitors
Magnetic forming	Coil
Electroforming	Electrolyte, electrodes, and molds
Powder metal forming	Punch and die arrangement, presses
Plastics molding	Molds, molding machine

Extruding

Material made from metal powders or plastic pellets that are forced through dies to produce long shapes must be extruded. Methods used for extruding depend on the characteristics of the material; some are extruded cold with a binder, while others must be heated to the desired extruding temperature. Generally, the powder is first compressed into a billet and then heated in a nonoxidizing atmosphere, or the pellets are melted before being placed in the press.

Rolling

Steel ingots that are not to be remelted and cast into molds are taken while still hot to the roll-ing mill and rolled into such intermediate shapes as blooms, billets, or slabs. A bloom has a square cross section with a minimum size of 6 inches on a side. A billet is smaller than a bloom and may have any square section from 1½ inches up to the size of a bloom. Slabs may be rolled from either an ingot or a bloom and have a rectangular cross section with a minimum width of 10 inches and a minimum thickness of 1½ inches.

Roll Forming

Roll forming is reducing and tapering short lengths of bar stock by passing them through rolls. Material in sizes from thin strips to bars ¾ of an inch thick can be worked in this way.

Drawing/Wire Drawing

Deep-drawn pieces, such as metal cans or cups, are formed from flat stock in a series of operations in which each die draws the piece nearer to its final form. These deep-drawn parts are usually produced by hydraulic presses, since a slow, steady pressure at a controlled rate is desired rather than a sharp blow. The method of using a series of dies, each smaller than the preceding one, to produce continuous wire from bar stock is called wire drawing.

Swaging

Swaging is cold working by a compressive force or impact causing the metal to flow in some desired shape determined by the design of the dies. The metal conforms to the shape of the dies, but it is not restrained completely and may flow at some angle in the direction in which the force is applied.

Bending

In bending, strips of sheet or plate metal are formed into angles, channels, tubes, or complex irregular shapes by passing them between a series of rolls, allowing the material to gradually go into the desired shape. Bending the metal into circular form, called roll bending, consists of passing sheets, plates, or bars of flat stock through a number of flat rolls (three is common), spaced so that they bend the material into a ring or curve of predetermined radius. A press brake can be used for making sharp bends in one operation.

Punching

In punching, metal plate or sheet is pierced with a die to stamp out a predetermined shape.

Shearing

Shearing is cutting metal by stressing it in shear above its ultimate strength between adjacent sharp edges.

Processes used to change the shape of material. *Top*: In the wire drawing operation shown here, there are a series of dies in the hydraulic press, each smaller than the last, that produce ever finer wire. *Bottom*: Many processes are available for cold working, and in making a selection features such as required capacity, type of drive, and power source must be considered. Punch presses, such as the one shown here, are used to manufacture small-to-medium size metal parts in large quantities. This particular press is used for the production of metal gaskets.

Spinning

Spinning is shaping thin metal into simple or complex round shapes by pressing it against a form while rotating. Generally, the blank is clamped against a wood, plastic, or metal pattern, and both are rotated in a lathe. The metal is gradually worked into shape against a pattern by means of simple hand tools or rollers, which are pressed against the rotating workpiece, or automatically operated tools in special lathes. Figures 6.1 and 6.2 demonstrate how metal spinning shear forming may be done.

Stretch Forming

In stretch forming, large sheets of thin metal are shaped into shapes by using a metal press. The process is a stretching one that causes the sheet to be stressed above its elastic limit while conforming to the shape of the die.

Figure 6.1   Metal-spinning operation

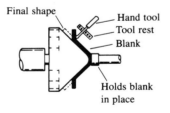

Figure 6.2   Progressive forming in a shear-forming operation in which a flat plate is formed into a conical shape

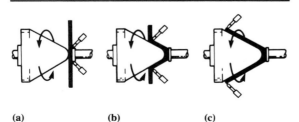

(a)   (b)   (c)

Torch Cutting

Torch cutting is cutting heavy steel by burning with a cutting torch. A simple hand torch for flame cutting differs from the welding torch in that it has several small holes for preheating flames to emerge and surround a jet of pure oxygen passing through a central orifice. When the preheating flames raise the temperature of the base metal at a given spot to a critical point, the operator turns on the jet of oxygen, which actually burns the heated metal, forming a hole through the work. As the torch is moved along the cutting line, a rough slot is made through the metal until the cut is completed. The method is economical but produces a rough edge and can consume an appreciable amount of material.

Explosive Forming

Explosive forming is a high-energy-rate forming process by which parts are formed rapidly under extremely high pressures created by explosives directed against the workpiece. Both low- and high-energy explosives are used in the various processes, depending on the mass of the part and the degree of change in shape required. The charges, whether exploded in air or in a liquid, set up intense shock waves that pass through the medium between the charge and the workpiece. Figure 6.3 provides four simplified drawings of different ways that explosive forming may be utilized.

Electrohydraulic Forming

Electrohydraulic forming is a process whereby electrical energy is directly converted into work. The operation is similar to explosive forming, but the pressure is obtained from a spark gap instead of an explosive charge. A bank of capacitors is first charged to a high voltage and then discharged across a gap between two electrodes in a suitable, generally noncon-

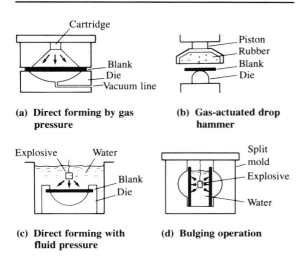

Figure 6.3 High-energy-rate forming

(a) Direct forming by gas pressure

(b) Gas-actuated drop hammer

(c) Direct forming with fluid pressure

(d) Bulging operation

ducting, liquid medium. This generates a shock wave that travels radially from the spark at high velocity, supplying the necessary force to form the workpiece to shape.

Magnetic Forming

In magnetic forming, electric energy is directly converted into useful work. In Figure 6.4 the schematic diagram of an electromagnetic forming circuit is shown. The system consists of a bank of capacitors, a voltage source, and a coil. Depending on the desired shape to be formed, the workpiece is placed in or near the coil. For

Figure 6.4 Electromagnetic forming circuit

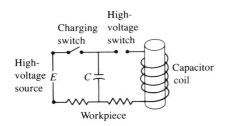

the process to begin, the charging voltage E is supplied from a high-energy source to the bank of capacitors. With the triggering of a high-voltage switch, a magnetic field is created. This field induces a current in the workpiece and thus creates a strong force acting on it. Since this force exceeds the elastic limit of the material of the workpiece, the desired permanent deformation occurs.

Electroforming

A special process for shaping metals, electroforming produces parts by electrolytic deposition of metal upon a conductive, removable mold or matrix. Electroforming is a very valuable process for fabricating thin-walled parts that require a high order of accuracy, internal surface finish, and complicated internal configurations that are difficult to core or machine.

Powder Metal Forming

Powder metal forming is the art and science of producing commercial products from metallic powders by pressure. Heat, which may or may not be used, is limited to keep the temperature below the melting point of the powder. The application of heat during or after the process is known as sintering and results in bonding the fine particles together, thus improving the strength and other properties of the finished product. Originally developed for use with metals, this process has been expanded to include ceramics. Very fine ceramic powders are subjected to high pressures to form the desired shape and then are fired to complete the bonding. One application is a ceramic bearing for pumps that may be machined to close tolerances much like its steel counterpart.

Plastics Molding

The term plastic is applied to all material that is capable of being molded or modeled. Modern usage of this word has changed its meaning to include a large group of synthetic, organic materials that become plastic by the application of heat and that are capable of being formed to shape under pressure. The various plastic compounds differ greatly from each other and lend themselves to a variety of processing methods, including compression molding, transfer molding, injection molding, rotational molding, and blow molding.

6.2 PROCESSES USED FOR MACHINING PARTS TO A FIXED DIMENSION

These secondary processes are necessary for many products that require close dimensional accuracy. In such operations, metal is removed from the parts in small chips by machine tools in which either the tool or the work reciprocates or rotates. Machining operations may be classified as traditional or nontraditional.

Traditional Chip Removal Processes

Table 6.2 lists 12 traditional methods of removing material by chipping or by a chip removal process. It also includes the machines and/or equipment used in each of the processes as well as some of the more prominent characteristics and/or applications. Each of these methods is briefly described in the text.

Turning

Turning is a process in which the workpiece rotates against a fixed tool. The most common turning machine is the lathe; and in this operation, chips are removed to generate a symmetrical form such that the cross section at any point on the axis is a circle.

Planing

Planing is the removing of metal by moving the work in a straight line against a single-edged tool held in the planer machine.

Shaping

Shaping is the moving of a reciprocating tool across the stationary work to create a plane surface.

Drilling

Drilling is producing a hole in an object by forcing a rotating tool, a drill, against it. The same can be accomplished by holding the drill stationary and rotating the work.

Boring

Boring is a machine operation for enlarging a hole that has already been drilled or cored.

Reaming

Reaming is slightly enlarging a machined hole to proper size with a very smooth finish and a close tolerance.

Sawing

Sawing is a reciprocating or continuous cutting motion operation using a toothed blade, performed on materials and bar stock in preparation for subsequent machining operations.

Broaching

Broaching is the removal of metal by an elongated tool having a number of successive teeth of increasing depth that cut in a fixed path. The original purpose was to produce holes having square or splined shapes, but broaching has since been adapted to machine external surfaces.

Milling

Milling is a widely used method of machining in which metal is removed by feeding the work against a rotating cutter. Its applica-

Table 6.2   Machines used in traditional chip removal processes

Process	Machines/Equipment	Applications/Characteristics
Turning	Speed lathe	Woodworking, metal spinning, polishing
	Engine lathe	Step-cone pulley, gear-drive, or variable speed control
	Bench lathe	Similar to, but smaller than, the speed lathe
	Toolroom lathe	Precision version of bench lathe
	Special-purpose lathe	As appropriate
	Turret lathe	Horizontal: ram or saddle type
		Vertical: single or multistation
	Automatic lathe	Single or multispindle machine allowing sequential and/or simultaneous operations
	Automatic screw machine	Single or multispindle machine for producing screws, bolts, and similar cylindrical products
Planing	Planing machine	Double housing, open side, pit-type, plate, or edge planer
Shaping	Shaper	Horizontal-push cut: plain (production work) or universal (toolroom work)
		Vertical: slotter or keyseater; special purpose
Drilling	Portable drill	Hand-held, general use
	Drill press machine	Bench or floor mounted
	Upright drilling machine	Light or heavy duty, gang drilling
	Radial drilling machine	Drill spindle travels along radial arm
	Turret drilling machine	Any one of several spindles may be selected; useful when a number of different sized holes must be drilled
	Multispindle drilling machine	Multiple drillings in a single operation
	Automatic-production drilling machine	Indexing table, transfer type
	Deep-hole drilling machine	Accommodates longer drills than standard machines
Boring	Boring machine	Jig boring, vertical boring, horizontal boring
Reaming	Reamer	Hand, chucking, shell, taper, expansion, adjustable, and special-purpose reamer
Sawing	Reciprocating machine	Horizontal: power hacksaw
		Vertical: sawing and filing
	Circular saw	Metal saw, steel friction disk, abrasive disk
	Band saw	Saw blade, friction blade, wire blade
Broaching	Broaching machine	Vertical single-slide, vertical pull-down, vertical pull-up; horizontal broaching, and continuous broaching machine
Milling	Column and knee milling machine	Hand miller, plain milling machine, universal milling machine, vertical milling machine
	Fixed-bed type	Simplex, duplex, and triplex milling machine
Grinding	Cylindrical grinder	Surface, centerless, and gear tooth grinding
Hobbing	Hobbing machine	For gear cutting
Routing	Routing machine	Shaping wood, plastic, and sometimes metal
		Portable and fixed machines

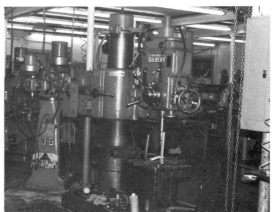

Processes used for machining parts to a fixed dimension. *Top left*: This precision toolroom lathe uses the turning process for woodworking, metal spinning, or polishing; the workpiece rotates against the tool, which removes material from it, giving it perfectly circular cross sections. *Top right*: Radial-arm drills, like the one shown here, facilitate the drilling of holes at different locations on large workpieces that cannot be readily moved and clamped on an upright drilling machine. Versatility can be obtained by allowing large workpieces to be fastened directly to the base of the machine, whereas small pieces can be mounted to a worktable attatched to the base. Vertical holes can be drilled, as can holes at an angle in a vertical plane. *Bottom left*: The horizontal load power saw combines the flexibility of a reciprocating power hacksaw and the continuous cutting action of a vertical handsaw. A machine such as this one can hold, clamp, and cut several metal or plastic pieces simultaneously and can be used to cut tubing, bars, and pipes of various sizes. *Bottom right*: Electrical discharge machining (EDM) is one of the most widely used of the new, nontraditional metal-working processes. A pulsating direct current creates a series of sparks between an electrically conducting metal tool and the workpiece to cut or erode the metal of the workpiece. The EDM is especially useful to form dies or tools of extremely hard metals.

tions are almost unlimited, ranging from surface finishing to gear cutting to grooving and fluting.

Grinding

Grinding is the removal of metal by a rotating abrasive wheel; either the grinder or the work may be fed to the other.

Hobbing

Hobbing is a process for generating gear teeth, consisting of rotating and advancing a fluted steel worm cutter against a slowly revolving blank.

Routing

In routing, a portable tool with a very small cutter, generally from ⅛ to ¾ of an inch in diameter, is driven at very high speeds of up to 40,000 rpm to make shallow carvings of numerous possible shapes.

Nontraditional Machining Processes

In more recent years, unusual equipment and unusual applications have been developed to machine various materials. In some cases these processes have merely improved manufacturing efficiency, but in others they have made possible the introduction of new materials or tolerances that were previously unachievable.

Table 6.3 is a listing of eight of these relatively new processes for machining materials, the equipment used, and the characteristics and/or applications as appropriate. As has been our practice, each of these methods is briefly described in the text.

Ultrasonic

The ultrasonic process is mainly used to machine hard and brittle materials. The system consists of an ultrasonic machining tool, a transducer, abrasive grains, and a carrying

Table 6.3 Machines and materials used in nontraditional machining processes

Process	Machines/Equipment	Applications/Characteristics
Ultrasonic	Tool, transducer, abrasive grains, a carrying fluid	Machining hard and brittle materials
Electrical discharge	Electric discharge machine (wire EDM)	Forming dies and molds; especially used for very hard materials that are difficult to machine
Laser	Ruby, gaseous state, liquid state, and semiconductor lasers	Computer controlled, for accurating metal cutting and welding
Electrochemical	Cathode, electric circuit, electrolyte	A depleting process; the tool is shaped like the desired design in the metal
Chemical milling	Alkaline and acid chemical reagent	Microprocessor chips, engraved chips. Alkaline for aluminum, acid for steel
Abrasive jet	Jet gun, air, CO_2, aluminum oxide, silicon, carbide powders, dolomite, sodium bicarbonate powders	Etching, cutting, shaping, drilling of fragile materials
Electron beam	Electron beam machine	Depositing thin film of metal, welding, or cutting
Plasma arc	Plasma torch	Very high temperature for cutting metals

fluid; the schematic diagram is shown in Figure 6.5. The toolholder carrying the tool is attached to the transducer, which produces elastic wave energy at a frequency of 20 to 30 kilohertz at an amplitude of 0.001 to 0.005 inch. This causes the toolholder to oscillate, resulting in the expansion and contraction of the normal length of the tool material. Because of the tool motion, the abrasive grains in the carrying liquid bombard the workpiece at a high velocity. The chipping pattern on the workpiece is controlled by the tool shape and contour. Boron carbide or similar materials of a 280 mesh size or finer are used as the abrasive grains. The process is mainly used to machine carbides, tool steels, gemstones, and synthetic crystals. Skilled workers are not required; this fact, the low tool cost, and the absence of thermal stresses are major advantages of this process.

Electrical Discharge

Electrical discharge machining is one of the fastest-growing and most widely used of all the new and more exotic metal-working processes. It is most useful for forming dies, molds, and other tools, especially in very hard materials that are difficult to machine by other means. Basically, the cutting or erosion of the metal in the workpiece is accomplished by the action of a series of sparks or electrical discharges between a shaped tool and the workpiece. The tool may be of copper, graphite, or any other electrically conducting material. A dielectric

liquid, usually a light mineral oil, is made to flow under pressure between the tool and the work to carry away the disintegrated metal and to prevent burning of the workpiece or the tool. The sparks are created by a pulsating direct current. The principal advantage of the electric discharge method is that it can be used in the fabrication of dies made from carbides and hardened tool steels, a machining operation that would otherwise be very difficult or impossible. Figure 6.6 is a schematic of a typical setup for this type of operation.

Laser

The laser is a device that provides a means for generating a narrow beam of monochromatic light of extremely high intensity in very short pulses. So far, its application has been largely in scientific and research projects and communication, but it is beginning to be used in manufacturing, principally for metal removal and welding. Because of the extremely narrow beam and high intensity, it is possible to perforate stainless steel 0.05 inch thick with a single pulse of energy.

One suggested application is in the field of balancing rotating equipment. The laser beam can be triggered by the stroboscopic light that is currently used in balancing equipment, and in this way it might selectively remove metal from the heavier parts of the rotating object until a perfect balance is obtained, the object never having to be stopped during the operation.

Figure 6.5   Schematic diagram of the ultrasonic machining process

Figure 6.6   Diagram for electrical discharge machining

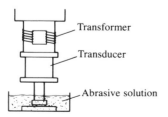

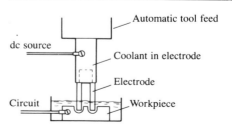

In welding, the laser has been limited to spot-welding thin material. Work is being done on extending the field of laser welding to produce butt welds in materials up to approximately ⅛ inch in thickness.

Electrochemical

Electrochemical machining is based on the same principles that are used in electroplating except that the workpiece is the anode and the tool is the cathode; therefore it is actually a depleting operation. Electrochemical machining performs stress-free cutting of all metals, has high current efficiency, and can produce complex configurations that are difficult to obtain by conventional machining methods.

Chemical Milling

Chemical milling is a controlled etching in which metal is removed to produce multifaceted patterns, lightweight parts, tapered-thickness sheets, and integrally stiffened structures. The initial step consists first of thoroughly cleaning the sheet or part to be etched and then masking, with a chemically resistant coating, those areas that are not to be affected by the etching process. (If the entire area is to be reduced, masking is unnecessary.) The part is then submerged in a hot alkaline solution in which metal in the unprotected area is eroded, the amount depending primarily on the time the part is in the heated solution. Finally, the part is rinsed, and the masking material is removed. In comparison with machine milling, the following advantages are claimed for this process:

· Material can be removed uniformly from all exposed surfaces.
· Material can be removed after parts are formed to shape.
· Highly skilled operators are not required.
· Operating and equipment costs are less than for milling machines.

Abrasive Jet Machining

The process of abrasive jet machining has found numerous applications in the metalworking and electronics industries. Fine abrasive particles (between 27 and 50 microns) are mixed with air or inert gas under pressure and blown against the workpiece with considerable force in a very fine stream or jet. The nozzle may be held by hand or mounted on the carriage of a machine. The abrasive action can be used for etching, cutting, shaping, or drilling fragile materials that would be extremely difficult to work by other methods. Examples of its application are cutting of external or internal threads on glass tubing, drilling and shaping of quartz crystals, and cutting thin sheets of titanium without cracking or breaking this very fragile material. Figure 6.7 provides an example of how abrasive jet cutting operates.

Electron Beam Machining

In this process, heat is generated by impelling high-speed electrons at the workpiece. At the point where the energy of the electrons is focused, the beam is transformed into sufficient thermal energy to vaporize the metal locally. The process must be carried out in a vacuum.

Plasma Arc Machining

In a plasma torch a gas is heated by a tungsten arc to such a high temperature that it becomes ionized and acts as a conductor of

Figure 6.7 Abrasive jet cutting with aluminum oxide particles

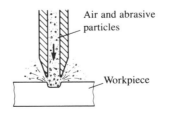

electricity. In this state the arc gas is known as plasma. The torch is generally designed so that the gas is closely confined to the arc column through a small orifice. Such a torch can be used to replace certain rough machining operations such as turning and planing. While it is effective in cutting all metals regardless of hardness, there is a resulting rough finish and possible surface damage due to oxidation and overheating.

6.3 PROCESSES FOR OBTAINING A SURFACE FINISH

These operations are used to produce a smooth surface, great accuracy, aesthetic appearance, or a protective coating. Table 6.4 is a listing of ten methods of obtaining surface finishes and the machines, equipment, and/or materials used with each. Each of these processes is briefly described in the text.

Polishing

Cloth wheels or belts coated with abrasive particles are used for polishing operations. It is not a precision metal removal process, but scratches and other minor surface imperfections can be eliminated. The amount of metal removed and the surface finish are controlled by belt speed, pressure, and grit size, as well as the characteristics of the material being polished.

Abrasive Belt Grinding

This operation, also termed high-energy grinding, is used for stock removal and surface preparation. It is performed by using a tensioned abrasive belt moving over precision pulleys at a high speed of between 250 and 6000 feet per minute. This method is used for preparing flats, tubing, and extrusions and for finishing partially fabricated stampings, forgings, and castings.

Table 6.4 Machines, equipment, and materials used for obtaining surface finishes

Process	Machines/Equipment/Materials
Polishing	Polishing wheels coated with abrasives
Abrasive belt grinding	Abrasive belts, belt sanders; metal or wood finishing
Barrel tumbling	Rotating barrel, abrasive medium (water or oil)
Electroplating	Electrolytes, electrodes
Honing	Honing machine
Lapping	Lapping machine
Superfinishing	Abrasive stones
Metal spraying	Metal-spraying guns
Parkerizing	Manganese dihydrogen phosphate solution
Anodizing	Sulphuric acid, oxalic acid, chromic acid

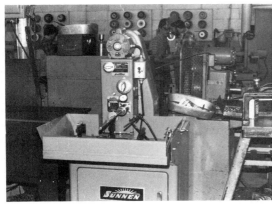
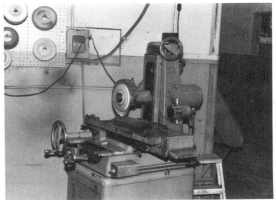

Processes for obtaining a surface finish. *Top left*: The vertical-spindle surface grinder uses a magnetic disk as a rotating table to hold a workpiece. The table slides horizontally until the part is in direct contact with the grinding wheel on a vertical spindle; it allows various sizes of workpieces to be machined at adjustable speeds. Machining speed is high as both table and grinding wheel rotate. *Top right*: Sanding removes minor surface imperfections by rubbing a metal surface against a wheel or belt coated with abrasive particles. Small-to-medium size metal parts can easily be handheld. *Bottom left*: Honing machines use fine abrasive stones to remove very small amounts of metal. Single-spindle machines such as this one are used to size and finish bored holes, remove tool marks left by grinding, and remove common boring errors such as tapers, waviness, and tool marks. The amount of metal removed is typically about 13 mm (0.005 inch). *Bottom right*: A horizontal-spindle surface grinder allows a reciprocating table to be utilized. The table enables the stroke length to be adjusted to the size of each workpiece, reducing the machining time for each part and providing versatility.

Barrel Tumbling

Barrel finishing or tumbling is a controlled method of processing parts to remove burrs, scale, flash, and oxides, as well as to improve surface finish. By this process a uniform surface finish, not possible by hand, is obtained. The parts are placed in a rotating barrel or a vibrating container along with an abrasive medium, water or oil, and usually some chemical compound. As the barrel rotates slowly, the upper layer of the workpiece slides toward the lower side of the barrel, causing the abrading or polishing action to occur. The same results can also be obtained with a vibrating unit in which the entire contents of the container are in constant motion. This process is the most economical method for surface cleaning large quantities of small parts.

Electroplating

Electroplating is the process of applying decorative and protective coatings, usually of nickel, chromium, copper, silver, zinc, gold, or tin, to other metals. The object to be coated is placed in a tank containing an electrolyte, the salts of the metal to be applied. A plate of pure metal is used as the anode, and the object to be coated is the cathode. When a d.c. voltage is applied across the circuit, the resulting current causes metal from the anode to replenish the electrolyte solution while ions of the coating metal are deposited on the workpiece in a solid state. The rate of deposition and the properties of the plated material depend upon the magnitude of the current, temperature of the electrolyte, condition of the surface, and properties of the workpiece material.

Honing

In this process, light spring pressure pushes slowly rotating blocks of very fine abrasive material against the work surface, such as the in-side of a cylinder, while they are moved back and forth laterally. The rates of rotation and of lateral movement are adjusted in some odd ratio so that an individual spot on the hone does not cover the same path on the work more than once. The amount of material removed is only 0.001 inch or less.

Lapping

The purpose of lapping is to produce geometrically true surfaces, correct minor imperfections, improve dimensional accuracy, or provide a very close fit between two contact surfaces. Lapping is quite similar to honing except that the lapping plate or block is of metal with a fine abrasive material in paste or liquid form used between the metal lap and the work surface. The amount of material removed is less than 0.001 inch.

Superfinishing

All finishing operations leave a surface that is coated with fragmented, noncrystalline metal, which results in excessive wear, increased clearances, noisy operation, and lubrication difficulties. Superfinishing is a surface-improving process that removes this undesirable fragmentation metal, leaving a base of solid crystalline metal.

Metal Spraying

Melted metal can be sprayed by air pressure, almost like paint or lacquer, to build up worn parts, protect the surface, or form molds for plastics or a soft metal casting. A gun similar to a paint spray gun is used with metal in wire or powder form. The metal is melted by an oxyacetylene or other gas flame at the nozzle of the gun and is sprayed in the form of a fine mist onto the work. Upon striking the surface, the minute drops of metal flatten out into flakes. If the workpiece has been cleaned and roughened properly, a good bond is obtained. Heavy coatings up to ¾ inch thick can be built up by spraying. Figure 6.8 shows a typical nozzle in cross section.

Parkerizing

Parkerizing is the process for producing a thin phosphate coating on steel to act as a base or primer for enamels and other paints or as a wear-in surface for high-contact pressure parts such as gears. The steel is dipped in a 190°F solution of manganese dihydrogen phosphate for about 45 minutes.

Anodizing

Aluminum can be anodized by a procedure that has some similarity to electroplating. A chemical reaction between the aluminum anode and an electrolyte of sulphuric, oxalic, or chromic acid is initiated by a direct current and results in surface oxidation of the aluminum. Since the coating is produced entirely by oxidation and not by plating, the oxide is a permanent and integral part of the original base material. Colored aluminum tumblers and pitchers are examples of products that are made by this process. A copper alloy of aluminum cannot be hard anodized.

Figure 6.8   Nozzle of a metal spray gun

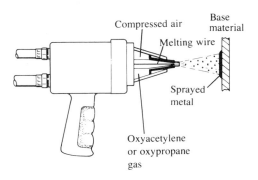

Compressed air

Base material

Melting wire

Sprayed metal

Oxyacetylene or oxypropane gas

6.4 PROCESSES USED FOR JOINING PARTS OR MATERIALS

The joining of two or more parts is usually accomplished by one of the processes listed in Table 6.5. Each of these methods is briefly described in the text.

Welding

Welding is the process of joining two pieces of metal by melting them directly together with or without the addition of more material of a similar nature and melting point. There are a few broad classifications depending upon the details of the process involved. These include electric arc welding, resistance welding, thermite welding, electric ray welding, pressure welding, and gas welding.

Soldering

Soldering is the uniting of two pieces of the same or different metals with a third metal (solder), which is applied between the two in a molten state at a temperature not exceeding 800°F. When the solder cools and hardens, it bonds the two parts together.

Brazing

Brazing is the introduction of a molten nonferrous alloy between pieces of metal to be joined and allowing it to solidify. The filler material, with a melting temperature of over 800°F but lower than the melting temperature of the parent metal, is distributed between the surfaces by capillary action.

Sintering

Sintering is the process of bonding metallic or other powders by subjecting them to elevated temperatures in an oven.

Table 6.5   Methods of joining parts or materials

Process	Equipment/Characteristics
Welding	Welding torch, arc welder, spot-welding machine; high heat generated to melt together two metal pieces
Soldering	Electric soldering iron; joining of metals by adhesion with a solder metal; solder metal has a melting temperature below 450°C
Brazing	Gas flame torch; joining of metals by adhesion with a brazing metal that has a melting temperature above 450°C but below that of the base metal
Sintering	Dies and ovens used; metal powders pressed into designed shape and heated
Riveting	Riveting hammer or press and a punch
Screw fastening	Screw or bolt
Adhesive joining	Adhesive agents; produces joints; cannot hold great loads
Pressing	Two parts, one with an inner diameter the same as the outer diameter of the other, are forced together

Riveting

In a riveting operation a solid or hollow rivet is placed through holes made in the parts to be fastened together, and the end is pressed to shape by a series of sharp blows delivered by a rivet gun or a hammer.

Screw Fastening

The most widely used method of joining parts is by driving a threaded shaft (a bolt or a screw) between the two parts.

Adhesive Bonding

Adhesive bonding is the process of joining parts by the application of suitable adhesive between them. Different types of adhesives are animal glues, vegetable glues, casein, sodium silicate, natural and synthetic rubbers, phenol formaldehyde, urea formaldehyde, resorcinol resins, alkyd resins, the acrylics, cellulose derivatives, melamine resins, polyester resins, polyurethane resins, epoxy resins, and silicon.

Pressing

When two mating parts are forced together, the process is known as pressing. The interior part should be slightly larger than the hole into which it fits. The forcing together of the parts can cause a displacement of material, and the joint will hold.

6.5 PROCESSES FOR CHANGING PHYSICAL PROPERTIES

The physical properties of a material can be modified by a number of processes, some of which are (1) heat treatment, (2) hot working, (3) cold working, and (4) shot peening.

Heat Treatment

In heat treating operations a metal is heated or cooled in its solid state to change its physical properties. Depending upon the procedure, steel can be made hard to resist cutting action, or it can be softened to allow further machining. If proper heat treatment is allowed, internal stresses may be removed, grain size reduced, toughness increased, or a hard surface produced with a ductile interior.

Hot Working

To be useful, an ingot of steel should be formed into a shape that can be machined. The ingot can be converted into a structural shape, bar stock, or sheet form by working it hot, above the recrystallization temperature.

Cold Working

Most ferrous metals are cold-worked after hot working to improve strength, machinability, dimensional accuracy, and surface finish. Cold working is done at a temperature below the recrystallization temperature of the metal, leaves residual stresses, and work hardens the piece.

Shot Peening

Shot peening is a method that is used to improve the fatigue resistance of a metal to compressive stresses set up in its surface by blasting or hurling a rain of small shot at high velocity against the surface to be strengthened. Small indentations, produced as a result of the striking shot, cause a slight plastic flow of the surface metal to a depth of a few thousandths of an inch. This stretching of the outer fibers is resisted by those underneath, and this resistance tends to return them to their original length, thus producing an outer layer having a compressive stress, while those below are in tension.

6.6 AUTOMATION

Many of the changes that have been made in manufacturing thus far in the computer age have been in terms of refinements of methods of operations and controls rather than changes in the fundamentals of the process itself. Prominent among the modifications are automation and numerical control.

Automation in manufacturing is the name given to an automatic system of production. An automated system may be considered to be made up of four distinct but closely interrelated parts: (1) the manufacturing system, (2) material handling, (3) sensing equipment, and (4) the control system.

We will study types of manufacturing systems and material handling in later chapters and will introduce sensory techniques, numerical control machines, and the robot and its applications at this point.

Sensing Methods

Sensing equipment plays a role similar to that of the physical senses. It observes what is happening and transmits the information to the control unit. The sensing techniques employ photoelectric cells, infrared cells, high-frequency electronic devices, and units making use of isotopes, X-rays, ultrasonics, and resonance. Sensing devices offer many advantages. Speeds of operation are many times faster than the maximum sensing that is possible by humans. There is no human fatigue problem, and absolute accuracy of inspection is assured within machine limits. Observations can also be made in places that are inaccessible to, or unsafe for, human beings.

Sensors are the eyes and ears of any automated process. One cannot operate a numerically controlled machine without knowing the position of the stock relative to the cutting tool. Feedback is needed to determine when the stock has moved to the proper cutting position or whether the cutting tool has retracted from the stock so that it can be positioned for the next cut. The controllers that perform the work must be guided by sensors that measure the critical parameters of the job being performed.

Sensors are classified as either discrete or continuous on the basis of the type of signal that they produce. Discrete sensors produce a bistate input corresponding to some bistate observation—for example, closed or open, full or not full, on or off, running or stopped. Since most discrete devices normally close or open a switch on the basis of the condition they observe, they are commonly referred to as contact sense inputs. Limit switches are a very common example of a contact sense input. For example, a numerically controlled machine can be inhibited from moving the stock until a limit switch closes, signifying that a drill bit has fully retracted. House thermostats are another type of discrete sensor. When the house temperature falls below a preset limit, a contact closure is made that turns on the heating system. Once the temperature rises, the system is again turned off by the thermostat switch opening. These home heating systems then are either fully on or all the way off, and there exists no way to heat the house at some intermediate level. Therefore raising the thermostat's temperature set point heats a house to a higher temperature, but not any faster. Even though temperature is a continuous variable, the sensor's output is bistate, making the device a discrete sensor.

By contrast, a continuous sensor can signal any intermediate value between 0 and 100 percent and is not limited to simple contact closure. The output of a continuous sensor is normally linear with respect to some observed variable. For example, a pressure sensor can be used to measure the level of a tank in which the pressure on the sensor is proportional to the height of the column of fluid above the sensor. The sensor output may be measured electrically as volts or milliamps and then scaled to the appropriate observed variable. At times the variable that exhibits linearity with the sensor output is not the variable of interest. This is typically the case in measuring flows, where the sensing device's output is linear with respect to the pressure difference between two points. The square root of this differential pressure must be obtained in order to have a signal proportional to the flow. The fuel tank measurement in a car is a common example of a continuous sensor. The sensor output is a measurement of the height of the gasoline in the tank, which, owing to the tank's geometry, is roughly linear with respect to the volume of gas remaining.

Sensor technology continues to advance as new types of measurements augment the old classics such as position, pressure, temperature, and flow. Many of these new sensors involve the measurement of infrared or visual light, radiation, or ultrasonic sound. These more exotic measurement techniques are used when the classical techniques fail or are inadequate. The measurement of coal in a metal hopper is an example of such an application. Since coal dust obscures any attempt to determine a hopper's fullness by light, a means is needed that makes this dust transparent. To accomplish this, a radiation source is aimed through the hopper to illuminate an ionization chamber (receiver). The signal level is then calibrated for the full-scale reading with no coal present so that the effect of the metal sides and air gap are removed from the reading. Any additional attenuation will therefore be caused by the coal and will be proportional to the mass of the coal between the source and the receiver. Hence dusting will cause a minimal drop in the signal level, whereas a solid mass will provide a more substantial change. This radiation is used to detect when the level in a coal hopper has fallen below the preset limit in order to restart the coal-feed mechanism for a power plant. Radiation is thereby used to produce a discrete contact sense output used to trigger a sequence of events. It should be further noted that this more sophisticated measurement is more costly, is more difficult to maintain, and requires special licensing in the handling and storage of the radioactive source. Simpler techniques should therefore be used whenever they are practical.

Numerically Controlled Machines

Use of computers in manufacturing can be marked by the introduction of numerically controlled (NC) machines to shop floors. These are machines that operate under a given set of instructions, as opposed to the manual manipulations that are used with traditional machines. NC machines are currently used in metal cutting for milling, boring, drilling, turning, grinding, and flame cutting and in metal joining for riveting and spot and continuous welding. In fact, the total number of NC machines installed for metal working almost doubled from 1978 to 1983 to about 103,000 units. Because of the growing use of NC machines, we will describe these in some detail.

The earlier development of NC machines is credited to the aircraft industry and the Massachusetts Institute of Technology (MIT). Many identical and complicated parts were needed in manufacturing; and to help avoid manual errors in the duplication of the parts, MIT was asked by the U.S. Air Force in the late 1940s to develop technology that would cut the required shapes from metal sheets automatically. A machine was developed that would read a series of numerical codes indicating the tool operations and act on these to produce a part. The same operations could be called upon for the same design the next time the same piece needed to be processed. Thus began the introduction of NC machines in manufacturing.

Components of NC machines

The three basic components in the operation of an NC machine are a set of instructions or software, a machine controller unit (MCU) or controller unit (CU), and machine tools.

Instructions: The software for NC machines consists of a set of commands instructing the controller units regarding operation of machine tools. These are very detailed step-by-step directions giving machine tools and their tables the next moves. The instruction code is a binary coded decimal (BCD) system. In a conventional NC machine this can be easily observed by the presence or absence of holes in the correct positions on the punched paper tape that is used to transfer instructions to the controller. In modern systems, instructions are read on floppy disks or entered through control keyboards of the machines.

Like using any other machine language for writing a program for a computer, composing instructions in BCD is a difficult job. Simpler languages have been developed that allow the format of the instructions to be more like commands in English. Automatically Programmed Tools (APT) is one such language developed by MIT. This is one of the most widely used languages today, and a variation (ADAPT) developed by IBM for the U.S. Air Force allows application of APT to small computers. Additionally, some manufacturers have developed languages for use with the machines they produce. Sundstrand Processing Language Internal Translator (SPLIT), developed for Sundstrand machines, is one example of this type of software. If it is intended to have more than one NC machine in a plant, it is desirable to make them compatible by assuring that all of them can use the same language.

Numerically controlled machines. The three basic components in the operation of an NC machine used in manufacturing are a set of instructions (software), a machine controller unit (MCU) or controller unit (CU), and machine tools. *Top left*: Here, the tape reader of the CU deciphers the step-by-step software commands that have been punched into paper tape. *Top right*: The CU converts the software commands into mechanical actions by tools, such as this NC machine used for milling small metal castings. *Bottom*: In industrial applications, like the one shown here, NC machines use commands from the MCU to control electronic board assemblies.

Control unit: It is the hardware, electronic controls, and servomotors that read the software instructions and convert them into mechanical actions by tools and their work table. Some earlier NC machines also had feedback controls that formed a closed loop system, ensuring the position of the table and workpieces with respect to the tool. However, experience has proven the reliability of NC machines, and such feedback controls are not used in most modern machines. The components of an individual NC machine may depend on the system of which it is a part (these systems are described next) and may include a tape reader, computers, servomotors, input/output signal controllers, a control panel, and a CRT monitor.

Machine tool: The machine tool is the third component of an NC machine; it performs the actual work. The tool consists of the work table, fixtures on the table, a turret, spindle(s), and tool(s) on the spindle(s). The tools are designed from materials that enable them to be wear resistant, as well as having toughness qualities (impact and heat resistance) combined with high strength.

Machine System

Four NC machine systems are currently in general use in industries. They range from a single, stand-alone NC machine to a combination of a large mainframe computer and a number of NC machines, each with its own small computer. The mode of instruction transfer and the associated flexibility, which vary from system to system, are described in the following paragraphs.

Conventional NC machine: Instructions for conventional NC machines are read by a tape reader (a part of the controller) from a one-inch-wide tape. The tape may be made of paper, a low-cost but not very durable medium, or Mylar-reinforced paper, aluminum strip, or even plastic. Manual instructions on the tape are produced by using a typewriterlike machine called the Flexowriter, while a computer-assisted tape (a tape produced by using the computer) is punched by a device called a tape punch.

A tape is fed through the tape reader every time a piece is to be produced by the conventional NC machine. Thus even for production of identical pieces the same tape of instructions must be read for each piece. The reader deciphers one instruction at a time, saves it in memory, and, while the controller is acting on this instruction, reads the next instruction. Actions of the controller are to control movements of the tool(s) and the work table according to the instruction. This mode of operation has some drawbacks. Storage and upkeep of the number of tapes needed to produce different parts could become a problem. Also, any change in part design, however small, can mean producing a new tape. The straightforward loading of the machining instructions does not allow adjustments in machining parameters in real time; hence changing the machine tool operation while the work is in progress is difficult at best.

Direct numerical control: In a direct numerical control (DNC) arrangement, as illustrated in Figure 6.9, a group of NC machines is connected to a mainframe computer or even a fairly large minicomputer, and each receives its instructions from this computer on a time-sharing basis. The paper tape instructions are no longer required, and the high speed of the computer allows each machine to interrupt and get its instructions on a continuing basis. A number of different programs can be stored on magnetic tape, disk, or main memory, ready to be recalled immediately when needed by any machine. Part changes can be incorporated in the program with very little effort. However, the system does have disadvantages. Any failure in the main computer affects all NC machines,

Figure 6.9   Direct numerical control

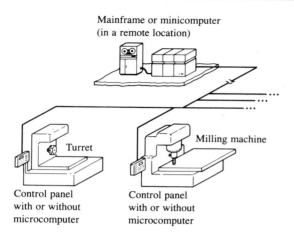

Mainframe or minicomputer
(in a remote location)

Turret

Milling machine

Control panel
with or without
microcomputer

Control panel
with or without
microcomputer

Figure 6.10   Computer numerical control

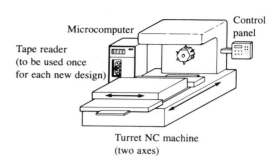

Microcomputer

Control
panel

Tape reader
(to be used once
for each new design)

Turret NC machine
(two axes)

perhaps resulting in a serious loss of production. Low-voltage transmission lines from the computer to the machines also occasionally make perfect transmission of instructions difficult, causing errors in manufacturing.

Computer numerical control (CNC): The next step in the development of NC machines is a minicomputer or a microcomputer attached to each NC machine rather than having them share one large mainframe computer. (See Figure 6.10.) The instructions are entered through a floppy disk or a keyboard associated with the computer on the machine. There is great flexibility here. Program modification can be completed almost immediately through the keyboard; in addition, instructions for the operator and machine data can be displayed on a cathode-ray tube (CRT) if needed.

Modern DNC: Recent DNC systems take advantage of a CNC configuration as it connects a group of CNC units to a central computer. New programs can be developed with the main computer and then transferred to the individual computer on the particular NC machine. It

also allows storage of large numbers of programs in the main computer, including very long programs or large numbers of short programs that are used infrequently, and then transfers them on an as-needed basis to individual machines. The main computer can also be used for other activities such as controlling inventory, scheduling production and labor resources, and producing work orders for maintenance; all of these lead to overall computerization of the plant, which is called computer-aided manufacturing (CAM).

Advantages of NC Machines

One of the greatest benefits of the NC machine is the accuracy with which the parts can be produced. Complex geometrical parts can be repeatedly produced to the same very close tolerances. The NC machine can accommodate design changes easily by merely changing the tape program instead of requiring expensive changes in jigs and fixtures. NC machines increase productivity by reducing setup and handling times. These machines have proven to be ideal where greater flexibility is required in terms of production quantity (for example, one unit or thousands of units), when the processing is expensive, and when the cost of a defective unit is considerable.

Classification of NC Machines

An NC machine is categorized on the basis of its control features, and the cost of a machine increases as its performance capability increases. The following list gives the categories in ascending order of complexity.

1. *Point-to-point machine:* capable of moving from one designated point to another. These points must be defined beforehand. (See Figure 6.11.)

2. *Straight-cut machine:* capable of moving continuously at a controlled speed parallel to one of the major axes (*x*, *y*, or *z*). (See Figure 6.12.)

3. *Contouring machine:* capable of following any contours or geometric shapes such

Figure 6.11　A point-to-point operation

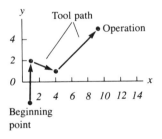

Figure 6.12　Absolute and incremental positioning

$x = 7$ and $y = 4$ for incremental positioning
$x = 10$ and $y = 6$ for absolute positioning

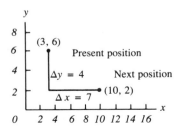

Figure 6.13　Straight-line cutting

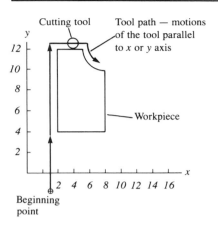

as a circle or a curve in two or three dimensions. (See Figure 6.13.)

4. *Adaptive control machine:* the next advancement, developed for the Air Force in 1962, is an adaptive control machine capable of varying its speed and feed based on the condition of the workpiece or tool wear. For example, if the hardness of a workpiece is reduced at a point, the cutting speed might increase. Even a small air pocket in the cutting path can mean easier cutting, again increasing speed and/or feed. These machines are used when processing time is a large part of the total time of production. (See Figure 6.14.)

Industrial Robots

The Robot Institute of America defines *robot* as "a programmable, multifunctional, manipular design to move material, parts, tools, or specialized devices through variable programmed motions for performance of a variety of tasks." A robot can be programmed and reprogrammed to perform various repetitive tasks. Robots are currently being used in loading and unloading production machines such as stamping, forging, metal cutting, and injection mold-

Figure 6.14   An adaptive control system

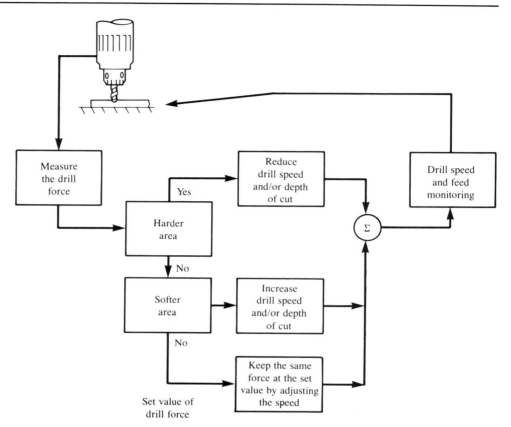

ing. The robot also performs spray painting, spot and continuous welding, and even adaptive welding, in which it scans the joint by laser and then follows it by welding the seam. A robot with vision or an "eye" has been effective in inspection processes.

Operating costs for robots can be economically attractive, being less than $5.00 per hour for some models. Purchase prices range from $25,000 to $200,000 but have been consistently decreasing owing to standardization and increased volume. Robots are reliable, following the same preplanned motions every time, and productive, working about 98 percent of the time. They can be strong, lifting heavy objects, and will work in hazardous conditions in which high temperatures, dangerous chemicals, or radioactive substances are involved.

Degrees of Freedom

To emulate human arm motion requires six degrees of freedom. As shown in Figure 6.15, the arm can move at the shoulder, elbow, and wrist. At the shoulder it can rotate about two axes, giving two degrees of freedom. At the elbow it has one degree of freedom, since it can move only up and down. At the wrist, up-and-down motion, side-to-side motion, and rota-

Figure 6.15 A robot with six degrees of freedom

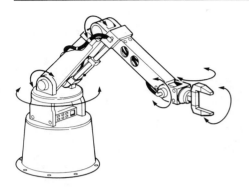

tional motion give three additional degrees of freedom. Robots are designed to use one or more of these motions on the job. Drives such as a rack and pinion, screws, and hydraulic cylinders are used to achieve transverse motion; angular or rotational motions are obtained by gear drives and timing belts.

The grasping of an object by a robot is equivalent to human finger motions. Vacuum cups, hooks, electromagnets, clamps, scoops, or even fingerlike mechanisms are used to achieve grasping.

Classification of Robots

Like NC machines, robots can be classified with respect to their abilities. A point-to-point robot can go only from one predefined point to another. A continuous path robot can go to a number of points in a path, but not necessarily continuously following the path. A controlled path robot can follow a predefined geometric path continuously, and a servo-controlled robot has the ability to sense the points on the path and feed back the information so that corrective actions can be taken by the robot controller.

Robot Programming

At present, over 90 percent of robots are taught by leading them through a sequence of operations. A robot in the learning mode remembers these operations and repeats the sequence when in the operating mode. The method is popular because an operator on the shop floor can show the robot what to do.

Robots can also be programmed like NC machines by giving them a set of instructions. Robot programming languages are as varied as the number of manufacturers of robots, since each tends to develop a language of its own. These languages are developed as modifications to existing programming languages by adding syntax or subroutines or by creating a new language. Some of the programming languages are WAVE, RAIL, VAL, PAL, AL, MHI, and HELP.

Selecting a Robot

A robot works as an integral part of a production system, and its selection is based on where and how it should function. Environmental conditions such as heat, humidity, and dust determine the surroundings in which the robot must work. The range of motions (less than one foot, from one to four feet, between four and ten feet, or ten feet and over) and the speed at which it must operate (low speeds of less than one foot per minute, medium speeds of from one to five feet per minute, or high speeds of greater than five feet per minute) determine the capabilities of the robot. The type or classification of controls and the needed sensitivity—such as detection of an object by proximity or touch, detection of holes and edges by simple vision, or recognition of objects by complex vision—determine the degree of refinement of the robot. Naturally, as the complexity of the robot increases, the purchase and installation costs increase correspondingly.

Some examples of robot use were mentioned earlier. The important features in selecting the robot for such applications can now be defined. For example, for material handling, a robot must be able to travel from one place to another, perhaps on wheels, rails, or guided

Industrial robots. *Top*: Robots can be programmed to accurately perform repetitive tasks such as welding in automobile manufacturing. *Bottom*: A robot with an ''eye,'' such as the one shown here, can detect leaks and take corrective action.

paths. For loading and unloading of machines a robot should be able to manipulate a unit and should also be able to travel, but this travel is mostly in terms of arm motions. Welding and spraying operations require robots that are good in manipulating and following contouring and continuous motions. Robots for machinery and assembly operations should be able to manipulate as well as sense the positions of workplaces. The need for sensing becomes more acute when the robots are used in inspection processes.

6.7 ECONOMIC EVALUATION OF PROCESSES

We have seen different processes and machines that can be used in manufacturing. Yet the question may still remain: Which method is best suited for the intended operation?

In most instances the selection of a process is dominated by design needs such as shape, material tolerance, and the required quantity of production. However, whenever alternative methods are available that will give the same quality product, economic evaluation may be performed to select the best alternative. An example illustrates such an approach.

A machine manufacturer requires close tolerance of ± 2 mils (0.002 inch) on the cams it produces. The manufacturer can perform the machining by using one of the following methods:

1. Cut the part on milling machine and then grind to the required tolerance on a grinder.
2. Use a wire-EDM that will produce the cam with the required tolerance.

The manufacturer expects to produce 1000 cams of various sizes per month and, on the average, estimates the information in Table 6.6 to be the necessary data. The interest rate is currently 12 percent per year.

Using the factors in Table 6.6, determine the process to use by calculating the monthly costs of the two alternatives.

The depreciation cost per month ($i = 1$ percent, $n = 120$, $A/P = 0.0143$) for each machine is as follows:

Milling machine: $20,000 \times 0.0143 = \$286.00$

Grinder: $15,000 \times 0.0143 = \$214.50$

EDM: $90,000 \times 0.0143 = \$1287.00$

The number of units that must be initiated per month to allow for defective units are

Grinder: $\dfrac{1000}{0.995} = 1005.0$

Milling machine: $\dfrac{1005}{0.97} = 1036.1 \approx 1037$

EDM: $\dfrac{1000}{0.995} = 1005.0$

Table 6.6   Data for machine manufacturing example

	Profile Milling Machine	Surface Grinder	Wire-EDM
Initial Cost	$20,000	$15,000	$90,000
Time for Setup Operation and Handling per Unit	10 min	8 min	9 min
Defective Percentage	3	1/2	1/2
Unit Cost of Rejects	$15.00	$18.00	$20.00
Operating Expense including Maintenance and Tool Replacement per Hour	$1.50	$2.00	$5.00
Useful Life in Years	10	10	10
Operator Cost per Hour	$15.00	$15.00	$20.00

Therefore expected number of rejects on each
machine are

Grinder: $1005 - 1000 = 5$

Milling machine: $1037 - 1005 = 32$

EDM: $1005 - 1000 = 5$

Assuming 80 percent efficiency in operation,
the average monthly time required on each process is

Milling machine: $\dfrac{1037 \times 10}{60 \times 0.8} = 216.04$ hours

Grinder: $\dfrac{1005 \times 8}{60 \times 0.8} = 167.50$ hours

EDM: $\dfrac{1005 \times 9}{60 \times 0.8} = 188.43$ hours

Now we can determine the cost of manufacture on each machine, which includes depreciation, rejects, labor, operation, and overhead
cost (overhead is assumed to be 150 percent of
the direct labor cost).

Thus the costs for milling are

Depreciation: $286

Rejects: $32 \times \$15 = \$480$

Labor: $216.04 \times \$15 = \$3240.60$

Machine cost: $216.04 \times \$1.50 = \$324.06$

Overhead: $216.04 \times \$15 \times 1.5 = \$4860.94$

Total = $9191.60

Similarly, the cost for grinding is

$$214.5 + 5 \times 18 + 167.5(15 + 2.0 \\ + 1.5 \times 15) = \$6920.75$$

And the cost for EDM is

$$1287.0 + 5 \times 20 + 188.43(20 + 5.0 \\ + 1.5 \times 20) = \$11,750.65$$

Therefore the cost of Process I is

$$9191.60 + 6920.75 = \$16,112.35$$

and that of Process II is \$11,750.65. It is ob-
vious that Process II, that of using EDM, would
be selected.

Further investigation showed that a CNC-
milling machine could be used that would
allow one operator to tend both milling and
grinding operations; however, the cycle time
would be extended to 11 minutes (see Section
8.3), and the overhead cost would change to
175 percent of the direct labor. The operator is
still paid \$15 per hour. The data for the CNC-
milling machine are as follows:

· Initial cost: 35,000
· Defective percentage: ½
· Operating expense, including mainte-
  nance tool/hour: 2.00
· Useful life (in years): 10

By following the same procedure as before, the depreciation cost per month is 35,000
$\times$ 0.0143 = \$500.5. The number of bad units
produced is $(1005/0.995) - 1005 = 1010 - 1005 = 5$.

Assuming 80 percent efficiency, the
monthly time required of CNC-milling is (1010
$\times$ 11)/(60 $\times$ 0.8) = 231.46 hours. The required time on the grinder would be (1005 $\times$
11)/(60 $\times$ 0.8) = 230.31 hours. Since both operations are to be carried out simultaneously,
the longer of the two times—231.46 hours—
will be used in our calculations as the cycle
time. Therefore the cost per month for Process
I is

$$500.5 + 214.5 + 5 \times 15 + 5 \times 18 \\ + 231.46(15 + 2 + 2 + 1.75 \times 15) \\ = \$11,353.56$$

This cost is less than that for EDM alone,
and the decision will now be to adopt the CNC-
milling and grinder combination; however, the
closeness of the costs of the two alternatives

would suggest a need for more careful data collection and analysis.

SUMMARY

A brief introduction to manufacturing processes is presented in this chapter. It involves a discussion of various operations that provide different degrees of tolerance, surface finishes, and strengths for the completed items. In the initial phase of changing the shape, for example, one might transform an ingot into a casting or a metal plate into stamped parts. The tolerance for such operations, however, is not very close, and to obtain a fixed dimension with better accuracy machining might be necessary.

Most manufacturing plants commonly use equipment such as lathes, milling machines, and drills; whereas others may utilize modern techniques such as laser beam, electrochemical milling, and ultrasonic machining. Need also arises for joining two parts or changing physical properties of material, and though the discussion in the chapter is brief, it serves as a good introduction to all these topics.

In manufacturing, automation plays an important part in improving productivity. This chapter discusses two of its components, sensors and numerically controlled machines and robots. Sensors are the eyes and ears of automated systems. They observe what is happening and relay the information to the control unit, which takes appropriate corrective action. Both discrete and continuous sensors have found a number of applications in production facilities.

Numerically controlled (NC) machines provide more flexibility and greater accuracy than their conventional counterparts in operations such as milling, turning, drilling, and grinding. These machines follow instructions provided in a binary decimal system or in a language that includes APT, SPLIT, and ADAPT, especially developed for NC machines. Instructions are entered into memory using punch tapes, floppy disk drives, or keyboards and are then converted by control units into mechanical actions by tools and worktables. When a plant has more than one NC machine, they are controlled by either an individual computer for each machine or a large and small computer combination. Many arrangements are possible and are designated as direct numerical control and computer numerical control. NC machines are classified according to their performance capacities, with point-to-point motion as the lowest classification and adaptive control as the most sophisticated.

Because of their reliability and ability to work in environments that may be hostile to humans, industrial robots are used in operations such as loading/unloading, spray painting, and welding. Freedom of motion is an important trait and is measured by a robot's ability to emulate human motions. Robots are programmed and classified in a manner very similar to that for NC machines.

With many different processes and methods available to perform a task, it is advantageous to make an economic evaluation between alternatives. The chapter concludes with an illustration of such an analysis.

PROBLEMS

6.1 What types of activities are involved in manufacturing processes? Give two examples of each type.

6.2 List the methods that are used to change the shapes of materials.

6.3 How has forging changed over the years?

6.4 Distinguish between blooms, billets, and slabs.

6.5 How is electrohydraulic forming different from explosive forming?

6.6 What is a plastic and how can it be molded?

6.7 Describe the following:
  a. Planing
  b. Drilling
  c. Grinding

6.8 Nontraditional machining methods are more expensive than their traditional counterparts, but they are often used nevertheless. Why?

6.9 What is the advantage of the electrical discharge process over grinding?

6.10 What are the advantages of chemical milling over machine milling?

6.11 What determines the surface finish in the polishing process?

6.12 How do welding, soldering, and brazing differ?

6.13 How can the physical properties of a material be modified?

6.14 List the manufacturing processes that would be used in the production of the following:
  a. A nail
  b. A small gold cross
  c. An automobile body

6.15 What is automation and what are the components? Give two examples of the types of facilities that can be automated and the degree of automation that can be achieved for each.

6.16 What is an NC machine? When and where is it preferable to a traditional machine?

6.17 Describe the basic components in the operation of an NC machine. How do they vary within different NC machine systems?

6.18 Distinguish between CNC and DNC systems.

6.19 Define *robot*. Where can one be used?

6.20 It is said that a robot provides flexible automation as opposed to hard automation. Explain this statement.

6.21 What is a degree of freedom for a robot? How many degrees of freedom are needed to emulate a human arm? Describe each.

6.22 How are robots classified? Identify at least two applications for each type of robot.

6.23 What are the attributes of a robot that should be considered before acquiring one?

6.24 A robot costing $50,000 is considered for use in a welding operation that is currently being performed by two welders. Each welder is paid $15 per hour and can produce 15 parts per hour. The robot can weld at a rate of 30 parts per hour and has an operating cost of $5 per hour (including the average cost for setups). If the required production rate is 200 units per day, how long will it take before the robot can "pay for itself"? Assume an interest rate of 12 percent.

6.25 A manufacturer of staplers has the choice of two processes to produce the housing of the stapler. Method I produces the part with a punch press and smooths it with an abrasive belt grinder. Method II produces the part by casting and smooths it with an abrasive belt grinder. The manufacturer expects to produce 10,000 units per month. The present interest rate is 12 percent. The factors on which the manufacturer will base the decision are shown in the accompanying table. Determine the monthly cost of each alternative.

	I		*II*	
	Punch Press	*Belt Grinder*	*Casting (Oven Mold Label)*	*Belt Grinder*
Initial Cost	$20,000	$2500	$7000	$2500
Time for Setup Operation and Handling per Unit	1 min	0.2 min	1.5 min	1.5 min
Percent Defective Units	1/2	1/5	5	1/5
Unit Cost of Rejects	$0.80	$0.60	$0.90	$0.70
Operating Expense includes Maintenance and Tool Repair per Hour	$0.20	$0.05	$0.08	$0.05
Useful Life in Years	8	5	8	5
Operator Cost per Hour	$12.00	$8.00	$11.00	$8.00

SUGGESTED READINGS

Manufacturing Processes

De Garmo, P.E., *Materials and Processes in Manufacturing,* Macmillan, New York, 1984.

Flinn, R.A., *Fundamentals of Metal Casting,* Addison-Wesley, 1963.

Manufacturing Engineering, Society of Manufacturing Engineers, Dearborn, Mich., 1985.

McGannin, H.E., *The Making, Shaping, and Treating of Steel,* 9th edition, United States Steel Corporation, 1971.

Niebel, B.W., and A.B. Draper, *Product Design and Process Engineering,* McGraw-Hill, New York, 1974.

Strauss, K., *Applied Science in Casting of Metals,* Pergamon Press, New York, 1970.

Automation

Asfahl, C.R., *Robots and Manufacturing Automation,* John Wiley and Sons, New York, 1985.

Groover, M.P., *Automation, Production Systems, and Computer-Aided Manufacturing,* Prentice-Hall, Inc., Englewood Cliffs, N.J., 1980.

Knebusch, M., "New Software Simplifies CNC," *Machine Design,* Aug. 1984, pp. 63–66.

Krouse, J.K., *What Every Engineer Should Know About Computer-Aided Design and Computer-Aided Manufacturing, The CAD/CAM Revolution,* Marcel Dekker, Inc., New York, 1982.

"A Look at NC Today," *Manufacturing Engineering,* Oct. 1975, pp. 24–28.

Winship, J., "CNC Lasers Cut Out Saw Blades," *American Machinist,* May 1984, pp. 92–94.

CHAPTER

7

Production Charts and Systems

Manufacturing of a product requires transformation of design ideas into reality, which ultimately results in a useful item. This is accomplished through the efficient completion of operations by a group of well-trained workers in a properly developed plant. Managers and engineers must carefully analyze different ideas regarding production of an item and then select the alternative that is most cost-effective and feasible under the limitations that the plant must operate. Consideration of several factors is warranted in many cases—for example, the production of one or of several items, volume of production of each item, preferred skills of the workers, and flexibility to be incorporated in the plant design for product and/or product mix changes. All of these factors contribute heavily in overall plant development.

The discussion that follows is intended to illustrate available means for converting production ideas into tangible results. It is divided into three phases. First, production charts are illustrated. These charts are the backbone of planning and improving activities. Next, the discussion on production systems leads to our appreciation of various types of production arrangements and their associated advantages and disadvantages. Lastly, we illustrate some manpower planning models that are used in improving efficiencies in workplaces.

7.1 PRODUCTION CHARTS

To illustrate the activities associated with production, engineers often use charts to represent the process graphically. Such an approach increases understanding of the course of action that is needed in manufacturing and helps to resolve many problems related to design of the production layout. Assembly charts, operation process charts, and process charts are important representations that significantly contribute to this goal. All three charts are drawn by using symbols standardized by the American Society of Mechanical Engineers in 1947.

Symbols and Descriptions

The symbols representing the five basic activities in manufacturing are shown below along with definitions that are commonly used throughout industry.

· ○: *Operation.* The item is intentionally changed in one of its characteristics. Examples are filling a bottle with a soft drink, bending a metal sheet, and writing a letter.

· →: *Transportation.* The item is moved from one place to another (except when the movement occurs as an integral part of an operation or inspection). Examples are moving an item on a conveyor belt between operations and moving an item to storage.

· □: *Inspection.* The unit at an inspection point is compared with the quality standard established for that point.

· D: *Delay.* The next planned action does not take place. Examples are delay in moving a unit and delay in performance of the next operation.

· ▽: *Storage.* An item is stored in a place such that its withdrawal requires authorization.

Assembly and Operation Process Charts

In the planning stages of developing a production system the layouts of machines and the plant are not yet known. One can only visualize, from parts and product designs, the operations that are necessary and the sequence in which they must be performed. Assembly charts and operation process charts are the graphical representations of such methods. An assembly chart gives a broad overview of how several parts manufactured separately are to be assembled to make the final product, for example, an internal combustion engine. It may also show a flow in reverse—how a product that is in unit form is disassembled and distributed in different processes, for example, raw milk in a dairy. Figure 7.1 is the assembly chart for the teakettle mentioned in Chapter 4.

Another important use for an assembly chart is in scheduling production, especially in a job shop. As is shown in Figure 7.2, such charts must be drawn to scale using some convenient unit of time along the horizontal. The graph is a set of bars, each representing the starting and completion times for the manufacture or installation of components and/or subassemblies. Knowing the promised delivery date, a scheduler can work backward and determine the time for processing component work orders.

Returning to the teakettle example, Figure 7.3 presents the operating process chart for that item. Preparing such a chart is essentially the task of detailing the assembly chart (Figure 7.1), showing every operation and inspection that each part needs as it progresses from raw material to the final assembly.

Figure 7.1

Assembly chart for teakettle

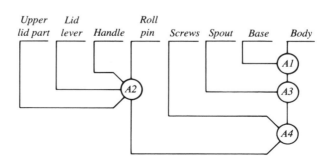

Figure 7.2 Assembly chart for scheduling production

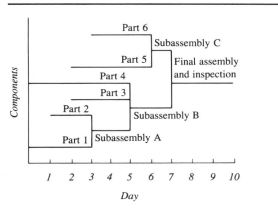

Time Estimates

For each activity (operation or inspection) in the operation process chart an estimate should be made of the time it will take to complete the activity. To develop such estimates is a complex task. The time to perform an operation depends on the machine and the jigs and fixtures used. Automatic machines, for example, would probably not take as long to finish a given job as would manually operated machines. Furthermore, the speed and feed of the machine also affect the time to perform the activity, and the degree of automation in the material-han-

Figure 7.3 Operation process chart for teakettle

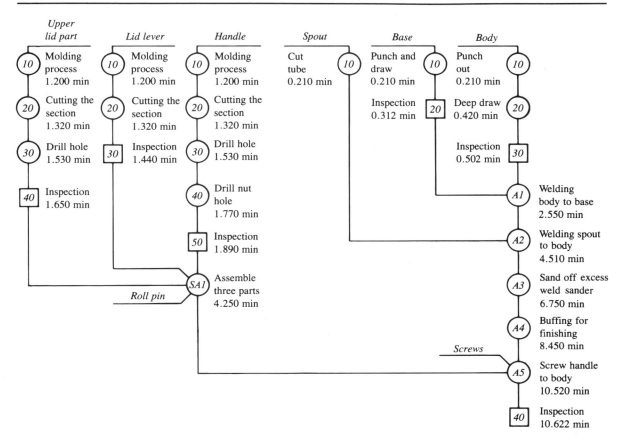

dling equipment should be reflected in the setup time.

In the case of an established manufacturer introducing a new product, the time estimates can be based on experience and records for the processes, machines, and procedures. For a new enterprise, however, such company data are seldom available. Standard predetermined time tables such as work factors, brief work factors, MTM, or element time data may be used to the extent possible in developing time estimates. These tables, however, give time values mainly for manual operations. One such effort, the use of brief work factors, is illustrated for a welding job in Figure 7.4.

The machine operation times could be established by using standard machine formulas obtained from any handbook on manufactur-

ing. For example, cutting time on a milling machine is given by the following equation:

Cutting time/piece in minutes
$$= \frac{\text{Table or tool travel per cut in inches/cut}}{\begin{array}{c}\text{Table or tool feed in inches/minute} \\ \times \text{ Number of cuts required per piece}\end{array}}$$

The total distance traveled by the table or tool is equal to the length of the milling surface plus overtravel on each side. The table feed might change somewhat from machine to machine; therefore the time estimate might have to be modified on the basis of the machine used.

Using the methods described above, one could develop the time values for most of the

Figure 7.4   Brief work factors table

PART NAME: Body and base assembly   **PART NUMBERS:** 1 and 2
OPERATION NAME: Welding   **PERSONAL ALLOWANCE:** 20%
TOOLS: Welder
FIXTURE: Body and base
ANALYST: Chong

ELEMENT		DESCRIPTION	TIME (0.001 MIN.)
1	P10	Pick up the body	10
2	MR10	Move to the right	10
3	MR10	Move hand to the welder	10
4	G15	Grasp the welder	15
5	P20	Pick up the welder	20
6	ML20	Move the welder to the left	20
7	15A40	Weld the body to the base	600
8	MR20	Move the welder to the right	20
9	R10	Release the welder in position	10
10	LR30	Lift the welded parts with care	30
11	MR30	Move the welded parts aside	30

```
Total: 775
1 unit = 0.001 minute
Total calculated time: 0.775 minute (120%)
Total operation time: 0.930 minute (100%)
Standard time: 0.016 hour
Units per hour: 64.516
```

manual and machine operations. For some other elements such as inspection the times might still have to be estimated. The accuracy of these time values will determine the initial operational setup. Of course, the industrial engineer will be asked to update these time values and perhaps redesign the operations once production has begun.

Routing Sheet or Production Work Order

The next step in planning is to draw a routing sheet (sometimes called route sheet or production routing). It shows how a part is to be produced, which machines are needed, the tools to use, estimated setup times for the machines, and production in terms of the number of units expected per hour from each machine. Figure 7.5 shows a routing sheet. One routing sheet is required for each part in the assembly.

The information collected from all the routing sheets (remember that there might be more than one product to manufacture) is extremely important and is used in many phases of future planning. As we will observe in the following chapters, this information is needed for determining the number and types of machines to be purchased to produce certain output rate(s), the number and skill of the employees needed, the production system to use, and indeed how the entire plant should be laid out. Routing sheets and the bill of materials form the major data base for future planning.

Other Charts

Some of the other charts used in manufacturing and control are now briefly described.

Left-Hand, Right-Hand Chart

The left-hand, right-hand chart lists the work performed simultaneously by both the left hand and the right hand of an operator at a specific work station. There are two columns, one for each hand. which list the sequential ele-

Figure 7.5   Production routing sheet

PRODUCT: <u>Teakettle</u>   PART: <u>U. lid part</u>   PART NO.: <u>6</u>
PREPARED BY: <u>Chong</u>   DATE: <u>Jan 28</u>   SHEET: <u>1</u> OF <u>1</u>

OPERATION		MACHINE	AUX. EQUIP.	SETUP TIME (HR)	HR/PC	PC/HR
NO.	DESCRIPTION					
10	Molding process	Injection molding machine		0.05	.02	50 sections of 6 parts
20	Cutting out of sections (combined with lower part)	Hand cutter			.002	500
30	Drill roll pin hole	Drill press		0.02	.0035	285
40	Inspection				.0017	580

ments of the task. There is an additional column for each hand that, by means of two symbols, indicates whether the hand is involved in transporting a part or is performing such actions as grasping, positioning, using, or releasing the item. A small circle designates transporting a part or moving the hand, and a large circle indicates an operation. Figure 7.6 is a small portion of a left-hand, right-hand chart for the assembly of a bolt and a nut.

The objective is to design the work station and the work sequence so that both hands are fully utilized working rather than doing less productive tasks such as holding, excessive delays, and transporting.

Gang Chart

The gang chart, also called the multiple-activity chart or the man-machine chart, depicts the simultaneous activities of all the members in a gang or team or the activities performed by a combination of one or more persons and one or more objects, such as people and machines, or one person and one machine. The purpose is to visualize all details of the

Figure 7.6   Left-hand, right-hand chart

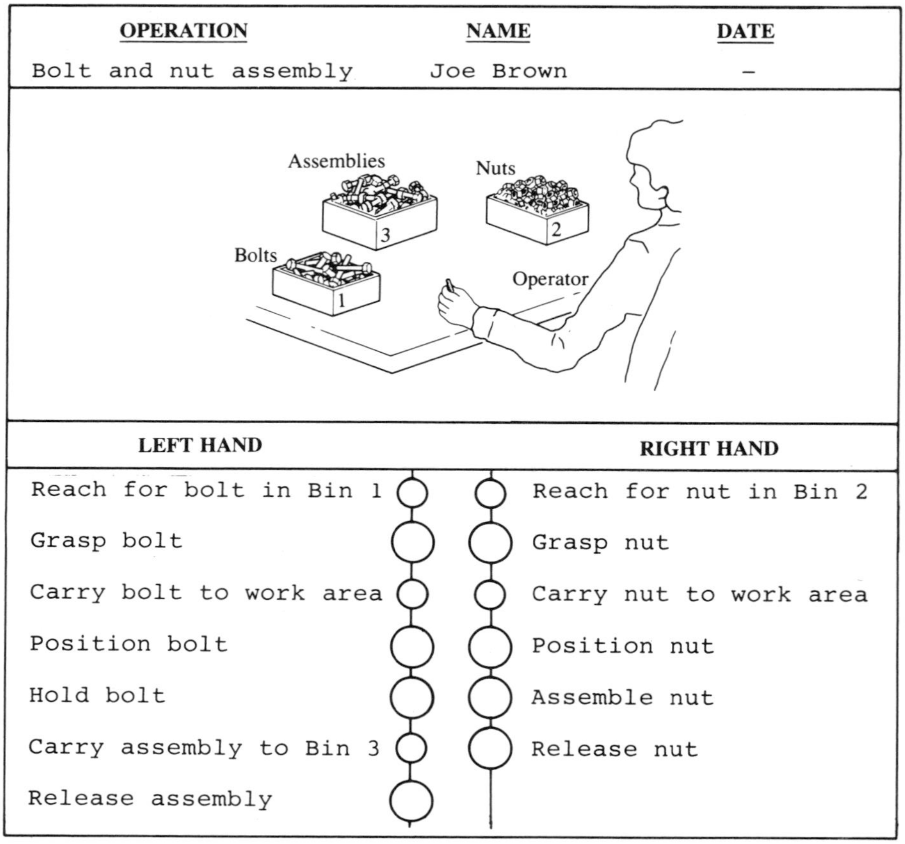

work being performed by a team to eliminate or minimize any nonproductive element for an individual and to obtain a good balance of work among all the team members. Figure 7.7 is a sample gang chart depicting the steps performed by a three-person team in completing a simple task.

Gantt Chart

In a Gantt chart, horizontal bars indicate the proposed (and perhaps actual) times spent in performing each activity or task. These times could be reduced if resource allocations, such as manpower, were increased. The objective is to assign the limited resources to different activi-

Figure 7.7 Gang chart

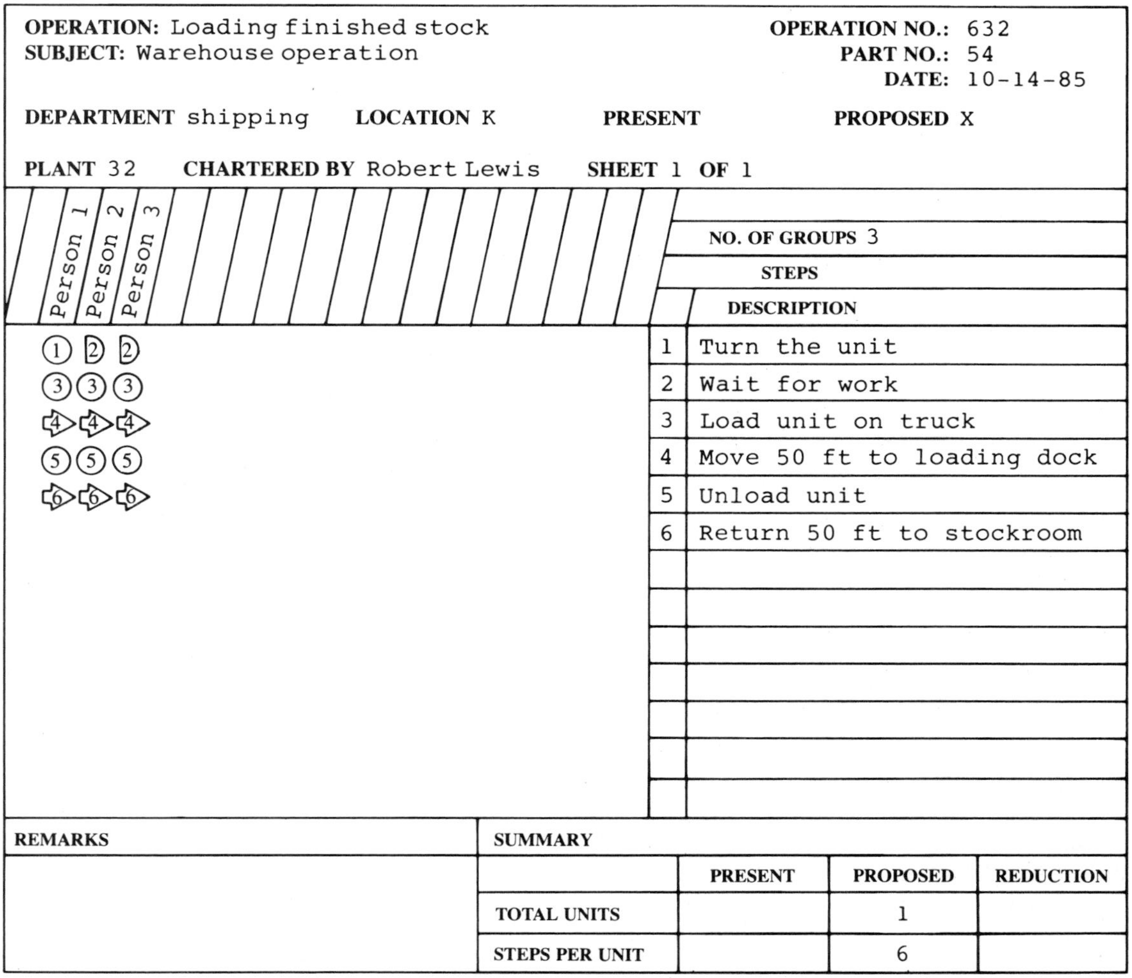
OPERATION: Loading finished stock
SUBJECT: Warehouse operation
OPERATION NO.: 632
PART NO.: 54
DATE: 10-14-85
DEPARTMENT shipping
LOCATION K
PRESENT
PROPOSED X
PLANT 32
CHARTERED BY Robert Lewis
SHEET 1 OF 1
Person 1
Person 2
Person 3
NO. OF GROUPS 3
STEPS
DESCRIPTION
1 Turn the unit
2 Wait for work
3 Load unit on truck
4 Move 50 ft to loading dock
5 Unload unit
6 Return 50 ft to stockroom
REMARKS
SUMMARY
PRESENT
PROPOSED
REDUCTION
TOTAL UNITS
1
STEPS PER UNIT
6

Figure 7.8

Gantt charts

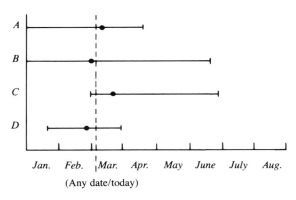

(Any date/today)

(a)

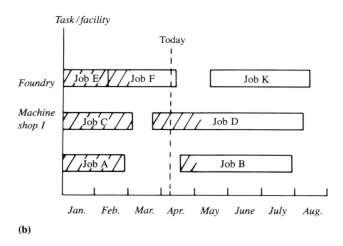

(b)

ties so as to minimize the total time required for completion of all activities. Figure 7.8(a) presents a basic Gantt chart for four activities involved in a particular contract. This could be for a major subdivision, or it could be the entire project. The dashed line is used to represent a sliding bar, which is moved horizontally to any date to help compare the progress of the project with the schedule of that date. The actual progress is marked by an individual slide for each task, shown in the figure as a solid dot on the line.

For example, as of the date shown, Tasks B and D are shown to be behind schedule.

Figure 7.8(b) represents another version of a Gantt chart in which the facilities are shown with the operations in the sequence in which they are scheduled. The bar length represents the planned time while the shaded area represents the actual progress on each activity.

Figure 7.9 Flow process chart

SUMMARY	PRESENT NO.	PRESENT TIME	PROPOSED NO.	PROPOSED TIME	DIFFERENCE NO.	DIFFERENCE TIME
○ OPERATIONS	5					
⇨ TRANSPORTATIONS	9					
□ INSPECTIONS	1					
D DELAYS	2					
▽ STORAGES	3					
Distance Traveled	1485 FT			FT		FT

JOB	
MANUFACTURE OF A TISSUE BOX	
□ OPERATOR	
☒ MATERIAL	
CHART BEGINS	RECEIVING (RAW MATERIALS)
CHART ENDS	SHIPPING (FINISHED PRODUCT)
CHARTED BY	T.P.C.

QUESTION EACH DETAIL	ANALYSIS	
	WHAT?	WHEN?
	WHY?	WHO?
	WHERE?	HOW?
DATE		
NUMBER		
PAGE 1 OF 1		

DETAILS OF (PRESENT / PROPOSED) METHOD	OPERATION	TRANSPORT	INSPECTION	DELAY	STORAGE	DISTANCE IN FEET	QUANTITY	TIME (min.)	POSSIBILITIES: ELIMINATE	CHANGES: COMBINE	CHANGES: SEQUENCE	CHANGES: PLACE	CHANGES: PERSON	IMPROVE	SAFER?	$ SAVED?	NOTES
1. RECEIVE RAW MATERIALS		⇨				50	5	7.1									5 cartons of 1000 sheets of cardboard
2. INSPECT			□					1.3									
3. MOVE BY FORKLIFT		⇨				40		3.1									
4. STORE					▽			12.3									
5. MOVE BY FORKLIFT		⇨				45		3.7									
6. SET UP AND PRINT	○							120.1						X		X	
7. MOVED BY PRINTER		⇨				120		5.2									
8. STACK AT END OF PRINTER	○							15.3									
9. MOVE TO STRIPPING		⇨				165		3.9									
10. DELAY				D				2.2	X								
11. BEING STRIPPED	○							18.9									
12. MOVE TO TEMP. STORAGE		⇨				150		10.8									
13. STORAGE					▽			20.1				X		X	X	X	
14. MOVE TO FOLDERS		⇨				200		15.4									
15. DELAY				D				6.3	X								
16. SET UP, FOLD, GLUE	○							210.5									
17. MECHANICALLY MOVED		⇨				90		3.8									
18. STACK, COUNT, CRATE	○							45.0									
19. MOVE BY FORKLIFT		⇨				525		22.8									
20. STORAGE					▽												

Flow process chart tabulates steps and moves in a process, in this case for the manufacture of a small box with a printed label. Standard American Society of Mechanical Engineers (ASME) process elements and symbols are used. The first column from the left is the step number, followed by a brief description of the activity. In the next column, the symbol that best describes the activity is used in tracking the step-by-step flow from top to bottom. Improvement opportunities can be noted in the columns headed "possibilities," and additional information can be entered under "Notes." A single chart can be used to track the flow of either an object or a person — but not both simultaneously.

The chart may be extended to that of the CPM and PERT networks, details for which may be found in Muther and Hale (1979).

Flow Process Chart

The flow process chart is another device that is very helpful in analyzing an existing flow layout. It graphically displays every step a unit follows in the plant, starting with raw material and continuing until the product is completed. All of the operations, transportations, delays, inspections, and storages are noted symbolically on the chart. The objective is to determine the method of producing the unit that uses the least number of such symbols. For example, it might be possible to combine or eliminate some operations, reduce transportation by relocating machines or finding alternative routes, and eliminate delays by more efficient scheduling. The analyst should ask the questions What? Why? When? Who? Where? and How? for each step to see whether any beneficial changes can be made. Each inquiry should lead to improved plant operations and/or layout. A flow process chart is prepared for keeping track of either a product flow or a person but not both simultaneously. Figure 7.9 (see page 117) illustrates a flow process chart (courtesy of the Material Handling Institute).

7.2 PRODUCTION SYSTEMS

The next logical step in the development of a manufacturing facility is to determine the type of production system that would be best utilized. We will therefore discuss here typical production systems and manpower planning methods. Production systems are classified according to the arrangement of machines and departments within manufacturing plants. The spectrum of production systems ranges from job shops with mainly manual operators to completely automated assembly lines. The number of different products a company makes, the types of orders (made to stock or made to order), the volume of its sales, and the frequency of reorders strongly influence what the most efficient production system would be for a given firm. How certain the demand is and how long the production run must be also play an important part in this decision. As a general rule, high volume favors automation, and low volume favors manual operations. Numerically controlled (NC) machines, however, can produce both high and low volumes with almost equal efficiency. They do need a high initial investment that could be hard to justify if the machines are not fully utilized. It is not unusual to have different modes of production in the same plant for different products because of the products' individual sales volumes. The four major categories of production systems are job shop production, batch production, mass and continuous production, and cellular and flexible manufacturing.

Job Shop Production

A job shop is a suitable mode for a company that produces many different products with a relatively small volume for each. The various items being produced to customer specifications require the collection of a variety of general-purpose equipment, highly skilled labor to operate this equipment, and general-purpose tooling and fixtures. Similar machines can be grouped together within the plant to form a department. (See Figure 7.10.) The product moves from one department to another on the basis of its operational sequence.

Many problems are encountered in this type of operation. The expense associated with the variety of tools and fixtures can be large; and to achieve machine-load balancing through proper job sequencing is very difficult. There is excessive material handling as parts are moved from one department to another; some might

Figure 7.10 Job shop or batch production floor plan

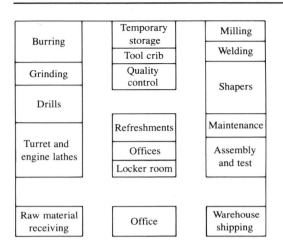

pass through the same department several times. Items might have to be held in temporary storage awaiting further processing, resulting in the need for a large area devoted to inventory. Each item or batch of parts requires its own production shop orders, which can result in excessive bookkeeping.

The productivity of a job shop—that is, the percentage of time spent in actual production—is low, mainly because of excessive setup time and material movement. The plant has a minimum of automation; however, it does attain a high degree of flexibility to produce a variety of different products. The system is very responsive to changes in the market. It is estimated that about 30–50 percent of the manufacturing systems in the United States are of the job shop type.

Batch Production

Batch production is a production system that falls between a job shop and an assembly line in its complexity. The system is suitable for a firm that has numerous items to produce but

not so large a variety as to require a job shop type of production. Unlike the situation in the job shop, the products to be made throughout the year are known, each having a stable and continuous demand. The production capacity of the plant is larger than the demand for an item, and hence the products are manufactured in batches. An item is produced once and stored at a preplanned inventory level to meet both present and future demand, and then the facility is switched to produce the next item in the production sequence. A product is rescheduled for production when its stock level has diminished to a predetermined level.

The production machinery and equipment used in the batch shop are somewhat specialized and of high speed, but the skills of the labor force might not be as highly developed as in a job shop. There is some automation within the plant, and the machines and equipment are grouped together to perform certain tasks, which may mean that different types of machines are placed together. The productivity in a batch shop is higher than that in a job shop.

Typical users of batch shop production systems are manufacturers of furniture, appliances, and mobile homes.

Mass Production

The mass production system is used for high-volume production. Often the entire plant and its equipment are dedicated to manufacturing a single product. The equipment is very specialized and fast, and the investment in special tools, jigs, and fixtures is large. The work content is broken down into many small groups; a high degree of efficiency is obtained by perfecting the tooling and method of work in each group. The overall labor skills required are minimized by such an approach, but the assigned tasks for the individual can become repetitious and in some cases, uninteresting. The productivity in mass production is very high and is achieved to a large extent by automation.

Production systems in manufacturing facilities. *Top left*: A job shop, like the one shown here, processes several different items in small numbers and has the flexibility needed in a tool room, which must make or obtain the various tools a plant needs as well as store and maintain them. *Top right*: In the assembly line operation here, partial assemblies of electronic boards are moved by conveyor belt from one work station to the next so workers can use specialized equipment to assemble the product in small steps. *Bottom left*: In cellular manufacturing, similar parts are made together in areas called cells, which can often be completely automated. Automated guided vehicles can perform material-handling tasks accurately and efficiently. Pictured here is the recharging area for such vehicles. *Bottom right*: Automated guided vehicles like this one can transport parts and/or material from robot-controlled storage systems to automated production machines without direct human control.

In mass production, controls of the line output and manpower scheduling can readily be achieved. One can control the output rate by the design of the line (stations) and/or the manpower assigned to these stations. (These concepts are explained further by an example later in the chapter.) Periodic checks of the inventory level and product demand might suggest a change in the output rate.

There are two ways of further classifying mass production systems: assembly line and flow line. An assembly line is used to produce a discrete product. Typically, partial assemblies are moved from one workstation to the next in sequence by a fast-moving material-handling system (such as conveyor belts) past each station, advancing the product toward its final assembly. Travel of the assembly within the stations is continuous or intermittent, depending on the nature of the job to be performed. The total work for a job is distributed among the work stations that form the assembly line such that all stations complete their assigned tasks in approximately the same time, called the cycle time. Automobile manufacturing is a prime example of an assembly line operation.

The term flow line is typically used to describe a continuous production process such as that of chemicals, liquids, gaseous products, and paper, as well as die operations such as wire manufacturing.

Cellular and Flexible Manufacturing

Cellular manufacturing is a system in which a large number of common parts are grouped together and produced in a cell consisting of all the machines that are needed to produce that group. When large quantities of each part are required, cells can be made almost completely automated, leading to the designation of that cell as flexible manufacturing.

The major advantage of the mass production system is that it lowers the unit cost of production, which is composed of two components: a fixed cost (for setup) and a variable cost (primarily labor and material). The portion of fixed cost assigned to a unit depends on the quantity produced with the same setup (same fixed cost). The larger the volume of production, the lower the share that each unit must carry. In mass production the quantity produced with a particular setup is large, and therefore the unit cost of production is much lower than it would be in a job shop with a small production lot. But what can be done if the required quantity for each item does not warrant the use of a mass production system? Group technology and cellular and flexible manufacturing might provide the answer.

Group technology refers to the concept of identifying similar parts and grouping them to take advantage of their common characteristics in design and production methods. For example, from 1000 parts that are produced in a plant, perhaps 20 groups could be formed, each needing similar machines, jigs, and fixtures so that the items within a single group could be produced with very little change in the setup. Thus the major setup cost is distributed among a large number of items, however small the quantity produced for each individual item may be.

Many different methods are used to develop such groups. The quickest, but perhaps a little more difficult and not so accurate as others, is simple observation, a process in which one tries to form groups by looking at the tasks that are most often performed by a number of machines. To aid this process, another method called production flow analysis (PFA) can be used. We will demonstrate this procedure by means of a very simple example.

The analysis begins with an examination of the production routing sheets for the items under consideration because the information provided by such traveler sheets includes the sequence of machines needed to manufacture the part. This important data will be used in the next step of developing the machine groups.

For example, suppose there are ten items to produce, and analysis yields the information presented in Table 7.1 regarding the machines needed in their production.

The data can be transformed into the matrix shown in Table 7.2, in which items are the rows and machines are the columns of the matrix.

For Item 1, entry 1-A indicates the use of Machine A in its manufacture; the next entries, 1-C and 1-D, complete the sequence. Similar entries are made for all other items in the table. The next step is to rearrange the items so that groupings of machines are obvious, but this is not as simple as might be thought and can involve several trials before the proper grouping is obtained. Table 7.3 presents the final grouping for the example at hand.

Table 7.1 Production schedule

Item	Required Machines
1	A, C, D
2	A, B, C
3	C, D
4	H, I
5	I
6	E, G
7	E, F, G
8	B, C, D
9	A, B
10	E, G

Table 7.2 Initial production matrix

Item	A	B	C	D	E	F	G	H	I
1	X		X	X					
2	X	X	X						
3			X	X					
4								X	X
5									X
6					X		X		
7					X	X	X		
8		X	X	X					
9	X	X							
10					X		X		

Machine (header spanning A–I)

Table 7.3 Final production matrix

Item	A	B	C	D	E	F	G	H	I
1	X		X	X					
2	X	X	X						
3			X	X					
8		X	X	X					
9	X	X							
6					X		X		
7					X	X	X		
10					X	X			
4								X	X
5									X

Machine (header spanning A–I)

In cellular manufacturing, the machines used to produce each of a group of items are clustered together. The matrix in Table 7.3 shows that three groups of cells (Machines A through D, E through G, and H and I) can be formed to most efficiently manufacture the ten items of our example. Remember that the products need not be identical in size and/or shape to make possible the reaping of benefits from cellular manufacturing. The similarity must be in the particular machines and setups required in their production.

A computer program that attempts to form such groups is described in Section 7.4. Figure 7.11 shows the input data for an example problem consisting of 20 jobs and 10 machines. Figure 7.12 presents the corresponding groupings obtained by the program. The program is only illustrative and forms the satisfactory groups in many cases; however, it is not a generalized program.

Small or medium-sized lots (varying from 1 to 200 units) of each item can be made with a minimum of changeover in setups. Anywhere from one to fifteen machines are grouped together, typically in a U-shaped layout, as shown in Figure 7.13. These machines may be general-purpose or highly automated. The U-shaped layout has many advantages. It provides flexibility in that the number of workers who are trained to perform a range of tasks can easily be increased or decreased on the basis of the amount of work assigned to that cell. Workers are made responsible for more than one machine, a situation called machine coupling, so that the utility of a worker is not limited by the utility of a machine. Close contact and cooperation of workers in a U-shaped layout can contribute to an increase in productivity by decreasing idle time, poor quality, and in-process inventory.

Flexible manufacturing is cellular manufacturing in highly automated form. The cells are made very efficient to produce large quantities rapidly by including NC machines, ro-

Figure 7.11

Data for the production flow analysis program

	A	B	C	D	E	F	G	H	I	J	K	L	M	N	O	P	Q	R	S
1	X	X				X													
2							X		X										
3							X	X	X										
4	X		X			X	X	X											
5							X	X	X										
6								X	X						X	X	X	X	
7	X	X	X		X	X													
8		X	X																
9			X	X	X														
10	X	X	X																
11				X		X													
12	X	X						X	X										
13			X	X				X	X										
14			X	X				X	X										
15	X	X	X		X	X				X		X							
16	X	X	X		X	X												X	
17									X	X	X	X			X	X			
18				X	X	X			X	X									
19															X	X			
20	X	X																	
21	X	X	X																
22				X			X	X	X										
23	X	X	X																
24				X	X	X													
25				X	X														
26	X							X		X									
27							X	X	X										
28		X		X		X													
29	X	X						X	X										
30	X	X	X					X	X										
31						X	X	X											
32	X	X						X	X										
33	X						X												
34	X	X	X					X	X										
35	X	X						X				X		X					
36			X	X	X			X	X										
37	X							X	X										
38			X	X				X											
39	X							X	X										
40			X					X	X										

124

Figure 7.12

Output of the production flow
analysis program

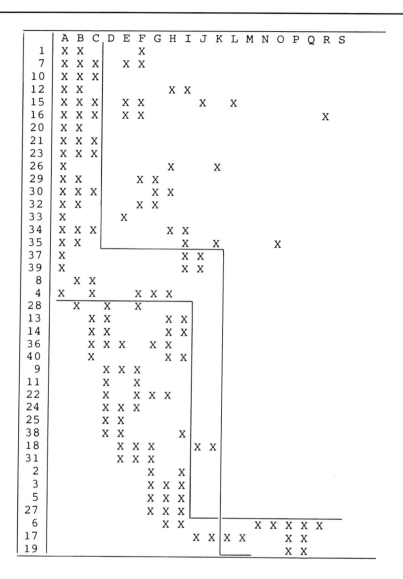

Figure 7.13 U-shaped arrangement

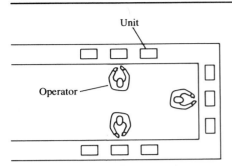

bots, and computer-controlled conveyor systems. A cell in this case might be arranged in a U-shape as before, or even in a straight line form, which can increase efficiency of material movement. This system of manufacturing is used mainly to produce a family of parts in lots of 200–1000 units, since the NC machines require almost no changeover in setups to produce the items in a family. A high degree of automation means high efficiency and very little need for manpower. It is not unusual to find only one or two workers controlling the entire production.

7.3 MANPOWER ASSIGNMENTS

Let us consider two prominent production layout types in which the manpower needed can change depending on how the total production is distributed among different stations. The production line, which is mainly a batch-processing arrangement, and the assembly line where the units are to be produced at a constant rate are examined next.

Manpower Assignment in Production Lines

In some job shops and batch shops, production facilities are referred to as production lines. Though similar to assembly lines, production lines differ from them primarily in the rate of flow. Units are processed through many different stations (work areas) before finally being transformed into complete assemblies. Within a given production line, each station might consist of a single machine and/or a worker, an entire department, a group of machines and their operators, or even a number of manual laborers assigned to perform a specific task. The rate of production in each station depends on the type of work assigned and quite often varies from station to station. The production rate can be controlled by the manpower allotted to that station; the more manpower provided, the greater the production, up to a certain limit. There might also be partially finished assemblies being left over from a previous production run of the same product, called inventory, in front of a station. For a specific station, such inventory consists of units that have received all of the operations by preceding stations and can be made into the final assembly by starting with the operations of the specific station. The problem is to determine the number of workers needed in each station to achieve the required output from a given production line. The following example illustrates this procedure.

Consider the production line shown in Figure 7.14, requiring 20 stations to complete the assembly. For convenience the stations are numbered from left to right in the direction of material flow. Subassemblies 1 and 2 are combined in Station 5, which is also the common input station for the three parallel branches 6–9, 7–10, and 8–11. These branches are required in this example because of limitations on space, equipment, and manpower available to work at the stations.

The output from the three branches—that is, from Stations 9, 10, and 11—goes into common storage, from which parts for Stations 12 and 13 are drawn. In Station 17, two subassemblies are joined in the ratio of 2:5. Finally, Station 18 supplies both Stations 19 and 20, where

Figure 7.14 Production line

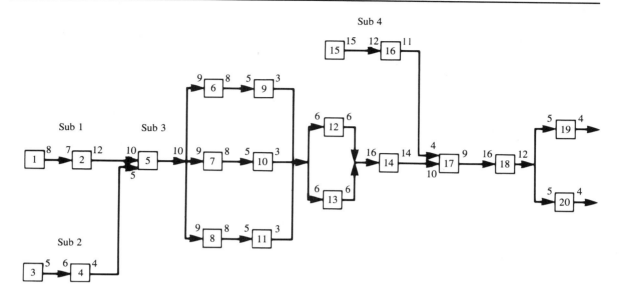

the final processing is accomplished. Figure 7.14 also shows the input/output requirements per worker at each station; for example, in Station 18 a worker will require 16 units of input and will produce 12 good units.

Table 7.4 presents the tabular approach. The information contained in Columns 1, 2, 3, 4, 5, and 10 is known for the given product. By way of illustration, consider Station 17. Production per worker at this location is 9 units, but the next work station, Station 18, requires 16 units per worker as its input. The maximum number of workers that can be assigned to Station 17 is 8.

Continuing with the analysis, we will now examine a few specific stations of interest in the development of Table 7.4. Station 18 provides the input for two stations, 19 and 20; and Column 5 for Station 18 lists the input requirement for each of these stations (19 and 20).

Stations 9, 10, and 11 provide the input to a common storage area. They are grouped to-

gether and are marked as such by a brace. Stations 12 and 13 also form a similar group.

Station 17 is the succeeding station for both stations 16 and 14; however, the units demanded are different, since two different subassemblies are involved. The appropriate input is shown for each station in Column 5.

In this example the required output is 16 units from Station 20 and 18 units from Station 19. After accounting for present inventories by completing Column 2, the procedure is to work backward from the last work area (Station 20), filling in all the columns for each station (or each group of stations). For example, for Station 17 the number of workers assigned to the succeeding station—that is, Station 18—is 4. The output requirement is therefore 64 (16 × 4) units. (Column 7 is Column 6 multiplied by Column 5.) Since there is an initial inventory of 7 units, the net production should be 57 (64 − 7) units. (Column 8 is found by subtracting Column 2 from Column 7.) The necessary

Table 7.4 Production line planning

Station (1)	Initial After-Station Inventory (2)	Production Rate of the Station (3)	Succeeding Station (4)	Input Required on Succeeding Station per Worker in that Station (5)	Workers Assigned in Succeeding Stations (6)	Output Required (6) × (5) (7)	Net Production Required (7) − (2) (8)	Minimum Manpower Required (9)	Limit on Manpower Assignment (10)	Workers Assigned (11)	Output (3) × (11) (12)	Final After-Station Inventory (12) − (8) (13)
20	0	4	—	—	—	16	16	4	6	4	16	0
19	0	4	—	—	—	18	18	5	6	5	20	2
18	5	12	19, 20	5, 5	5, 4	25 + 20 = 45	40	4	8	4	48	8
17	7	9	18	16	4	64	57	7	8	7	63	6
16	10	11	17	4	7	28	18	2	4	2	22	4
15	15	15	16	12	2	24	9	1	4	1	15	6
14	34	14	17	10	7	70	36	3	4	3	42	6
13 }	5	6	14	16	3	48	43	8	4	4	24	5
12 }		6	14	16					5	4	24	
11 }	1	3	12, 13	6, 6	4, 4	24 + 24 = 48	47	16	8	5	15	1
10 }		3	12, 13						8	5	15	
9 }		3	12, 13						10	6	18	
8	6	8	11	5	5	25	19	3	4	3	24	5
7	4	8	10	5	5	25	21	3	4	3	24	3
6	7	8	9	5	6	30	23	3	4	3	24	1
5	10	10	6, 7, 8	9, 9, 9	3, 3, 3	27, 27, 27	71	8	10	8	80	9
4	15	4	5	5	8	40	25	7	9	7	28	3
3	5	5	4	6	7	42	37	8	8	8	40	3
2	0	12	5	10	8	80	80	6	8	6	72	8
1	5	8	2	7	6	42	27	5	7	5	40	3

manpower is calculated by dividing the net production required by the rate of production per worker in this station. (Column 9 results when Column 8 is divided by Column 3.) Since this must be a whole number and cannot be negative, the minimum value is zero, and fractions are rounded off to the next higher integer. For Station 17 we need a minimum of 7 workers (Column 9) to operate the machines; the maximum number of workers that can be assigned is limited to 8 (Column 10). Suppose the production manager decides to assign 7 workers (Column 11). The total output from this station then becomes $(7 \times 9) = 63$ units (Column 12 = Column 3 $\times$ Column 11). The final remaining inventory is what is produced minus what is needed (Column 17 = Column 12 − Column 8), or 6 in this case.

Now consider some other stations of interest. For Station 18, Column 6 contains the number of workers assigned to Stations 19 and 20. Thus the output required (Column 7) is the total output necessary to supply both stations.

For the group consisting of Stations 9, 10, and 11 the succeeding work areas are Stations 12 and 13. For each of the latter the input requirement is 6 units per worker assigned. Therefore the total output required from Stations 9, 10, and 11 is 48 units; but with the initial inventory of 1 unit, the net production required is 47 units. Each worker in Stations 9, 10, and 11 can produce 3 units; therefore we will need a minimum of 16 workers (Column 9). These workers are distributed 5, 5, and 6, respectively, among Stations 9, 10, and 11, keeping within the maximum manpower assignment limit. In the common storage area, units arrive from all three stations (a total of 48 units), leaving one unit on hand at the end of the production run.

The analysis is complete when the entries associated with all the stations have been made. In this example, Station 1 is the last station for which the data are entered. When the analysis is finished, the method tells what the manpower assignments should be and what the final inventory will be for each station.

Manpower Planning

Quite often, each station represents a stage in a production process; but with a limited crew it becomes impossible to work on all stages simultaneously. Thus there is a need to develop a manpower scheduling system that will indicate which station is to receive how many workers in what time period, perhaps daily.

One approach would be to complete all work for an entire order at a particular station before moving to the next station in the sequence. This method would almost certainly create excessive in-process inventory. The other extreme is to process a single unit in one station and then move the worker to the next station. This requires costly setup changes and worker movements that might be difficult to plan practically. It is quite difficult, for example, to keep a worker four minutes in one station, six in the next, and so on. (In fact, the worker might spend more time changing stations than working.)

One method of making assignments would be to follow the rules stated below:

1. Move the workers from one station to another only after the end of a period (a day in our example).
2. Assign a worker to the succeeding stations in the production sequence if the input to these stations is sufficient to meet the target production for the period.

We leave it as an exercise to the reader to develop the daily manpower schedule for the example in the preceding section.

Assembly Line Balancing

An assembly line is a method of production in which the parts are assembled and made into the final product as the unit progresses from station to station. To develop a balanced assembly line requires logical planning and involves the distribution of the total job among the work stations so that all stations can complete their designated tasks at approximately the same time. If the line were perfectly balanced, the time required in each station would be identical. However, such balance is rarely possible, and the longest station time dictates the actual cycle time for the entire assembly.

Various techniques are available for developing an assembly line, but the basic requirements for all are the same: break the total job content into small tasks or work elements and determine the precedence relationships, that is, decide which tasks must be performed before the next one begins. These relationships also indicate the tasks that can be performed simultaneously. For example, in paintbrush manufacturing, glue must be poured into a clip before the bristles can be inserted; however, at the same time, someone (or a machine) could polish and stamp the handle. Next, determine the time that will be required for each work element. Any other restrictions such as the requirement that none of a group of work elements may be placed together in the same work station, called negative zoning, or the requirement that some tasks must be placed in the same work station, called positive zoning, should also be considered and clearly stated.

Mathematical methods are available to balance assembly lines in terms that are mainly the application of optimization techniques such as integer programming and branch-and-bound methods. However, these exact methods are very tedious and cumbersome to apply to a large assembly line containing a multitude of tasks. We will show two heuristic methods that are popular because of their simplicity.

Balancing Procedures

The methods are illustrated by applying them to the following sample problem. The ele-

Figure 7.15   Precedence diagram

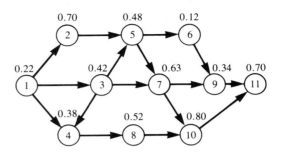

ments are represented in Figure 7.15 by nodes, and the precedence of each is indicated by the arrows connecting the nodes.

In the diagram the element (task) times are listed above their corresponding nodes. The same data are also presented in Table 7.5.

The total time needed to complete all tasks is 5.31 minutes.

Procedure I—largest candidate rule: The procedure determines the minimum number of stations necessary to obtain the desired cycle time, often referred to as *cycle time* and defined as the total available time per period divided by the required production per period. The objective is to distribute the work among the stations. The steps involved in the procedure are as follows:

1. List the tasks in decreasing order of magnitude of task times, the task requiring the largest time being first. Also list the corresponding immediate predecessor task(s) for each.

2. Designate the first station in Step 1 as Station 1 and number the remaining stations consecutively.

3. Beginning at the top of the task list, assign a feasible task to the station under consideration. Once the task is assigned, all reference to it is removed from the predecessor task list. A task is feasible only if it does not have any predecessors or if all predecessors have been

Table 7.5 Predecessor list

Element	Time	Immediate Predecessor(s)
1	0.22	—
2	0.70	1
3	0.42	1
4	0.38	1, 3
5	0.48	2, 3
6	0.12	5
7	0.63	3, 5
8	0.52	4
9	0.34	6, 7
10	0.80	7, 8
11	0.70	9, 10

Table 7.6 Task time

Task	Task Time	Immediate Predecessor(s)
10	0.80	7, 8
11	0.70	9, 10
2	0.70	1
7	0.63	3, 5
8	0.52	4
5	0.48	2, 3
3	0.42	1
4	0.38	1, 3
9	0.34	6, 7
1	0.22	—
6	0.12	5

deleted. It may be assigned only if it does not exceed the cycle time for the station, and this condition can be checked by comparing the cumulative time of all the jobs so far assigned to that station, including the task under consideration, with the cycle time. If the cumulative time is greater than the cycle time, the task under consideration cannot be assigned to the station. If no task is feasible, proceed to Step 5.

4. Delete the task that is assigned from the first column of the task list. If the list is now empty, go on to Step 6; otherwise, return to Step 3.

5. Create a new station by increasing the station count by one. Return to Step 3.

6. All jobs are assigned, and the present station number reflects the number of stations required. The procedure also shows the job assignments for each station. The largest of the cumulative times for the individual stations is the true cycle time.

We will demonstrate application of the above procedure to the sample problem presented in Figure 7.15 and Table 7.5, with a planned cycle time of one minute or with a production rate of 400/day, assuming that there are 400 working minutes in a day shift.

In following the first step we list the tasks

(elements) in the order of decreasing task time (Table 7.6). Also listed are the immediate predecessor(s) of each task. This information is needed in Step 2 and can be obtained either from the data table or from the precedence diagram.

The procedure begins with the element listed first in Table 7.6, Task 10. The accumulated task time would be equal to 0.80, which is less than the cycle time; but since neither Task 7 nor 8 is completed, Task 10 cannot yet be considered. The same procedure is repeated for the next element in the list, Task 11, with the same result. As we go down the list, Task 1 is the first feasible task to be encountered, and it is assigned to the station list. We delete all reference to Task 1 in the predecessor list. In reality these operations are performed on the same table; but for clarity, we will show the results in Table 7.7.

Next we return to the top of the list; that is, 10, 11, ... Task 2 is the next feasible task found with a cumulative time of 1.0 or less, namely, 0.22 + 0.70 = 0.92. It is therefore included with Task 1 in Station 1. Inasmuch as no other task can be added to that station without exceeding the cycle time, we proceed to Step 5 and create a new station. Task 3 has had its preceding task scratched, and its time is less than the cycle time; therefore it is assigned to Station 2. The procedure continues until all the

Table 7.7   Task time after assignment of Task 1

Task	Task Time	Immediate Predecessor(s)
10	0.80	7, 8
11	0.70	9, 10
2	0.70	X̶
7	0.63	3, 5
8	0.52	4
5	0.48	2, 3
3	0.42	X̶
4	0.38	X̶, 3
9	0.34	6, 7
X̶	0̶.̶2̶2̶	—
6	0.12	5

Table 7.8   Final task assignments

Station	Element	Task Time	Cumulative Task Time
1	1	0.22	0.22
	2	0.70	**0.92**
2	3	0.42	0.42
	5	0.48	0.90
3	7	0.63	0.63
	6	0.12	0.75
4	4	0.38	0.38
	8	0.52	0.90
5	10	0.80	0.80
6	9	0.34	0.34
7	11	0.70	0.70

tasks have been assigned. The final result is shown in Table 7.8.

We need seven stations with this arrangement, and the actual cycle time is 0.92, the time required by Station 1. One measure of efficiency (e) is $(1 - p) \times 100$, where p is defined as

$$\frac{\text{(Total worker time required/unit)} - \text{Job time/unit}}{\text{Total worker time required/unit}}$$

and is known as balanced delay.

In this case, seven stations are utilized, each worked by one person, with a cycle time of 0.92 minute. Therefore efficiency is

$$e = 1 - \left[\frac{(7 \times 0.92) - 5.31}{7 \times 0.92}\right] \times 100$$
$$= 82.5\%$$

Procedure II—ranked positional weighted method: In the previous method the tasks were listed in decreasing magnitude of time requirements. In this method they are ranked according to their importance to the completion of all the tasks that depend on them. The importance is measured by the ranked positional weight (RPW) of each element, which is the sum of the times for all elements that directly follow it in the precedence diagram plus the time for the particular task itself. For example,

the RPW for Element 5 is the sum of the times for Elements 6, 7, 9, 10, and 11 plus the time for Element 5, that is, $0.12 + 0.63 + 0.34 + 0.80 + 0.70 + 0.48 = 3.07$.

Once the elements have been listed in descending order of RPW along with their predecessor elements (as shown in Table 7.9), from there on the procedure for developing an as-

Table 7.9   Ranked positional weight listing

Task	RPW	Task Time	Immediate Predecessor(s)
1	5.31	0.22	—
3	4.39	0.42	1
2	3.77	0.70	1
5	3.07	0.48	2, 3
7	2.47	0.63	3, 5
4	2.40	0.38	1, 3
8	2.02	0.52	4
10	1.50	0.80	7, 8
6	1.16	0.12	5
9	1.04	0.34	6, 7
11	0.70	0.70	9, 10

sembly line is exactly the same as before (Steps 2–6).

Carrying out Steps 2–6 of Procedure I (largest candidate rule) results in the final assignments listed in Table 7.10.

Seven stations are still required, but the cycle time is increased to 0.98 minute, the cumulative time associated with Station 4. The efficiency of this arrangement is

$$e = 1 - \frac{(0.98 \times 7) - 5.31}{7 \times 0.98} \times 100$$
$$= 77.4\%$$

The present production rate is $400/0.98 \approx 408$ units/day.

It should be noticed that these heuristic procedures develop task assignments that might not be very satisfactory, as is the case here with an efficiency of 77.4%. One way to increase the efficiency is to try the procedure again with a different cycle time. For example, the cycle time of 0.9 minute results in the task distributions shown in Table 7.11.

With an efficiency of 88%, the production

Table 7.10   RPW assignments

Station	Element	Task Time	Cumulative Task Time
1	1	0.22	0.22
	3	0.42	0.64
2	2	0.70	0.70
3	5	0.48	0.48
	4	0.38	0.86
	6	0.12	0.98
4	7	0.63	0.63
	9	0.34	0.97
5	8	0.52	0.52
6	10	0.80	0.80
7	11	0.70	0.70

Table 7.11   Final task assignments

Station	Elements Assigned to the Station	Cumulative Station Time
1	1, 3	0.64
2	2	0.74
3	5, 4	0.86
4	7, 6	0.85
5	8, 9	0.86
6	10	0.80
7	11	0.70

rate is now increased to $400/0.86 \approx 465$ units/day.

It is also possible to increase the cycle time, thus reducing the production rate but increasing the efficiency by decreasing the number of required stations.

Parallel Grouping of Stations

In some instances the required production rate can necessitate having a cycle time that is even less than the time required for the completion of one of the tasks. For instance, in the previous case, if the output needed were 800 units per day, the cycle time would have to be $400/800 = 0.50$ minute. Tasks 10, 11, 2, 7, and 8 require more time than that for their completion. In such an instance the previous methods, in which the stations were operated in series only, would have to be modified to allow for a parallel setup.

When two or more stations are in parallel and all are assigned to perform the same tasks, the permitted time for completion of the tasks in a station is longer than the cycle time. As an example, suppose there are two stations in parallel, as shown in Figure 7.16(a), each requiring time t to perform the tasks assigned to it; then two units are produced per t time interval. Other stations in series may therefore operate with $t/2$ as their cycle time. Similarly, if there are n stations parallel to each other, the cycle

Figure 7.16   Parallel stations

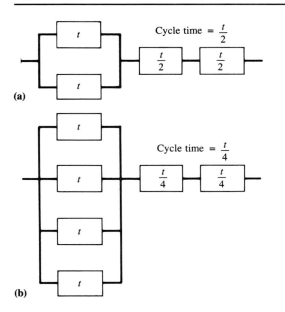

(a)

(b)

time for the remaining stations could be t/n, where the cycle time for the stations in parallel is t.

A question might be raised as to why we do not have six parallel stations in our example each performing all operations. This arrangement should obtain a production rate that is better than the one per minute needed and at the same time would reduce the manpower requirement from seven to six. There are several reasons why such a grouping may not be possible. An additional task assigned to a work station might mean additional investment at that station for the machines and tools that are necessary to perform the task. In multiple stations the same group of equipment would have to be available at each station, an investment that could be prohibitive. The task time could increase because as the work area is enlarged, the worker must move farther to perform more tasks. Workers with more experience and training would be needed to perform all tasks. A

larger floor space would generally be required if the same tasks were to be repeated in each station.

It should be recognized that in a practical layout of an assembly line, two parallel stations need not be physically parallel. They can, for example, be two successive stations fed by a common conveyor. It is possible to dedicate alternate units on the same conveyor to each of the stations in various ways, from using a sophisticated mechanical, push/pull arrangement to simply marking alternate positions on the conveyor by differently colored carriages such as trays.

Computerized Assembly Line Programs

Numerous computer programs have been designed for assembly line balancing. In most cases, each was developed for a particular situation, but all are based on one of the following techniques.

- Enumeration of all possible sequences
- Selecting the best from randomly chosen sequences
- Selecting a sequence that is similar to one already used in the company
- Applying a mathematical programming technique such as linear programming, integer programming, or the branch-and-bound method
- Using heuristic rules, such as the two we have just seen
- Applying rules that are specifically developed for the problem, possibly including, among other items, restrictions on tasks and assignments among the tasks

Applicable Programs by Others

It is beyond the scope of this chapter to go into the details of all known computer assembly line balancing programs. The following are some of the more popular.

· NULISP: Nottingham University Line Sequencing Program, developed by Nottingham University in 1978
· ALPACA: Assembly Line Planning And Control Activity, developed by General Motors in 1967
· GALS: Generalized Assembly Line Simulator, developed by ITT Research Institute in 1968
· COMSOAL: Computer Method of Sequencing Operations for Assembly Line, developed by Chrysler Corporation in 1966

7.4 COMPUTER PROGRAM DESCRIPTION

Three programs for this chapter are included on the program disk. For assembly line balancing the programs include one on the largest candidate rule and one on the ranked positional weighted method. The third program is associated with forming the groups for flexible manufacturing systems. All programs are in interactive mode, and data input is displayed on the screen.

Job Grouping

The program is written for a maximum of 19 machines and 40 jobs.

1. *Input lines.* For the first line, enter the number of jobs (JB) and the number of machines (MC) using format 2I3. (Note that for each job, a new line is required.) For the second and subsequent lines, indicate by 'X', the machine that the job utilizes using format 19A1. Each column represents one machine—for example, 'X' in columns 3 and 5 indicates that the job uses machines C and E.

2. *Output.* The output of the program shows a robust job grouping, and the analyst may apply further subjective judgment for additional improvements.

Assembly Line Balancing (in GW-BASIC)

Largest Candidate Rule

The input is in interactive mode. The questions asked are as follows.

1. *What is the required cycle time?* (Input the desired cycle time.)

2. *How many tasks are to be assigned?* (Input the total number of tasks.)

3. *What is the time for task I; number of predecessors for task I?* (Enter individually the task time and the number of predecessors for each task.)

4. *Task J must be preceded by the following task; give the # 'K' predecessor for task J.* (Enter individually each predecessor task number for task J.)

The output consists of (1) each station's task assignment, duration of each assigned task, and total station time; (2) the cycle time for the balanced assembly line; and (3) the efficiency of the line.

Ranked Positional Weighted Method

The data can be entered into the program interactively or by data statements. An answer to the first question below determines the mode.

1. *If you want interactive mode, enter number 1.*
2. *If you want data mode, enter number 0.*

If the data are to be entered interactively, use the same information and procedure as in the case of the largest candidate rule. If the information is entered in data mode, it should be typed as data statements in the same sequence as one would in interactive mode.

SUMMARY

Working with production charts is a manager's way of visualizing a manufacturing process. The charts allow one to design a sequence of operations that will produce a unit efficiently and economically. These charts evaluate the process in terms of movement and length of time required for a task and permit examination of each operation to see whether it is functioning efficiently. With a questioning mind, an analyst may make the necessary improvements should the operation be deemed unsatisfactory. Various charts are commonly used in industry, but among them the most common are assembly and operation process charts; routing sheets, left-hand, right-hand charts; gang charts; Gantt charts, and flow process charts.

A production arrangement within a plant may be designed to give different degrees of flexibility and production rates. A job shop arrangement is highly flexible and is suitable when a plant produces many different products, each in relatively small quantity. The efficiency in a job shop is, however, relatively low. The other extreme is an assembly line arrangement where each operation is carefully designed and balanced to provide maximum production rate. The sequential stations in an assembly line perform the next required operation on the job as the item moves from one station to the next on a fast-moving material-handling system such as a conveyor belt. An assembly line is generally designed for a single product and often has very high efficiency, but very little flexibility. A batch production system falls between a job shop and an assembly line in its complexity and efficiency.

With cellular and flexible manufacturing, a production system can be developed that has the production rate of an assembly line, yet the flexibility of a job shop. Application of group technology is a key ingredient in forming appropriate machine cells for cellular manufacturing, and production flow analysis is very helpful in this regard. Arranging the cell in a U-shape form has many advantages; it provides flexibility for worker movement and provides the opportunity for machine coupling. Flexible manufacturing is cellular manufacturing that is highly automated, using NC machines, robots, and a computerized material-handling system, and employing minimum manpower. Flexible manufacturing requires a fairly large initial investment.

The chapter concludes the analysis of systems by describing two manpower assignment models, one for a production line and another for an assembly line. An efficient tabular method presented makes it easy to determine the required manpower at each station when the scheduled production rate is known.

Two methods are described for balancing an assembly line: (1) the largest candidate rule and (2) the ranked positional weighted method. Any problem should be solved by both of the methods, and the solution giving the greatest efficiency should be selected for planning. Occasionally, cycle times and production rates may be adjusted to increase the efficiency. Parallel stations are needed when a cycle time is less than the sum of the task times; however, there may be a practical limit to the number of parallel stations we may have.

PROBLEMS

7.1 Identify the following symbols:
　a. ○
　b. ▽
　c. □

7.2 Develop an assembly chart for making a pot of coffee.

7.3 How are time estimates used?

7.4 Use standard data tables to estimate a standard time for:
　a. Replacing a headlight in your car
　b. Sharpening a pencil

7.5 Describe the left-hand, right-hand chart. What is the objective in using the chart to design a work station?

7.6 Develop a left-hand, right-hand chart for putting on and tying a tennis shoe.

7.7 How are production systems classified? What factors influence a production system?

7.8 Define the following:
　a. Job shop production
　b. Batch production
　c. Mass and continuous production
　d. Cellular and flexible manufacturing

7.9 Using the production schedule in the accompanying table, develop a final production matrix and cell arrangement.

Item	Required Machines
1	A, C, E, G
2	B, D
3	D, F, G
4	B, C, D
5	F, G
6	C, F
7	A, B, C, E, G
8	D, F
9	A, B, C
10	C, D, E
11	D, E, F
12	A, B, C, D
13	C, D, E, F
14	B, C, D, E
15	A, B
	B, G

7.10 Develop an assembly line to produce one chair every 20 minutes for the job described by the schematic diagram in Figure P7.10. The time in minutes required for each element is indicated under the element.

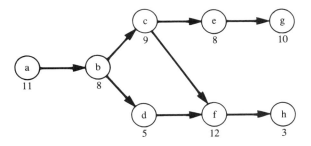

Figure P7.10

7.11 What factors does the rate of production depend on?

7.12 An electronics plant is to produce 500 calculators at 85% efficiency by one line. On the basis of the accompanying table, develop the balanced work station configurations required to meet this production goal.

Task	Time (in minutes)	Immediate Predecessor Task(s)
1	1.10	—
2	0.45	—
3	0.38	1
4	0.25	2, 3
5	2.50	4
6	0.50	5
7	0.10	5
8	0.05	6, 7

7.13 Plan a balanced assembly line for a spring scale production department on the basis of the accompanying table. The plant averages 92% efficiency and would like to produce 750 units per day.

Element	Time	Immediate Predecessor(s)
1	0.15	—
2	0.27	—
3	0.32	—

Element	Time	Immediate Predecessor(s)
4	0.45	1, 2
5	0.30	2, 3
6	0.19	3
7	0.24	4
8	0.38	4, 5
9	0.17	5
10	0.42	5, 6
11	0.21	6
12	0.35	7, 8
13	0.25	9
14	0.44	10, 11
15	0.33	11
16	0.20	12, 13
17	0.40	14, 15
18	0.22	16, 17

7.14 Using Figure P7.14,

    a. Develop the manpower requirement to produce 12 units each from Stations 7 and 8. Assume the initial inventory after each station from the accompanying table.

Station	1	2 & 3	4	5	6
Inventory in Units	5	8	4	3	6

    b. Develop a daily work schedule if the number of workers is limited to 5.
    c. Develop a daily work schedule if the number of workers is limited to 2.

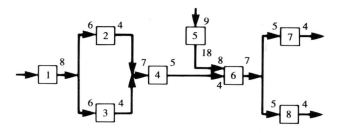

Figure P7.14

SUGGESTED READINGS

Charts

Galgut, P.E., *Production Planning and Control,* Elsevier, New York, 1987.

Muther, R., and Hale, L., *Systematic Planning of Industrial Facilities,* Vols. 1 and 2, Management and Industrial Research Publications, Kansas City, 1979.

Production Systems

Industrial Engineering, Vol. 15, No. 11, Nov. 1983.

Kusiak, A., *Modeling and Design of Flexible Manufacturing Systems,* Elsevier, New York, 1986.

Manpower Assignments

Barnes, R.M., *Motion and Time Study Design and Measurement of Work,* John Wiley and Sons, New York, 1980.

Bowman, E.N., "Assembly-Line Balancing by Linear Programming," *Operations Research,* May–June 1960, pp. 385–389.

Held, M.; Karp, R.M.; and Shareshian, R., "Assembly Line Balancing Dynamic Programming with Precedence Constraints," *Operations Research,* May–June 1963, pp. 442–459.

Sadowski, R.P., "Manpower Scheduling Method Simplifies Production Line Assignments Through Graphics," *Industrial Engineering,* Vol. 13, No. 8, 1981, pp. 34–41.

Sule, D.R., "Manpower Assignments on a Production Line Using Tabular Approach," *Industrial Management,* Vol. 25, No. 1, 1983, pp. 11–13.

Talbot, F.B., and Patterson, J.H., "An Integer Programming Algorithm with Network Cuts for Solving the Assembly Line Balancing Problem," *Management Science,* Vol. 30, No. 1, Jan. 1984, pp. 85–99.

Tonge, F.M., *A Heuristic Program of Assembly Line Balancing,* Prentice-Hall, Inc., Englewood Cliffs, N.J., 1961.

CHAPTER

8

Requirements and Selection of Machines and Labor

$\mathbf{M}$achines and labor are the backbone of an industry, and their selection requires careful evaluation of what is needed and what is available. This chapter describes the systematic steps involved in specifying machine and labor requirements. The description is divided into three main topics: machine selection, labor selection, and machine coupling.

8.1 MACHINE SELECTION

Machines are an integral part of manufacturing, yet seldom do we find an engineer developing production arrangements who also possesses expertise in machines. Some basic data such as types of machines, the names of suppliers, the range of costs, and the associated capacities are needed to begin the machine selection process. Where can we obtain this information? How can we use it? These are the kinds of questions we answer in the following sections.

Make-or-Buy Analysis

Before one proceeds to select the equipment that is essential for manufacturing and assembling the designed product, it is necessary to perform a preliminary make-or-buy analysis to determine what can and should be produced in the plant. The analysis is preliminary because the data required for the decision are based for the most part on experience and not necessarily on complete production knowledge. At some later date, further analysis might be made that will reflect more reliable data along with a better understanding of the manufacture and sale of the product itself.

The bill of materials (see Figure 5.9) provides the information on the parts that are necessary to make a finished product. The questions the engineer should ask are which of these parts should be made in the plant and which ones should be bought from outside vendors. Two determining factors might suggest the answers: expertise and economics.

Parts that require expertise in manufacturing technology other than what the firm possesses should be bought from vendors who have technical know-how in such fields. It is a common practice in the automobile industry, for example, to buy tires from tire manufacturers. Bottling plants that dispense soft drinks buy their bottles from glass manufacturers. Table lamp manufacturers hardly ever attempt to produce the lightbulbs without which their products will not function.

The economics of manufacturing also plays an important part in this decision. Production in a small quantity generally results in a larger unit cost than if the same parts are produced in large volumes. It might be cheaper to purchase 2000 units of ½-inch springs than to buy the equipment and employ the necessary labor to produce them in the plant. An economic analysis such as the one shown in the following example illustrates this point further.

Example: Component Part Source Selection

A prospective kitchen blender manufacturer has a design that requires hard plastic connecting gears between the electric motor and the cutting blade assembly. As shown in Table 8.1, there are three alternatives for obtaining such parts:

Alternative A: A molding specialty house can supply the parts for a price of $500 per thousand. This price includes the cost to design and build the tools necessary to manufacture the gears; however, the minimum-order quantity is 20,000 units. The company must also spend $2000 on an engineering effort to review the design before allowing the supplier to begin production.

Alternative B: Plant engineers can design, build, and perform initial testing of a single-cavity tool for $50,000. The gears can then be manufactured in the plant on a small automatic mold press at a cost of $200 per thousand. The unit costs include all of the variable costs: labor and material, as well as all normal overhead operating costs prorated per unit of output.

Alternative C: It is also possible to design and build a multiple-cavity tool for $100,000. This tool would be designed to run on a larger automatic mold press at a cost of $150 per thousand.

Determine the preferred alternative, given a specific requirement level. Assume that the time period over which the production will be required is short enough to eliminate the need to consider the time value of money.

Neglecting the time value of the money, the break-even point (Y_1) for purchasing the parts versus molding them with a single-cavity tool can be computed from the above data:

$$\$2000 + \$500 \times Y_1 = \$50,000 + \$200 \times Y_1$$
$$\$300 \times Y_1 = \$48,000$$
$$Y_1 = 160 \text{ thousand parts}$$

The break-even point (Y_2) for molding the parts in a single-cavity tool versus a multiple-cavity tool is

$$\$50,000 + \$200 \times Y_2 = \$100,000 + \$150 \times Y_2$$
$$\$50 \times Y_2 = \$50,000$$
$$Y_2 = 1000 \text{ thousand}$$
$$(1 \text{ million}) \text{ parts}$$

Table 8.1 Data for the three alternatives

Alternative	Initial Cost	Cost per 1000
A: Purchase mold tool, minimum order 20,000	▸ $2,000	$500
B: Manufacture with a single-cavity tool	$50,000	$200
C: Manufacture with a multiple-cavity tool	$100,000	$150

It is cheaper to purchase quantities of parts up to 160,000 units from a supplier. From 160,000 to 1,000,000 parts it is better to build a single-cavity tool and mold the parts in the factory; to produce more than 1,000,000 parts, a multiple-cavity tool should be used. (See Figure 8.1.)

Another important aspect of the analysis is the availability of initial capital. If it is not possible to raise $100,000 immediately, Alternative C is not feasible. In that case, even though production of more than 1,000,000 parts would be cheaper with Alternative C, only the single-cavity mold, Alternative B, could be used.

Sources of Information

Once the manufacturer has decided which parts are to be produced in the plant, the next question concerns which machines should be used in the production. This complex and important chore in machine selection will determine how efficiently and economically we can produce the unit. Machine cost contributes a large portion of the fixed cost that cannot be easily changed. Where and how can we obtain the necessary information about the machines to make a selection that meets our plant and product needs? Fortunately, there are many sources of such information.

Books and Periodicals

Experienced engineers have a broad background in dealing with suppliers, processes, and materials and generally know where to go for additional information to expand, modify, or build new facilities. Buyers, production clerks, maintenance personnel, and fellow engineers can frequently furnish the names of current and recent suppliers of different types of equipment. In most cases this knowledge is acquired from on-the-job training, associating with professional societies, and visiting trade shows. Trade journals and professional magazines that feature advertisements by suppliers provide considerable information about processes, materials, and equipment. Managers generally have access to information presented by current and new suppliers who are trying to demonstrate that their latest products are better than the one that is currently being used.

Figure 8.1

Total cost versus production level (molded plastic parts)

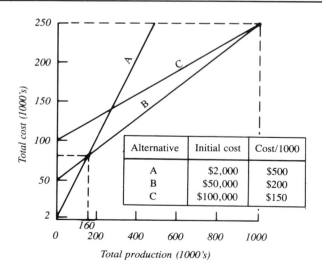

Alternative	Initial cost	Cost/1000
A	$2,000	$500
B	$50,000	$200
C	$100,000	$150

Total cost (1000's)

Total production (1000's)

The following discussion is intended to provide a means of locating a few additional records in many fields: information on professional societies, technical data on material and processes, and extensive listings of suppliers.

One of the best-known listings of suppliers is the *Thomas Register of American Manufacturers and Thomas Register Catalog File,* which is published annually by the Thomas Publishing Company, One Penn Plaza, New York, NY 10001. It contains several different listings, and the 1987 edition comprises 21 volumes. The register has an alphabetical listing of products and services of almost 21,000 pages in 12 volumes. (Some advertising is also included in this section.) A list of company names, addresses, telephone numbers, brand names, and trademarks is included in the next two volumes. The last seven volumes contain catalogs from some of the companies listed in the first section. In addition, Thomas has recently added a section in the last volume that provides considerable information on how to import products using the most cost-effective methods. In total, the *Thomas Register* has listings of 140,000 U.S. companies, 5000 Canadian companies, and 500 trademarks.

Several other directories and indexes are available that list publishers rather than manufacturers. These are of value when the concern is for sources of information about a product that is relatively new. These publications allow one to find what information is available about the product, who is trying to sell the equipment and materials, and what technical societies are involved in the field.

The IMS 1986 *Ayres Directory of Publications* contains a comprehensive listing of U.S. publications and a small number of foreign publications. The directory has several cross-indexes, including one of over 100 pages that lists trade and technical publications. This directory has been published annually for many years and is available from IMS Press, 26 Pennsylvania Avenue, Fort Washington, PA 19034.

Ulrich's International Periodicals Directory is a listing of domestic and foreign periodicals that contains over 1000 pages of titles and subjects along with the current price of each. The cross-index allows one to guess at the name of a publication or topic and, if successful, find a cross-reference to all of the other publications on the same topic. *Ulrich's* also lists two other sources that may be of value in researching information: Abstracting and Indexing Services and Micropublishers (suppliers of microfilm data bases). *Ulrich's Directory* is revised on a frequent basis by R. R. Bowker Company, 1180 Avenue of Americas, New York, NY 10036.

Another index of periodicals is the *Applied Science and Technology Index,* which is published annually by H. W. Wilson Company, 950 University Ave., Bronx, NY 10452. Like the *Ayres Directory,* it primarily refers to domestic periodicals, and it also includes the most recent subscription prices for the periodicals.

All of these indexes and directories make it possible to take advantage of the research work that has already been done by others and to direct a manufacturer to information on the product of interest. Through the articles and advertisements in technical journals, one can gain access to the most current information on major aspects of the business, since these technical journals are one of the first places that suppliers announce their new offerings. The magazines also list all of the regional and national trade shows where additional information can be obtained.

One of the features of *Ulrich's Directory* is the listing of professional society journals. In addition to the informative articles and listings of suppliers, they provide opportunities to make contact with other individuals who might have similar interests.

Computer Data Bases

Computer data bases are a source that, although currently unpublicized in most of the professional literature, will probably become the principal avenue for obtaining information in the near future. These data bases provide diverse technical data on almost any topic, including those of interest here: selecting locations, suppliers, processes, and equipment.

Dialog Information Services is one company that currently provides computerized data base systems. Located at 3460 Hillview Avenue, Palo Alto, CA 94304, the company has access to over 200 data bases containing more than 100 million items. Some of these provide a "Yellow Pages" service for trade and industry.

The companies that supply computer data bases, such as Dialog, can be connected by telephone lines through a modem, making it feasible for a person with even a small terminal costing less than $300 to have access to such data bases on a continuing basis. A large amount of information is currently available that includes references to published and unpublished literature, articles, newswires, financial data, and directory listings. While only a small portion of the information is currently of interest to engineering, it is expected that the electronic data bases will expand rapidly in the next few years to include most of the important references. A sampling of data bases that are currently available follows:

- Economic Literature Index and Trade and Industry Index are useful in market research and the management areas of business, industry, and economics.
- Donnelly Demographics can be utilized in the area of statistical and demographic data.
- Encyclopedia of Associations and Ulrich's International Periodicals provide directories of technical associations and periodicals.
- Compendex (engineering and technological literature abstract) and Scisearch supply scientific and technological information.

Production Arrangement

The next concern in machine selection is the production arrangement. Even though the final layout of the plant might be unknown at this point, the anticipated production arrangement should be decided upon, since it has a major bearing on machine selection. If a single product is to be produced in mass quantity with an assembly line arrangement (sometimes referred to as product arrangement), highly specialized machines will be needed, each capable of performing a specific task with great speed and accuracy. The other extreme is the job shop arrangement (sometimes referred to as process arrangement), in which machines capable of working on multiple jobs are used to achieve flexibility.

The number of products assigned for production and their required production rates might also dictate the choice of plant layout and ultimately determine the machine selection.

Cost Consideration

The decision as to whether to undertake labor-intensive or capital-intensive production is another problem that must be reconciled. At the present time, individual robots and groups of machines with fully automated material-handling systems are being considered as attractive alternatives for reducing labor costs. Before purchasing expensive equipment to replace manual operations, however, a manufacturer should perform a careful analysis of real costs associated with automatic equipment: costs of design, purchase-or-build, prove-in, and maintenance. Most automatic processes take longer than expected to produce quality products and are expensive to modify if design changes are necessary.

In most cases, machines are available in different capacities with different operating speeds and different feed rates. The available options also vary from one model to the next, and the initial investment and operating cost can differ considerably among the available alternatives. The initial investment is especially important, since it is money committed that cannot be easily recovered. Again, an economic analysis can suggest the best alternative.

Available Capacity

Machines can seldom be used at full capacity throughout the production period. There are several reasons why this is so.

The need for setup, preventive maintenance, tool sharpening, and unpredicted failures and repairs reduce the time available for production. Older machines are more susceptible to breakdown resulting in lower productivity than are newer machines.

Operator-related factors such as absence, personal time, and time spent in bookkeeping, machine adjustments, and material preparations reduce the machine availability.

As a result of quality requirements in some processes, a certain loss of production is inherent even though the equipment is properly set up and is being operated correctly. Some production is also lost as units are scrapped when a machine malfunctions and produces parts of undesirable quality.

At times an operator is assigned the responsibility for many machines, a situation called machine coupling. This is made possible by dividing the total cycle time into two parts, machine time and man time. Machine time is the time when the machine is operating without any assistance from the operator, and man time is the time the operator spends with the machine when the machine is not working—for example, the time for load. If the machine time is large in comparison to the man time, it might be possible for the operator to load another machine while the first machine is in its machine time. One operator therefore can frequently be assigned responsibilities for several machines.

In such a system a problem with one machine in the group can at times affect the production of all the other machines. An operator who is engaged in identifying and correcting the problem might have to neglect the other machines needing his or her time in their natural cycle of production.

Available time for production does not increase in the same proportion as the number of shifts; the addition of a second shift does not add another eight hours of operation. The loss of production time increases because the free time available for repairs decreases. For example, in a single-shift operation a machine that broke down almost at the end of the shift would normally be repaired during the second (maintenance) shift if the required repair time were expected to be long, say four hours or more. Because the machine is now scheduled for operation, however, such repairs must be accomplished during time scheduled for production in the second shift. Often factors such as a shortage of material, absence of technical help, and absenteeism of production workers also affect the overall performance in the second shift. In general, more workers are absent from the second shift than the first, and more are absent from the third shift than the second.

Official holidays such as Labor Day, New Year's Day, and Christmas also reduce the available working hours during the year.

The loss of useful time is accounted for by many different allowances; setup time is measured or estimated for each type of operation and is incorporated into the time necessary to produce the units; the personal allowance, ranging from 5 to 15 percent, depending on the physical strength and dexterity required to perform the job, is given to the operator to accommodate his or her personal needs. Generally, this is accomplished by including the allowance in calculating the standard time for the job. Shrinkage allowance is established to account for the loss of production as a result of rejection of parts due to poor quality even though the machine is set and operating correctly.

All other factors contributing to lost time are grouped together and measured against the efficiency of operation. For example, if one-shift operation is 95 percent efficient, then the time available for production, excluding the allowances we have considered before, is 456 minutes (0.95 × 480). In multishift operation it is a normal practice to indicate the efficiency of the additional shift. The terminology commonly used is frequently misleading. It might be said that in our example the addition of a second shift would add 75 percent to the efficiency of the plant. This does not mean that the first shift would continue at 95 percent and the second shift would add 75 percent for a total of 170 percent efficiency. It indicates that the two shifts together will operate at a rate of 1.7 times that which one shift would if it were at 100 percent efficiency. In our example the first shift might drop to 90 percent, and the second would be at 80 percent, for the total of 170 percent.

Required Capacity

The capacity needed depends on the operations the machine is assigned to perform and the required production rate for each item. The following example illustrates this point.

On the basis of the production routing sheets of the items manufactured in the plant,

it is decided to perform Operations 10, 12, and 15 of Item A, Operation 101 of Item B, and 157 of Item C on the new machine. The estimated time for completion of an operation on the machine is noted along with the estimated time for setup, and a personal allowance is assigned to compensate for personal needs. The quality requirement for the parts and the tolerance at which the machine and process can operate give an estimate of shrinkage allowance per operation. The required production per day is obtained from the scheduled deliveries. The data are recorded in Table 8.2.

The entries in Column 4 are the numbers of units that must be processed to obtain the required number of good parts per day when shrinkage is considered. An entry is obtained by dividing the parts required, Column 2, by the yield or 1 minus shrinkage allowance. The time allowed per operation, Column 7, is the estimated time of Column 5 multiplied by the quantity 1 plus the personal allowance. The total time, Column 9, needed to produce the items is Column 4 multiplied by time allowed per operation, Column 7, plus the setup time, Column 8, for that operation. The grand total is the time the new machine must operate each day, in this case 1308.7 minutes.

Table 8.2   Production data

(1) Item and Operation	(2) Units Required/ Day	(3) Shrinkage Allowance	(4) Items to Produce/ Day = (2)/(1 − (3))	(5) Time per Operation (min)	(6) Personal Allowance (%)	(7) Time Allowed = (5) (1 + (6))	(8) Setup Time/ Day (min)	(9) Total Time (min/day)
A-10	1000	0.03	1031	0.13	5	0.1365	15	155.73
A-12	2000	0.00	2000	0.15	5	0.1575	18	333.00
A-15	1000	0.01	1011	0.08	5	0.0840	12	96.92
B-101	100	0.10	112	02.2	5	2.3100	10	268.72
C-151	300	0.08	327	01.3	5	1.3650	8	454.35
							Total	1308.72

Numbers of Machines Needed

Once the required capacity and the available productive time per machine have been established, the calculation for the required number of machines is quite straightforward. In the above example, if the plant is to operate on one shift with 95 percent efficiency, the available time per machine is 456 minutes (480 × 0.95). The total capacity needed is 1308.72 minutes, necessitating having 2.87 (1308.72/456) or 3 machines.

It should be noted that the data in the table were applicable to the machine under consideration. If a different type of machine were to change the data (e.g., setup, processing time, or shrinkage allowance), a new required capacity would have to be calculated, perhaps giving a different answer to the problem.

Yield and Cost Models

Let us now go one step further and determine the quantity that should be processed at each stage as a product proceeds through a series of stations in its manufacturing sequence. We will illustrate the calculations by considering an example. Suppose production of a part requires a three-stage operation with a shrinkage rate for each stage as indicated in Figure 8.2. Assume further that 100 good units are required from this setup.

The input to any stage must allow for rejects in that stage. For example, only 95 percent of the products in Stage 3 are good, and therefore 100/0.95 = 105.26 ≈ 106 units must be started in that stage. If we proceed backward to Stage 2, to obtain 106 good units the input to Stage 2 must be 106/0.97 = 109.3 ≈ 110 units. Similarly, in Stage 1, 110/0.94 ≈ 117 units must be started. By using appropriate input values, the time requirement for the product in each stage can be calculated by using the procedure described earlier. As the reader has perhaps observed, the numbers are rounded up at

Figure 8.2   Three-stage sequential production process

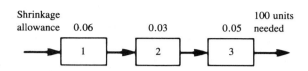

each stage; this is mainly to make certain that at least the required number of good units are produced in the three stages.

Extensions of Variations on Stage Analysis

To make the formulation general, let P_i denote the percent defectives in State i. Then if N is the number of units at the beginning of the process, the number of good units after the first state is $[N(1 - P_1)]$ where the brackets stand for integer portion of the result only. The number of good units after the second stage is $[[N(1 - P_1)](1 - P_2)]$.

One can extend the analysis to all the stages. It is also possible to determine the value of N if the required number of final good products is known. A fairly good approximation of the input/output relationship is given by calculating the yield of the line. Such calculations ignore stage-by-stage integerization to give

$$\text{Line yield} = (1 - P_1)(1 - P_2)(1 - P_3) \cdots (1 - P_n)$$

and input quality N = required output/yield of the line. In the previous example the line yield is $(1 - 0.06)(1 - 0.03)(1 - 0.05) = 0.866$ and therefore to produce 100 good units would require starting production with $100/0.866 = 115.4 \approx 116$ units altogether.

Another interesting aspect of this analysis can be investigated by calculating the cost per good unit produced. Let C_i be the cost of processing a unit in State i. Then the number of units processed at each stage, multiplied by the cost of processing each unit in the stage, gives

Figure 8.3   Recycling of rejected units to backstream

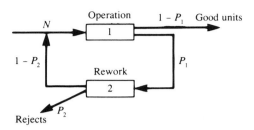

the cost of each stage. Accumulating the individual stage cost and dividing it by the total number of good units produced gives cost per good unit. In our example, if $C_1 = \$5$, $C_2 = \$8$, and $C_3 = \$10$, the cost of a good unit is given by

$$\frac{5 \times 117 + 8 \times 110 + 10 \times 106}{100} = \$25.25$$

The cost can be approximated by adding the cost of each serial stage and dividing it by the yield. In our case it would be $(5 + 8 + 10)/0.866 = \$26.56$.

The analysis may be further expanded to include nonserial systems formed by rework and/or repair operations. For example, consider a system such as is shown in Figure 8.3, in which a portion of the units rejected in Operation 1 can be reclaimed in Operation 2 and again become an input to Operation 1.

If P_1 and P_2 denote the percent defectives at each stage, respectively, then for N units of input the number of good units produced would be

If we denote $a = P_1(1 - P_2)$ and since both P_1 and $P_2 \leq 1$, a is also ≤ 1, we have

$$N(1 - P_1)[1 + a + a^2 + \cdots]$$

Since the terms in the brackets follow the form of a geometric series, the expression transforms to $N(1 - P_1)/(1 - a)$. Hence the yield of the system is $(1 - P_1)/(1 - a)$.

The value of N can be determined for a required number of good units. For example, if $P_1 = 0.2$, $P_2 = 0.1$, and 100 good units are required, then

$$a = 0.2(1 - 0.1) = 0.18$$

and the yield $= (1 - P_1)/(1 - a) = 0.975$. Therefore $N = 100/0.975 = 102.56 \approx 103$ units.

The total cost of processing N units is the sum of the costs at each stage multiplied by the number of units processed at that stage. Since we have a recycling process, the terms in the sum are infinite. However, they are decreasing in value successively:

$$
\begin{aligned}
\text{Total cost} = {} & C_1 N + C_2 N P_1 + C_1 N P_1 (1 - P_2) \\
& + C_2 N P_1 (1 - P_2) P_1 \\
& + C_1 N P_1 (1 - P_2) P_1 (1 - P_2) \\
& + C_2 N P_1 (1 - P_2) P_1 (1 - P_2) P_1 \\
& + \cdots \\
= {} & C_1 N[1 + P_1 (1 - P_2) \\
& + P_1^2 (1 - P_2)^2 + \cdots] \\
& + C_2 N P_1 [1 + P_1 (1 - P_2) \\
& + P_1^2 (1 - P_2)^2 + \cdots]
\end{aligned}
$$

Since $a = P_1(1 - P_2)$

$$\underbrace{N(1 - P_1) + NP_1 (1 - P_2)(1 - P_1)}_{\substack{\text{units reworked} \\ \text{once}}} + \underbrace{NP_1 (1 - P_2)P_1(1 - P_2)(1 - P_1)}_{\substack{\text{units reworked} \\ \text{twice}}} + \cdots \quad = N(1 - P_1)[1 + P_1 (1 - P_2) + P_1^2 (1 - P_2)^2 + \cdots]$$

$$\text{Total cost} = \frac{C_1 N}{1 - a} + \frac{C_2 N P_1}{1 - a}$$

$$= \frac{N(C_1 + C_2 P_1)}{1 - a}$$

Suppose that in the previous illustration $C_1 = \$5$ and $C_2 = \$3$. The cost to produce 100 good units would be

$$\frac{103[5 + (3 \times 0.2)]}{1 - 0.18} = 703.41$$

or $7.03 per unit.

The recycling could also be of the form in which the unit is as good after repair as a unit coming from the operation itself. (See Figure 8.4.)

The total good units produced would be the sum of good units produced in Operation 1 and in Operation 2, that is,

$$\text{Number of good units} = N(1 - P_1)$$
$$+ NP_1(1 - P_2)$$

The system yield is $(1 - P_1) + P_1(1 - P_2)$, and the total cost of the operation is $NC_1 + P_1 NC_2$, from which the cost per good unit of output can be determined.

System Decomposition

A system consisting of a complex structure can be broken down step by step into a familiar pattern, which can then be analyzed with the result illustrated in the previous section.

Figure 8.4 Recycling of rejected units to upstream

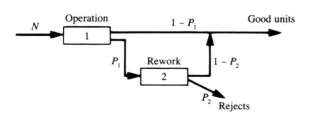

The example below demonstrates this approach.

Example: A Composite Production Line

Consider the following production pattern with the data as given in the accompanying table.

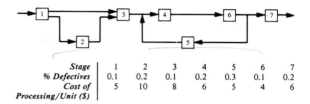

Stage	1	2	3	4	5	6	7
% Defectives	0.1	0.2	0.1	0.2	0.3	0.1	0.2
Cost of Processing/Unit ($)	5	10	8	6	5	4	6

The objective at this point is to reduce the system into a serial system. We shall combine different stages into a single stage and determine its equivalent percent defectives and cost.

Combine Operations 4 and 6 into a stage, and denote it as Stage 8.

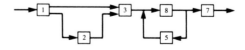

Since Operations 4 and 6 form a serial system, the yield is $(1 - P_4)(1 - P_6) = 0.8 \times 0.9 = 0.72$ or percent defectives $P_8 = 1 - 0.72 = 0.28$. The unit cost consists of the costs for the operation in Stage 4 on all units plus the operation in Stage 6 on the good units coming from Stage 4. That is,

$$\text{Unit cost, } C_8 = \$6 + (1 - 0.2)4 = \$9.20$$

The resulting system is as follows:

Now combine Stage 8 and Stage 5 into one operation, Operation 9. We have

$$a = P_8(1 - P_5) = 0.28(1 - 0.3) = 0.196$$

and hence

$$\text{Yield} = \frac{1 - P_8}{1 - a} = \frac{1 - 0.28}{1 - 0.196}$$
$$= 0.895$$
$$\% \text{ defectives, } P_9 = 1 - \text{yield}$$
$$= 1 - 0.895 = 0.105$$

and

$$\text{Cost/unit, } C_9 = \frac{C_8 + C_5 P_8}{1 - a}$$
$$= \frac{9.2 + 5(0.28)}{1 - 0.196} = \$13.18$$

The resulting system is as follows:

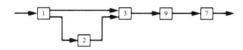

Now let us combine Stages 1 and 2 into Stage 10.

$$\text{Yield of Stage 10} = (1 - P_1) + P_1(1 - P_2)$$
$$= 0.9 + (0.1 \times 0.8)$$
$$= 0.98$$

Therefore

$$\% \text{ defectives, } P_{10} = 1 - 0.98 = 0.02$$

and

$$C_{10} = (C_1 + P_1 C_2)/1 - P_{10}$$
$$= \frac{5 + (0.2 \times 10)}{.98} = \$7.14$$

The resulting system is as follows:

Since it is in serial arrangment, we have

$$\begin{aligned}\text{Yield of}\\ \text{the line} &= (1 - P_{10})(1 - P_3)(1 - P_4)(1 - P_7)\\ &= 0.98 \times 0.9 \times 0.895 \times 0.8\\ &= 0.6315\end{aligned}$$

and

$$\begin{aligned}\text{Estimated}\\ \text{cost/good unit} &= \frac{C_{10} + C_3 + C_9 + C_7}{\text{yield}}\\ &= \frac{7.14 + 8 + 13.8 + 6}{0.6315}\\ &= \$54.35\end{aligned}$$

Machine Specifications

A decision to buy a machine must now be documented for further action by the shop supervisor and the maintenance department. For example, appropriate locations for placement of the machines must be determined, the required utilities must be provided, and foundations must be planned so that when the machines do arrive, they can be installed without further delay. The models developed in Chapters 2–4 can be utilized profitably to make the placement efficient. Tables 8.3 and 8.4 display one of the more common forms of specifications.

Table 8.3   Forming press specifications

Item	*Specification*
Machine	150-ton forming press
Manufacturer	Hess
Model	SC2
Condition	New
Cost	$240,000
Crew	1
Foundation	8-inch steel-reinforced concrete
Motor horsepower	20 hp at 1200 rpm
Voltage	440 V at 60 Hz, 3-phase
Dimensions	50 × 123 inches
Work rate	35 strokes/minute
Die change time	3 hours
Number required	1
Usage	Model C—Operation 330: plate forming

Table 8.4 Drying oven specifications

Item	Specification
Machine	Paint-drying oven
Manufacturer	Monroe Industrial Machinery
Model	Custom built
Condition	New
Crew	1 — 1 hour/day
Cost	$98,000
Foundation	6-inch steel reinforced concrete
Motor horsepower	20 hp at 1800 rpm
Fuel	Natural gas
Requirement	1,850,000 ft³/hour*
Voltage	440 V at 60 Hz, 3-phase
Dimensions	40 × 60 feet
Work rate	8 fpm
Number required	1
Usage	Bake enamel paint onto parts

*Overall for paint system.

Table 8.5 Air compressor specifications

Item	Specification
Machine	Air compressor
Manufacturer	Redman
Model	500 — H Type SSR
Condition	New
Cost	$27,500*
Foundation	Gravel with treated wood runners
Motor horsepower	133 hp; 152 amps at full load
Voltage	460 V at 60 Hz, 3-phase
Dimensions	110 × 69 × 58 inches
Capacity	125 psi each 500 cfm
Number required	1
Usage	(1) Air for assembly work
	(2) Reserve for paint system

*This includes the compressor, outdoor protection package, air dryer, tank, cooler, etc.

In most cases the necessary information can be obtained from the supplier of the machine. It should be remembered that even though the supplier will provide such information as the utility and foundation requirements, the ultimate responsibility for installing and running the machine normally rests with the user.

Auxiliary Equipment

Air compressors, tool sharpeners, and water treatment facilities are other machines and equipment that are not directly related to production but that are necessary to achieve the production objectives. These are referred to as auxiliary equipment. An analysis similar to the ones performed for the main equipment should be conducted to determine their required capacities and the number of units needed. Table 8.5 shows a specification for an air compressor.

Material-handling equipment can also be categorized as auxiliary equipment. More detailed descriptions of these types are given in the chapter on material handling, Chapter 10.

8.2 LABOR REQUIREMENT AND SELECTION

Few plants can operate very long without a qualified work force that can do its job well. A careful analysis is required to obtain a good match between job and worker. Inappropriate assignments generally lead to problems of inefficiency in production, poor-quality parts, disgruntled employees, and low morale.

The employees in a manufacturing facility either are engaged in production or provide the necessary services to facilitate production. The people involved in immediate production activities such as operating the machines, working at assigned stations, or operating a forklift truck or other material-handling equipment are referred to as direct labor. The people who provide the supporting services such as maintenance, janitorial services, and cafeteria operation are termed indirect labor. We shall study indirect labor further in Chapter 9; however, whether the labor is direct or indirect, an appropriate job evaluaton should be conducted to determine the duties and responsibilities of the job and who can best perform them. In the following sections we will concentrate on job evaluation, which defines the factors that should be considered in hiring new employees, and on the job description, which indicates what tasks are assigned to an employee.

Job Evaluation

One of the tasks in developing a production design is to determine the number and qualifications of workers needed in each step of production. The production arrangement described in Chapter 7 plays an important part in this decision. For example, in an assembly line arrangement, in which a worker is mainly responsible for work in one station, job skills and responsibilities are not as demanding as they would be in a job shop environment. Each position should be evaluated in terms of the level of skill, effort, responsibility, working conditions, and safety involved in the job content. This will enable a manager to hire a person with the right qualifications and pay him or her the salary that is appropriate for the job.

The process of job evaluation is begun by comparing each attribute of the job in question with corresponding standards established within the company and then assigning a point value to each attribute to indicate the degree of its variation from the standard. Since not all of the attributes are equally significant, the relative importance of each is reflected by its allowed point spread. The total point value, obtained by adding the points assigned to each attribute, reflects the job level and even perhaps the step in that level for the job.

A company should develop a work factor point scale similar to that shown in Table 8.6.

In each category, Level 1 corresponds to simple, routine duties, Level 2 indicates well-defined duties that may require some decision making, Level 3 factors require working independently and devising or modifying methods based on policy, and Level 4 is assigned to highly technical or complex work involving independent decisions.

Job evaluation allows management to group all jobs requiring the same types of skills together; thus an employee with a given set of skills might be qualified to work in a variety of jobs. This provides information that permits greater flexibility in manpower assignments. Management can also achieve a uniform salary structure throughout the plant by comparing point values of each job with the pay scale. Besides assisting in matching a worker's qualifications to the job, the evaluation provides information for training, promotions, transfers, and performance-rating evaluations.

Sample Job Evaluation: Machine Shop Supervisor

We will illustrate the technique of job evaluation by applying it to a machine shop supervisor for the company mentioned before, with a work factor chart as in Table 8.7. For each factor the level of difficulty for the job is judged, and corresponding points from the table are assigned to that category.

The total points assigned to a job indicate its relative importance. For example, a job worth 280 points would be twice as important to a company as one worth 140 points.

Sample Job Evaluation: Salary Determination

Uniform salary determination is another benefit of job evaluation. For example, the possible points assigned by this company range from 217 to 384. These numbers can be used to assign salaries for the firm. The graph in Figure 8.5 shows the relationship between the point value of a job and the corresponding salary.

Table 8.6   Job evaluation point scale

	Level			
Work Factor	*1*	*2*	*3*	*4*
I. Skill				
1. Job Knowledge	40	60	80	100
2. Experience	30	35	40	45
3. Education	15	18	21	24
4. Dexterity	8	10	12	14
II. Effort				
5. Mental Effort	25	30	35	40
6. Physical Effort	20	24	28	32
7. Fatigue	5	7	9	11
8. Monotony	2	3	4	5
III. Responsibility				
9. Equipment	17	20	23	26
10. Others	20	25	30	35
11. Errors	12	15	18	21
12. Coordinating Activities	19	20	21	22
IV. Working Conditions				
13. Hazards	3	4	5	5
14. Environment	1	2	3	4
Total	217	273	329	384

One might note that as the number of points assigned to a job increases, the salary also increases. In the job evaluation for the machine shop supervisor, the various work factors were found to have a total value of 283 points. The graph equates this to a salary of $25,000, which in practice could be the midpoint of the salary range for the supervisor, allowing for periodic raises based on performance.

Standard Job Analysis

The *Dictionary of Occupational Titles* (DOT) is a 1400-page listing of 20,000 occupations that is published by the U.S. Department of Labor. The latest edition was issued in 1977, and a list of revisions and additions is printed each year. The occupations are grouped into classes based on the interrelationship of tasks and require-

Table 8.7   Job evaluation (machine shop supervisor)

Work Factor	*Level*	*Points*
1	2	60
2	3	40
3	1	15
4	2	10
5	2	30
6	3	28
7	2	7
8	1	2
9	2	20
10	3	30
11	2	15
12	2	20
13	2	4
14	2	2
Total		283

Figure 8.5

Salary
determination

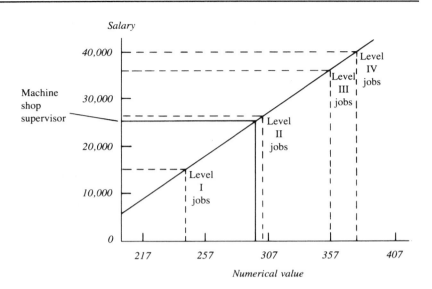

ments. The dictionary is designed to aid in matching job requirements with worker skills. It begins by listing the occupational categories as a quick reference for identifying the numerical classifications. The major part of the book lists the many different occupations in detail according to the numerical classifications, categorizing the jobs and giving brief descriptions. The last section is an alphabetical listing of the jobs that refers back to the numerical listing.

The DOT is an example of job analysis on a large scale. All aspects of jobs are broken down into three broad categories: data, people, and things. Each of these categories is subdivided further, and each division is assigned a point value according to intensity. The lower the number, the more time and know-how are required by that job trait. For example, instructing is assigned a 2, whereas supervising is assigned a 3 because teaching requires more intense interacting with people. Likewise, setting up has a lower "things" value than tending a machine, and computing has a lower "data" value than copying information.

The DOT begins by breaking down jobs into nine primary occupational categories. Numbered from 1 to 9, they are as follows: Professional and Technical; Clerical and Sales; Service; Agricultural, Fishery, Forestry, and Related; Processing; Machine Trades; Bench Work; Structural Work; and Miscellaneous Occupations. The numbers identifying these categories are the first numerals in the code for each job, and the two that follow indicate one of the many divisions within each category.

The "Data, People, and Things" categories are each divided and numbered accordingly. For Data the subdivisions range from 0 to 6 as follows: Synthesizing, Coordinating, Analyzing, Compiling, Computing, Copying, and Comparing. For People, Mentoring, Negotiating, Instructing, Supervising, Diverting, Persuading, Speaking-Signaling, Serving, and

Taking Instructions-Helping are 0 to 8. The categories 0 to 7 for Things are Setting up, Precision Working, Operating-Controlling, Driving-Operating, Manipulating, Tending, Feeding on/off Machines, and Handling.

The last three digits of the code are assigned to distinguish between jobs for which the first six digits are identical. When there is no other job with the same first six numerals, the last three are 000. If the first six digits are the same, then the last three digits are assigned to categorize in alphabetical order the different jobs in steps of four, beginning with 010. For example, the Fruit Farmworker, Fig Caprifier, Fruit Harvest Worker, and Vine Pruner all have 403.687 as the first six digits. The last three digits assigned to distinguish the four jobs are 010, 014, 018, and 022, respectively.

In using the number code, consider the job of Export Manager, which is listed as 163.117-014. The 1 in the first part identifies it as a professional/technical job, and the 63 classifies it as a sales and distribution management occupation. The second set of numbers, 117, corresponds to Data, People, and Things; the first 1 shows that the export manager must be able to analyze and coordinate data, the second 1 indicates that he or she must instruct and negotiate with people, and the 7 is indicative of the fact that the export manager has very little to do with things.

Another example is the occupation listed as 559.382-054. The 5's are processing occupations, and all of the 559's are occupations in the processing of chemicals, plastics, synthetics, rubber, paint, and related products. The number 054 identifies this occupation as a soap maker. By looking at the 382, we find that the soap maker must be able to compile data (3), take instructions (8), and operate and control machinery (2). Following the number classification is a brief description of the job, emphasizing some of the major qualifications and responsibilities.

There are several advantages to using this system. It can be considered a standard for job analysis, allowing different companies with similar jobs to compare these jobs, such that identical jobs would result in the same analysis and description. The DOT can be used as a basis for job evaluation by any company, saving the considerable amount of time and money required to start from the beginning in gathering information that is already available and contained in the DOT. Finally, the numerical codes can be of help in developing pay scales. The lower the numbers in a particular category, the higher the value of the job. A pay scale may be based on the emphasis a particular company places on the three traits: Data, People, and Things.

Job Definition and Description

Job definition and evaluation go hand in hand. A job cannot be evaluated unless it is well defined. Different analysts might very well develop different tasks as they divide the total work required to manufacture a product into jobs. An analyst may assign one worker per station, assign one operator per machine, or make a group of workers collectively responsible for a large task or for operating a group of machines. In a teamwork concept, a job can be defined in a manner that allows for crossover and cooperation between the workers to complete the total task. The initial job description might have to be modified as one gains work experience, leading to the concepts of job reevaluation and job enlargement.

In all cases the job description is important. It gives a complete listing of an employee's duties and responsibilities and his or her line of command. It helps form an understanding between an employee and an employer as to what is expected from the employee and what the employer will pay in return. Table 8.8

Table 8.8   Job description

(1) Title (DOT)	(2) Salary	(3) Educational and Experience Requirements	(4) Job	(5) Reports to
Supervisor, Plastics 556.130-010	$19,500/year	2 years of college, 8 years experience in injection molding	Supervises activities of workers in molding and plastics operations. Trains workers in job duties and production techniques.	Plant Engineer
Supervisor, Metal Parts Line 619.130-030	$19,500/year	2 years of college, 8 years experience in press line management	Supervises activities of workers engaged in punch and form press operations. Trains workers on metal parts assembly line.	Plant Engineer
Injection Molding Machine Tender 556.685-038	$16,500/year	High school diploma, 3 years experience on injection molding machine	Tends injection molding machine. Fills hopper of molding machine. Starts up machine to liquefy material to be shot into mold. Observes gages and controls temperature to ensure good molding.	Supervisor, Plastics
Automatic Punch Press Operator 615.482-026	$16,000/year	High school diploma, 2 years experience on automatic punch and forming presses	Sets up and operates power presses that will automatically feed rolls of sheet aluminum into the punch. Positions and clamps feed guides.	Line Supervisor
Welder 811.684-014	$7.30/hour	Trade school, 1 year welding experience	(1) Welds base to body. (2) Welds spout to body.	Supervisor, Metal Parts Line

156

illustrates an example from a job description chart.

8.3 MACHINE COUPLING

The concept of machine coupling was explained in Chapter 7. Recall that in machine coupling a single operator is made responsible for operating a group of machines. In production planning it is necessary to decide whether machine coupling is feasible so that an appropriate job description for the operator can be written.

The use of man–machine charts plays an important part in this development. The chart graphically displays on the same scale the time when the person is working or idle and when the machine is working or idle. The person could be made responsible for more than one machine if he or she has sufficient idle time during which it would be possible to attend to another machine. The following example illustrates this concept. It should be noted that machine coupling is possible only if the machine, when it is working, is capable of performing its task without any assistance from the operator.

Example of Machine Coupling

Suppose we have three machines for which machine coupling is to be investigated, one A and two B's. The loading, processing, and cycle times (the sum of loading and processing) in minutes for the parts scheduled for production on these machines are given in Table 8.9.

The company estimates that the cost of the operator is \$6.00/hour, while the costs for Machines A and B are \$12.00/hour and \$18.00/hour, respectively. It will take about 30 seconds for the operator to move from one machine to another.

Table 8.9 Data for an illustrative example

	Machine	
	A	*B*
Loading	2	5
Processing	8	7
Cycle time	10	12

Alternative A

Let us begin by assigning one operator per machine. As shown in Figure 8.6, in each chart the working time and the idle time for both man and machine are plotted simultaneously. The time is represented by the vertical distance, the bar representing the cycle time for the corresponding machine. The load time, which is the integral part of an operation, is needed to load and unload a unit from the machine and as such is not considered part of the idle time for the machine. The symbols used are M/C—machine, L—load, W—work, I—idle, and M—move.

Figure 8.6 Man–machine chart showing three operators (one worker per machine)

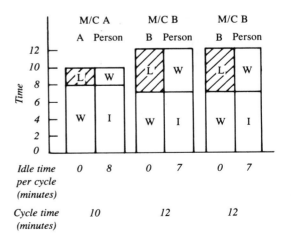

The total cost consists of two components, machine cost and operator cost. In making comparisons between the alternatives, we will include the cost of idle time for the machine as one component along with the operator cost. This is because when machines are in use, they are performing desirable, productive activities; however, in the idle state they are nonproductive and an expense. The idle condition can be reduced by increasing the number of operators, thus reducing the machine component of the cost while increasing the operator cost. We are seeking a balance that will minimize the total cost. The total cost of the present alternative is

$$0 + 3(\$6) = \$18.00/\text{hour}$$

Alternative B

Now let us examine the time distribution when one operator is assigned to Machines A and B and one to the remaining Machine B. The time bars are illustrated in Figure 8.7. The minimum cycle time when Machines A and B

are tended by the same operator is the maximum of cycle times for either Machine A or B.

$$
\begin{aligned}
\text{Total cost/hour} &= \text{Idle cost of M/C A} \\
&\quad + \text{Cost of operators} \\
&= (2/12)\$12 + (2 \times \$6) \\
&= \$14.00
\end{aligned}
$$

Alternative C

If one operator is assigned to Machines B and B and one to Machine A, the time bars are as shown in Figure 8.8. The cycle time on two B machines is 12 minutes; on machine A it is 10 minutes. There is no idle machine time, and so the total cost/hour consists of only the cost of the operators, that is,

$$
\begin{aligned}
\text{Total cost/hour} &= 2 \times \$6 \\
&= \$12.00
\end{aligned}
$$

Figure 8.8 Man–machine chart showing two operators (one worker for Machines B and B and one worker for Machine A)

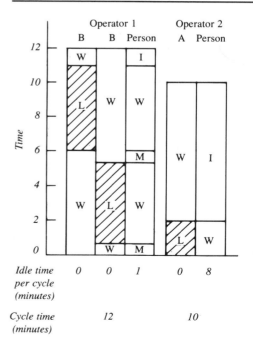

Figure 8.7 Man–machine chart showing two operators (one worker for Machines A and B and one worker for Machine B)

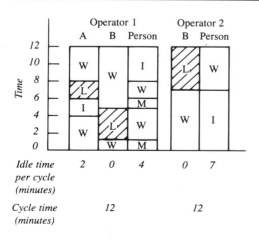

Figure 8.9 Man–machine chart showing one operator (one worker for Machines A, B, and B)

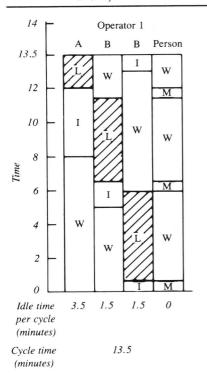

Idle time per cycle (minutes): A 3.5 B 1.5 B 1.5 Person 0

Cycle time (minutes): 13.5

Alternative D

If one operator is assigned for all three machines, the time bars are as shown in Figure 8.9. The minimum cycle time must allow the setup of each machine and the time for an operator to go from one machine to the next in the job sequence. There are three "walks," one setup for A and two setups for B, giving a minimum cycle time of $3(\frac{1}{2}) + 2 + 2(5) = 13\frac{1}{2}$ minutes.

Total cost/hour
$$= \text{Cost of idle time on each M/C} + \text{Cost of operation}$$
$$= \frac{3.5}{13.5} \times 12 + \frac{(1.5 + 1.5)}{(13.5)} \times 18 + 6$$
$$= \$13.11$$

Alternative C has the lowest cost of all the alternatives; therefore it is the preferred alternative. One operator should be made responsible for two B machines and one operator for Machine A.

Machine coupling is increasingly used with a robot working with a number of automated machines. The cycle times can be controlled to very precise values, and the repetitious tasks of loading and unloading each machine are assigned to the robot.

8.4 TOTAL PERSONNEL REQUIREMENT

The next step is to list all the personnel requirements in the plant. The number of assembly stations and the amount of manual effort required, along with the machines and their associated operators, determine the number of direct factory workers needed. It should be understood that more than one machine might be assigned to an operator through machine coupling, or it might be necessary to assign more than one operator to a machine as is the case in a paper-manufacturing facility. Organization and factory layout determine the number of direct supervisory personnel required. For example, in a job shop environment consisting of a large number of machines and operators in each department, a foreman might be appointed for each department such as gear cutting, heat treatment, and milling. One foreman might be sufficient for the entire machine shop if it consists of a much smaller-scale operation, such as only 8–10 machines and 15 total operators.

Though management theorists claim that assigning 10–12 people per supervisor is most efficient, in an assembly line environment the number of workers reporting to one supervisor may be considerably higher. It is not uncommon to see a ratio of 25:1 in an assembly line in which there are similar work patterns, more manual work than machine operations, and a layout that permits the supervisor to easily see all workers.

If the plant is spread over a large area, the ratio of 12:1 might have to be reduced accordingly. Similarly, if in addition to supervision the person has other responsibilities such as long-range planning, new product development, or sales analysis, then the time available for direct supervision is limited, and the number of personnel assigned per supervisor should be modified correspondingly.

A proposed or current organization chart also plays a role in defining responsibilities and developing the number of required personnel. (More about organization charts in Chapter 9.) For example, the maintenance and quality control departments might be directly supervised by the plant manager in a small operation; or a technician could be hired having broad responsibilities for process control, quality control, and incoming and outgoing inspections in a medium-sized plant. A large plant might require a number of technicians and supervisors to perform similar tasks. Matrix organization, regardless of size, requires an independent manager for each product as well as for each department.

The number of staff appointments in such areas as production and material planning, accounting, computer operations, purchasing, trucking and transportation operations, labor relations, and material handling will depend on the size of the company. For example, each administrator might hire his or her own secretary if the clerical work involved justifies it or if the confidential nature of the work demands it. Whenever possible, a secretary is shared between departments if the floor plan permits. With today's technology and the availability of desktop computers, many companies allow technical people to develop their own letters and memos. The secretaries are assigned broad responsibilities in communications, computer

Table 8.10   Direct labor table

Position	Salary	Number Needed	Total Salary
Rip saw operator	$12,000/year	2	$24,000/year
Radial arm saw operator	$12,000/year	2	$24,000/year
Variety saw operator	$13,000/year	2	$26,000/year
Belt sander operator	$11,000/year	1	$11,000/year
Machine sander operator	$11,000/year	4	$44,000/year
Drill press operator	$11,000/year	3	$33,000/year
Cutter	$11,000/year	2	$22,000/year
Upholstery sewer	$11,000/year	2	$22,000/year
Chair padder	$11,000/year	2	$22,000/year
Spray varnisher	$12,000/year	5	$60,000/year
Inspector	$13,000/year	4	$52,000/year

Table 8.11   Support personnel table

Position	Salary	Number Needed	Total Salary
Plant manager	$40,000/year	1	$40,000/year
Bookkeeper	$17,000/year	1	$17,000/year
Secretary	$13,000/year	1	$13,000/year
Maintenance person	$19,000/year	2	$38,000/year
Cleaner	$11,000/year	1	$11,000/year
Plant foreman	$23,000/year	1	$23,000/year
Forklift operator	$15,000/year	3	$45,000/year
Utility worker	$13,000/year	2	$26,000/year
Packager	$13,000/year	3	$39,000/year

operation and data entry, and computer report generation. On occasion a secretary might handle the clerical needs of three different departments with a total of 50–60 engineering personnel.

The number of work shifts has an obvious impact on the requirements for both direct and indirect labor. It is a common practice, however, to limit indirect personnel on the second and third shifts. Only the indirect personnel who are responsible for production-related operations are generally working during second and third shifts—for example, a shop foreman and a skeleton maintenance crew.

The direct and indirect personnel requirements for a small firm working a one-shift operation are listed in Tables 8.10 and 8.11. Note that the plant manager has been given major responsibilities in different areas such as personnel management, sales, and planning along with overall supervision of the plant.

SUMMARY

To begin manufacturing, we require, among others, two important resources: appropriate machines and a good labor force. The judicious selection for both can be a challenging task.

Make-or-buy evaluations and capability or expertise studies help identify the components that could be produced within a plant, but then proper machines must be selected to make this possible. This selection is influenced by needed capabilities, capacities, and prices of the various machines, as well as the mode of operation within the plant. In order to make this determination, we must first gather the information on all available machines in the market that would satisfy our needs. Within the plant, good sources for such information are the people with experience—fellow workers and engineers, buyers, maintenance personnel, and supervisors. We may also acquire this information from outside sources such as technical periodicals, the *Thomas Register, Ayers Directory of Publications,* and *Ulrich's Directory.* Computer data bases are also becoming more economical and popular. Information services like Dialog, for example, provide access to over 200 data bases containing more than 100 million items.

Once the machines are identified, the next step is to determine the capacities and the number of machines that would be needed. For that, we must first perform a detailed analysis of all production routing sheets and augment the calculations by taking into account the shrinkage factor for each stage in the entire production facility. Such analysis establishes the estimate of the total required capacity for each type of machine operation; knowing the capability of each machine, their required numbers can be easily established. Economics plays an important part in determining the final selection.

All machines, including the auxiliary equipment, should be specified in detail so that both buyer and supplier are aware of what the requirements are. These specifications are also helpful to plant engineers as they develop the floor plan and install utilities for the new machines.

Proper labor selection is another important element that should be critically reviewed. Each job should be evaluated for factors such as required skills, efforts, responsibilities, and working conditions. Such an evaluation goes a long way in hiring a person who is suitable and qualified for the job. Job evaluation also helps in determining the salary range for the position, which is compatible with the salary structure within the plant. The *Dictionary of Occupational Titles* (DOT) provides standardized job analysis that can save a considerable amount of time and effort by eliminating individual job evaluation within the plant.

Coupling machines, thus allowing one worker to tend more than one machine, is a technique that can be profitably used to reduce the total work force. Finally, constructing a total labor chart, listing all jobs and indicating for each its title, salary range, and duties, leads us to realize the relative importance and value of each job. It also displays the total labor force and total labor cost for the plant.

PROBLEMS

8.1 What basic data are needed to begin the machine selection process?

8.2 How do expertise and economics affect the make-or-buy decision?

8.3 In performing central nervous system research a laboratory is using microelectrodes, which it believes will last one year. Microelectrodes can be made in the laboratory or purchased. The equipment to produce the electrodes would cost $800.00 and could be used for six months. The research technician could manufacture the electrodes and is paid $12.00 an hour. A research assistant could be trained to produce the electrodes and is paid only $6.80 an hour. However, the assistant would require two weeks of training, during which time the electrodes would have to be bought from an outside source. The cost of purchasing the electrodes is $25.00 apiece. The technician could produce ten electrodes a day, as could the assistant after proper training. The material cost for each electrode is $4.00. If the laboratory will need 500 electrodes in total, determine how they should be acquired.

8.4 A toy factory is planning to make an additional item, which would be processed in four different departments. The company is proposing to produce 600,000 units per year, working 250 days per year. The estimates of times made by an industrial engineer are shown in the accompanying table. The parts are produced within the departments in a sequential manner (from department 1 to 2 to 3 to 4).
   a. Determine the number of units that must be produced each day in each department if all the defective units are scrapped.
   b. Determine the number of machines needed in each department to meet the production goals.

| | Departments | | | |
Times	1	2	3	4
Average processing time (seconds/part)	8	13	10	50
Average maintenance time/day (in minutes)	20	30	20	15
Average daily setup time (in minutes)	20	15	30	20
Average bookkeeping time/day (in minutes)	10	8	12	5
Machine adjustment time/day (in minutes)	15	10	15	15
Defective items (in % of production)	5	6	4	4

c. Determine the efficiency of each departmental operation if only one shift is operated daily from 7:30 A.M. until 4:30 P.M., with a 40-minute break for lunch and two 10-minute coffee breaks.

d. Adding a second shift would reduce the efficiency of the first shift by 10%, and the efficiency of the second shift would be 5% less than that of the first shift. How many machines of each type would be required?

8.5 In a small machine shop the manger must decide how many and what types of machines to buy. She has collected data, as shown in the accompanying table, on an NC machine and a traditional milling machine on which she plans to produce two products per day. All of item 1 will be produced first, and then all of item 2 will be manufactured.

	NC Machine	Traditional Milling Machine
Cost	$15,000	$2300
Useful Life	5 years	3 years
Manufacturing/Operating Cost/Year	$500	$300
Labor Cost	$10/hour	$10/hour
Product 1		
Output Rate/Hour	25 units	10 units
% Defective	0.1%	6.0%
Cost of a Defective Unit	$10	$10
Setup Time	5 minutes	25 minutes
Needed Output/Day	80	80
Product 2		
Output Rate/Hour	30 units	12 units
% Defective	0.1%	5.0%
Cost of a Defective Unit	$8	$8
Setup Time	10 minutes	30 minutes
Needed Output/Day	100	100

    a. Assuming 8 hours of working time per day and 250 days of operation each year, determine how many NC machines would be needed if all the production were done on NC machines.

    b. How many traditional milling machines would be needed if all the work were done on such machines?

    c. Calculate the manufacturing cost associated with each good unit of each product (excluding materials) produced in this station. Assume an interest rate of 12%.

8.6 At your library, select a recent issue of an engineering periodical such as *Material Handling* or *Plant Engineering.* List five equipment-related articles and/or advertisements in that issue. Note the information supplied on the equipment.

8.7 Refer to the *Thomas Register* and list the companies that manufacture forklift trucks.

8.8 Check the literature on forklift trucks and then write a specification for forklift trucks capable of picking up one thousand 1500-pound cartons per day.

8.9 What information related to facility design can be found in a computerized data bank such as Dialog?

8.10 Distinguish between a product arrangement and a process arrangement.

8.11 List the more important advantages and disadvantages of automated equipment.

8.12 Why can machines not be utilized at their full capacities during production? What actions should a plant manager take to increase machine utilization?

8.13 What is a job evaluation? What are some advantages in developing such a system?

8.14 How does job evaluation within an organization help determine salaries?

8.15 Using Table 8.6 first, determine the point value for a supervisor of a machine shop and then decide the supervisor's salary based on Figure 8.5.

8.16 How is a job analysis related to job evaluation?

8.17 Develop descriptions of the following jobs and compare them with those listed in the *Dictionary of Occupational Titles:*
    a. News Editor
    b. Maintenance Supervisor
    c. Tack Welder

8.18 What are some advantages to using the DOT for job evaluations?

8.19 Using DOT number codes, develop a point system pay scale for a general machine shop operation.

8.20 A bank would like to develop a job evaluation point plan for the president, teller, and loan officer.

Miller distributor.

→ ~~Boots~~ choice brands,
US 165 north
monroe.

Dont go and get at teh, already there Professors, know that we
are working some where, so every one gets fucked up. If something happens bad.
I think its better to go and get in monroe, bk
change your hair style lil bit, so that if someone comes to your station
ok also, they cant identify you, and keep your car at back.

*

<u>Practice</u>

<u>6 6 , 6·7</u>

$$\frac{500}{92} - \frac{250}{46} \quad \frac{125}{\backslash 23} \quad 21 \quad \frac{5}{125}$$

$$\frac{5}{115} \quad \underline{5} \quad \frac{115}{100}$$

a. Write the job specifications for each.
b. Construct a list of ten factors to be considered in the evaluation.
c. Determine the value and assign points to each factor.
d. Evaluate the three jobs.
e. Develop the salary scales.

8.21 a. Machine operators cost $9.50 an hour. The cost to run the machines in the accompanying table is $10, $12, and $15, respectively. If it takes 30 seconds to get from one machine to the next, investigate the possibility of machine coupling.

| | Machine | | |
Process	A	B	C
Loading	8	5	10
Processing	14	10	12

b. If two more machines are added with operating costs of $18 per hour, investigate coupling with all five machines. The setup and processing costs are as given in the accompanying table.

| | Machine |
Process	D, E
Loading	6
Processing	15

8.22 Production of a unit follows the schematic diagram shown in Figure P8.22. P_i, the probability of generating a defective unit in Stage i, and C_i, the unit cost of processing in Stage i, are given in the accompanying table.

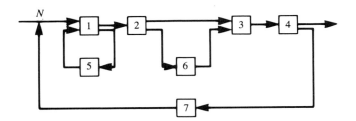

Figure P8.22

Stage	1	2	3	4	5	6	7
P_i	0.2	0.1	0.1	0.15	0.3	0.4	0.3
C_i	5	8	6	4	5	4	15

Determine the cost of producing 100 good units and the number of units that must be started (i.e., the value of N) in Stage 1 to obtain 100 good units.

8.23 Product A and by-products B and C can be produced with the production line shown in Figure P8.23. The average percentage of the input product that will be channeled into a branch is shown by the numerical value above that branch. The cost data for each unit of input in each stage are given in the accompanying table.

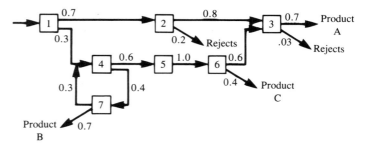

Figure P8.23

Stage	1	2	3	4	5	6	7
Cost ($)	10	30	15	18	12	13	5

Determine the average cost of each unit of products A, B, and C.

SUGGESTED READINGS

Machine Selection

Applied Science and Technical Index, The H.W. Wilson Company, New York, 1985.

The IMS 1984 Ayres Directory of Publications, IMS Press, Fort Washington, Penn., 1984.

Thomas Register of American Manufacturers and Thomas Register of Catalog File, Thomas Publishing Company, New York, 1980.

Ulrich's International Periodicals Directory, R.R. Bowker Company, New York, 1980.

Labor Requirement and Selection

Dictionary of Occupational Titles, U.S. Department of Labor, Washington, D.C., 1977.

Lowdy, F.J., and Trumbo, D.A., *Psychology of Work Behavior,* The Dorsey Press, Homewood, Ill., 1980.

CHAPTER

9

Building, Organization, Communications, and Selected Support Requirements

This chapter discusses a variety of topics that contribute to the overall design and working of a production facility. The manufacturing unit must be housed in a building, and Section 9.1 discusses a few traits of more popular industrial buildings. The facility must be operated efficiently, which requires an organization structure such as that described in Section 9.2. As we have seen, the nature of an organization affects the total manpower resources that are needed in the plant. It also influences the office and plant layouts detailed in Chapter 13. Section 9.3 is a short statement on the communications needed to operate a plant coherently. The mode of communication also defines the facilities required in the plant. For example, storing documents in the quality control room might require a stacked cabinet, or the documents could be stored on a floppy disk if a desktop computer is available; written memos could be transmitted on paper or through minicomputers connected in a network if there are minicomputers in every office. Section 9.4 discusses the services and facilities that are necessary in any production operation, including shop offices, inspection and maintenance departments, locker rooms, lavatories, a cafeteria, and even a parking area. Essentially, this chapter provides information on integrated plant design.

9.1 BUILDING

More and more industrial engineers, along with architects and civil engineers, are getting involved in the basic phases of designing and constructing industrial buildings. An industrial engineer is concerned with the efficient opera-

tion of a plant and therefore is responsible for arranging machines, departments, and material-handling facilities within it. Many times the engineer can obtain a superior arrangement if the building is constructed to fit the layout rather than a layout being developed to accommodate the existing building. Layout considerations must also play an important part in enlarging or revising an existing structure. We will now study some of the building features in industrial plants.

Conventional Building Characteristics

The following are the major components of an industrial building, their properties, and requirements.

Structure

Structural considerations of a plant building are influenced by the activities to be conducted within the building. For heavy manufacturing, which includes iron smelting, steel fabrication, ship building, and automobile manufacturing, the support structure is large and heavy. This is mainly due to the demands placed on the structure by gantries, cranes, and other heavy equipment.

Light industrial buildings are generally single-story (unless the land is very expensive) and are constructed of steel frame. The spacing of columns and bays within the building plays an important role in overall layout. It is common practice to define major aisles and partition walls for various departments along the columns. The sizes of "bays" (between columns) range from 20 × 20 feet to 35 × 160 feet. The longer the span, the more expensive it is to construct the building; however, the long span also reduces the number of columns that are necessary to support the entire structure. It is estimated that each column effectively uses up between 9 and 16 square feet of floor space, an area that could be used for productive activity. Long spans also require deeper steel frame construction, and this permits the overhead space within the frame to be used for compressed-air piping, air-conditioning ducts, and electrical conduits. A square building is preferred over a rectangular shape, being more flexible for expansion. The clear height of the building differs depending upon the products, generally ranging from 11 feet to 38 feet, 18–20 feet being most common.

Walls

Outside walls furnish protection and security to the plant. In conventional structures, most walls are masonry or concrete, which provides good fire protection and presents a neat appearance. Such walls are not moveable, however, and therefore expanding or revising the plant is not easy. Temporary walls of corrugated aluminum or sheet metal with fiberglass insulation are normally used when expansion is envisioned.

Inside partition walls are generally non-load-bearing, and they can be prefabricated or built on-site using various material. In a food plant, for example, walls might be tiled for a smooth and sanitary interior.

Floor

The floor of a plant is almost always a concrete slab. Such a floor can be made strong enough to support heavy equipment loads and can be leveled to an accuracy ranging from ±4 mm in 2 m (normal) to ±1 mm in 2 m (superflat). The flatness is important, since it relates to the vehicle dynamics and mast deflection of a floor-riding material-handling device such as a forklift truck.

In general, the floor and its covering should be durable, impact- and vibration-resistant, sound-absorbent, nonskid, nonsparking, and easy to clean. It should not be affected by changes in temperature or humidity or by accidental spills of chemicals, oils, or other industrial substances.

Roof

The choice of roof shape from the several shapes that are generally used (see Figure 9.1) can give distinct character to the building. A

Figure 9.1 Roof types

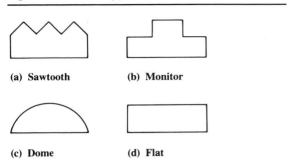

(a) Sawtooth **(b) Monitor**

(c) Dome **(d) Flat**

sawtooth roof with translucent glass on one side of the teeth allows natural light to penetrate into the plant without causing glare. A sawtooth roof can also allow an opening for fresh air and ventilation. A monitor roof is useful when an overhead crane is needed at the center of the plant. A dome shape encloses a large volume when a great height in the center is desired or essential. A flat roof is most common in modern buildings because of its low cost of construction as well as of installing insulation and acoustic ceilings.

The material used in roofing can vary from long spans of aluminum or steel, costing as much as $300 per 100 square feet, to a combination of asbestos and cement sheeting, costing about $100 per 100 square feet. A flat, reinforced concrete roof is often covered outside with an insulating material similar to cork to reduce temperature variations in the plant.

Interior

A pleasant interior environment can add considerably to the worker's morale and productivity. It is the responsibility of engineering and layout personnel to provide such an environment.

Temperature in the plant should be maintained between 68° and 72°F at a relative humidity of 50%. Air-conditioning and heating units might be necessary to prevent the ambient temperature from varying much. A flow of air within the facility is desirable to avoid a feeling of stuffiness; and dust, odor, and the vapors and fumes generated by machines or workplaces should be controlled. This can be accomplished by providing vacuum ducts and tubing that transport the air and particles to a centralized location away from the work spaces. Depending upon the nature and amount of the contamination, any of numerous collecting devices might be employed; or if no hazard is involved, the air could be discharged to the atmosphere. Appropriate drainage should also be provided whenever necessary.

Noise within the plant should be minimized by isolating the sources (if possible) or controlling them. Proper maintenance and sound-absorbing tiles on high ceilings and walls contribute greatly toward noise reduction. Hearing protection (earplugs or muffs) should be worn by people who work in an environment where an unsafe, high-noise level cannot be reduced. The Occupational Safety and Health Administration (OSHA) has issued regulations governing the maximum noise level permitted.

Good illumination within the plant is important, and between 50 and 60 foot-candles of light should be maintained throughout the facility. At times a work station for delicate or close manual work might need much more illuminiation, perhaps as much as 100 foot-candles or more.

Interior colors also contribute toward making the workplace enjoyable. Dark colors on the walls might hide dirt, but they also absorb light, making the place look dull. Cream, peach, and light shades of green and blue are favorite colors for walls, and white is most common for ceilings. Yellow is used to indicate the need for caution (low overhead pipes, for example) and other problem areas that might need special attention. Black and white stripes generally define traffic areas, and aisles are marked with white borders. Red is associated with danger and is used to indicate fire protection apparatus, stop signs, traffic lights, and dangerous areas where admission might be restricted.

Prefabricated Buildings

Many sources are now available for prefabricated buildings of almost any desired shape. Prefabrication is a cost-effective and time-saving method of construction, mainly using a corrugated or flat-channel steel body shell. The modular approach to the building also allows expansion or alteration. Interior walls or offices can be moved from one location to another with ease to modify the layout within an existing plant.

9.2 ORGANIZATION

Proper organizational planning eases the coordination of all personnel in a manufacturing concern by showing how their efforts contribute to a common purpose. It also helps in facility planning to outline the number and types of personnel who will be working in the organization and their relationships with each other. The method of structuring is determined by the aims of the company, what resources are available, and the amount of time the organization has to accomplish the task.

Employees within a business who share one or more important common factors should be grouped together. Some of these factors are skills, processes, geographical locations, types of customers, and goals. The chain of command should facilitate decision making at the lowest possible level at which the required information is available, but such decisions should be reviewed, and modified if necessary, by people at higher levels in the organization. The structure should be as simple as possible to avoid overlap, redundancy, and stifling of creative efforts. Required coordination time and communication difficulties increase drastically with each added layer of supervision.

Organizational Concepts

Classical U.S. and European organizational theories are based on principles derived from military and religious orders. In contrast, the typical Japanese manufacturing corporation does not even have an organization chart, often employing project teams whose managers may readily be moved from one area to another. The individual worker not only feels responsible for performing his or her assigned task, but also sees to it that all tasks assigned to his or her project team are carried out. Japanese workers view employment as a lifetime arrangement and co-workers as lifetime colleagues; as a result, they make every effort to get along well together. Japanese factory workers usually tend to be generalists trained in most aspects of a manufacturing process.

In the United States, on the other hand, workers are more frequently specialists, and businesses are designed with as little redundancy as possible. Management attempts to make lines of communication easily identifiable, tends to push employees toward specialization, and compartmentalizes functions. A well-defined organization helps to decrease operating costs, reduces organizational friction, and consequently minimizes the number of management problems.

Organization Charts

How an organization is structured at a particular time can be depicted by a chart that shows the formal relationships among functions and provides the titles (and frequently the names) of the people who are responsible for those activities. Some companies avoid using such charts because they feel that organization charts can cause inflexibility; other companies like them because they show who is responsible for what function in the lines of direct authority. The charts can help a company prevent duplication of effort or having a responsibility not be properly assigned. The charts can also be used in expansion planning and to show employees who reports to whom throughout the organization.

Organizational Structures

The structure of an industrial plant can be set up in one of several ways depending on the size of the company, the type of business being carried out, and the complexity of difficulties encountered in day-to-day operations. The structure can range from a straight line to a multidimensional matrix organization.

Line Organization

The oldest and simplest format is the line form. The straight line organization most often exists in small companies in which authority extends from the highest to the lowest level of the concern. A simple line organization chart is shown in Figure 9.2. The head of the company oversees all functions such as personnel, sales, engineering, and purchasing. An organization of this type has a unity of command in which each person reports to, and receives orders from, only one other person. It has the advantage of having easily understood relationships at all levels. Conversely, it has the disadvantage of requiring personnel at most levels to be capable in several functional areas. This may result in some duplication of effort; for example, each supervisor might find it necessary to keep a set of equipment catalogues. As such a company increases in size, managers become overwhelmed by the details that must be handled. In this situation the organization will probably evolve into a staff-line organization.

Staff-Line Organization

A staff-line organization evolves when the need for specialists in different areas becomes so great that managers cannot handle the burden of details. Figure 9.3 shows an example of a typical staff-line organization. It is similar to a line organization, but specialists have been added to contribute technical expertise in particular areas. In theory, authority and responsibility still lie within the line, but the staff can exert considerable influence while acting in its advisory capacity. A modification of this type of structure is the line and functional staff organization in which the specialists have full authority over matters relating to their particular areas of expertise. Staff functions are usually one or a combination of service, control, advisory, or coordination types.

Product Organization

When a business becomes a multiproduct company, organizing it into a product-type structure might prove more efficient. This is especially true if the products are easily distinguishable, in either form or market, from each other. Figure 9.4 shows how the chart of such a firm might look. The disadvantage of this arrangement is that several people with the same general type of skill might be required, but this may be overcome in part by centralizing some

Figure 9.2   Line organization

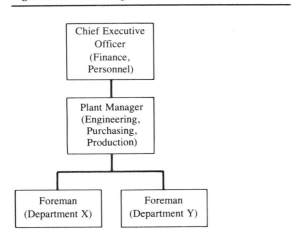

Figure 9.3

Staff-line
organization

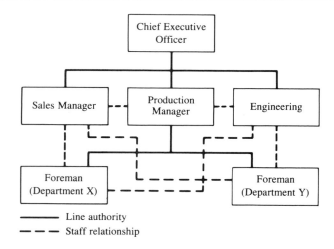

Figure 9.4

Product
organization

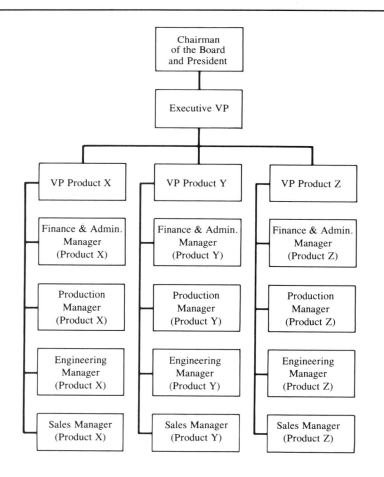

Figure 9.5   Matrix organization

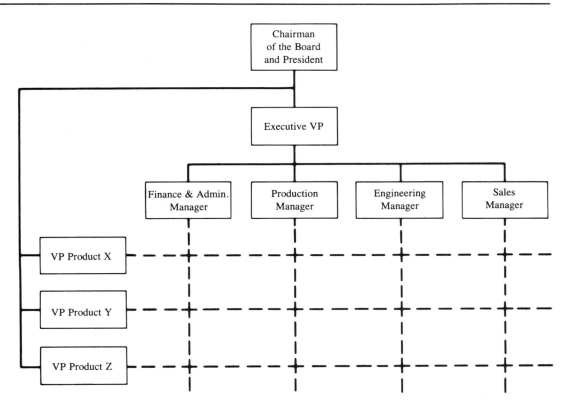

functions such as engineering and administration. As the business grows, this classification becomes quite complex, and a more modern concept, the matrix organization, might be used to alleviate the problems.

Matrix Organization

In a matrix chart, such as that in Figure 9.5, activities are shown vertically, and products are shown horizontally. In this arrangement, financial, engineering, and other methods can be standardized through function managers who can assign personnel and resources as needed to the different products. (In practice, the function managers are actually staff managers.) However, the matrix organi-

zation does have some inherent disadvantages. The more significant ones are that it violates the principle of unit of command and that it makes communications and coordination even more critical, both vertically and horizontally.

Selection of Organization Structure

Of the organizational forms discussed above, choosing the one that is best for a particular business depends on a thorough understanding of (1) the goals of the organization, (2) the product or service to be offered, and (3) the capabilities and willingness of the individuals who work in the organization. In reaching this understanding, one must determine and analyze the types of manufacturing facilities and departments required for efficient and profitable operation. Once the appropriate structure is selected, implementation will depend in part upon effective communication throughout the organzation.

9.3 COMMUNICATION

The methods of communications are changing in modern plants and offices, and so are the necessary facilities. It is therefore beneficial to understand the need and means of communication as we develop the facilities.

 Communication is an essential function because it is the foundation for transmitting management objectives and provides the means for understanding those objectives. Communication can have one or more of several purposes:

 · To furnish information or transmit data
 · To persuade
 · To give directives or direction
 · To gather information or receive input

 In order that the people who receive a communication take the desired action or retain the desired knowledge, they must accept and understand the message. Communications can be written or spoken; each method has inherent advantages and disadvantages.

Written Messages

It is often difficult to maintain attention to a written communication that is very large. The media of radio and television have made modern people less receptive to written messages than to spoken communication. Confirmation of the recipient's understanding of a document is more difficult and less timely. However, a written communication has the advantages that the same message is received by everyone in-

Methods of communication. Here diagrams on a blackboard provide written communication to supplement an oral presentation.

volved, it is easily disseminated, and it provides a permanent record to which one can refer at a later time.

Spoken Messages

Spoken communications offer the chance for immediate feedback to clarify or confirm understanding of the message. Psychologically, however, people tend to remember less of what they hear than of what they see. If a communicator has a poor delivery, even a good message might be improperly received and accepted.

Methods of Presentation

A message might be intended for one, two, or many people. When information is to be given general distribution, one of several means can be used. Group meetings can be held in a conference room with the speaker present, or the message can be delivered live via a teleconference or a broadcast. Delayed presentations can be made via video recordings, films, or tape recordings. One of the more common methods of in-plant mass communication is use of bulletin boards. In plants with large work forces, employee periodicals, handbooks, and policy manuals are effective means of using mass communication. Although the manuals might not be read by everyone, they do serve as written records of the information made available.

Person-to-person communications can be carried out face-to-face, by telephone, or via interactive computer terminals. Rapid advancements in modern communication technology will continue to increase the speed with which communications can be transmitted. The number and sophistication of the methods of communicating will also continue to increase.

Documentation

We stated earlier that the design and operation of a manufacturing plant are a team effort. Each member of the team is an expert in one field of operation, yet coordination between all the members is essential to accomplish overall integration in the plant. Documentation is a means of achieving a common objective without duplication, confusion, and chaos.

Three principal factors contribute to the need for documentation in achieving integration of work. They are the number of workers (people) involved, the physical distribution of the workers and their activities, and the time span over which the activities are performed.

For a small organization with one-person supervision and a small area of operation, oral communication might serve well; but for most other organizations, written rules, instructions, and records are required. The red tape created by paperwork is not meant to impede the activities of a creative individual but to aid in the team effort by maintaining orderliness in the information flow and by reducing mistakes. Care must be taken to avoid overburdening employees with preparing and reading unnecessary reports. That is, such documents should not be required simply for the sake of having reports. Ideally, the system will provide everyone with accurate, timely, and useful informa-

tion required for efficient operation of the organization.

When workers are located in widely separated areas (different rooms or buildings, for example), oral communication requires the assembly of all workers in a conference room, or the speaker must telephone or travel from one employee or group to the next. All of these alternatives are time consuming and not cost-effective, especially if only routine matters are involved. A memorandum or a message transmitted by computer network is preferable.

This also resolves problems associated with remembering instructions over an extended period, as well as ensuring that all people concerned receive the message even if some are not present when the message is transmitted.

Printed forms provide a convenient means of recording standard information. They prevent wordiness and display information in the same manner every time, making it easy to read and recognize. They also help a person who is new in the job to understand information that must be collected and recorded.

Engineers also frequently record their calculations, notes, and communications on a notepad, which becomes a legal document that should be readily available to other authorized personnel. While working on a project, engineers should carefully record all calculations, assumptions, instructions, and dates. They are often asked to produce these records in support of their presentations of project reports.

At times, owing to an emergency or other critical event, people might have to deviate from normal procedures by "cutting the red tape." An organization in which such deviation is possible without causing confusion, hard feelings, or disorder has two principal traits: trust between the employees who will help each other and people who are very knowledgeable about the workings of the system. Any shortcuts taken by one person can add to the time demanded from another; in a well-organized and well-run plant this procedure should not become a common practice.

An example of good documentation is shown in Appendix B. The instructions are for a quality control operation in a spring-manufacturing facility.

9.4 SUPPORT FACILITIES AND REQUIREMENTS

The facilities and services that are not directly necessary for production but are still required for the operation of the business are termed support facilities and services. Their availability and timeliness contribute greatly to improving worker satisfaction and productivity. We will discuss some of the more important support services in this section; some others, including utilities, will be discussed in Chapter 15. We will exclude managerial activities such as supervision, personnel management, accounting, sales, and advertising. These activities, for the most part, are self-explanatory with respect to their purposes.

Shop Offices

In its operation the manufacturing facility is normally divided into several shops or departments, each managed by a foreman and supporting staff. Though many support functions such as payroll, engineering, and personnel are directed through the central office, an area must be set aside for office space for the people who are directly responsible for supervising the activities of a department.

Inspection

There might be one centralized location for all inspection activities when the products are small and can be easily transported, or there might be multiple inspection stations, perhaps more than one for each department if the product is large or inspection frequency demands it. The necessary space for the inspectors and their equipment must be planned. The personnel should be trained in statistical methods of quality control, they should be familiar with the production processes and characteristics of the product, and they must be able to communicate effectively, both orally and with written reports.

Maintenance

All production plants need good maintenance facilities. There is a large investment in the building, equipment, and material-handling system, and it is important to obtain continuous and well-planned use from them. Any failure in the conveyor system, for example, can halt production throughout the plant. The maintenance department anticipates such potential failures and takes action through preventive maintenance programs. It also makes emergency repairs to avoid excessive loss of production when failures occur. Maintenance personnel are generally skilled in a variety of trades. They are trained in electrical wiring, hydraulic and pneumatic systems, controls, heating and air-conditioning, welding, and working with specific machines. The primary skills required vary from plant to plant depending on the equipment and processes installed. Most maintenance departments include one or more people with specialized training such as electricians, welders, and mechanics. In small plants, one person might be trained in more than one trade. This is especially true when maintenance of the building is involved: the same individual might be required to paint and clean and do plumbing, electrical, and air-conditioning work.

For highly skilled work such as the maintenance of computers, or when the demand for certain skills is infrequent, it might be better to contract the maintenance work to outside specialists. Routine maintenance, such as lubrication and general inspection, can be assigned to the operator of an individual machine.

Facilities for minor or emergency repairs might be provided in each shop; a centralized location might be needed for major repairs and overhauls. At times, heavy machines require service within the plant, and the maintenance department must be located near the production area. The shop should contain hand and power tools, workbenches, storage areas and racks, and light machines such as small lathes.

Tool Room

The purpose of a tool room is to procure (or make), store, and repair tools that are commonly used in the plant. Typically, the tools consist of mold dies, punch press tools, milling cutters, drills, and cutting tools for lathes, as well as jigs and fixtures. As part of a normal maintenance program, tool room workers inspect tools, molds, and jigs and fixtures to see that they comply with the specifications and drawings. When necessary, they restore, clean, grind, sharpen, polish, and repair the items. A tool room is a job shop operation; depending on the activities in the plant, it might contain high-precision equipment such as a Bridgeport drill press, boring machines, planers, a surface grinder, a lathe, and various hand tools. Some tool rooms are also equipped with specialized steel stock used in making tools, a heat treatment furnace, and high-speed cut-off equipment. A tool room should have all the utilities that are available in the plant, such as compressed air, 220-V/110-V electric outlets, and lubrication facilities. Material handling in a tool room is generally achieved by using hand carts and monorail and jib cranes. The layout of a typical tool room is shown in Figure 9.6.

Tool Crib

Necessary tools are issued to the workers in a plant through a tool crib, which serves as a storage center for all tools and accessories. A tool crib might consist of cabinets and racks for holding such items as emery cloth, steel wool, sander belts, socket sets, drill bits, milling tools, cam followers, grinding wheels, special clamps, and general-purpose dowel pins. Some security must be provided for a tool crib area; it generally consists of an 8–12′ area with a 10-gage chain link fence around it, with a window open-

Figure 9.6

A typical tool room layout

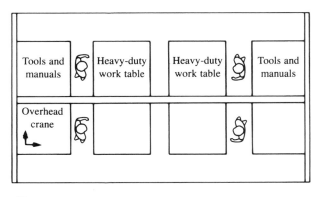

= Mechanic

Support facilities. *Top left*: Tools that workers need to do their jobs are kept in a tool crib like the one shown here. This secure enclosure has racks and cabinets that hold anything from simple wrenches to special drilling or grinding tools. *Top right*: In plants that involve messy work, washing facilities like this foot-operated hand-washing facility are located conveniently near the work area and can serve several people at one time. *Bottom left*: In-plant food services are beneficial to both workers and the company. Workers can get good, inexpensive meals without having to leave the plant, and the company avoids problems of drinking and absenteeism. *Bottom right*: Coffee service in a work area contributes to employee morale, and cart delivery saves employee travel time during breaks.

Figure 9.7 Tool crib

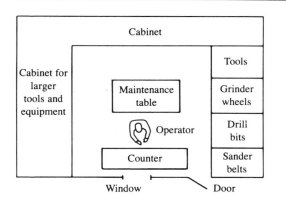

ing for issuing the tools. The layout of a typical tool crib is shown in Figure 9.7.

Lockers/Changing Area

Each employee might be provided with a locker to store his or her belongings. If a uniform is required or if the work environment makes protective clothing necessary, a changing area should be provided in the plant. Layouts for locker rooms are shown in Figures 9.8 and 9.9.

Lavatories

In most plants, locker rooms and lavatories are combined into one area. However, in a large plant there might be a need for lavatories in several different locations; to minimize time spent away from the job, they should be close to the work space. Individual states and local authorities might have regulations defining the number of lavatories required for a specific number of workers. OSHA has regulations regarding the requirements for toilets, lavatories, and showers in an industrial setting. Depending on the number of employees, the minimum number of toilets required is specified as given in Table 9.1.

For dividing the toilets on the basis of male/female population, OSHA has the following regulation: "Where toilet rooms will be occupied by no more than one person at a time, can be locked from the inside, and contain at least one water closet, separate toilet rooms for each sex need not be provided." Otherwise, "toilet facilities, in toilet rooms separate for each sex shall be provided. . . ."

An average of 4.5 feet per employee for washroom space is recommended; generally,

Figure 9.8

Locker room plan I

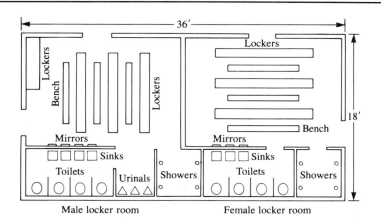

Figure 9.9

Locker room plan II

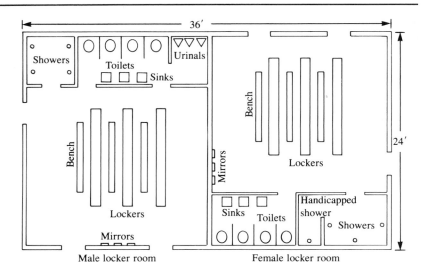

Male locker room   Female locker room

1.2–1.5 percent of the working area should be set aside for lavatories.

Lavatories are required in all places of employment; showers are required when certain materials such as paints, herbicides, and insecticides are handled or produced. One shower is required for each ten employees who will be using such a facility. Both hot and cold water should be provided.

Janitorial and Custodial Services

Upkeep and cleaning of the facility are the responsibility of the janitorial group. Most of the work must be performed at night after normal working hours so as to not disrupt activity during the heavy working time. Some companies prefer to subcontract these services to businesses that can obtain labor at less cost than an

Table 9.1   Number of toilets required

Number of Employees	Minimum Number of Toilets
1–15	1
16–35	2
36–55	3
56–80	4
81–110	5
110–150	6
Over 150	1 additional fixture for each additional 40 employees

industrial plant's contract with its labor union might permit.

A storage area must be provided for custodians to maintain brooms, mops, cleaning powders, fluids, and other needs. This area is nothing more than a small closet in many cases.

Eating Area

A plant should provide an area where employees can relax and eat during their breaks, the size obviously depending on the number of employees. Such places might provide anything from a few tables and benches, vending machines, and a microwave oven to a full-size cafeteria serving a variety of hot lunches. See Figures 9.10, 9.11, and 9.12 for various cafeteria layouts.

A rough estimate of the area needed for a lunchroom is 17 square feet per employee for those who will be eating at one time. For example, if it is expected that at most 200 people will eat at one time, the area assigned to the cafeteria would be 17 × 200 = 3400 square feet.

Figure 9.10

Cafeteria plan I

Figure 9.11

Cafeteria plan II

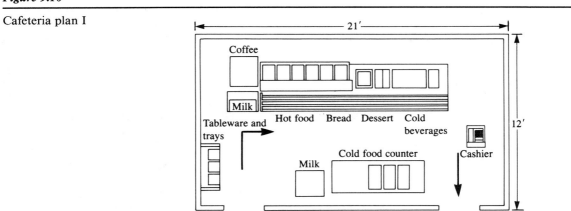

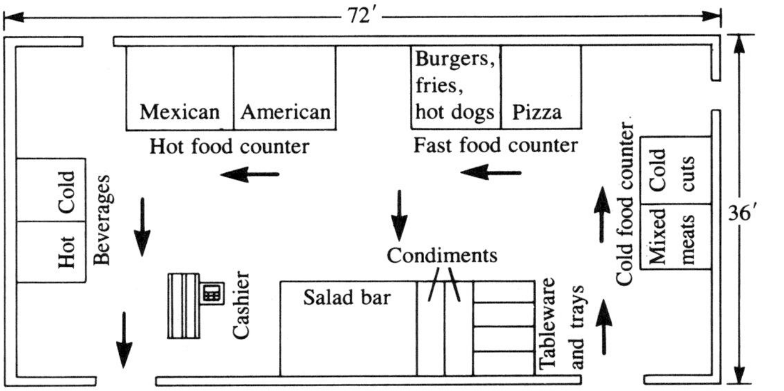

Figure 9.12 Cafeteria plan III

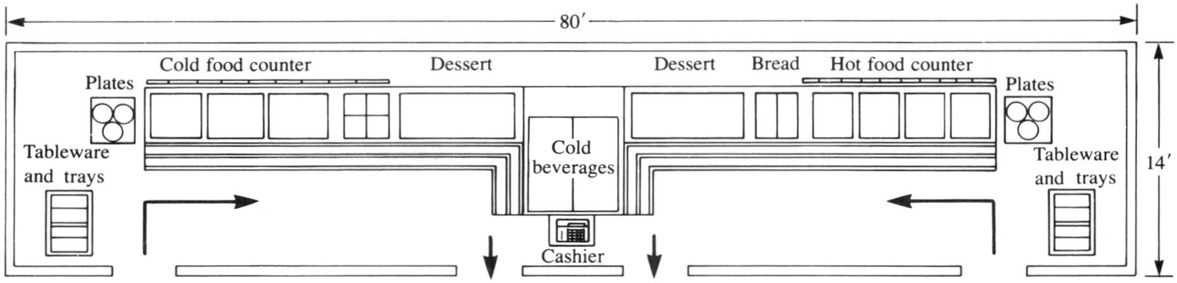

This includes space for tables, aisles, counters, and serving tables.

Kitchen area allocation depends on the number of meals to be prepared and served. Table 9.2 provides an estimate of the area required for kitchen activities.

If the company does provide a cafeteria, associated workers such as cooks, helpers, and dishwashers are also needed. Some companies choose to subcontract the entire cafeteria activity. In many instances the meals provided are subsidized by the firm as a fringe benefit to its workers.

A wide range of food can easily be distributed by vending machines. Soup, sandwiches, hot food entrees, desserts, ice cream, milk, soft drinks, tea, coffee, fruit, and candy are some of the most popular items that are thus sold. The machines range in dimension from $1'4'' \times 1'9'' \times 3'10\frac{1}{2}''$ for a small hot chocolate dispenser to $3'1'' \times 2'6'' \times 5'2''$ for a candy bar dispenser. One or more microwave ovens in vending areas can quickly warm the food if desired. Eating areas can be provided adjacent to the machines and might include tables, chairs, and benches. See Figure 9.13 for a typical layout.

A snack bar or sandwich line can be accommodated in a small area to provide a variety of quick hot and cold food such as hamburgers, hot dogs, potatoes, a variety of fresh salads, and drinks. The equipment area might include a refrigerator, coffee maker, stove, drink dispenser, sink, and counter with a cash register. The sitting area might include stools, tables, and chairs. Though a facility with a sandwich line requires more space than do vending machines, it is considerably more compact than a cafeteria. It also provides more flexibility in food preparation for an individual than vending machines offer.

Certain benefits can be reaped from providing food services within the plant. It will no longer be necessary for employees to leave the facility. The company still maintains supervision over the employees, reducing or even eliminating the problems of drinking, returning late, or not returning at all. Of course, it might not always be feasible economically or otherwise for a plant to have such facilities.

Table 9.2 Kitchen area requirements

Number of Meals	Kichen Area (square feet)
200	1000
400	1800
800	2800
1000	3000
2000	5000
3000	6000

Figure 9.13

Eating area plan

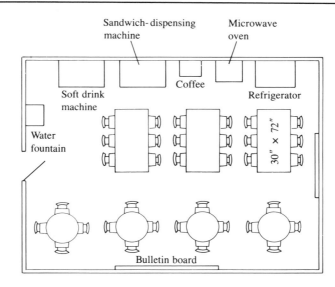

Security Force

Plant security requirements might necessitate employing a full-time security force. This could be because of danger present within the plant, such as producing chemicals, or because of the nature of the products—for instance, secret defense materials. Most plants, however, require nothing more than a night watchman or an alarm system.

Many plants issue passes to all of their employees. A pass usually contains the person's picture and must be presented to gain entrance to the plant. A guard or receptionist might be used to check each pass. As an alternative, automatic doors and gates are available that can be opened only by using a coded magnetic strip on the pass.

If the plant employs a security force, worker areas must be provided for the people involved. These might be a guardhouse at each entrance and/or a security office within the plant.

Parking Lot

A parking area is needed whenever public parking is not available in the immediate vicinity; many states require it by law. However, a parking area can mean considerable investment, and an analysis should be performed to determine the details of its development. For example, lots with parking spaces at 90 degrees or 60 degrees are quite popular; however, each requires a different average area per car. To provide parking for full-size cars, the stall dimensions frequently used for a 90-degree arrangement are 9 × 19 feet (though any width between 8.5 feet and 10.0 feet is common), with a 24-foot-wide driveway. Thus an average of 300 square feet of parking area is needed per car (including driveway allocation). If 60-degree angular parking is chosen, the stalls become elongated parallelograms with sides of 10.5 and 21 feet. The width of the driveway can safely be reduced to 18 feet, but the total allocation of space per automobile increases to 360 square feet. The advantages of the 60-degree arrangement over the 90-degree setup are that parking and backing out are easier and safer, and a narrower lot can be used for the same number of rows of stalls. Figure 9.14 illustrates parallel, 60-degree, and 90-degree parking areas.

Small cars require much less area than large cars. It is therefore possible to obtain many additional parking spaces in the same area if the lot is divided into two segments: one for small cars and one for large cars. Such a division should be based on the proportion of the small and large cars expected in the total automobile population.

Many physical characteristics of a parking lot must be taken into consideration. The parking area should be firm and built with a weather-resistant surface such as asphalt or concrete. There should be adequate lighting in the area and security provided if needed. Parking spaces and the direction of travel should be clearly marked on the surface. Entrance, exit, and speed limit signs should be posted.

The number of parking spaces needed might be dictated by state and local regulations. Generally, one parking space for every one to three employees is required. Locations that have adequate public transportation will not need as many spaces as plants in areas where all the employees must drive or ride in private cars.

If shipping and receiving is by truck, additional area must be available for parking and maneuvering the trucks. The exact area required is determined by the type, size, and number of trucks that the plant will handle at one time. The surface used for trucks must be much stronger than that for automobiles. Concrete with heavy reinforcing bars or rods is the most widely used paving material. Loading and receiving docks should be properly located with regard to other traffic in the plant vicinity. The opening to the plant should be of sufficient area to enable mechanical equipment such as forklift trucks to operate. Adjustable lift docks might be necessary when the sizes of trucks

Figure 9.14

Two parking lot layouts

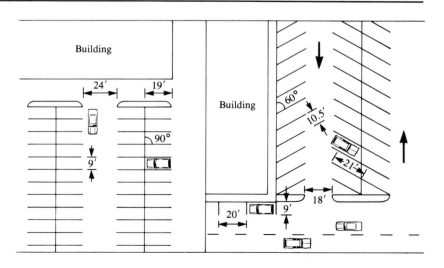

vary through a broad range. (More about loading docks in Chapter 12.)

Medical Facilities

Medical facilities in a plant might range from strategically located first aid kits to an infirmary capable of handling minor medical emergencies. In any case the company should provide the necessary personnel and equipment for handling sickness and accident. If a major medical facility is located close to the plant, a well-stocked first aid kit and someone trained in administering first aid might be all that is necessary. When the environment is particularly dangerous or hazardous materials are being handled, the availability of medical treatment should be increased accordingly.

Noise Exposure

Exposure of employees to high noise level or noises over a long period of time can result in permanent hearing loss. Any noise over 90 decibels is considered to be a noise hazard, and hearing protection is strongly suggested. Table 9.3 lists the permissible noise exposures over a

Table 9.3   Permissible noise exposures

Duration per Day (in hours)	Maximum Sound Level (in dBA)
8	90
6	92
4	95
3	97
2	100
1.5	102
1	105
0.5	110
0.25 or less	115

period of time that OSHA has developed for various intensities of noise. All manufacturing facilities are required to adhere to these limits.

SUMMARY

This chapter provides information for integrated plant operation and discusses different topics that contribute to the design and workings of a production facility. The building in which the plant is located can have many different traits. The structure of the building may differ depending on whether heavy or light industry is to be accommodated. The building spans can vary from 20 × 20 feet to 35 × 160 feet. The walls can be masonry for permanent buildings or corrugated aluminum for temporary use. Though plant floors are almost always concrete, their smoothness depends on the material-handling equipment that is to be installed. The roof can be of many different shapes, such as dome, sawtooth, or flat, each having its own characteristics. The interior of a building should be pleasant and should have temperature and noise control, as well as proper illumination.

Organization within a plant dictates the command structure, which, in turn, may influence plant and office layouts. The chapter briefly discusses various organizational forms such as line, staff-line, product, and matrix. Each form has both benefits and drawbacks.

Communication is important in operating a plant efficiently, and a brief discussion of the subject is presented. Facilities available for communication can also influence the design and development of a plant.

Support facilities are necessary for the smooth operation of a plant, even though they require valuable space within it. For example, offices are needed for departmental foremen and support staff; inspection and maintenance departments require a place to work; employees need lockers, lavatories, and eating areas; room for automobile parking is essential and must be well planned. The basic information and nec-essary data for the various facets of planning support activities are provided in this chapter. The analysis performed using this information is essential in developing a total plant layout.

PROBLEMS

9.1 List the basic structural considerations that would be called for in a building constructed for:
   a. Heavy equipment use
   b. Light equipment use

9.2 Discuss the factors of importance in building the:
   a. Walls
   b. Floors
   c. Roof

9.3 What are the advantages of the following types of roofs?
   a. Sawtooth
   b. Dome
   c. Flat

9.4 A pleasant environment should be maintained for the comfort of workers. Discuss this in terms of:
   a. Temperature
   b. Ventilation
   c. Noise
   d. Lighting
   e. Colors

9.5 In the interior of a manufacturing plant, the colors white, yellow, and red are used to indicate certain features. Specifically, what do they point out?

9.6 What is "good" illumination within a plant? How does the type of work affect this requirement?

9.7 Contrast U.S. and Japanese organizational theories.

9.8 Use an organization chart to show the chain of command in the College of Engineering at a state university.

9.9 What are the advantages and disadvantages of line organization? What benefits does a matrix organization have over a line organization?

9.10 What is the purpose of communication? Why should a facility planner consider the need for communication?

9.11 A manager of a canned food company has a list of things to do. He needs to request a financial report from accounting, make inquiries concerning a new piece of equipment the company might be interested in purchasing, distribute information to the workers under his supervision concerning a change in policy, and confirm his appointment calendar with his secretary. As forms of communications, how should each of these matters be handled?

9.12 What are support facilities and why are they necessary? Describe each of the following support facilities, indicating why it is necessary and how it should be planned.
   a. Eating area
   b. Maintenance
   c. Parking
   d. Custodial service

9.13 How is the number of spaces required in the parking lot determined? How does transportation by trucks affect the parking lot requirement? (See Chapter 11 also.)

9.14 a. A company employs 350 people, 105 of whom are women. One quarter of the employees work in conditions in which they need to shower and change clothes before leaving work. How many showers and toilets should be provided?
   b. The same company has a cafeteria that serves lunch. There is only one lunch shift, and approximately 90 percent of the employees eat in the cafeteria. How much space should be provided for the kitchen and the lunchroom?
   c. Design parking lots for each of the following conditions for the same company if 75 percent of the employees drive to work. Include all dimensions. First, use 90-degree parking spaces. Then use 60-degree parking spaces. Finally, consider 40 percent of those cars to be small cars requiring spaces one foot less in length and width than the standard. Use either a 60-degree or a 90-degree arrangement.

SUGGESTED READINGS

Building

Estall, R.C., and Buchanan, R.D., *Industrial Activity and Economic Geography,* Hutchinson Co., London, England, 1980.

Lewis, B.T., and Marron, J.P., *Facilities and Plant Engineering Handbook,* McGraw-Hill, New York, 1973.

Spradlin, W.M., *Walker's Building Estimator's Reference Book,* The Company, Chicago, USA, 1986.

Organization

Miller, R.W., "IE's Face Challenges in Managing Human Resources to Improve Declining Productivity," *Industrial Engineering,* Vol. 17, No. 1, Jan. 1985, pp. 72–79.

Shannon, R.E., *Engineering Management,* John Wiley and Sons, New York, 1980.

Support Facilities and Requirements

Occupational Health and Safety Administration, *OSHA Safety and Health Standards,* U.S. Department of Labor, Washington, D.C., 1983.

Communication

Borman, E.G.; Howell, W.S.; Nichols, R.G.; and Shapiro, G.L., *Interpersonal Communication in the Modern Organization,* Prentice-Hall, Inc., Englewood Cliffs, N.J., 1982.

Fallon, W.K., *Effective Communication on the Job,* AMACON, New York, 1981.

Verderber, R.F., *Communicate!,* 4th edition, Wadsworth, Belmont, Calif., 1984.

Williams, F., *The New Communications,* Wadsworth, Belmont, Calif., 1984.

Material Handling: Principles and Equipment Description

One of the most important aspects in new plant development or in modification of an existing plant is thorough analysis of the material-handling system. Material handling can account for 30–75 percent of the total cost, and efficient material handling can be primarily responsible for reducing a plant's operating cost by 15–30 percent. How material is handled can determine some of the building requirements, department arrangements, and time needed to produce a unit. When an employee handles an item, he or she adds nothing to the product's value but does add to its cost. Planning the handling, storage, and transportation associated with manufacturing can reduce the cost of material handling considerably. In an assembly line system, for example, properly designed equipment will space the production along the assembly lines, bringing material to an operator at a steady rate and sending the part or subassembly to the next station after the operator has finished the task.

In this chapter an introduction to material handling is presented. The purpose of the chapter is twofold: to help the reader understand the relationship between material handling and plant layout as well as the complexity of designing such a system and to describe the more commonly used material-handling equipment. The information serves to illustrate the basics of material handling as applied to different environments such as a warehouse or a manufacturing plant.

10.1 DEFINITION OF MATERIAL HANDLING

Several definitions are available for material handling. The most comprehensive is the one provided by the Material Handling Institute

(MHI), which states: "Material handling embraces all of the basic operations involved in the movement of bulk, packaged, and individual products in a semisolid or a solid state by means of machinery, and within the limits of a place of business."

Even a cursory examination of the statement reveals that material handling involves much more than just moving the material by

---

This chapter has been written with contributions from Mohsen M.D. Hassan.

using machinery; several additional functions are implied in the system.

First, material handling involves the movement of material in a horizontal (transfer) and a vertical (lifting) direction, as well as the loading and unloading of items. Second, specifying that the movement of materials is "within a place of business" implies that the movement includes raw materials to work stations, semifinished products between work stations, and removal of the finished products to their storage locations. It also distinguishes material handling from transportation; the latter involves moving materials from suppliers to places of business or from places of business to customers.

Third, the selection of handling equipment is another activity in designed material-handling systems. Fourth, the term "bulk" indicates that the materials are to be moved in large, unpackaged volumes such as sand, sawdust, or coal. And fifth, using machinery for handling material is the preferred method even though the initial cost might be high. The use of human beings on a continuous basis is inefficient and can be costly; material-handling equipment soon pays for itself, especially in societies in which the cost of labor is high.

10.2 OBJECTIVES OF MATERIAL HANDLING

The need for study and careful planning of a material-handling system (MHS) can be attributed to two factors. First, as was mentioned before, material-handling costs represent a large portion of production cost. Second, material handling affects the operations and design of the facilities in which it is implemented. These then lead us to the major objective of MHS design, that of reducing production cost through efficient handling or, more specifically:

· To increase the efficiency of material flow by ensuring the availability of materials when and where they are needed.
· To reduce material-handling cost.
· To improve facilities utilization.
· To improve safety and working conditions.
· To facilitate the manufacturing process.
· To increase productivity.

This chapter will clarify how these objectives can be met.

10.3 MATERIAL-HANDLING EQUIPMENT TYPES

The backbone of an MHS is the handling equipment. A wide variety of equipment is available, each having distinct characteristics and cost that distinguish it from the others. All such equipment, however, can be classified into three main types: conveyors, cranes, and trucks. Each type has its own advantages and disadvantages, and some equipment is more suitable for certain tasks than others. This is mainly based on the characteristics of the material, the physical characteristics of the workplace, and the nature of the process using the equipment.

A brief discussion of the main equipment types is given here. A detailed description of the equipment is postponed to the latter sections of the chapter.

Conveyors

Conveyors are used for moving materials continuously over a fixed path. Examples of different types of conveyors are roller, belt, and chute conveyors.

Material-handling equipment. *Top left*: In this interior view of a covered conveyor system, which is 12′ in diameter and 500′ long, coal and waste wood are being moved from storage areas to boilers without spillage. *Top right*: The bridge crane shown here removes large logs directly from railroad cars and places them on trucks. *Bottom left*: Forklift trucks like this one are used in warehouse stores to stack merchandise efficiently and economically. *Bottom right*: Some trucks are customized for special uses in industrial applications. One example is the log-handling truck shown here. The jaws of this truck can move short wood logs from outside storage piles to a plant's conveyor system.

Advantages of Conveyors

· Their high capacity permits moving a large number of items.
· Their speed is adjustable.
· Handling combined with other activities such as processing and inspection is possible.
· They are versatile and can be on the floor or overhead.

· Temporary storage of loads between work stations is possible (for overhead conveyors in particular).
· Load transfer is automatic and does not require the assistance of many operators.
· Straight line paths or aisles are not required.
· Utilization of the cube (entire volume of the workplace) is feasible through the use of overhead conveyors.

Disadvantages of Conveyors

· They follow a fixed path serving only limited areas.
· Bottlenecks can develop in the system.
· A breakdown in any part of the conveyor stops the entire line.
· Since conveyors are fixed in position, they hinder the movement of mobile equipment on the floor.

Cranes and Hoists

Cranes and hoists are items of overhead equipment for moving loads intermittently within a limited area. Bridge cranes, jib cranes, monorail cranes, and hoists are examples of this basic equipment type.

Advantages of Cranes and Hoists

· Lifting as well as transferring of materials is possible.
· Interference with the work on the floor is minimized.
· Valuable floor space is saved for work rather than being utilized for installation of handling equipment.
· Such equipment is capable of handling heavy loads.
· Such equipment can be used for loading and unloading material.

Disadvantages of Cranes and Hoists

· They require heavy investment (especially bridge cranes).
· They serve a limited area.
· Some cranes move only in a straight line and thus cannot make turns.
· Utilization may not be as high as desirable since cranes are used only for a short time during daily work.
· An operator has to be available for operating some types, such as bridge cranes.

We find cranes being used in places such as shipyards and heavy equipment production facilities.

Trucks

Hand or powered trucks move loads over varying paths. Examples of such trucks include lift trucks, hand trucks, fork trucks, trailer trains, and automated guided vehicles.

Advantages of Trucks

· They are not required to follow a fixed path of movement and therefore can be used anywhere on the floor where space permits.
· They are capable of loading, unloading, and lifting, in addition to transferring material.
· Because of their unrestricted mobility, which allows them to serve different areas, trucks can achieve high utilization.

Disadvantages of Trucks

· They cannot handle heavy loads.
· They have limited capacity per trip.
· Aisles are required; otherwise the trucks will interfere with the work on the floor.
· Most trucks have to be driven by an operator.

· Trucks do not allow handling to be combined with processing and inspection as other types of equipment do.

10.4 DEGREES OF MECHANIZATION

A material-handling system can be completely manual or fully automated; different degrees of mechanization also exist between these two extremes. Classification of a handling system according to its level of mechanization is based on the source of power for handling and the degree of involvement of humans and computers in operating the equipment. The levels of mechanization can be classified as follows:

1. *Manual and dependent on physical effort.* This level also includes manually driven equipment such as hand trucks.

2. *Mechanized.* Power instead of physical effort is used for driving the equipment. Some trucks, conveyors, and cranes fall into this level. Here operators are needed for operating the equipment as opposed to providing the power.

3. *Mechanized complemented with computers* (an extension of the second level). The function of the computers is to generate documents specifying the moves and operations.

4. *Automated.* Minimal human intervention is used for driving and operating the equipment, and most of these functions are performed by computers. Examples include conveyors, automated guided vehicles, and AS/RS (automated storage/retrieval system). The equipment usually receives instructions from keyboards, push buttons, and tape or card readers.

5. *Fully automated.* This level is similar to the fourth level, but computers perform the additional task of on-line control, thus eliminating the need for human intervention.

The cost and complexity of designing the system increase as the degree of mechanization increases. However, efficiency of operations and labor savings can result.

The advantages of using mechanized and higher level systems include an increase in speed of handling operations, which in turn may decrease the overall production time; a reduction in fatigue and an improvement in safety; better control of material flow; lower labor cost; and better record keeping regarding inventory status of the material.

There are also some disadvantages as the degree of mechanization increases. For example, mechanization requires a high investment cost, training of operators and maintenance personnel, and specialized equipment and personnel, which reduces flexibility. It is therefore necessary to weigh advantages and disadvantages carefully before deciding which system to use. Changing from one mode of operation to another is always expensive and time consuming.

Unit loads used in the plant play an important part in defining the items to be moved. The degree of mechanization affects the unit load; conversely, the defined unit load also influences the degree of achievable mechanization.

10.5 THE UNIT LOAD CONCEPT

The unit load concept depends on the fact that it is more economical to move items and material in groups than individually. A unit load is defined as a number of items arranged such that they can be handled as a single object. This can be accomplished by palletization, unitization, and containerization.

Palletization is the assembling and securing of individual items on a platform that can be moved by a truck or a crane. Unitization is also the assembling of goods, but as one compact load. Unlike palletization, additional materials are used for packaging and wrapping the items as a complete unit. The unit load can be handled by trucks, conveyors, or cranes depending on its size and weight. Containerization is the assembling of items in a box or a bin. It is most suitable for use with conveyors, especially for small items.

Each unit load type is most suitable for certain situations. For example, a pallet is most suitable for stacking similar items that have regular shapes. Items that have different shapes and sizes can be grouped inside a container. In general, the factors that influence the selection of the unit load type are the weight, size, and shape of the material; compatibility with the material-handling equipment; cost of the unit load; and the additional functions provided by the unit load such as stacking and protection of material.

Using unit loads has both advantages and disadvantages. Among the advantages are the following. Using unit loads allows for moving large quantities of material, which reduces the frequency of movement and therefore reduces the handling cost. The ease of stacking helps achieve better space and cube utilization and promotes good housekeeping. There is greater speed in loading and unloading and a corresponding reduction of handling time. Protection against material damage is provided.

The disadvantages of using unit loads are as follows. Costs of the unit load can be high if a large number are required, especially if the containers are not reusable. Loading and unloading equipment that is different from what is available might be required. When used in shipping to suppliers, there is the problem of returning the empty pallets and containers if they are reusable.

10.6 PRINCIPLES OF MATERIAL HANDLING

Designing and operating a material-handling system is a complex task because of the numerous issues involved. There are no definite rules that can be followed for achieving a successful material-handling system. There are, however, several guidelines that can result in reducing the system cost and in enhancing its efficiency. These guidelines are known as the principles of material handling. They represent the experience of designers who have been working in the design and operations of handling systems. The twenty principles of material handling are listed in Table 10.1.

These principles can also be compiled slightly differently to suggest how the objectives are to be achieved. For example, to lower the handling cost, one should reduce unnecessary handling by properly planning material movement; by delivering units to the required place the first time without stopovers; by using proper material-handling equipment such as forklift trucks, pallets, boxes, and conveyors; by replacing obsolete equipment with new and more efficient systems when the savings justify it; and by reducing the ratio of dead weight (pallets, boxes) to payload. One can also use unit loads and move as many pieces at one time as possible.

One can increase productivity by minimizing machine operators' waiting time by delivering raw materials and subassemblies when needed and by maintaining a steady movement of work at a rate that will match the machine operator's rate.

One can make all workers more productive by eliminating nonproductive activities associated with proper material handling, by removing damaged parts from assembly lines before they reach the work station, by not using nonstandard and unique equipment, and by co-

Table 10.1 Principles of material handling

Principle	Description
1. Planning	Plan all material-handling and storage activities to obtain maximum overall operating efficiency.
2. Systems flow	Integrate as many handling activities as is practical into a coordinated system of operations covering vendor, receiving, storage, production, inspection, packaging, warehousing, shipping, transportation, and customer.
3. Material flow	Provide an operation sequence and equipment layout optimizing material flow.
4. Simplification	Simplify handling by reducing, eliminating, or combining unnecessary movements and/or equipment.
5. Gravity	Use gravity to move material whenever practical.
6. Space utilization	Make optimum utilization of the building cube.
7. Unit size	Increase the quantity, size, or weight of unit loads or flow rate.
8. Mechanization	Mechanize handling operations.
9. Automation	Provide automation to include production, handling, and storage functions.
10. Equipment selection	In selecting handling equipment, consider all aspects of the material handled, the movement, and the method to be used.
11. Standardization	Standardize handling methods as well as types and sizes of handling equipment.
12. Adaptability	Use methods and equipment that can best perform a variety of tasks and applications when special-purpose equipment is not justified.
13. Dead weight	Reduce the ratio of dead weight of mobile handling equipment to load carried.
14. Utilization	Plan for optimum utilization of handling equipment and manpower.
15. Maintenance	Plan for preventive maintenance and scheduled repairs of all handling equipment.
16. Obsolescence	Replace obsolete handling methods and equipment when more efficient methods of equipment will improve operations.
17. Control	Use material-handling activities to improve control of production, inventory, and order handling.
18. Capacity	Use handling equipment to help achieve the desired production capacity.
19. Performance	Determine the effectiveness of handling performance in terms of expense per unit handled.
20. Safety	Provide suitable methods and equipment for safe handling.

Courtesy of The Material Handling Institute, Inc.

ordinating material movement throughout the plant.

One can reduce floor space use by using material-handling equipment and production schedules that will require a minimum amount of stock on the floor; by storing material in spaces that do not hinder production (for ex-

ample, stock material should not be piled so close to a machine that it interferes wtih the operator's ability to perform); and by arranging the plant layout so that it will permit smooth flow of material between stations.

Accidents can be reduced by using material-handling equipment that has appropriate safety features for lifting and moving heavy materials and by using gravity to move the material whenever possible.

How do we apply these principles? Some applications are easily recognizable. For example, to apply the gravity principle, use chutes.

To apply the safety principle, reduce or eliminate manual handling that causes injuries. For the space utilization principle, stack items and use overhead equipment. According to the unit size principle, use containers and pallets to move groups of items. To apply the utilization principle, select equipment that is capable of performing several handling tasks in different areas, thus preventing it from being idle.

Compatibility of the Principles

The principles of material handling are compatible with each other and with the objectives of material handling. Achieving some of the principles will help achieve others. As examples, consider the following.

When applied properly, equipment selection and adaptability principles will help achieve the utilization principle, since in this case only the necessary equipment will be acquired, and thus will rarely be idle.

The mechanization principle will reduce manual handling, thus minimizing injuries and helping to achieve the safety principle.

Unit size and equipment selection principles help in achieving the space utilization principle. Using a unit load will permit stacking items, thus reducing the floor space required. Selecting overhead equipment will free some space on the floor for other purposes.

Difficulties in Applications of the Principles

The designers of MHS are usually advised to follow these principles. However, in some cases they might not be able to apply them to the full-

est extent because of factors such as the limitation on capital, physical characteristics of the building, and capability of the equipment.

Lack of the desired capital might prevent one from realizing a high degree of mechanization or from applying the maintenance principle. Building characteristics such as ceiling height, location of columns, and number and width of aisles can influence the gravity as well as the space utilization and material flow principles. Capability and type of equipment may affect application of the space utilization, unit sizing, and utilization principles.

10.7 MATERIAL-HANDLING COST

The main costs involved in designing and operating a material-handling system are:

· Equipment cost, which is comprised of the purchasing of the equipment and auxiliary components, and installation.
· Operating cost, which includes maintenance, fuel, and labor cost, consisting of both wages and injury compensation.
· Unit purchase cost, which is associated with purchasing the pallets and containers.
· Cost due to packaging and damaged material.

Reducing such cost is one of the primary objectives of a handling system. There are several ways of achieving this goal. For example, one can minimize the idle time of the equipment. High utilization of equipment will eliminate the need to acquire extra units. One can minimize rehandling of material and backtracking, thus reducing operating cost. One can arrange closely related departments near each other to result in material being moved only short distances. One can prevent excessive repairs by planning maintenance activities in advance. One should use proper equipment to reduce material damage and use unit loads whenever possible. The gravity principle should be used whenever possible, since it can reduce operating cost. One should eliminate unsafe practices by employees such as lifting heavy items; this will reduce injuries and consequent worker compensation. One can minimize the variations in equipment types, thus eliminating the need for an inventory of a variety of spare parts and their associated costs. One can replace obsolete equipment with new and more efficient ones when the savings justify it. As the reader probably realized by now, we are again applying the twenty principles of material handling.

10.8 RELATIONSHIP BETWEEN MATERIAL HANDLING AND PLANT LAYOUT

In a manufacturing system, no two other activities affect each other as do plant layout and material handling. The relationship between the two involves the data required for designing each activity, their common objectives, the effect on space, and the flow pattern. Specifically, plant layout problems require knowledge of the equipment operating cost in order to locate the departments in a manner that will minimize the total material-handling cost. At the same time, in designing a material-handling system the layout should be known in order to have the move length, move time, and source and destination of the move. Because of this dependency, many designers stress the need to solve the two problems jointly. However, the only feasible way is to start with one problem, use its solution for solving the other, then go back and modify the first problem on the basis of the new information obtained from the second, and so on until a satisfactory design is obtained.

Plant layout and material handling have the common objective of cost minimization.

The material-handling cost can be minimized by arranging closely related departments such that the material moves only short distances. Determining the flow pattern is a common concern in both problems and will be treated in a later section.

Additionally, material handling and plant layout influence one another in terms of space requirements and utilization. Trucks that are compact in size and have the capability of side loading do not require wide aisles. Overhead equipment does not occupy any space on the floor of the layout. Stacking items as high as possible by using the appropriate unit load will help to reduce the space taken by these activities and best utilize the cube. Trucks that have the ability to lift to a high level will assist in achieving cube utilization. It can also be achieved by using mezzanines, carousels, and high-rise storage for storing material.

Finally, the physical characteristics of the building such as aisle width, ceiling height, and columns will affect equipment selection and its routing.

10.9 MATERIAL-HANDLING SYSTEM DESIGN

In production systems, developing an MHS is one of the most crucial and difficult tasks a designer faces owing to its impact on the cost and efficiency of operations. There are no standard steps that can be followed, although the general procedures of designing a system can be adopted. The procedure is iterative; the analyzer has to go back and forth between the different steps until a satisfactory design has been obtained and can be implemented. Experience and sound judgment are indispensible during this process.

Developing a material-handling system involves selection of material-handling equipment, selection of a unit load, and assignment of the equipment to the moves and determining their routes. These three components of the design have been expressed in what is known as the material-handling equation. This equation is a nonquantitative relationship between the three components and is expressed as follows:

Material + moves = methods

Questions concerning the materials involve their type, size, shape, quantity, and weight. Those concerning the moves involve their origin and destination, length, frequency of movement, and duration of the move. (This part of the design is closely related to plant layout.) Determining the material-handling methods—that is, equipment and unit load—depends on the information obtained from studying the material and the moves. Because of the heavy investment associated with equipment that render irreversible the decisions that have been made and implemented, selection of the equipment is usually the most crucial and difficult task of the design.

10.10 DILEMMA OF AN ANALYZER

The difficulties in designing a material-handling system, particularly in equipment selection, are numerous and can be attributed to several factors associated with subjective analysis and/or application of analytical models to the design. Most important among these factors are the following.

1. The interrelation between material handling and plant layout as was explained previously is highly significant.

2. The existence of a large variety of equipment with different capabilities and limitations that renders evaluating and comparing all of them an impossible task. Undoubtedly, this will result in ignoring some good alternatives.

3. Some of the characteristics of the equipment are difficult to quantify, either for comparison between different types of equipment or for inclusion in an analytical model. Examples of these characteristics are equipment flexibility and maneuverability and the effort required to move the material.

4. While the most common objective of the design is to ensure a system with minimum cost, other objectives have to be considered. Notable among these are maximum utilization of the equipment, reducing the variation in equipment types, and safety of employees.

5. Application of an analytical model to the design problem might require an inordinate amount of calculations with no guarantee of an optimal system design. For example, equipment selection, when performed analytically, will result in a maximum of pq combinations to evaluate, where p and q are the number of equipment types and moves, respectively. If there are many moves and/or several types of equipment to be considered, pq can get quite large.

6. Some of the data required for designing the system, such as move time and equipment-operating cost, cannot be known exactly unless the system is already operating. Therefore those data must be estimated, which is not an easy task, and a poor estimate can result in several modifications after the system has been installed and is operating.

10.11 SPECIFICATIONS OF THE DESIGN

Generally speaking, the steps to be followed in designing a handling system are as follows, assuming the layout has already been prepared.

1. State the intended function of the handling system—whether it is for a warehouse whose function is storing, packaging, inspec-

tion, and shipping to customers or for a manufacturing system where the function is to move items or partial assemblies from station to station. Knowing the type of manufacturing system (product, process, group technology, or any other type) is very helpful here.

2. Collect the necessary data about the material such as its characteristics and the quantities involved. Data regarding quantity may be summarized in the form of a chart.

3. Identify the moves, their origin and destination, their path, and their length.

4. Determine the basic handling system to be used and the degree of mechanization desired. Here, an idea is to establish whether a conveyor, a truck, or a crane will be best suited for the situation.

5. Perform an initial screening of suitable equipment and select a set of candidate equipment among them. Evaluate the candidate equipment on the basis of such measures as cost and utilization. Always match the equipment with the material characteristics.

6. Select a set of suitable unit loads and match them with material and equipment characteristics.

These steps are more easily stated than followed and must be repeated several times until compatibility between the components of the material-handling equation is assured. Several factors influence the designer in making decisions at each of these steps. Important factors are stated briefly here.

· Costs of equipment and unit loads and availability of funds. This factor will affect the degree of mechanization achieved in the design.
· Physical characteristics of the building and the available space. Aisle width and number will be affected by the available space, which in turn will influence decisions regarding mobile equipment. Over-

head equipment may or may not be considered, depending on the height of the ceiling.
· Management attitude toward safety and employee welfare, which will affect the degree of involvement of material-handling personnel in manual handling.
· Degree of involvement between handling and processing.

During the design of the system, the objectives and the principles of material handling should be kept in mind. Achieving as many of the objectives and principles as possible will result in a satisfactory and efficient design. While there are no benchmarks against which to measure the quality of the design, a good material-handling design should possess most or all of the following characteristics:

· Well planned
· Handling combined with processing whenever possible
· Mechanical whenever possible
· Minimum manual handling
· Minimum handling by production personnel
· Safe
· Protection of material provided
· Minimum variation in equipment types
· Maximum utilization of equipment
· Minimum backtracking, handling, or transferring
· Minimum congestion or delay
· Economical

10.12 ANALYZING AN EXISTING MATERIAL-HANDLING SYSTEM

To analyze an existing material-handling system means to determine whether it is functioning efficiently without creating any bottlenecks or excessive inventories and is transporting the units when and where needed. The problems in an existing material-handling system will be evident if one can observe one or more of the following symptoms in the system (courtesy of The Material Handling Institute, Inc.):

· Backtracking in material flow path
· Built-in hindrances to flow
· Cluttered aisles
· Confusion at the dock
· Disorganized storage
· Excess scrap
· Excessive handling of individual pieces
· Excessive manual effort
· Excessive walking
· Failure to use gravity
· Fragmented operations
· High indirect labor costs
· Idle machines
· Inefficient use of skilled labor
· Lack of cube storage
· Lack of parts and supplies
· Long hauls
· Material piled up on the floor
· No standardization
· Overcrowding
· Poor housekeeping
· Poor inventory control
· Product damage
· Repetitive handling
· Service areas not conveniently located
· Trucks delayed or tied up
· Two-person lifting jobs

It is also a good idea to examine the entire material-handling system in the plant with a checklist similar to the one shown in Table 10.2 and identify the problems. Again, the checklist is developed by MHI; however, one can develop a more specific checklist for the particular plant in question.

Once the problem areas have been identified, they must be reexamined for possible improvements. In performing a study, some basic questions must be asked, such as: Why? What? Where? When? How? Who? A process chart,

described in Chapter 7, is very helpful in making such an analysis.

Why is this activity taking place? For example, moving an incoming unit from receiving to inspection and then back to receiving could be avoided if the final quality assurance inspection takes place at the supplier and we can rely on the supplier's data.

The *What* question is associated with understanding the type of material to be handled. The material to move and its frequency could define the type of equipment and accessories that might be used in making such material moves. For example, small discrete units could be handled by a conveyor or by a forklift by unitizing, while large bulky units such as pressure vessels might need a crane. Parts and material lists and production schedules are good sources of this information.

Where describes the data associated with the move. The travel path, distance, and any physical limitations should be noted. Process charts, flow diagrams, and scale models of the plant are some of the sources from which these data can be gathered.

When defines the time at which the material is to be moved and within what time span the move must be made. Its answer provides the speed and frequency with which the material-handling system must operate.

How indicates the method for the move. By analyzing the method, inefficient and expensive aspects are revealed. Operations charts and time study data are useful in this analysis.

Who refers to the person responsible for material handling as well as the manpower required for such a task. Sufficient manpower is necessary for efficient working; and though the degree of mechanization determines the manpower, the material-handling workers should also realize their responsibility and importance in the smooth operation of the plant.

Asking these questions about the material-handling system, especially the problem areas, reveals the nature of overall difficulties.

Table 10.2   Material-handling checklist

— Is the material-handling equipment more than ten years old?
— Do you use a wide variety of makes and models that require a high spare parts inventory?
— Are equipment breakdowns the result of poor preventive maintenance?
— Do the lift trucks have to go too far for servicing?
— Are there excessive employee accidents due to manual handling of materials?
— Are materials weighing more than 50 pounds handled manually?
— Are there many handling tasks that require two or more employees?
— Are skilled employees wasting time handling materials?
— Does material become congested at any point?
— Is production work delayed owing to poorly scheduled delivery and removal of materials?
— Is high storage space being wasted?
— Are high demurrage charges experienced?
— Is material being damaged during handling?
— Do shop trucks operate empty more than 20 percent of the time?
— Does the plant have an excessive number of handling points?
— Is power equipment used on jobs that could be handled by gravity?
— Are too many pieces of equipment being used because their scope of activity is confined?
— Are many handling operations unnecessary?
— Are single pieces being handled where unit loads could be used?
— Are floors and ramps dirty or in need of repair?
— Is handling equipment being overloaded?
— Is there unnecessary transfer of material from one container to another?
— Are inadequate storage areas hampering efficient scheduling of movement?
— Is it difficult to analyze the system because there is no detailed flowchart?
— Are indirect labor costs too high?

Courtesy of The Material Handling Institute, Inc.

In most instances the answers suggest the mode of improvement and/or provide guidance as to where further in-depth study would be appropriate.

10.13 PRODUCTIVITY RATIOS

Every system within a manufacturing plant must work efficiently to reduce the production cost and be competitive. Several typical productivity ratios are used as indicators of the performance of a system. Some such ratios are listed below; however, the list is by no means exhaustive, and an organization often develops its own ratios that more clearly identify the fac-tors with which it is concerned. These ratios are monitored periodically to be sure that their values are within acceptable tolerances. Any wild fluctuation from period to period is a clear signal that the system is not working smoothly and needs further investigation and control. Though some ratios do not specifically address productivity associated with material handling, all the ratios are stated here for completeness. They should be reviewed when we discuss topics such as storage and energy consumption later in the text. Most factors listed below are defined by MHI and require little or no explanation. They are listed here courtesy of The Material Handling Institute, Inc. Collection of the required information is likely to present problems, at least initially. These can probably be overcome in a given plant once the guidelines are established to identify and quantify the information necessary to calculate each ratio used. Since several departments might be involved, interdepartmental cooperation is essential in forming the guidelines and collecting the data.

Material Handling–Labor (MHL) Ratio

$$MHL = \frac{\text{Personnel assigned to material handling}}{\text{Total operating personnel}}$$

The personnel assigned could be measured in terms of a number of full-time workers or a dollar expenditure. If a person is not performing a material-handling task full time, a percentage of the person's work associated with material handling must be estimated and used in the above calculations. The ratio should be less than 1, and a reasonable value would be less than 0.30 in a plant, while in a warehouse a higher value should be expected.

Handling Equipment Utilization (HEU) Ratio

$$HEU = \frac{\text{Items (or load weight) moved per hour}}{\text{Theoretical capacity}}$$

Ideally, the ratio should be close to 1.0; however, equipment breakdown, poor scheduling, poor housekeeping, and building geography can reduce the load movement.

Storage Space Utilization (SSU) Ratio

$$SSU = \frac{\text{Storage space occupied}}{\text{Total available storage space}}$$

If the storage areas, such as bins or racks, are only partially full, then the percent of utiliza-

tion should be estimated and included in the calculation. A value close to 1 indicates assignment of appropriate space for the storage activities.

Aisle Space Percentage (ASP)

$$ASP = \frac{\text{Space occupied by aisles}}{\text{Total space}}$$

The ASP should have a value between 0.10 and 0.15.

Movement/Operation (MO) Ratio

$$M/O = \frac{\text{Number of moves}}{\text{Number of productive operations}}$$

The ratio indicates the amount of material handling performed. The moves involved may consist of material moved from receiving, from storage to an operation and back to storage, and so on. A high value indicates room for improvement.

Manufacturing Cycle Efficiency (MCE)

$$MCE = \frac{\text{Time in actual production operations (machine time)}}{\text{Time in production department}}$$

Time not spent in production could be caused by delays in material movement, poor scheduling, machine failure, and storage limitation, among other items. For increasing machine utilization, the delay should be eliminated or at least minimized. The performance index should be observed over a time period for consistency.

Damaged Loads (DL) Ratio

$$DL = \frac{\text{Number of damaged loads}}{\text{Total number of loads}}$$

The ratio measures the quality performance of

material-handling personnel. Damage to the loads during receiving, in-process movement, and shipping should be minimized.

Energy Ratio (ER)

$$\text{Energy ratio} = \frac{\begin{array}{c}\text{Total BTU consumption}\\\text{in the warehouse}\end{array}}{\text{Warehouse space}}$$

The energy ratio measures the efficiency of heating and cooling operations. Some of the ways in which this ratio can be improved are reducing heating or cooling of an unmanned portion of the warehouse, turning lights off when not needed, and using lights on moving vehicles rather than permanent lighting.

10.14 EQUIPMENT USED FOR MATERIAL HANDLING

A wide variety of equipment is available for material handling. Which equipment to use under what conditions depends on the responsible person's judgment and knowledge of the machines and costs associated with the task of moving the materials. An economic analysis can be performed to justify the selection of a particular machine, and we will illustrate this approach by examples in Chapter 11.

Equipment Types

The equipment can be characterized by the area it is intended to serve:

1. Between fixed points over a fixed path

   a. Belt conveyor
   b. Roller conveyor
   c. Chute conveyor
   d. Slat conveyor
   e. Screw conveyor
   f. Chain conveyor
   g. Overhead monorail conveyor
   h. Trolley conveyor
   i. Wheel conveyor
   j. Tow conveyor
   k. Bucket conveyor
   l. Cart-on-track conveyor
   m. Pneumatic tube conveyor

2. Over limited areas

   a. Hoists
   b. Overhead cranes
   c. Hydraulic scissors lift

3. Over large areas

   a. Handcart/truck
   b. Tier platform truck
   c. Hand lift truck/pallet jack
   d. Power-driven handtruck
   e. Power-driven platform truck
   f. Forklift truck
   g. Narrow-aisle truck
   h. Tractor-trailer train
   i. Material lift
   j. Drum truck
   k. Drum lifter
   l. Dolly
   m. Automated guided vehicle system

Equipment Description

Each of the equipment types listed above is discussed in some detail in this section. A brief description is provided, followed by a listing of the equipment's principal characteristics and/or applications.

Belt Conveyor

A belt conveyor is an endless belt, driven by power rollers or drums at one or both ends and supported by flat beds or rollers. These rollers can produce a flat conveyor belt or a trough conveyor. The belt can be made of rubber, woven wires, metal, or fabric, depending on the load to be carried. On special occasions, to carry ferrous metal or to separate ferrous from

Equipment used for material handling between fixed points over a fixed distance. *Top left*: An overhead rail conveyor can be used to move material between work stations and provides temporary storage and a time delay, allowing cooling and drying of the material during transport. *Top right*: In a powered wheel conveyor, the wheels are mounted on a shaft to transport light loads. Here, circuit boards are transported between work stations. *Middle left*: A belt conveyor carries partial assemblies between work stations in an assembly line. *Middle right*: A power roller conveyor can carry heavy loads in warehousing and storage systems. *Bottom left*: Here, a trolley conveyor carries shelves loaded with parts. Height is adjusted so that a worker seated on a stool can reach a workpiece in one of the trolleys. *Bottom right*: A chain conveyor transports logs. It can lift bulky material up an inclined trough.

other types of metals, the belt may have a magnetic bed. The belt can be vibrated to feed assembly parts, position items, and deliver small amounts of bulk material.

Characteristics/Applications:

- The belt can operate at the horizontal or on an incline of up to 30 degrees.
- A flatbed belt can be used to carry light objects in assembly lines.
- Roller belts can be used to carry heavy boxes, bags, or other containers in warehousing and storage operations.
- Trough rollers can be used to carry bulk material such as coal and raw materials.
- Belt speed can be adjusted from 2' to 300' per minute.
- Belt width can be from 12" to 36" with a capacity of 300–1500 pounds per linear foot.

Roller Conveyor

A roller conveyor consists of rollers attached to side rails supported by a steel frame. The load is carried on the rollers, each of which rotates about a fixed axis. The type of roll (steel, rubber, or wood), its shape (cylindrical or wheel, the latter sometimes called a wheel conveyor), and its spacing depend on the load being carried. The conveyor can be either gravity-operated or power-driven. The gravity-operated conveyor has a slight downward slope, allowing material to move because of gravitational force. On the power-driven conveyor, some of the rollers are driven by chains or belts to provide the motion for the material on the conveyor.

Characteristics/Applications:

- Material can be moved between work stations.
- The height can be adjusted to the level of the work area.

- Loads must have a firm, even base.
- Uneven and fragile objects can be carried in boxes, containers, or pallets placed on the conveyor.
- The width can be between 7" and 51", with a capacity of 460–25,000 pounds per linear foot. Conveyors can be bought in sections of 5' to 10'.

Chute Conveyor

A chute conveyor is a slide, generally made of metal, that guides materials as they are lowered from a higher-level to a lower-level work station. The shape of the chute can be straight or spiral to save space.

Characteristics/Applications:

- Materials (boxes, packages) are moved a short distance because of gravity.
- The chute can have a door that controls the flow of the work items (usually in batches).
- This type of conveyor is very inexpensive and provides an efficient means of connecting conveyors at different levels.
- The chute can occasionally be jammed if the design is not proper for the size and shape of the objects being transported.
- The diameter of spiral can be between 18" and 48", and the length can be customized.

Slat Conveyor

A slat conveyor is a moving surface that is made up of slats attached to power-driven chains. The slats move over rollers at each end of the conveyor to form a closed loop.

Characteristics/Applications:

- Heavy, uneven loads can be placed directly on the slats.
- Operation can be horizontal or can be inclined as much as 40 degrees.

· Slat conveyors can be used as a series of platforms in assembly operations.
· Sanitation is better than for other conveyors. Slat conveyors are mainly used for bottling and canning owing to ease of cleaning.
· Slat size can be 3¼" to 7½" for plastic slats and 3¼" to 12" for steel slats, with slats arranged ½" to ⅞" apart.

Screw Conveyor

A screw conveyor is a large spiral or screw contained in a trough or tube. The rotating movement of the screw moves the material in a spiral along the path of the trough or through the tube.

Characteristics/Applications:

· It can be used horizontally or on an incline.
· Because of the small volume, it can be used in cramped space.

Chain Conveyor

A chain conveyor is an endless chain directly carrying loads, sometimes located at the bottom of a trough.

Characteristics/Applications:

· It is useful in moving tote boxes and pallets.
· A chain conveyor can pull bulk material along a trough.
· Generally, the length can be between 10' and 100' with a capacity of 300–3000 pounds per linear foot.

Overhead Monorail

An overhead monorail is a track to transport carrying devices such as trolleys and hooks. The track itself can form a closed loop, and each trolley can be either powered or manually operated.

Characteristics/Applications:

· Trolleys can be independently powered and controlled by a hand device.
· The monorail can be designed to carry heavy objects.
· Monorails are often used in transporting units to a spray paint booth or a baking oven where a uniform rate of travel is necessary with the entire unit suspended in the air.
· Overhead conveyors are generally 8' to 9' from the floor.

Trolley Conveyor

A trolley conveyor is a closed-loop, overhead track with an endless chain carrying uniformly spaced trolleys that support the loads.

Characteristics/Applications:

· Loads can move horizontally or at a significant incline.
· This type of conveyor can be used for overhead storage, thereby saving floor space.
· Degreasing, spray painting, assembly, and packaging are the most common applications.
· The conveyor can have a free, unpowered track to which trolleys can be shifted when accumulating is important.
· Conveyors are generally 8' to 9' from the floor and can be functional in vertical, horizontal, or inclined positions.

Wheel Conveyor

The wheel conveyor is similar in function to the roller conveyor; but instead of long rollers, wheels are mounted on a shaft.

Characteristics/Applications:

· The spacing of the wheels depends on the load to be transported.

- Wheel conveyors are more economical than roller conveyors.
- This type of conveyor is used for light loads.

Tow Conveyor

A tow conveyor is a tow line and the trolleys, trucks, or dollies it pulls over a fixed path.

Characteristics/Applications:

- The tow line can be overhead or in the ground.
- The lines can be installed so as to permit automatic switching from one tow line to another.
- This type of conveyor is generally used in frequent trips of long distances.
- Carts are 3' × 5' or larger.

Bucket Conveyor

In a bucket conveyor, buckets are evenly spaced on a chain moving between levels. The buckets automatically tip at the top or bottom of the conveyor to deliver the material.

Characteristics/Applications:

- This type of conveyor can be used for moving raw materials.
- The chain can be vertical or inclined.

Cart-on-Track Conveyor

In a cart-on-track conveyor, carts are spaced along a track with a tube beneath the carts to move them along by rotating. A drive wheel rotates the tube.

Characteristics/Applications:

- Carts can be accumulated when necessary.
- Independent control of each cart is possible.

Pneumatic Tube System

A pneumatic tube system consists of a cylinder in which messages or small items are carried over a predetermined path by compressed air or vacuum.

Characteristics/Applications:

- Such a system is designed to transport lightweight objects (small tools, dies, gages, money, and messages) rapidly to and from stations within toolrooms, banks, offices, lumberyards, and similar locations.
- The requirement for a constant pressure or vacuum results in high operating and maintenance costs for large, complicated systems with multiple branches and stations.

Hoist

A hoist is a lifting device attached to monorails, cranes, or a fixed point. A hoist can be powered manually or by electric or pneumatic motors. Strictly speaking, a hoist is the lifting device itself; however, by general usage it is frequently named by the kind of crane to which it is attached. There are three major types.

1. *Chain hoist:* A hoist that serves a fixed spot directly beneath the hoist. A hand chain hoist has a capacity lift from 250 pounds to 30 tons with a standard lift of 7' to 10'. An electric chain hoist has a capacity of 500 pounds to 50 tons with a lift of 10' to 35'.

2. *Monorail hoist:* A hoist that is free to move along an overhead rail serving any spot under the track.

3. *Jib hoist:* A hoist that serves any area circumscribed by the jib in 360-degree rotation (a boom extending from, and free to pivot around, a fixed vertical post). The height can be adjusted from 4' to 30'.

Equipment used for material handling over limited areas. *Top left*: In a jib hoist, the boom that lifts the material may pivot 360 degrees around a fixed post. Here, heavy subassemblies are being unloaded from pallets and placed on conveyors. *Top right*: A bridge crane is mounted on a pair of tracks and can lift and transport items up, down, sideways, and forward and back. This particular crane is used to change mold-dies in the machines. *Bottom left*: Gantry cranes are most suitable for heavy outdoor lifting. This crane is used for loading and unloading large steel beams. *Bottom right*: Tower cranes like this one can reach great heights and are commonly used in building construction.

Characteristics/Applications:

· Hoists are primarily intended for transferring moderately heavy objects a short distance.
· Hoists are also used to suspend a workpiece while various operations are being performed.

Overhead Traveling Crane/Bridge Crane

This is an overhead handling unit resembling a bridge that is mounted on a pair of tracks, traveling lengthwise. Within the bridge is a cable and hoist that can be positioned at any point along the bridge.

Characteristics/Applications:

· Such a crane covers all the area within the rectangle over which it travels. It provides three-dimensional coverage while moving up and down, sideways, and lengthwise.
· With various accessories such as buckets, electromagnetic disks, and chains, it can handle almost any material from light tools to heavy, flat metal plates.
· Such cranes can have capacities from ¼ ton to 50 tons and can lift from 10′ to 30′.

Variations of bridge cranes are the following.

1. *Stacker crane:* A crane that, instead of hoisting, uses platforms with forks. It is used mainly in storing and retrieving a unit load (palletized materials, boxes, or other containerized loads).

2. *Tower crane:* A crane that is used mainly on large construction projects. It consists of a hoist that travels on a horizontal boom attached at one end to a vertical post. The other end of the boom is supported by a guy line to the top of the post. The boom itself can be rotated 360 degrees around its post.

3. *Gantry crane:* A large traveling crane supported by towers or side frames running on parallel tracks. It is used mainly in heavy outdoor activities such as loading and unloading on shipping docks, shipbuilding, and the like.

4. *Jib crane:* A crane that can travel on a horizontal boom, which may be mounted on a column, or mast. The mast can be fastened to the floor or roof, or the boom can be fastened directly to the wall brackets or rails on the wall.

Jib cranes can rotate 360 degrees, are inexpensive and versatile, and are used in loading and unloading individual work stations or the material-handling carriers.

Hydraulic Scissors Lift Equipment

Hydraulic scissors lift equipment consists of scissor legs and hydraulic cylinders with a platform on top to travel in a vertical direction from the ground up to about 10′.

Characteristics/Applications:

· This equipment is used for raising, lowering, and supporting heavy loads during loading and unloading.
· It is primarily permanently located but may be moved short distances.
· It is designed to serve in transfer operations employing forklift trucks or other powered equipment, or as an industrial heavy-duty work positioner.
· It can also be used as an adjustable loading dock.

Handcart/Truck

A handcart or handtruck is a wheel-mounted platform with handles to manually push or pull the unit.

Characteristics/Applications:

· It is the simplest and most inexpensive method of transporting a load.

Equipment used for material handling over large areas. *Top left*: A handtruck is the simplest, least expensive way to move small loads short distances. *Top right*: A forklift truck, which can lift and carry heavy loads on pallets, is especially useful in warehouse storage. *Bottom left*: A high-rise forklift truck is adapted to hydraulically lift heavy loads for high-level storage. *Bottom right*: In this automated guided vehicle system, the computer-controlled vehicle follows a buried guide path according to preprogrammed instructions or coordination by a more sophisticated computer.

· It is used to move material a short distance with frequent stops for loading and unloading.
· It can be used for storage of materials between operations.
· It requires smooth and mainly level floors.

· The platform can have a capacity from 200 to 10,000 pounds.

Tier Platform Truck

A tier platform truck is a handtruck with one or more additional platforms stacked vertically.

Characteristics/Applications:

- It is used to carry light parts or materials in large quantities.
- Additional platforms increase both storage and handling capacities.
- There are from one to five trays with a capacity of 150 to 1800 pounds.

Hand Lift Truck/Pallet Jack

A hand lift truck or pallet jack is a hand-operated truck that can raise loads hydraulically or mechanically to clear the floor before transporting them to the desired destination(s).

Characteristics/Applications:

- Operators can transport heavier objects more easily than with a handtruck.
- Applications are similar to those of a handtruck or tier truck.
- It is used mainly in moving material into and out of storage areas on pallets or skids.
- Its capacity can range from 500 to 1000 pounds, with the load center from 8″ to 15″ from the frame. Vertical movement is from 64″ to 136″.

Power-Driven Handtruck

A power-driven handtruck is similar to the hand lift truck, except that it is driven by a battery-operated electric motor.

Characteristics/Applications:

- It is used when the distances are greater than can be conveniently negotiated by hand lift trucks, about 150′ to 300′.
- It can be used with a slightly inclined floor.
- It provides greater productivity than hand-driven trucks.

- The operator walks behind the truck.
- Its capacity is from 500 to 6000 pounds, with a lift range from 0″ to 54″ and a speed of 4–6 miles per hour. It can operate on 12-V or 24-V batteries.

Power-Driven Platform Truck

A power-driven platform truck is a much larger device than the power-driven handtruck. It carries both load and operator. Power is supplied by a diesel or gasoline engine or by an electric motor.

Characteristics/Applications:

- It can carry a much heavier load for a longer distance; about 400′ to 500′ is the normal application.
- It is often used in maintenance and heavy storeroom work.
- Its capacity varies from 200 to 2000 pounds.

Forklift Truck

A forklift truck is an operator-ridden, power-driven truck with forks in front that lift and carry heavy loads on skids or pallets.

Characteristics/Applications:

- It can carry loads weighing thousands of pounds; a load of 100,000 pounds is not unusual.
- It can lift loads on skids as high as about 25′.
- It can be used to load and unload trucks or railroad boxcars.
- It is used extensively in storage and warehousing.
- It generally requires an aisle 10′ to 12′ wide.
- Its capacity is from 1000 to 100,000 pounds.
- It can be operated on gas or electricity.

Narrow-Aisle Trucks

Narrow-aisle trucks are variations of industrial trucks that are specifically designed for aisles, in which regular industrial trucks are too wide to operate.

Characteristics/Applications:

- Less aisle space is necessary (5′ or 6′).
- Narrow-aisle trucks are more maneuverable than regular industrial trucks.
- Electricity or gas powers them.
- Their capacity is from 1000 to 100,000 pounds, with a lift height up to 30′.

Variations of the narrow-aisle truck are the following.

1. *Side-loader truck:* A fork truck with forks on the side rather than the front.

2. *Straddle truck:* A fork truck with outriggers to balance loaded trucks.

3. *Reach truck:* A fork truck with telescoping forks to reach loads that are set back.

4. *Order-picker truck:* A truck with a platform that lifts the operator to the desired level to choose the load desired. It is used mainly when only part of a pallet is desired.

5. *Turret truck:* A fork truck with forks that can rotate left or right to place or pick up a load without the truck having to turn in the aisle.

Tractor-Trailer Train

A tractor-trailer train is a series of carts pulled by a self-propelled tractor.

Characteristics/Applications:

- A tractor-trailer train is used for long-distance moves.
- It allows for a variety of load movement, such as platforms, bins, and racks.

- It allows many trains to be driven by the same tractor. The tractor can be disconnected and used to pull another train while the first is being loaded or unloaded.
- It can carry relatively heavy loads.
- It is mainly used for stop-and-go operations carrying loads from different points and delivering them to various destinations.
- It is usually used where the traveling distance is between 200′ and 300′.

Material Lift

A material lift is a handtruck with a winch or hydraulic lifting mechanism for relatively small loads.

Characteristics/Applications:

- It is manually moved.
- It has a variety of uses in warehouses, storerooms, offices, and shops where materials are lifted and moved from one location to another.
- It is easily maneuverable.
- It can be used to stack small items.
- Its lifting capabilities vary from 2′ to 6′.
- Its capacity is from 20 to 1000 pounds, with a lift height up to 78″. The width can vary from 15″ to 24″.

Drum Truck

A drum truck carries bulk like a handtruck, but it is specifically designed for drums. It has a rectangular, upright, open-type frame with two or four wheels. It has nose tips at the bottom for placing under drums.

Characteristics/Applications:

- It is manually moved.
- One person can easily handle a full drum.
- It is used on smooth, level surfaces.
- Four wheels allow the truck to stand with-

out support when either loaded or unloaded.

· It can move drums onto or off of pallets.
· It may have a winch or hydraulic lift to lift the drum.
· It has a capacity of up to 1000 pounds.

Drum Lifter

A drum lifter is a drum-handling unit with gripping arms that slide onto the forks of a lift truck.

Characteristics/Applications:

· It allows the lift truck operator to pick up, transport, and deposit drums without assistance and without leaving the operator's seat.
· It provides no-tilt lifting of closed-head or open-head drums, with or without the tops in place.

Dolly

A dolly is a horizontal platform or open-type frame with wheels attached to the underside. It is used for transporting relatively light weights and low volumes short distances.

Characteristics/Applications:

· It is manually operated.
· It is primarily for use on level surfaces.
· It is highly maneuverable.
· Dollies come in various shapes and sizes for specific uses such as moving pallets, drums, cabinets, and heavy machinery.
· It can be designed to pry heavy objects.
· It is inexpensive.
· It provides easy one-person movement of objects.
· It is stable.
· There are two-wheel and four-wheel models with capacities from 500 to 14,000 pounds. Dollies can weigh from 6 to 127

pounds and may be made of aluminum, plastic, or steel.

Automated Guided Vehicle System

The use of computer controls has been extended to material-handling tasks. A system operator at a station can control the vehicles by causing them to move along a predetermined path and to perform certain duties. Such a setup is known as an automated guided vehicle (AGV) system.

The guided vehicle can take any of several forms including a flat load carrier similar to a dolly, pallet trucks, automatic forklifts, side-loading vehicles, and robot arms. These machines can load themselves, travel to their destinations, unload, and return to the place of origin or any other desirable location. For example, the pallet lift could be programmed to pick up a pallet in storage, bring it to the shipping dock, and return for another load. Human interaction is held to a minimum, thereby reducing labor costs, human error, and the possibility of injury.

AGV systems capable of handling up to 10 tons of weight are commonly available today. In addition, a recently developed system that floats on a microthin film (about 0.003") of air between vehicle and floor can carry heavy loads of up to 35 tons. The frictionless design permits movement of loads on rough floors and can eliminate the use of cranes and bridge structures, which otherwise would be necessary for moving such heavy loads. These vehicles can also be maneuvered manually, requiring only about 1 pound of force for each 2000 pounds of carrying load.

There are two basic types of automated guided vehicles: "dumb" and "smart." They are categorized on the basis of how much control is given the guide paths and other elements outside the vehicle and how much control is in the vehicle itself. A system that has the entire control outside the vehicle is built on a zone arrangement. The routes are divided into zones, and the vehicle is not allowed to enter a region unless other traffic in the zone is cleared. A vehicle can be made smart by using on-board microprocessors and radio transmitters. Different levels of communication between outside control and vehicles can be developed. For example, the destination can be punched into the vehicle's microprocessor, and it will travel along the route, communicating with other vehicles by radio to avoid collision. A more sophisticated system can have a host computer controlling the movement of all vehicles. The computer is instructed as to what is to be moved, and then it selects the available vehicle closest to the job and issues orders to the vehicle to perform the task. The net result is good coordination and utilization of vehicles as well as good record keeping on the movement of the goods.

Ray Kulwiec (1984) lists the following specific advantages of an AGV system.

1. *Material control:* More accurate accounting of the use and transfer of material is possible as it is transferred between designated points, reducing a need for large safety stock.

2. *More efficient use of personnel:* Many forklift truck operators are no longer required; this reduces the overall number of material-handling personnel and the necessary coordination between them.

3. *Efficient work environment:* AGVs allow loading and unloading of each station independently of other stations, thus permitting each operator to work at his or her own pace and not limiting the operator to the speed of the conveyor. Many AGVs have movable platforms that allow the work to be delivered at the desired height.

4. *Flexibility:* The routes can be changed and new ones added with considerably greater ease than can be accomplished with systems of fixed bed conveyors.

5. *Better use of floor space:* Buried guide lines in the floor do not permanently occupy floor space as do the conveyors fixed on the floor.

6. *Adaptability to automation:* AGVs can operate efficiently with other automated and computer-controlled systems such as robots, automatic storage and retrieval systems, conveyors, elevators, doors, strappers, wrappers, and automatic production machines.

7. *Integration within plant:* AGVs can provide a link between different cells of automation within the plant, thereby integrating the overall operation.

8. *Adaptability to existing facilities:* A new AGV system can be installed within an existing plant with a minimum of structural changes in the plant building. The guide paths require no more than slots in the floor compared to heavy support structure for cranes and substantial floor or structural modifications for conveyor systems

Accessories

A variety of accessories is currently available to facilitate material handling economically. Investment in these accessories can easily become quite large. Therefore it is necessary to carefully analyze each potential purchase before a decision is made to obtain a certain type. Listed next are some of the most widely used accessories.

Pallet

A pallet is a platform on which material can be stacked in unit loads and handled by lifting equipment such as the forklift truck. The pallet and its load can then be moved from one place to another by such equipment.

Pallets are made of wood, plastic, metal, or a combination of the three and are available

Accessories. *Left*: Tote pans hold small parts such as electronic subassembly parts for transport between work stations. *Right*: Pallets are portable platforms, often made of wood, on which boxes or other items are stacked for transport as unit loads.

in various sizes, the more popular being the following:

24″ × 32″	42″ × 42″
32″ × 40″	48″ × 40″
32″ × 48″	48″ × 48″
36″ × 36″	48″ × 60″
36″ × 42″	48″ × 72″
40″ × 48″	

The size selected depends on the size and weight of the unit load; the equipment used in moving pallets; the sizes of aisles, doors, and other spaces; and the cost of the pallet and whether it is to be saved or delivered and not returned.

Pallets are constructed so that the forks of the forklift truck can enter from either two or four sides. Types of pallets are the following:

1. *Single-faced pallets:* Pallets that consist of two or more stringers to which are attached the three or more planks composing the upper deck.

2. *Double-faced pallets:* Two or more stringers with attached upper and lower decks.

These pallets may be reversible and require very little care in storage when not in use.

3. *Box pallets:* Single- or double-faced pallets with a framework that forms a box. They are used mainly to contain and stack non-uniform or bulk loads to convenient heights.

Box

A box is a portable container in which parts or material can be stored in unit loads. Boxes are made of cardboard, wood, plastic, or metal and are available in various sizes. Some types may be folded for easy storage. Boxes can store and move parts or material that is small or varying in size without crushing or dropping the units. Boxes can be moved by either hand-operated or power-driven equipment. The size can vary from 11½″ × 2¾″ × 2¾″ to 71″ × 18″ × 19″ (length × width × height).

Tote Pan

A tote pan is a portable container that is smaller in size than a box. It is used to carry small parts. Tote pans are made of plastic, metal, or wood and are used to transport small parts from one work station to another. Their size can vary on the basis of the job requirements. They can be moved either by power-driven or hand-operated devices. Their size can vary from 16¾″ × 10¾″ × 3″ to 46″ × 34″ × 33″ (length × width × height).

Skid

A skid is similar to a pallet, except that the construction does not permit stacking of loaded skids on top of each other. Skids are made mainly of metal or heavy wood and are used to store and move heavy and/or bulky materials. They can be moved by power-driven or hand-operated devices and can be made portable by attaching two wheels on one end and a carrying dolly at the other.

Optical Code Reader

An optical code reader is a hand-held device that can read an optical code to identify the product or handling device on which the code is affixed. Optical code readers are small and easily carried between work stations. They can be used to keep track of inventory or products as items are moved from station to station. For example, an operator can acknowledge receipt of material by using an electric pencil to read the optical code affixed to the container (skid, box, etc.) and identify himself or herself by punching a personal code or reading his or her special wristband with the same electrical pencil. This identifies the material, the operator, and the time when the material was received. A similar procedure is followed to signify completion of work on the material by the operator and its dispatch to the next station.

The type of scanner can be a gun, wand, pistol, pen, or magnetic swipe. The light source can be laser, LED, infrared LED, or incandescent. The scan rate is up to 100 scans per second. The depth of the field can be from 0″ to 25″, while the height of the scan can be from 0″ to 12″. The life of the light source is between 3000 and 100,000 hours, The memory capacity can vary from 2K to 512K.

SUMMARY

The chapter presents basic principles for designing a good material-handling system within a plant. The objectives are to increase efficiency in material flow, reduce material-handling cost, and make the overall system more productive and safe. The equipment available for such purposes can be broken down into three major categories: conveyors, cranes, and trucks. Conveyors are popular when material is relatively light and is to be moved over a fixed path. Cranes also travel a fixed path but are generally used to handle much heavier loads. Trucks, on the other hand, can travel over varying paths but, as with all other systems, have some advantages and disadvantages.

The degree of mechanization can influence the cost of material handling. The level of sophistication may vary from a manual handling system to a completely automated system. The initial required investment and operating cost in each case are of course considerably different. Moving material in unit loads is one way to reduce the frequency of trips, and therefore development of a unit load is an important facet in overall planning.

There are twenty basic principles that serve as guides in designing and operating a material-handling sytem. The chapter lists them all and suggests ways of applying them. Many of these principles are compatible with one another, the application of one being able to achieve fulfillment of another.

Specifying a material-handling system for a new plant and developing its layout have an interesting interrelationship. If the layout is known, then the point-of-origin and destination for each material move is known, along with the floor plan; hence, one can design a material-handling system to suit the layout. But, it is also possible to develop a layout around a known material-handling system. To obtain the maximum operating efficiency, both the layout and the material-handling system should be considered simultaneously. This may require a procedure that goes back and forth between two developments.

The existing material-handling system should also be audited frequently to be sure that it is functioning efficiently. The chapter lists the symptoms of an inefficient material-handling system and provides a checklist that may suggest ways for improvements. Productivity ratios can also be used to trace inefficiencies and to incorporate further controls.

A wide variety of equipment is available for material handling. The chapter lists the major types of conveyors, cranes, and trucks. The description of commonly used equipment in a material-handling system should help the reader to understand the advantages, characteristics, and applications of various alternatives. Not all equipment and their features can be described in detail in the limited space available here. The reader should refer to the following monthly publications for recent developments and applications in material handling:

- *Modern Material Handling.* Available from Cahners Publishing, 1350 E. Touhy Ave., Des Plaines, IL 60018.
- *Plant Engineering.* Available from Technical Publishing Corp., P.O. Box 1030, Barrington, IL 60010.
- *Materials Handling Engineering.* 614 Superior Ave., Cleveland, OH 44113.

More information on each product can also be obtained by writing directly to the manufacturer, the address for which can be found in the above-mentioned publications or in the *Thomas Register of American Manufacturers* referred to in Chapter 8.

PROBLEMS

10.1 Define material handling. State at least four principles involved, and indicate how you would apply them.

10.2 Discuss the significance of material handling in industry. Give some examples of material handling in a light industrial plant that you might have visited.

10.3 What are the three basic types of material-handling systems? Give three advantages and three disadvantages of each type.

10.4 What is a unit load? What factors influence the formulation of a unit load?

10.5 What are the benefits of automation in material handling?

10.6 The principles of material handling given in Table 10.1 can be remembered by the following mnemonic:

The Galactic EMPire Called SPAM was Attacking Uno Over the Seven Stars. Spam's Mother Ship returned UnDestroyed.

or by a more down to earth phrase:

GUSS M.D. has A CAMP to teach POSSUM to be C.E.S.

Identify each principle by the letters capitalized in the above mnemonics.

10.7 Give examples for each principle listed in Problem 10.6.

10.8 Select an establishment in your community. Discuss its material-handling methods and machinery.

10.9 For the establishment chosen in Problem 10.8, go through the questioning process for one of its material moves.

10.10 Identify the productivity ratios that are important to you if you are:
  a. Plant manager
  b. Warehouse supervisor
  c. Industrial engineer of the plant
  d. Designing a plant layout

10.11 Discuss pros and cons associated with replacing a belt conveyor connecting two stations with an automatic guide vehicle. What volume of production justifies such replacement? Will the outlook change if ten stations distributed within the plant are to be connected by either a conveyor system or an AGV system?

10.12 Determine the appropriate pallet size required to load packages with a base of $5'' \times 12''$ so as to minimize wasted space.

10.13 Differentiate among the following material-handling equipments, giving at least two advantages and disadvantages of each:
  a. Roller conveyor and slat conveyor
  b. Belt conveyor and wheel conveyor
  c. Forklift truck and trolley conveyor
  d. Handtruck and forklift truck
  e. Jib hoist and birdge crane
  f. Automated guided vehicle and forklift truck

10.14 What factors must be considered in choosing the size of pallets to be used in a plant?

SUGGESTED READINGS

Material Handling

Apple, J.M., *Material Handling Systems Design,* John Wiley and Sons, New York, 1972.

Basics of Material Handling, The Material Handling Institute, Inc., Charlotte, N.C., 1973.

"Equipment—The Big Goal: More Work at Lower Cost," *Modern Material Handling,* May 1975, pp. 66–69.

Kulwiec, R., "Trends in Automatic Guided Vehicle Systems," *Plant Engineering,* Oct. 1984, pp. 66–73.

"Labor—Productivity Is Where the Savings Are," *Modern Material Handling,* May 1975, pp. 52–59.

Rose, W., *Logistics Management: Systems and Components,* William C. Brown Company Publishers, Dubuque, Iowa, 1979.

Operations Research

Ravindran, A.; Phillips, D.T., and Solberg, J.J., *Operations Research—Principles and Practices,* John Wiley and Sons, New York, 1987.

Taha, H., *Operations Research,* 3rd edition, Macmillan, New York, 1982.

CHAPTER 11

Material Handling: Equipment Selection, Flow Lines, and Packaging

Three important topics are discussed in this chapter. First, the discussion of equipment selection shows different methods for determining the type and/or number of units required to perform the necessary material moves. The examples involve algebraic methods as well as the more advanced approach of a heuristic solution procedure.

The second section concerns development of a material flow pattern. In a continuous manufacturing system, conveyors are extensively used, and this section illustrates a few concepts for basic conveyor analysis. The examples include the use of operations research techniques such as queuing theory and simulation.

The last three sections discuss packaging. Packaging can aid in the selling of the product in addition to influencing the material-handling cost. These sections discuss the need for packaging, the equipment available, and several means of reducing packaging costs.

11.1 BASICS OF EQUIPMENT SELECTION

There are many activities in a production plant where material is moved. For example, the most critical and prominent activities in warehousing are receiving, storing, and reissuing of materials, partial assemblies, and final products. At each stage considerable material handling is involved; to a large extent, the efficient operation of a warehouse depends on the proper selection of equipment for such handling. Some types of equipment have already been described; others will be discussed later.

In this section we shall concentrate on methods of equipment selection that are dependent on the following factors:

- *Material to be moved:* The type, weight, volume, shape, and size of the raw material and other assemblies that are to be moved.
- *Movement:* The frequency, path, aisle space, and loading/unloading mechanism.
- *Storage:* The area, volume, shape, and size of storage facility, columns, and other obstacles; the spacing arrangement of

shelves and racks; and company policies regarding storage and issuing.

- *Costs:* The investment and operating expenses of the equipment, interest rate, depreciation, and useful life of the equipment.
- *Other factors:* Flexibility to perform multiple tasks and to work on many different products and obsolescence of equipment.

The following examples illustrate some of the economic analysis needed in evaluating material-handling equipment requirements.

Example: Equipment-Operating Cost per Unit Distance

A forklift truck initially costs $20,000 and has an expected life of 5 years. The fuel cost is $10.00 per 8 hours of operation, and the maintenance is $1.50 an hour. If the truck travels an average of 10,000 feet per day, determine the cost per foot. Assume that the truck operates 360 days a year and that the operator is paid $10.00 an hour (including fringe benefits).

Solution: First determine depreciation. Using the straight line depreciation method, we have

$$20,000/5 \text{ years} \times 1 \text{ year}/360 \text{ days} \\ \times 1 \text{ day}/8 \text{ hours} = \$1.39/\text{hour}$$

The distance traveled per hour is

$$10,000 \text{ feet/day} \times 1 \text{ day}/8 \text{ hours} \\ = 1250 \text{ feet/hour}$$

$$\text{Total cost per hour} = \text{Maintenance} \\ + \text{Fuel} \\ + \text{Depreciation} \\ + \text{Operator costs}$$

$$= 1.50 + 10/8 \\ + 1.39 + 10.00$$

$$= \$14.14/\text{hour}$$

$$\text{Operating cost per foot} = \frac{\text{Cost/hour}}{\text{Feet/hour}}$$

$$= \frac{14.14}{1250}$$

$$= \$0.0113/\text{foot}$$

Example: Equipment Selection

The warehouse receives daily loads of Items A, B, and C from the plant, and shipments are made once a week. Information about these items is given in Table 11.1. The company is considering purchasing a tractor that pulls four trucks, a forklift truck, or a handtruck for use in transporting. Pertinent sizes and costs for these are listed in Table 11.2. The loading and unloading cost includes not only the cost for loading and unloading for one trip, but also the cost for making the return trip to pick up another load. Determine the least expensive method for transporting the goods.

Table 11.1 Material characteristics

Item	Volume (L × W × H) (inches)	Distance from Receiving (feet)	Distance to Shipping (feet)	Units Received per Day	Units Shipped per Week
A	12 × 6 × 6	525	125	118	590
B	48 × 36 × 24	225	375	165	825
C	24 × 24 × 24	400	200	121	605

Table 11.2   Equipment specifications

Equipment	Maximum Volume (cubic inches)	Loading and Unloading Cost ($)	Cost/ Foot ($)
Tractor truck	$4(60 \times 27 \times 72)$	1.20/trip	0.010
Forklift truck (pallet)	$48 \times 48 \times 48$	0.05/trip	0.007
Handtruck	$60 \times 27 \times 72$	0.40/trip	0.005

Solution: The number of units that can be moved per trip is given by the capacity of the piece of equipment divided by the volume of the unit to be moved.

For Item A using the tractor truck, the capacity of each truck is the volume of the truck divided by the volume of the item.

$$\text{Capacity} = \frac{60}{12} \times \frac{27}{6} \times \frac{72}{6}$$
$$= 5 \times 4 \times 12$$
$$= 240 \text{ units/truck}$$

Because parts of units cannot be transported, only a whole number answer for each dimension division is used. For example, $27 \div 6 = 4.5$; and since half a unit cannot be transported, 4.5 becomes 4.

Four trucks are pulled by the tractor.

$$\text{Total capacity} = 4(240)$$
$$= 960 \text{ units per trip}$$

Each day, 118 units are moved. The tractor can move 960 units at a time, so only one trip is necessary. At shipping, 590 units are moved, still requiring only one trip.

For the forklift truck,

$$\text{Capacity} = \frac{48}{12} \times \frac{48}{6} \times \frac{48}{6}$$
$$= 4 \times 8 \times 8$$
$$= 256 \text{ units/pallet}$$

With 590 units to be moved at shipping, three trips will be required.

The handtruck is the same size as one of the trucks for the tractor; therefore it can carry 240 units per trip and will require three trips to move the 590 units at shipping.

Next, using this information and that from Tables 11.1 and 11.2, we can determine the cost for transporting Item A using each piece of equipment.

Receiving cost = Loading and unloading cost per week + Travel cost per week

= (Loading and unloading cost per trip) (trips per day) (5 days) + (Cost per foot) (Number of trips) (feet per trip) (5 days)

Shipping cost = Loading and unloading cost per order + Travel cost per order

= (Loading and unloading cost per trip) (Number of trips) + (Cost per foot) (Number of trips) (feet/trip)

Total cost = Receiving cost + Shipping cost

The following abbreviations will be used:

T = Tractor

F = Forklift truck

H = Handtruck

RC = Receiving cost

SC = Shipping cost

TC = Total cost

LT = Loading and unloading trip

Tractor:

RC (T) = (5 days) (Loading and unloading cost per trip) (1 trip per day) + (5 days) (Cost per foot) (feet per trip) (1 trip per day)

= (5 days) (1.20 per trip) + (5 days) (0.01 per foot) (525 feet) = $32.25

SC (T) = (Loading and unloading cost per trip) (1 trip) + (feet per trip) (1 trip) (Cost per foot)

= 1.20 per trip + (125 feet) (0.01 per foot) = $2.45

TC (T) = 32.25 + 2.45 = $34.70

Forklift truck:

RC (F) = (0.05 per trip) (5 days) + 5 (525 feet) (0.007 per foot) = $18.63

SC (F) = 3 trips (0.05 per trip) + 3 trips (125 feet) (0.007 per foot) = $2.78

TC (F) = $21.41

Handtruck:

RC (H) = (0.40 per trip) (5 days) + 5 days (525 feet) (0.005 per foot) = $15.13

SC (H) = 4 trips (0.4 per trip) + 4 trips (125 feet) (0.005 per foot) = $4.10

TC (H) = $19.23

Similar calculations are made for Items B and C, and the results are shown in Table 11.3. The tractor truck is the least expensive and is therefore selected.

Work Volume Analysis

We illustrate here a basic tabular method called work volume analysis, which is used to determine the number of handling units needed to perform a material-handling task. It consists of measuring the volume, characteristics, and handling requirements of work that must be moved from one work center to another. The measurements may be divided on the basis of the type of work such as receiving, shipping, storage, and in-process movement. The data are analyzed to determine type and number of handling units that would be needed to accomplish the task. Table 11.4 shows one example of work volume analysis, taken mostly from *Basics of Material Handling* (1973). It shows an analysis of receiving docks for a plant that is being supplied by both rail cars and trucks.

Column 1 identifies the location where the analysis is performed. Columns 2 and 3 identify the commodity and mode of transportation used. Columns 4, 5, and 6 estimate the

Table 11.3 Final cost values (in dollars)

Equipment	Item A	B	C	Total Cost
Tractor truck	34.70	583.05	239.20	856.95
Forklift truck	21.40	1779.16	338.20	2138.77
Handtruck	19.23	*	203.60	*

* = Not feasible; item is too heavy.

Table 11.4 Work volume analysis

(1) Dock Identity	(2) Commodity	(3) Transport Type	(4) Average	(5) Peak	(6) Selected Mean	(7) Unload to	(8) Load Quantity	(9) Handling Unit	(10) Handling Units per Load	(11) Estimated Handling Minutes per Handling Unit	(12) Handling Hours per Load	(13) Dock Hours per Load, Including Allowances	(14) Operating Hours per Day	(15) Dock Spot Capacity per Day	(16) Calculated Number of Dock Spots Required	(17) Actual Number of Dock Spots Required	(18) Handling Units per Day	(19) Handling Hours per Day	(20) Number of Trucks Required
Dock H	Bulk A	R.R.	.25 c/l	1.0	1.0	Open bin	100,000 lb	800 lb, Grab Hook	125	3	6.3	7.0	16	2.2	.5	1	125	6.25	
Dock F	Bulk A	R.R.	.1 c/l	1.0	.5	CT 626	100,000 lb	400 lb, shovel	250	3	12.5	13.0	16	1.2	.5		125	6.25	
	Loose A	R.R.	2.1 c/l	5.0	3.5	CT 626	100,000 lb	600 lb, shovel	167	3	8.4	9.0	16	1.7	2.1		585	29.25	
	Package A	R.R.	.2 c/l	1.0	.75	CT 540	20 pallets	1 unit load	20	3	1.0	2.0	16	8.0	.1		15	.75	
	Package B	R.R.	2.1 c/l	5.0	3.5	CT 610	Loose 20 containers	1 container full	20	10	3.3	4.0	16	4.0	.9		70	3.50	
	Total		4.5 c/l	12.0	8.25										3.6	4	795	39.75	2.5
Dock M	Bulk B	Truck	.8 t/l	5.0	4.0	Undergrnd tank	3000 gal			1000 gph	3.0	3.5	16	4.5	.9	1			
Dock C	Loose B	Truck	1.25 t/l	3.0	3.0	CT 610	30,000 lb	1500 lb per cont.	20	10	3.3	4.0	8	2.0	1.5		60	3.0	
	Package A	Truck	.5 t/l	2.0	1.25	CT 540	10 pallets	1 unit load	10	3	.5	1.0	8	8.0	.2		13	.7	
	Package B	Truck	6.5 t/l	10.0	8.0	CT 610	21,000 lb	1500 lb per cont.	14	7	1.6	2.0	8	4.0	2.0		112	5.6	
	Total		8.25 t/l	15.0	12.25										3.7	4	185	9.3	1.2

loads per day of the transportation unit that will need services, the value in Column 6 being taken as the design parameter. Column 7 indicates where the unloaded material should be stored. Column 8 indicates the load quantity per unit of the transportation (one rail car, one truck, etc.). Column 9 identifies the material-handling unit that will be used to unload and its capacity per operation. Column 10 calculates how many complete operations (strokes) of the handling unit are required to unload the quantity listed in Column 8; that is, Column 10 = Column 8/Column 9. Column 11 is an estimate of the time to perform one stroke of the handling unit; the information is used to calculate the time (in hours) needed to unload one transportation unit; that is, Column 12 = Column 11 × Column 10/60. Column 13 modifies the time in Column 12 by including allowances. Column 14 states the working hours of the plant (one shift = 8 hours, two shifts = 16 hours). The dock spot capacity, Column 15, is calculated by dividing Column 14 by Column 13. This value indicates how many times during a day's work a single delivery unit (rail car, truck) can be emptied on a single dock. Since the average number of daily loads is given in Column 6, the number of dock spots needed, Column 16, is obtained by dividing Column 6 by Column 15. The actual number of dock spots required, Column 17, must be an integer equal to or greater than the value calculated in Column 16.

To calculate the number of forklift trucks necessary, the unit used in some docks, we must first calculate the handling loads per day, Column 18. This is obtained by taking the product of Column 6 and Column 10. Column 19 indicates the expected number of hours a forklift will work. The numbers are obtained by dividing entries in Column 18 by 20, the estimated loads a forklift will handle per hour. The number of trucks required is given by Column 20, which is obtained by dividing the total hours required, given as the total in Column 19, by the number of operating hours per day, Column 14.

Example: Application of Operations Research to Material-Handling Problems

Operations research models have been applied to the design and operations of material-handling systems. These applications involve using mathematical programming, simulation, queuing theory, and network models. Some areas of material handling have benefited more than others from these models. A few examples of the applications are

- Conveyor systems
- Pallet design and loading
- Equipment selection
- Dock design
- Equipment routing
- Packaging
- Storage system design

In this section a heuristic procedure is presented, first proposed by Hassan et al. (1985), that can assist designers in the selection of material-handling equipment and its assignment to departmental material-handling tasks (moves). Such selection is applicable after an initial screening has been performed by the designer to determine the most promising candidates, from which the final selection is to be made analytically.

In such circumstances a set of candidate equipment types is generally available. Each move can be performed by most or all of the candidate equipment. Thus for each move there are different values for the operating cost and time based on the equipment used. (See Table 11.5.) The problem requires selection of equipment among the candidate set and assigning them to the moves such that a move is not made by more than one item of equipment unless they are of the same type (that is, each move is assigned to only one equipment type) and all moves assigned to a piece of equipment

Table 11.5   Candidate equipment and move data

| Move | Equipment | | | | | | | |
| | 1 | | 2 | | 3 | | ... P | |
	Cost (W_{1j})	Time (h_{1j})	Cost (W_{2j})	Time (h_{2j})	Cost (W_{3j})	Time (h_{3j})	Cost (W_{pj})	Time (h_{pj})
1	200	0.3*	300	0.2	M†	—		
2	500	0.5	250	0.7				
$\vdots$	$\vdots$	$\vdots$	$\vdots$	$\vdots$	$\vdots$	$\vdots$	$\vdots$	$\vdots$
q	W_{1q}	h_{1q}	W_{2q}	h_{2q}	W_{3q}	h_{3q}	W_{pq}	h_{pq}

*Assuming that the total operating time of an equipment unit is 1, the operating time of a move can be expressed as a fraction.

†Move 1 cannot be performed by equipment 3, which is shown with a very large operating cost denoted by M and an associated operating time denoted by —.

can be performed in the available time on the equipment.

The primary objective of the problem is cost (operating and initial) minimization. There are also some secondary objectives such as maximum utilization of equipment and minimum variation in the selected types, but they are most often compatible with our primary objective. The problem can be expressed mathematically as follows:

$$\text{Minimizing } z = \sum_{i=1}^{p} \sum_{j=1}^{q} W_{ij} X_{ij} + \sum_{i=1}^{p} \lambda_i K_i$$

subject to

$$\sum_{i=1}^{p} a_{ij} X_{ij} = 1 \qquad j = 1, 2, 3, \ldots, q$$

$$\sum_{j=1}^{q} h_{ij} X_{ij} \leq \lambda_i H_i \qquad i = 1, 2, 3, \ldots, p$$

$X_{ij} = \{0 \text{ or } 1\}$ for all ij

$\lambda_i \geq 0$ and integer for all i

where

$a_{ij} = 1$ if equipment type i can perform move j

= 0   otherwise

h_{ij} = Total operating time required by equipment type i to perform move j

H_i = Available operating time of one unit of equipment type i

K_i = Capital cost of one unit of equipment type i in the same time unit as the operating cost (e.g., 1 year)

q_i = Number of moves that can be assigned on equipment type i

q = Total number of moves to be assigned

p = Number of candidate equipment types

W_{ij} = Total operating cost of performing move j by equipment type i in the same time unit as K_i (e.g., 1 year)

$X_{ij} = 1$ if equipment type i is assigned move j

= 0   otherwise

λ_i = Number of units of equipment of a selected equipment type i that are required

The algorithm presented here for an equipment selection problem is based on an analogy to both the knapsack* and loading problems† (see Hassan et al., 1985). Assuming that an equipment type has been selected, then an assignment to one unit of that equipment is analogous to allocating items to a knapsack. (The analogy is illustrated in Table 11.6.) When a second unit of the same equipment type is required for a move assignment, the situation becomes analogous to the loading problem.

The algorithm considers the equipment types one at a time. Moves are assigned to a unit of the selected equipment until it is fully utilized or no other move can be assigned. A selection of the second unit or another type is then made, and the moves are assigned until that second unit or type is also fully utilized or no further assignment is possible. The algorithm terminates when all moves are assigned. Both equipment selection and move assignment are performed in a manner that helps in cost minimization. The steps of the algorithm are as follows.

1. For each equipment type, calculate the number of units that would be needed if the equipment performs all the moves as follows:

$$\text{Let } Y_i = \sum_j \frac{h_{ij}}{H_i}$$

If the division is an exact integer, then

$$\lambda_i = Y_i$$

---

*The one-dimensional knapsack problem involves selecting some items from a set of candidates. Each item has a specific weight (or volume) and a measure of merit or value associated with it. The objective of the problem is to maximize the total value of the selected items such that the capacity of the knapsack is not exceeded.

†In the loading problem, some items, each with a certain volume, require allocation to boxes with equal capacities such that the total number of boxes or their value is minimized.

If the division is not an exact integer, then

$$\lambda_i = [Y_i] + 1$$

where the quantity in brackets is the integer portion of Y_i. Usually, H_i is a set equal to 1, and h_{ij} is expressed as a fraction of H_i.

2. Calculate the total cost of material handling for each equipment type as

$$Z_i = \lambda_i K_i + \sum_{j \in E_i} W_{ij}$$

E_i is the vector of the moves that can be performed by equipment type i, and the number of these moves (used in the next step) is q_i.

3. Calculate the average cost for each equipment type per move as

$$\overline{Z}_i = \frac{Z_i}{q_i}$$

4. Select the equipment having the smallest $\overline{Z}_i$ first. Resolve ties by selecting the equipment with the smallest Z_i. If ties persist, resolve them by selecting in order of ascending $\lambda_i K_i$.

5. For the selected equipment type, arrange the moves that can be performed by it in increasing order of operating cost.

6. Assign the moves to the selected equipment starting with the move having the smallest operating cost. After each assignment, check to see whether the sum of h_{ij} is equal to H_i or within a tolerance E_i of it. If the sum of h_{ij} is equal to H_i, go to the next step; otherwise, check either of the following two cases:

a. If the moves are the only remaining moves, or they cannot be assigned to another piece of equipment, leave the assignment as it is.

b. If the sum of h_{ij} is greater than H_i (or a multiple of H_i depending on the number of units required of the equipment

Table 11.6　Analogy between the knapsack and move assignment to a piece of equipment

Equipment	Knapsack
Moves	Items
Operating time of a move	Volume or weight of item
Available time on equipment	Capacity of knapsack
Operating cost of the moves	Value of the items

so far), check the difference between the least integer multiple of H (making it greater than the sum of h_i) and the sum of h_{ij}. If the difference, which represents idle time, is less than or equal to E_2 (a specified acceptable idle time), leave the assignment as it is. If the difference is larger than E_2, remove moves from the equipment, starting with the last assigned move, until the acceptable utilization level is achieved.

7. Delete the moves assigned from consideration for the remaining moves, calculate a new value for Z_i as before, and repeat the steps until all the moves are assigned.

To demonstrate the procedure, we use the following example. Four candidate types of equipment have the data shown in Table 11.7. H_i is assumed to be equal to 1 for all i, and E_1 and E_2 are 0.1 and 0.2, respectively. Table 11.8 shows the initial calculation. Capital costs for one unit of equipment types 1, 2, 3, and 4 are 5000, 2777.77, 3000, and 4000, respectively.

Table 11.8 illustrates the basic calculations of Steps 1–3.

The smallest $\overline{Z}_i$ is that of equipment type 4; hence type 4 is selected first, and the moves are arranged according to their operating costs as in Table 11.9. Move 7 is assigned first, and the sum of $h_{4j} = 0.4$. Move 8 is assigned next, and the sum of h_{4j} is now $0.4 + 0.4 = 0.8$.

Table 11.7　Cost per year and move time data, cycle I

Move	*Equipment Type*							
	1		*2*		*3*		*4*	
	W_{1j}	h_{1j}	W_{2j}	h_{2j}	W_{3j}	h_{3j}	W_{4j}	h_{4j}
1	400	0.4	M	—	200	0.5	M	—
2	600	0.6	400	1	900	0.8	M	—
3	400	0.5	500	1	900	0.9	M	—
4	500	0.4	400	1	800	0.9	M	—
5	100	0.3	300	1	M	—	M	—
6	200	0.7	900	1	M	—	M	—
7	M	—	400	1	400	0.7	200	0.4
8	M	—	100	1	300	0.6	200	0.4
9	M	—	200	1	800	0.7	900	0.3
10	M	—	100	1	M	—	500	0.2
Total	2200	2.9	3300	9	4300	5.1	1800	1.3

*Data for the problem are from the paper by Webster and Reed (1971).

Table 11.8   Calculations for cycle I

Equipment Type	Total Number of Possible Moves, q_i	Total Operating Time, h_{ij}	Number of Equipment, λ_i	Total Capital Cost, $\lambda_i K_i$	Total Operating Cost, ΣW_{ij}	Total Cost Z_i, (5) + (6)	$\overline{Z}_i$, (7)/(2)
(1)	(2)	(3)	(4)	(5)	(6)	(7)	(8)
1	6	2.9	3	15,000	2200	17,200	2866
2	9	9.0	9	25,000	3300	28,300	3144
3	7	5.1	6	18,000	4300	22,300	3185
4	4	1.3	2	8,000	1800	9,800	2450

Move 10 renders the sum of $h_{4j} = 1$; therefore this iteration is terminated, and one unit of equipment 4 is fully utilized.

With the assignment of moves 7, 8, and 10 to equipment 4, the capacity of that unit is utilized to the extent possible (in this case, completely). Moves 7, 8, and 10 are deleted from the set of moves.

For the next iteration, the cost and move time data are shown in Table 11.10. The moves already assigned are not included in this table.

The calculations for $\overline{Z}_i$ based on the data in Table 11.10 are shown in Table 11.11. Equipment 1 has the smallest $\overline{Z}$; therefore it is the next one to be selected.

Equipment 1 has the smallest $\overline{Z}_i$. Table

11.12 lists the moves that can be made by this equipment in ascending order of their move cost (W_{ij}). Moves 5 and 6 are assigned to one unit of equipment 1.

If we continue in the same manner, the

Table 11.9   The ranked moves and their parameters for a selected unit of equipment 4, iteration I

Move	W_{4j}	h_{4j}	Σh_{4j}
7	200	0.4	0.4
8	200	0.4	0.8
10	500	0.2	1.0
9	900	0.3	

Table 11.10   Cost and move time data, iteration II

	Equipment Type							
	1		2		3		4	
Move	W_{1j}	h_{1j}	W_{2j}	h_{2j}	W_{3j}	h_{3j}	W_{4j}	h_{4j}
1	400	0.4	M	—	200	0.5	M	—
2	600	0.6	400	1	900	0.8	M	—
3	400	0.5	500	1	900	0.9	M	—
4	500	0.4	400	1	800	0.9	M	—
5	100	0.3	300	1	M	—	M	—
6	200	0.7	900	1	M	—	M	—
9	M	—	200	1	800	0.7	900	0.3
Total	2200	2.9	2700	6	3600	3.8	900	0.3

Table 11.11   Calculations for iteration II

Equipment Type	q_i	h_{ij}	λ_i	$\Sigma W_i K_i$	W_{ij}	Z_{ij} (5) + (6)	$\overline{Z}_{ij}$ (7)/(2)
(1)	*(2)*	*(3)*	*(4)*	*(5)*	*(6)*	*(7)*	*(8)*
1	6	2.9	3	15,000	2200	17,200	2866.66
2	6	6	6	16,666.6	2700	19,366	3227.66
3	4	3.8	4	12,000	3600	15,000	3900
4	1	0.3	1	4,000	900	4,900	4900

final assignment of the moves to the candidate equipment is the following:

Move	Equipment
1	1
2	1
3	1
4	1
5	1
6	1
7	4
8	4
9	2
10	4

The required number of units of each equipment type is 3 of type 1, 1 of type 2, and 1 of type 4. The resulting cost is $24,300. Table 11.13 is a summary of the iterations performed in this example.

The algorithm presented is of the con-

struction type. An improvement algorithm has also been developed by Webster and Reed (1971). (For a discussion of construction and improvement types, see Section 14.4.) This algorithm first assigns moves to equipment on the basis of cost alone. The initial solution obtained is then improved by changing the assignment of moves to other equipment in an attempt to improve equipment utilization and thus reduce cost.

Example: Determining Required Number of AGVs

Suppose an automated guided vehicle (AGV) system is chosen for delivering small parts from storage to four shops in a plant, as shown in Figure 11.1. (Most guided vehicles can carry a load of 1000 to 4000 pounds.) Determine the number of vehicles needed, given

Table 11.12   The ranked moves and their parameters for a selected unit of equipment 1, iteration II

Move	W_{1j}	h_{1j}	Σh_{1j}
5	100	0.3	0.3
6	200	0.7	1.0
1	400	0.4	
3	400	0.5	
4	500	0.4	
2	600	0.6	

Table 11.13   Summary of the iterations for the example problem

Iteration	Smallest $\overline{Z}_i$	Equipment Selected	Moves Assigned
1	2450	4	7, 8, 10
2	2866	1	5, 6
3	2975	1	1, 3
4	3050	1	4, 2
5	2977	2	9

Figure 11.1 Departmental layout

the information in Table 11.14. (Assume that each vehicle can carry parts to and from only one shop at a time.) The AGVs under consideration can travel at a rate of 200 feet per minute. (200–260 feet per minute is a normal speed for an AGV.)

Solution: The steps involved in the procedure are as follows. First, determine the total time needed to travel from storage to each destination (round trip). To this we must add the time for loading and unloading. Using the frequency of trips, determine the total vehicle use time. Divide this number by the traffic congestion factor, generally 0.85, to determine the total required time in minutes. Divide the total load by 60 minutes per hour to obtain the number of vehicles necessary.

As shown below, the time to travel to and from a shop to storage is calculated by dividing the total distance between them by the average speed of the vehicles, to which we must add loading and unloading times to obtain the necessary time per trip.

Table 11.14 AGV distribution system data

Shop Number	Distance from Storage (feet)	Number of Trips (per hour)	Loading/ Unloading (minutes)
1	300	14	2
2	500	16	3
3	600	5	3
4	800	8	2

Shop	*Travel Time Calculations*	*Minutes Of AVG Use Per Hour*
1	$\dfrac{600 \times 14}{200} + 14 \times 2 \times 2 =$	98
2	$\dfrac{1000 \times 16}{200} + 16 \times 3 \times 2 =$	176
3	$\dfrac{1200 \times 5}{200} + 5 \times 3 \times 2 =$	60
4	$\dfrac{1600 \times 8}{200} + 8 \times 2 \times 2 =$	96
Total		Total = 430

Adding all these times results in the total vehicle use time of 430 minutes. With 0.85 as the traffic congestion factor, the total load is 430/0.85 = 505 minutes. The number of AGVs required would be 505/60 = 8.43 ≈ 9.

11.2 FLOW PATTERN IN ASSEMBLY LINES

Assembly lines and process lines are both developed with a continuous and smooth material flow in mind. Each work station is arranged in a sequence based on the production flow to perform the next necessary operation. Conveyors play an important part in delivering material to and carrying it from each station. Additional automation should be sought to make material handling as easy as possible within the work stations. In the following sections we discuss in some detail how job stations might be arranged in different patterns to obtain a continuous and smooth flow of material among them.

Serial and Modular Conveyor Systems

As illustrated in Figure 11.2, an assembly line can be developed along a single conveyor or

Figure 11.2

Serial systems

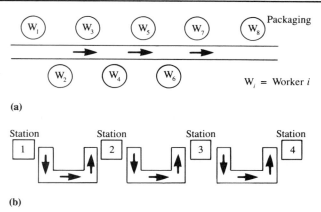

(a)

(b)

consist of a number of separable segments connected together, physically or functionally. There is a great deal of flexibility for the changing floor plan, machine arrangement, and operation sequence when the segments are thus connected.

A set of conveyors can also be arranged to obtain a modular manufacturing system (MMS). An MMS uses general-purpose and even portable equipment within it, allowing for production of many different items in the same physical setup.

For instance, as is shown in Figure 11.3, by including a loop and thus providing a return path within the system, jobs requiring the ma-

chining sequence of 2, 1, 3, 4 can be as easily produced as those requiring a sequence of 1, 2, 3, 4. A robot and conveyor combination, shown in Figure 11.4, can also achieve an MMS. The robot can be programmed to select different loading/unloading sequences on the machines, based on the job requirements.

In an assembly operation requiring highly skilled workers, the unit completion time can vary widely, and roller conveyors are frequently used to connect the stations. The units are moved manually from one station to the next by pushing. Banks, which will be discussed in the next subsection, are a must for efficient operation of such a system.

Figure 11.3   A modular manufacturing system

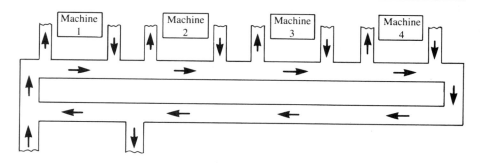

Figure 11.4   Another formation of MMS

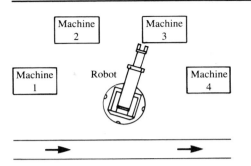

Figure 11.5   Intermittently operated serial conveyor

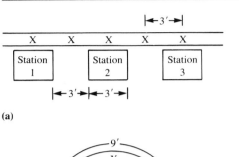

(a)

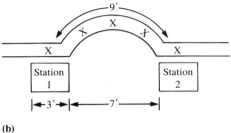

(b)

A power roller, belt, chain, or slat conveyor is often used in a self-paced assembly line. The conveyor serves stations by supplying an incoming unit or material and by carrying the finished item to the next station. The final operation on the line in many instances is that of packaging.

Conveyor speeds can be adjusted to the desired level; however, there are two types of movement, intermittent and continuous.

In a simple serial system in which the conveyor moves intermittently, the speed of the conveyor and the spacing of the items on the conveyor are adjusted to allow a part or subassembly to remain at a station for a cycle time. For example, with three work stations, each 3 feet long, and with a cycle time of 1 minute, the conveyor belt feeding the input and carrying the units to the next station requires a speed of 3 feet per minute. If the stations cannot be placed immediately adjacent to each other, the conveyor length should be 3 feet, or some integral multiple of that, between adjacent stations to permit a smooth and continuous material flow. The units should be placed 3 feet apart on the conveyor. Figure 11.5 shows how the system might be arranged. Figure 11.5(a) has one buffer item between work stations, and Figure 11.5(b) has three buffer items. For a continuous flow conveyor, such strict spacing might not be necessary as long as stations can work out of phase with each other; that is, stations can start and complete the work at times that might not be the same for each station.

Banking

Quite commonly, buffer spaces or banks are provided in conveyor systems to separate or decouple stations, allowing for the absorption of fluctuations in production rates within the workplaces. Such an arrangement also permits the cycle time to be the average production time of the slowest station rather than the slowest cycle time of the slowest station. In addition, a bank provides a space where in-process inspection between stations can be performed.

Banking can be achieved by either of two methods: by changing the product flow or by providing for overlapping operations. Each method has certain advantages and disadvantages, as will be indicated in the following discussion.

Figure 11.6   Change of product flow

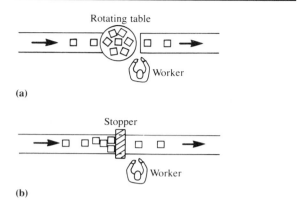

(a)

(b)

Figure 11.6 shows two methods of providing on-line space. In the first case, a rotating table serves as a temporary storage space on the conveyor. The operator selects a unit at random from the table, performs operations, and releases the unit on the other side of the conveyor (downstream). In the second case, a dam is created by placing across the conveyor a barrier such as a wooden plank (2 × 4) or a steel bar, which stops the incoming units. The operator picks up a unit, completes his or her task, and places the item on the downstream side, allowing it to proceed to the next station.

Other means for providing banking are by off-line product flow. Three alternatives are illustrated in Figure 11.7. A U-shaped arrangement provides storage in the curve of the U, the accumulation line furnishes auxiliary conveyor storage, and a completely detached line separates the input and output areas and generates storage at each point.

Supplying such banks increases the time a unit is available for the required operation as well as the number of units available at any one time. On-line banks reduce minor shock due to employee absenteeism, machine breakdown, or other problems in the prior station. Off-line banks require more floor space, increase material handling, and perhaps require more complex scheduling.

Another method of providing banking is utilizing overlapping operators by either one of the following procedures.

1. *Moving operator:* One operator, generally the supervisor, helps out in a group wherever and whenever assistance is required. In the auto industry, for example, it is common practice to provide seven operators for a six-station assembly line. (See Figure 11.8.) The seventh person is used for relieving an operator, helping as needed, and training operators on the six sta-

Figure 11.7

Off-line banking

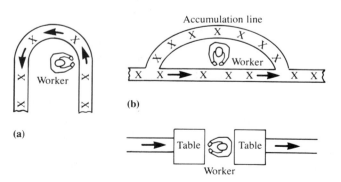

(a)

(b)

(c)

Figure 11.8   A moving operator

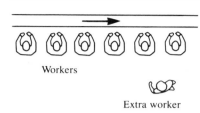

Workers

Extra worker

tions in addition to making minor repairs and doing paperwork.

2. *Overlapping operation:* An example of overlapping operations, shown in Figure 11.9, occurs when the duties of a station are increased substantially and two additional workers are assigned there. The middle operator is responsible for the overall operation of the station; however, the first operator assists on the upstream portion of the task, while the last operator helps on the downstream portion. The overlap of duties permits cooperation between operators and smooths out differences in their working speeds. The increase in operator cost is offset by an increase in efficiency and a reduction in the downtime of the assembly line. In addition, by performing the tasks on the conveyors themselves, the operators require no increase in working space. The costs are normally monitored periodically to determine whether the arrangement continues to be efficient.

Queuing Models

Example 1: Rotary Table Capacity

Items arrive at a station following a Poisson distribution with a mean of two units per

Figure 11.9   Banking by overlapping operators

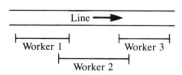

Line

Worker 1

Worker 3

Worker 2

minute. The station can service these units at a rate of three per minute. The incoming items are stored on a rotary table when the worker is busy. We wish to determine the required capacity of the table so that the probability of not being able to accommodate a unit on the table is less than 3 percent.

The situation can be analyzed as a single-channel queuing model. We refer the reader for a detailed discussion on queuing theory to any standard operations research book (Ravindran, 1987 and Taha, 1982). Appendix C supplies a few well-known formulas, which are used here in our analysis.

For a Poisson arrival and exponential service with the number of services being 1, we have

$$P_0 = 1 - \frac{\lambda}{\mu}$$

$$P_n = \left(\frac{\lambda}{\mu}\right)^n P_0$$

In our example, $\lambda = 2/\text{minute}$ and $\mu = 3/\text{minute}$.

The values of $P_0, P_1, P_2, \ldots, P_n$ are listed in Table 11.15. Each indicates the probability of being in that state. For example, $P_0 = 0.333$ implies that the probability of not having a unit on the table is 33.3 percent.

Table 11.15   Calculations of state and cumulative probabilities for each state

State	Probability	Cumulative Probabilities
0	$P_0 = (1 - 2/3) = 0.333$	0.333
1	$P_1 = (2/3)^1 (0.333) = 0.222$	0.555
2	$P_2 = (2/3)^2 (0.333) = 0.149$	0.704
3	$P_3 = (2/3)^3 (0.333) = 0.099$	0.803
4	$P_4 = (2/3)^4 (0.333) = 0.066$	0.869
5	$P_5 = (2/3)^5 (0.333) = 0.0439$	0.9129
6	$P_6 = (2/3)^6 (0.333) = 0.0293$	0.943
7	$P_7 = (2/3)^7 (0.333) = 0.0195$	0.9625
8	$P_8 = (2/3)^8 (0.333) = 0.0130$	0.9755

The cumulative probabilities indicate the percent of time the number of units in the system would be less than or equal to the state value. For example, for state 4 the cumulative probability is 86.9, implying that the number of units in the system would be less than or equal to four 86.9 percent of the time.

To accommodate all units at least 97 percent of time, the minimum size of the table should be sufficient to place eight units, since the associated cumulative probability is 0.9755.

It might be of interest to find, on the average, how many units would be on the table. This is given by $L = \Sigma_i i P_i$, which is equal to 1.7165 units.

Example 2: Rotary Table Capacity

Now suppose that in the previous example the machine periodically requires adjustment that stops the work for at most 3 minutes. What is the probability of not accommodating an incoming unit if the table capacity is set to 8?

For the time the machine is down, there is no service, and therefore the model has only the arriving pattern. The probability of having n units in the system in time period t is then given by

$$P_n(t) = \frac{(\lambda t)^n}{n!} e^{-\lambda t}$$

Here $\lambda t = 2 \times 3 = 6$; and with the assumption of no units being on the table the likelihood that the table would be full is the probability that we would have eight arrivals in 3 minutes. It can be calculated as

$$P_8(3) = \frac{(6)^8}{8!} e^{-6} = 0.1032 \quad \text{or } 10.32\%$$

Since on the average there are 1.7165 units on the table, the probability that the table would be full is the probability that we have $8 - 1.7165 = 6.28 \approx 6$ new arrivals. The value

of P_6 is

$$P_6(3) = \frac{(6)^6}{6!} e^{-6} = 0.1606 \quad \text{or } 16\%$$

If this is too high for the plant under consideration, the table capacity should be increased accordingly.

Example 3: Accumulation Line

Two workers receive units for operation from an accumulator roller conveyor. The main conveyor feeds the accumulator as long as it has room to accommodate a unit. Figure 11.10 shows the arrangement.

If the accumulator is full, the units continue on the main conveyor and are processed later by a relief worker. However, to the extent possible, we prefer not to have any unit going past the accumulator. If units arrive at a rate of three per minute and each worker can process two units per minute, determine the size of the needed accumulator.

The problem can be analyzed by applying the queuing model with Poisson arrival, exponential service, and two workers (Appendix C). The values of probabilities for having n units in the system—that is, $n - 2$ units on the accumulator—are listed in Table 11.16. Also listed are the values of the corresponding effective arrival rates.

Figure 11.10   Layout of the work area

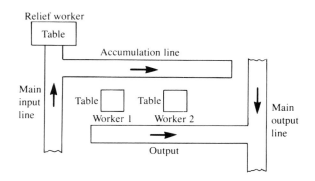

Table 11.16   Calculations for accumulation line example

Number of Units in System, n	Probability, P_n	Effective Arrival Rate, λ_e
2	0.3103	2.069
3	0.1888	2.434
4	0.1240	2.628
5	0.0851	2.745
6	0.0600	2.820
7	0.0430	2.871
8	0.0313	2.906
9	0.0229	2.931
10	0.0170	2.949
11	0.0130	2.961
12	0.0090	2.973

It is apparent that if the accumulation line is made long enough to accommodate eight units ($n = 10$), the probability that the system will be full is less than 2 percent ($P = 1.7\%$). In effect, the two workers can process 2.949 units per minute, leaving about $(3 - 2.949) \times 480 = 24.48$, approximately 24 or 25 units per day, to be served by the relief worker. This number being reasonable, the decision is made to provide for an accumulator with an eight-unit capacity.

Example: Simulation Model

We demonstrate the use of simulation in conveyor bank design by means of an example.

The final phase of production of a microwave oven consists of two operations. The first station finishes the assembly, and the second tests the product. There is a proposal to arrange them in the sequence as shown in Figure 11.11.

Station 1 receives units on skids, and a gravity roller conveyor is to be used between Stations 1 and 2. The conveyor will also serve as a temporary storage for the units. The service times required at Stations 1 and 2 follow uniform distributions with the following parameters:

· *Station 1:* 10 ± 3 minutes
· *Station 2:* 12 ± 5 minutes

Since, on the average, Station 2 takes two minutes longer than Station 1, it has been decided to assign an operator to work on Station 1 for 400 minutes and then transfer to work with the operator on Station 2 for the remaining 80 minutes of regular time. It is a policy of the company to finish testing all the units that are on the feeder conveyor before allowing workers to go home (even if it means overtime) because work held over would create a backlog that could not easily be eliminated the next morning. Station 1 requires essentially no setup time, whereas Station 2 takes about 5 minutes before testing can begin. Thus there is a small built-in time delay already in the system.

Simulation results for one day of operation are presented in Tables 11.17, 11.18, and 11.19. Table 11.17 shows the time of completion of a unit in the first station. Service times are generated by using a uniformly distributed variable generator, namely, $X_1 = a + b(\text{RAN})$, where a is the lower limit, b is the range of the distribution, and RAN is a random number between 0 and 1. (In our case, $a = 7$ and $b = 6$.) The numbers are rounded to the nearest minute because that unit of time is the largest that will allow the desired accuracy. The actual time of release of a unit on the conveyor is shown in the third column.

For Station 2, the arrival time of a unit is the release time of the unit from Station 1 (ne-

Figure 11.11   End stations arrangement

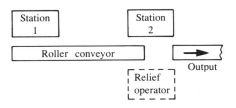

Table 11.17   Simulation of Station 1 (first 400 minutes of operations)

Random Number (RAN)	Time for Service, $X_1 = 7 + 6 \times (RAN)$	Time of Release
.25	$8.5 \approx 9$	9
.59	$10.54 \approx 11$	20
.19	$8.14 \approx 8$	28
.91	$12.46 \approx 12$	40
.61	$10.66 \approx 11$	51
.14	$7.84 \approx 8$	59
.15	$7.9 \approx 8$	67
.92	$12.52 \approx 13$	80
.45	$9.7 \approx 10$	90
.82	$11.92 \approx 12$	102
.50	$10.0 \approx 10$	112
.72	$11.32 \approx 11$	123
.76	$11.44 \approx 11$	134
.53	$10.18 \approx 10$	144
.26	$8.56 \approx 9$	153
.85	$12.1 \approx 12$	165
.39	$9.34 \approx 9$	174
.41	$9.46 \approx 9$	183
.80	$11.8 \approx 12$	195
.33	$8.98 \approx 9$	204
.46	$9.76 \approx 10$	214
.48	$9.88 \approx 10$	224
.60	$10.6 \approx 11$	235
.75	$11.5 \approx 12$	247
.68	$11.08 \approx 11$	258
.03	$7.18 \approx 7$	265
.98	$12.88 \approx 13$	278
.88	$12.28 \approx 12$	290
.01	$7.06 \approx 7$	297
.42	$9.52 \approx 10$	307
.36	$9.16 \approx 9$	316
.21	$8.26 \approx 8$	324
.20	$8.2 \approx 8$	332
.87	$12.22 \approx 12$	344
.16	$7.96 \approx 8$	352
.64	$10.84 \approx 11$	363
.54	$10.24 \approx 10$	373
.29	$8.74 \approx 9$	382
.91	$12.46 \approx 12$	394
.94	$12.64 \approx 13$	407*

*Data not used since it exceeds 400.

glecting time of travel on the conveyor). The service times, X_2, are again uniform and are generated by using $X_2 = 7 + 10(RAN)$. Column 2 in Table 11.18 indicates the random numbers used, and Column 3 shows the corresponding service time for the unit. Column 4 is the departure time obtained by adding the service time to the arrival time, or the previous unit release time, whichever is greater. If the arrival time is greater than the previous unit release time, the operator is waiting on the unit; and if the condition is reversed, the unit is waiting on the operator. For example, the unit that arrived at time 59 must wait until the server completes his or her task on the previous unit that arrived at time 51. This completion is scheduled at 71. Hence the unit will begin to be serviced at 71 and, needing 9 minutes of service time, will be released at 80. Since the server is busy until 80, all of the units that arrive after 59, through and including 80, must wait in the queue. There are two units in this category: one arriving at 67 and another at 80. Therefore the queue length for the time period $71 - 80$ is 2 as indicated in Columns 5 and 6 of the table.

At time 400, the operator at Station 1 ceases work on that station and transfers to help the operator at the second station. Table 11.19 indicates the disposition of the units that have accumulated on the conveyor at time 400. It shows the service time for each unit, which operator services the unit, and how long that operator is then occupied. For example, the unit that arrives at time 332 needs 9 minutes of service, and it is serviced by Operator 2, who then is occupied until time 416. The number of units in storage at that point is 6. The operators continue to work until all of the units arriving on that day are serviced.

The maximum queue length is nine units; therefore the conveyor must be long enough to accommodate at least nine units. Simulation being a sampling technique, we might obtain different queue lengths if other random numbers are used. From a practical standpoint, the

Table 11.18 Simulation of Station 2 (first 400 minutes of operations with 1 operator)

Arrival (1)	RAN (2)	Time for Service, $X_2 = 7 + 10 \times (RAN)$ (3)	Departure Time (4)	Time (5)	Number in Queue (6)
9	.67	13.7 ≈ 14	23	20–23	1
20	.70	14.0 ≈ 14	37	23–37	1
28	.30	10.0 ≈ 10	47	37–47	1
40	.46	11.6 ≈ 12	59	47–59	2
51	.52	12.2 ≈ 12	71	59–71	2
59	.16	8.6 ≈ 9	80	71–80	2
67	.80	15.0 ≈ 15	95	80–95	2
80	.65	13.5 ≈ 14	109	95–109	3
90	.95	16.5 ≈ 17	125	109–125	3
102	.96	16.6 ≈ 17	143	125–143	3
112	.38	10.8 ≈ 11	154	143–153	3
123	.33	10.3 ≈ 10	164	153–164	3
134	.96	16.6 ≈ 17	181	164–181	4
144	.11	8.1 ≈ 8	189	181–189	4
153	.76	14.6 ≈ 15	204	189–204	5
165	.66	13.6 ≈ 14	218	204–218	5
174	.12	8.2 ≈ 8	226	218–226	5
183	.76	14.6 ≈ 15	241	226–241	5
195	.37	10.7 ≈ 11	252	241–252	5
204	.62	13.2 ≈ 13	265	252–265	6
214	.24	9.4 ≈ 9	274	265–274	5
224	.91	16.1 ≈ 16	290	274–290	6
235	.41	11.1 ≈ 11	301	290–301	6
247	.23	9.3 ≈ 9	310	301–310	6
258	.64	13.4 ≈ 13	323	310–323	6
265	.98	16.8 ≈ 17	340	323–340	7
278	.83	15.3 ≈ 15	355	340–355	8
290	.56	12.6 ≈ 13	368	355–368	8
297	.95	16.5 ≈ 17	385	368–385	9
307	.97	16.7 ≈ 17	396	385–402	9
316	.03	7.3 ≈ 7			
324	.12	8.2 ≈ 8			
332	.16	8.6 ≈ 9			
344	.07	7.7 ≈ 8			
352	.92	16.2 ≈ 16			
363	.48	11.8 ≈ 12			
373	.88	15.8 ≈ 16			
382	.52	12.2 ≈ 12			
394	.73	14.3 ≈ 14			

Table 11.19 Simulation of Station 2 (after 400 minutes with 2 operators)

Arrival	Service Time	Operator 1	Operator 2*	Number in Queue
307		402		9
316	7		407	8
324	8	410		7
332	9		416	6
344	8	418		5
352	16		432	4
363	12	430		3
373	16	446		2
382	12		444	1
394	14		458	0

*Starts work at 400 minutes.

trials should be repeated several times to get a better estimate of the maximum queue length. In this case we will add a safety factor of about 20 percent and increase the storage space to 11.

Closed-Loop Conveyor Systems

A basic closed-loop conveyor is a nonreversible, continuous-operating system with carriers such as hooks, trays, and clamps attached to the conveyor that are not removed during normal operations. As shown in Figure 11.12, the parts with their fixtures are loaded at one station, and the final assemblies are unloaded at the same station or in its very close vicinity. Other stations in the assembly process are placed along the conveyor and perform their tasks as the conveyor is moved at a constant speed. The carriers are uniformly spaced, and the distance between them is controlled by the size of the family of parts that is scheduled for production. The spacings must be large enough to accommodate the largest part and its fixture.

The closed-loop conveyor has the following advantages and characteristics:

· Material handling is reduced considerably by depositing the parts fixtures at the starting point, conveniently allowing the operators to use them in reloading the conveyor.
· Workers are paced by the speed of the conveyor as they work off it.
· Material handling between the stations by operators is eliminated.
· The speed of the conveyor can be adjusted to suit different production rates and various products.
· The conveyor speed can be adjusted to compensate for the learning curve effect

Figure 11.12 Closed-loop conveyor system

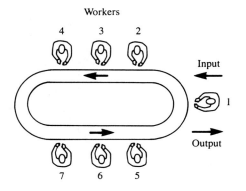

on the line, allowing it to run more slowly initially and at an increased speed as learning takes place.

· Conveyors can be (and often are) used as a temporary storage.
· In the case of job changeover, the conveyor speed is the slower of the two job speeds until the old job is completely replaced by the new job on the line.

For a basic closed-loop conveyor system, three principles must be followed (Kwo, 1960).

1. *Uniformity principle:* Suggests that the conveyor be loaded and unloaded uniformly over the entire length of the loop.

2. *Capacity principle:* States that the unit spacing (capacity) on the conveyor should be large enough for the number of units to be stored on the conveyor.

3. *Speed principle:* Determines the permissible speeds of the conveyor. The lowest speed is defined by the highest of the loading and unloading rates of the stations on the conveyor; the maximum speed is governed by the mechanical/electrical conveyor equipment or the initial and final stations' stock input or clearance rates.

A variation of the closed-loop conveyor is one in which each operator removes the material (unit) from the conveyor, places the work at the station in front of him or her, performs operation(s), and loads the unit back on the conveyor. (See Figure 11.13.) Carousel or tray conveyors are most often used in this arrangement, each level being reserved for the unit in a particular state of completion.

Another extension of the closed-loop conveyor allows for multiple input/output stations. (See Figure 11.14.) This is especially useful if an input/output station is a bottleneck because of its low speed of operation or lack of operating space.

Figure 11.13   Closed-loop conveyor with banks

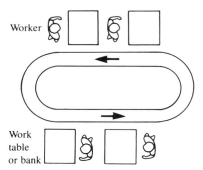

Example: A Closed-Loop Conveyor Design

In a closed-loop conveyor system, the conveyor is 90 feet long, and the carriers are placed 3 feet apart. Thirty-eight feet of the floor space along the conveyor is available for installing assembly stations. Each operator is comfortable working within a maximum of 6 feet of the work station, 3 to 4 feet being the preferred space. There is only one station for the input/output operation, and it takes a total of 0.4 minute to load and unload the new part that is being considered. The precedence relationships and time data for work elements are given in Figure 11.15.

Figure 11.14   Closed-loop conveyor with multiple input/output stations

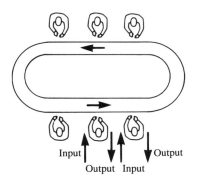

Figure 11.15   Time and precedence relationships

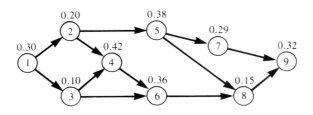

If a production rate of 100 units per hour is required, determine the conveyor speed, work station length, number of work stations, and load distribution.

The production rate of 100 units per hour indicates a cycle time of $60/100 = 0.6$ minute. If each station is made 4 feet long, and each item must travel from the beginning to the end of a station in 0.6 minute, the speed of the conveyor must be $4/0.6 = 6.67$ feet per minute. The maximum possible number of stations, n, on the conveyor with at least 2 feet of clearance between each station is

$$4'n + 2'(n - 1) \leq 38' \quad \text{or} \quad n = 6$$

Hence the part must be produced with six or fewer stations.

By applying the largest candidate rule, we obtain the task distribution shown in Table 11.20.

Table 11.20   Task distribution (cycle time 0.6 minute)

Station	Tasks	Total Time on Station (minutes)
1	1, 2, 3	0.6
2	4	0.42
3	5	0.38
4	6, 8	0.51
5	7	0.29
6	9	0.32

Table 11.21   Task distribution (cycle time 0.45 minute)

Station	Tasks	Total Time on Station (minutes)
1	1, 3	0.4
2	2	0.2
3	4	0.42
4	5	0.38
5	6	0.36
6	7, 8	0.44
7	9	0.32

Since there is room for six stations, the required production can be achieved.

Now suppose the production rate is to be increased to 133.33 units per hour, requiring a reduction of the cycle time to $60/133.33 = 0.45$ minute. This results in the task distribution shown in Table 11.21

Since there is room for only six stations, the seventh station cannot be accommodated unless the station dimensions are changed.

With a 4-foot working space per station, the velocity of the conveyor can be fixed to $4/0.45 = 8.88$ feet per minute. Station 2, however, requires only 0.2 minute to perform its task, and therefore the working space at that station could be reduced to $8.88 \times 0.2 = 1.78$ feet ≈ 2 feet. Similarly, the required dimensions for other stations could be recalculated as shown in Table 11.22.

Table 11.22   Space reallocation

Station	Total Time	Working Space Required
1	0.4	$3.55 \approx 4$
2	0.2	$1.78 \approx 2$
3	0.42	$3.73 \approx 4$
4	0.38	$3.37 \approx 3.5$
5	0.36	$3.19 \approx 3.5$
6	0.44	$3.90 \approx 4$
7	0.32	$2.84 \approx 3$

The total is 24 feet. To this we must add 2 feet for clearance between each two stations; there are six such gaps involved, giving the total required length of 24 + 12 = 36 feet. Since the available working space is greater than the required space, the assembly line can be formed with the length of each station set to the value calculated before.

Automated Controls and Transfers

Productivity in operations can be improved greatly by using automatic control and transfer mechanisms. Automatic controls on conveyor systems allow units to be counted; to be separated on the basis of quality, size, shape, weight, color, and desired destination; and to be collected from multiple conveyors onto one conveyor. The sensors discussed in Chapter 6 play an important part in developing the desired system of operation.

Accessories such as diverters, pushers, turntables, and positioners permit a load to change directions and/or be oriented for a machine tool, as well as change its position while still on the conveyor. (See Figure 11.16.) Such accessories are quite common in most conveyor systems.

Automatic transfer mechanisms not only transport a unit from one station to the next but also position it for the next operation with very little, if any, human interface. There are many ways of achieving such transfer, including the use of robots and AGVs. We will describe here very briefly two of the most commonly used methods.

Automated Transfer Machines

A transfer machine performs almost all operations to manufacture a specific product to a certain degree of completion. The machine consists of several stations, each performing a distinct operation. The part is automatically transported between the stations, positioned, and clamped, and then the next operation is carried out.

There are two general configurations of automated transfer machines: the in-line type and the rotary type. The in-line arrangement has its work stations in a straight line, whereas the rotary configuration has a circular flow of

Figure 11.16

Use of automatic controls

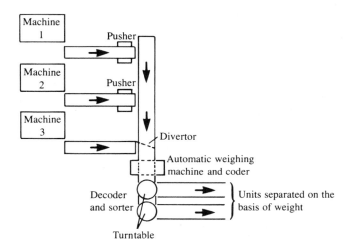

work, usually around a rotating table. Basically, the difference is only in the physical grouping and not in the method of operation.

Automated transfer machines offer many advantages, such as improved part quality and accuracy, reduced labor requirements, reduced floor space requirements, and reduced in-process inventory.

A common problem associated with automated transfer machines, however, is the low working efficiency of the line compared to that of the individual machines. This is mainly because of the linkage between the stations; if one station breaks down, then all stations must be shut down.

The problem can be partially overcome by providing buffer spaces between the different stations. If a station is then forced to shut down, the preceding station can continue to work until the buffer space is full. Also, the succeeding station can continue to operate if its buffer contains any materials.

Monorails

With the ability to automatically pick up, transfer, and place loads, automated monorails are becoming popular in U.S. industries as another means for automated transfer. Computerized monorails operating on lightweight tracks generally handle from 1000 to 2000 pounds per carrier. These monorails are often powered by current-carrying conductor bars designed into the tracks. Several manufacturers offer systems with a pair of channels that allow loads of up to 3000 pounds per carrier.

The prompt delivery and accurate positioning of material by monorails provide productivity boosts in parts fabrication, finishing, and assembly operations.

Self-powered carriers can automatically transport, lift, lower, pick up, and place loads. These systems are also capable of sorting loads and holding them in buffer storage. Monorails can easily be interfaced with automated guided vehicle systems, automated storage and re-

trieval systems, and other automated equipment.

Some advantages of monorails are the following. They are quiet, clean, and individually powered and controlled. They can receive and/or change commands while en route. Furthermore, monorails can operate at slow or high speeds (up to 500 feet/minute) and are capable of quick acceleration. Finally, monorails provide greater flexibility for expansion and layout changes because of overhead track systems with a single track and no chains.

Horizontal and Vertical Flow

A question might arise as to how the assembly line or main production flow should be laid out in a plant. The answer depends mainly on the physical structure and location of the receiving and shipping departments. For example, in a narrow building with facilities for receiving at one end and shipping at the other, a straight line flow would be suitable. A U-shape allows the shipping and receiving departments to be on the same side of the building. The circular-shape allows the same crew to receive and ship, whereas a serpentine flow allows for an assembly line of many stations. Other flow shapes could easily be developed to suit particular needs. Some of the more popular are shown in Figure 11.17.

A plant in a multistoried building presents a special problem. Material must flow vertically as well as horizontally on each floor. Again, many flow patterns could be developed, some of which are shown in Figure 11.18.

Vertical conveyors can play an important role in transporting material from one level to the next. One type, known as a vertical reciprocating conveyor, carries a car up and down in its guides. The car is specifically designed to carry a particular type of load (not people) by providing an appropriate bed (floor). For example, a roller conveyor bed can be used for a flat load, a chain conveyor bed might be appropriate to carry a truck or trolley, and a V-

Figure 11.17 Horizontal material flow lines

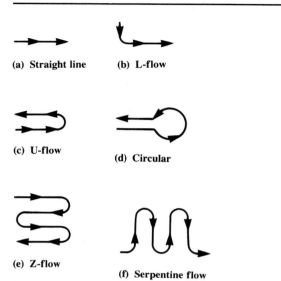

(a) **Straight line**   (b) **L-flow**

(c) **U-flow**

(d) **Circular**

(e) **Z-flow**

(f) **Serpentine flow**

shaped bed might be appropriate to carry a round load.

As shown in Figure 11.19, the arm tray or shelf conveyor can also be used to carry a number of relatively small loads continuously between floors. Arms, trays, and shelves may fold when traveling in the reverse direction, providing economy in space utilization.

The vertical flow pattern can be especially challenging when old, multistoried buildings are converted to productive use. Previous tax incentive schemes encouraged such conversions, and many companies were prompted to remodel old vacant buildings and develop manufacturing plants within them.

The existing walls, many of which are load-bearing and not easily modified, create the need to develop an efficient production facility within it. Figure 11.20 shows several possible arrangements of conveyor systems with lifting or turning tables. (These systems were designed

Figure 11.18

Vertical material flow lines

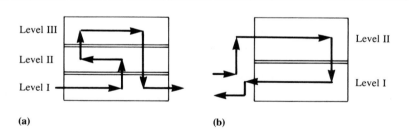

(a)   (b)

Figure 11.19

Vertical conveyors

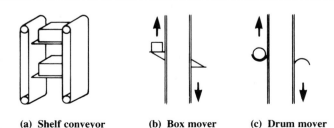

(a) **Shelf conveyor**   (b) **Box mover**   (c) **Drum mover**

Figure 11.20   Possible conveyor arrangements (redrawn courtesy of
PFLOW Industries, Inc., Milwaukee, WI)

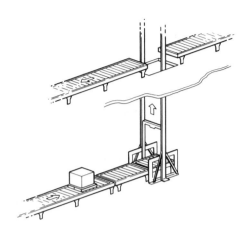

**(a) System 1: Power infeed to and on lift,
gravity discharge second level.**

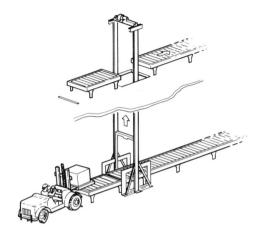

**(b) System 2: One pallet power infeed both
levels and multiple pallet power discharge
both levels.**

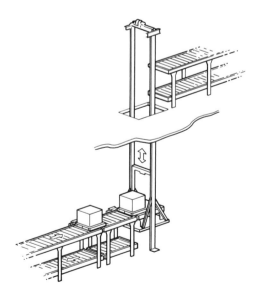

**(c) System 3: Floor space saving system. Upper
power infeed and lower gravity discharge
both levels.**

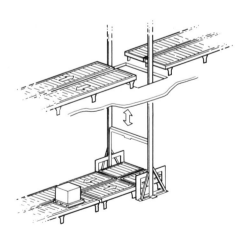

**(d) System 4: Double-width conveyor
carriage. Each side handles pallets in
one direction.**

Figure 11.20 Continued

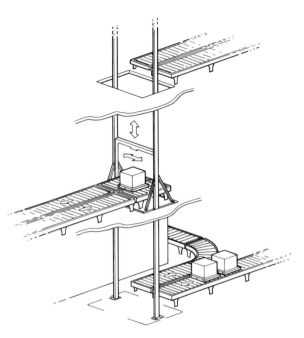

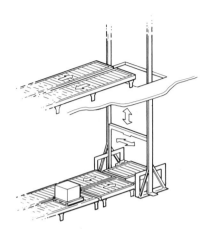

(f) **System 6: Double-width carriage with a single-width sliding conveyor. Allows infeed and discharge from the same side and level.**

(e) **System 5: Double-width conveyor carriage with drag chain transfer. Allows unit to handle two pallets at one time and still feed and discharge from the same side and level. Simultaneous feed and discharge is also possible.**

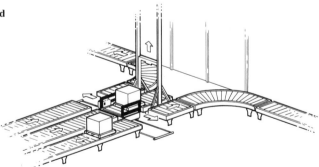

(g) **System 7: Single pallet conveyor lift with traveling transfer cart on first floor. Allows one lift to service many conveyors and locations on the same floor.**

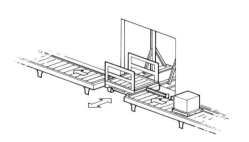

(h) **System 8: Multi-travel-carriage (MTC) system. Carriage extends to load or discharge and retracts to move to another level.**

continued

Figure 11.20   Continued

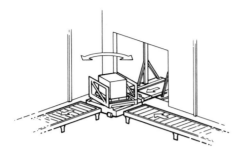

(i) System 9: MTC system incorporating a turntable on carriage for through-wall two-direction corner application.

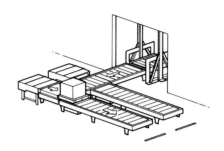

(j) System 10: Single-load conveyor lift discharging to 90° pushoff points. Used in systems handling different sizes or types of materials needing multiple unloading locations.

by Pflow Industries, Milwaukee, and are presented here through their courtesy.)

Material Flow in Job Shops

In a job shop environment there is no single fixed path over which all parts and materials are moved. This is because the work sequences of jobs may differ from one another, requiring different machines in a variety of sequences of operations. It is therefore not possible to define a single material flow pattern that would be ideal in all job shop plants. Trucks and cranes frequently play an important part in material moves that define the flow of patterns. A reduction in material handling is frequently possible, however, by planning the floor layout so that the departments with the most frequent interchange of parts are placed next to one another. Since we will study plant layout in detail in Chapter 13, there will be no further discussion of it here.

11.3 PACKAGING

Packaging is an important aspect in overall production and material handling and requires input from engineering, production, graphics, and advertising personnel. The specifications for the product package very much depend on the product design, and any change in design can cause a significant change in package requirements. It is therefore essential to consider packaging in the designing, production, and material-handling phases of the product.

Packaging also has a big role in a consumer's decision to purchase. If there are several varieties of the same basic product for about the same price, it is most likely that the one purchased will be the one that stands out the most. The size, shape, and colors of the packaging can be very instrumental in product sales. Even for industrial or commercial products, packaging plays an important role in delivering the product intact at minimal additional cost.

Functions

Packaging mainly serves to protect a product from damage caused by handling or exposure to environmental conditions involving heat, moisture, light, and even electronic interference and radiation. It allows a manufacturing firm flexibility in locating its facilities in a site

Packaging. *Top left*: A strapping machine wraps a pallet in metal straps to load-lock the units, forming a compact, secure package and a unit load. *Top right*: Automatic stretch film wrapping of palletized units prevents load shifting and allows easy handling by automated equipment. *Bottom left*: Packaging units in formed plastic trays such as are shown here provides protection against breakage while adding a minimum of weight. *Bottom right*: The pallets in the bin at right are fed one at a time into the palletizer. When a unit load has been collected, the full pallet is moved to the left and removed by a forklift truck.

that is most suitable in terms of production-oriented factors such as labor, raw materials, and utilities without having to be concerned with whether the finished product can be delivered safely to its customers. The type of packaging also contributes to formation of the unit load, which is necessary in the selection and use of the type of material-handling equipment.

There are three major categories in packaging: consumer, industrial, and military. Consumer packaging, which can be subdivided into retail and institutional, is characterized by small units of products handled in large numbers. When the packaging is for retail purposes, its appearance should be emphasized. For institutional use, protection, cost, and convenience are much more important than appearance. Quite commonly, large-sized units of a product indicate industrial packaging. Military packaging is specified by the government.

The important aspects of a package include its structure, aesthetic appeal, style, ability to communicate information to the user, and adherence to legal specifications. The development of a package follows steps similar to those of product design. First, the design should determine whether the packaging is for industrial or retail use to get a sense of the appropriate size and weight of a single package. Then the pallet size for shipping and how high the pallets can be stacked without damage will dictate a load.

The packaging personnel must be very familiar with the product to develop a proper package. This includes its physical specifications, how it is to be used, and details of its promotional information. They must also maintain high ethical standards by not using deceptive labeling and should pay attention to consumer needs, which can be identified through market research.

The type of material used for packaging is controlled by the protection needed for the product, which in turn depends on factors such as the sensitivity of the product (electronic instruments are very sensitive, while refrigerators and appliances are moderately rugged), the weight of the product, the method of shipping and handling, the desired shelf life of the material, and whether the packaged material is to be stored indoors or out. There are different materials to be used depending on the protection desired, for example, protection against breakage, moisture, or heat.

In designing the individual package, an existing design that fills all the packaging needs could be used, or the package could be designed entirely from scratch if no suitable modification to the existing design can be made. In any case, customer appeal, the packaging budget of the company, proper product labeling information, and the universal product code number should all be considered.

When the package is ready to be put into use, several production aspects must be kept in mind. The product manufacturing rate must be the minimum rate of packaging. The procedure should therefore be evaluated to determine the number of machines and personnel that would be needed to achieve this balance.

Protection

Shock from handling and transportation can be damaging to products, especially fragile objects. Formed plastic trays or styrofoam molds, which are lightweight and can be shaped to fit the object, can be helpful. Foam-in-place is very versatile, a light and inexpensive method of packaging that can be partly or fully automated if the volume justifies. Packaging materials such as styrofoam chips, thermofoam, polyethylene and polyurethane foams, paperboard partitions, air cushion mats, and die-cut corrugated inserts are other means of protecting against shock damage. Packaging the product in large units can help hold each individual unit in its place. Human error in handling that results in damage to the product can be reduced through training or the use of automated handling.

Federal and state regulations and company ethics require that packaging methods enhance the safety of the consumer. Potentially hazardous materials should be properly packaged and handled. In 1970 the Poison Prevention Packaging Act allowed for the formation of the Consumer Product Safety Commission. Among the services provided by this commission is the publication of a list of products requiring child-proof packaging. Products containing dangerous chemicals or even radioactive substances should be packed to ensure that no leakage will occur during the roughest handling.

Design and Material Considerations

Two major aspects in designing the package are careful consideration and evaluation of all available materials that could be used. Of particular importance are the static electricity, humidity, temperature, and barrier qualities; the material should keep out water and moisture, greases, oils, gases, and odors while holding in the product. The most common types of packaging materials are glass, steel, aluminum, paper, cardboard, wood, and petrochemical products such as synthetic rubber and plastics. Various chemicals, adhesives, inks, and solvents are also used in developing the final package. We discuss a few of these materials in the following paragraphs.

Glass

One of the oldest packaging materials is glass. It is formed by melting sand with limestone and soda ash. Glass has the advantage of being strong so as to securely hold the product, but it is relatively heavy for handling and is fragile, breaking easily upon dropping or bumping. Packages formed from glass include bottles, jars, tumblers, jugs, carboys, and vials.

The bottle is the most popular form of glass container. It is generally characterized by its narrow neck and mouth. The bottle is used primarily for holding liquids, and the small mouth minimizes the overall size of the closure. Numerous bottles can be spotted in the average home, storing such items as medicines, carbonated beverages, juices, and spices. A bottle with a wide mouth is classified as a jar, which is commonly used for food in a viscous, semisolid, or granular state. Common kitchen items stored in jars include instant coffee, jelly, mayonnaise, peanut butter, and pickles.

The tumbler and jug are two more glass containers that are variations of the bottle. A tumbler is an inexpensive drinking glass that is frequently used for packaging jams, jellies, and fruit preserves. The top is pressed on instead of being screwed on as in the case of a jar. When the original contents are consumed, the tumbler may become a drinking glass. A large bottle with a handle and a screw-on cap is called a jug and is frequently used to store liquid chemicals and foods.

Carboys and vials are so named because of the relative thickness of the glass. A carboy is a bottle made of very heavy glass used to hold liquid industrial chemicals, whereas a vial is a small, thin, tubular glass container used for expensive and sensitive drugs.

Metal

Steel and aluminum are the most commonly used metals because they are readily formed into cans and drums. Cans are constructed in two or three pieces. In two-piece construction, the metal is formed into the shape of a cup, and the top is added to seal the can. For three-piece construction, the metal is formed into a tube; the top and bottom are produced separately and are then secured to the cylinder to make the can. The inside of the can is frequently coated with tin or lined with a plastic sealant. The light weight of these cans and their resistance to chemical reaction with the product make them attractive, but they dent easily; and if made of steel, they are susceptible to rusting.

Paper Products

Cartons and bags are two major paper products used in packaging. They are very common because of their availability and ease of manufacturing, storing, and labeling.

Cartons: The folding carton is the most popular form of cardboard packaging because it is economical in terms of both the cost of the material and the cost to produce the finished carton. Their collapsibility makes them easy to ship, since they can be folded flat and may be stacked. Cardboard cartons are versatile, allowing for different styles and numerous printing and labeling methods. Lightweight cartons are relatively flimsy, not giving much protection to the contents; but heavy corrugated cardboard boxes rival those made of wood.

Bags: Bags are a form of packaging that can be made resistant to moisture, is easy to fill and empty, and has low shipping costs. Bags are light and can be folded and stacked. The most common bag materials are paper, plastic, and textiles. Bags have some disadvantages as packages; they are not supportive of the product, and their durability is only average. There are four basic types of bags:

· Pasted open mouth
· Sewn open mouth
· Pasted valve
· Sewn valve

The term valve indicates that the bag is secured or sealed at both ends once it is filled.

Examples of these are shown in Figure 11.21.

Plastic is obviously not a paper product, but plastic film can be formed in shapes similar

Figure 11.21   Types of bags used in packaging

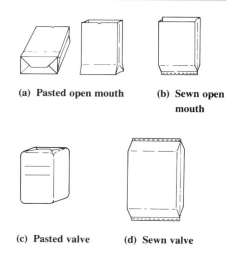

(a) Pasted open mouth   **(b) Sewn open mouth**

(c) Pasted valve   **(d) Sewn valve**

to those of paper bags. The ends can then be closed by sealing or by using a wire tie.

The most common textile bags are made of burlap or cotton and are sewn together and stitched closed. Burlap bags are commonly used for grains such as oats, while cotton sacks may be filled with finer substances such as flour. Textile bags have the advantages of being reusable.

Wood

Boxes and pallets made of plywood and wooden boards are used for packaging and carrying heavy, bulky material. They are comparatively strong and easy to palletize.

Consolidation and Palletizing

Protection for individual units can be increased through consolidation or combining several packages into a compact package in which the units support and protect each other. The number of units per container is generally defined by the sales department (for example, 24 bars of soap in a box), and the packaging department must then develop the size of the pallet.

The method of shipping also plays an important part in palletization. Shippers often limit the overall dimensions and weight of shipping containers. Air and ground parcel services quite often do not allow palletization because of size and weight limitations.

Several methods exist for consolidation of units. Adhesives can be used to hold them together, or units can be encased in shrouds, or wraps of stretch or shrink plastics. Boxes, bins, trucks, and rail cars can be used to transport the product in bulk. The large boxes and bins for this purpose are constructed of cardboard, wood, plastic, or metal. Groups of products can also be consolidated by securing them on pallets.

An additional consideration in palletizing is load-locking, which is a method used to prevent the various levels stacked on a pallet from sliding apart. One solution is to use a nonslip application on the exterior of the package; however, this can complicate the shipping process. Adhesive can be applied between the levels by spraying, brushing, or rolling. Enclosing the pallet in stretch or shrink wrap can also load-lock the units.

Shrink/Stretch Film

A common industrial method of holding a pallet of units together is the use of shrink wrap or stretch wrap. The wrap is used with the pallet to prevent shifting of loads and to facilitate the use of automated handling equipment. Shrink wrap requires a heat source to shrink the material around the units. Stretch wrap requires no heat source and therefore has a lower operating cost, but it does not conform as well to irregularly shaped loads as does shrink wrap.

Many different materials are currently in use as stretch and shrink wraps. Table 11.23 lists the most common films in use and their advantages and disadvantages.

Different densities of wraps are commercially available to fill the needs of the company. As density is increased, surface hardness and the heat seal temperature increase; flexibility is lowered; impact strength decreases; and moisture, gas, and grease resistance increase. Low-density materials are characterized by a lack of stiffness and high-impact strength. The high-density materials are stiffer and provide better barriers.

Strapping

Another way to palletize a group of items is to use steel strapping, in which bands of steel, plastic, or other strong materials are wrapped around the units on pallets. The efficiency is greatly improved if such a method can be automated rather than performed manually. However, a large volume of packaging is needed to justify the initial cost of the expensive and sometimes specialized equipment that is needed in such an operation.

Labeling

Once the package has been designed and manufactured, its proper labeling must be considered. The function of the label is to relay information to the consumer, information that includes a brief description of the product along with a list of ingredients or contents, instructions for use, warnings, the net weight, and/or the volume, and any other pertinent items. The label should also instruct the user regarding precautions and ways the product should not be used. The manufacturer's name and address should be somewhere on the package, as should the applicable universal product coding.

Final Step

A concluding inspection should be made to ensure that all items such as spare parts, mounting screws, and maintenance supplies as well as all necessary documents are packed. The latter might include the warranty, installation instructions, wiring diagrams as appropriate, shipping papers, and a consumer research questionnaire.

Table 11.23   Commonly used wrapping material

Material	Advantages	Disadvantages
Cellophane	Stretchable, gas barrier	Expensive, tears, short shelf life
Cellulose acetate	Transparent	Poor barrier for gases and moisture
Ethyl acetate	Strong	Expensive
Methyl cellulose	Oil barrier	Soluble in water
Nylon	Moisture barrier	Expensive, poor gas barrier
Polyester	Strong, transparent	Noisy, expensive, average barrier to moisture and gases
Polyvinyl chloride (PVC)	Cheap, transparent, good barrier	Retains odors
Rubber hydrochloride	Transparent, strong, oil barrier	Retains odors, average gas barrier
Polyvinylidene chloride (PVDC) (trade name: Saran)	Excellent barrier	Expensive
Styrene	Transparent	Poor barrier, develops static electrical charge
Polypropylene	Transparent, excellent barrier to gases and moisture	Average strength
Polyethylene	Low cost, soft, good moisture barrier	Poor gas barrier, slight haze

Packaging Equipment

Once a product is ready for shipment, it must be securely placed in the appropriate package and sealed for shipping and handling. Box flaps and bag openings have to be folded closed and sealed in preparation for palletization. Full pallets and irregularly shaped objects might require strapping to secure them. Bottles must be packed to prevent breakage during shipping. When the production volume is low, packaging can often be done by hand, but large companies with high volumes of production depend on semiautomatic equipment for sealing products.

The following sections list some of the equipment available.

Automatic Adhesive Sealers

An automatic adhesive sealer is a conveyor with an adhesive applicator and a compression unit for folding down and holding the glued flaps of a box. It features top or bottom sealers or simultaneous loader seals. Feeding of boxes can be manual or automatic.

Automatic Tape Sealers

An automatic tape sealer is a conveyor with a tape dispenser and a compressor. Tape is measured to the proper length, wetted if necessary, placed on the flaps of the box, and compressed in place to hold the flaps down. The tape sealer can be combined with an adhesive sealer to glue and tape the box at the same time. It features manual or automatic feeding of boxes.

Stitchers

In a stitcher, a coil of wire is fed through, cut to proper length, formed into a staple, driven into the box or bag, and tightly clinched in place to secure the flaps. Feeding of boxes can be manual or automatic.

Staplers

In a stapler, preformed staples are held in a magazine, pressed into a box or bag, and tightly clinched in place to secure the flaps. The bar-type stapler has a bar on which the box is placed to provide a surface for clinching the staple after it is driven through the box. The anvil-type stapler has two anvils that are driven through the box with the staples to provide a surface for clinching. Staplers can be manually or automatically operated.

Strappers

In a strapper, round wire or flat strapping is drawn around the box, bundle, or object and tightened. The ends are then secured together to close, strengthen, or hold the unit to be shipped. The unit can be passed through the strapper on a conveyor. Semiautomatic operations might require directing the strapping by the operator. Strapping is secured by twisting or clipping the ends together. Strappers are used primarily to secure pallets.

Wrappers

A wrapper is a conveyor-type machine that wraps the unit in paper or film and seals the ends. It is supplied by a roll of paper or film and can combine loading and wrapping.

Palletizers

A palletizer is an automatically controlled machine that is capable of stacking a unit load on a pallet. High-speed palletizers (e.g., those used in breweries) are capable of stacking 150 cases per minute, but they are expensive (frequently costing over $100,000) and require a large floor space (about 150 square feet). Low-speed palletizers (12–20 cases per minute) cost starting around $45,000 and require less than 100 square feet. However, there is a drawback; their stack height is limited to 80 inches or less, while high-speed units can stack to about 150 inches. If the cases being handled are lightweight (under 40 pounds), robotic palletizers, ranging in cost from $30,000 to $60,000, are available that operate at speeds of 5–15 cases per minute. They can operate at speeds of 5–15 cases per minute. They can operate in limited spaces and can be programmed for many different patterns.

Advantages in the control units used on palletizers have made it possible to integrate palletizers into existing automated systems. In recent years the use of programmable controllers has made it possible to increase the number of available patterns for palletizing from less than 10 with punched tape to about 50. The programmable controller also allows the loading pattern to be changed extremely rapidly.

11.4 REDUCING PACKAGING COSTS

Factors affecting the cost of packaging are similar to those influencing the cost of manufacturing a product. These elements are associated with the design, material, and production phases of packaging. We will discuss a few ways in which each factor can be controlled.

Design

To the extent possible, select a square shape for a package. This results in minimizing the required surface area, which in turn reduces the material needed for the package. For example, $2 \times 2 \times 2$ gives 8 cubic units of volume with a surface area of 24 square units. While $1 \times 1 \times 8$ also provides 8 cubic units of volume, it has a surface area of 34 square units. As a case in point, a physical fitness company decided to ship three weights in a box instead of two (when the order permits), which resulted in a square package and reduced the packaging cost by 5 percent.

Determine how a customer intends to unload the product and, if possible, design the package to accommodate the customer. This might mean developing different packages for different customers, but an increase in sales volume could justify such individualized attention. When an electrical company that supplies products primarily to manufacturers of electronic equipment changed its packaging to suit the automated opening and unloading machines used by such manufacturers, its sales and revenue increased greatly. This offset a small increase in packaging cost, and the net result was increased profit.

Make basic changes in design if they will result in a reduction in cost. Controlling factors could be reduction in the weight of a package, reduction in the number of components in the package, and redesign of the product itself (for instance, by eliminating indented rings around the cans, a soup manufacturer improved its packaging by eliminating wasted space between the cans in packing).

Material

The material used in packaging could perhaps be changed to lower the cost. A syrup manufacturer changed its bottles from glass to plastic and realized a savings of 20 percent, mainly in material handling because of the large reduction in total weight requiring transportation. It might be possible to obtain the necessary material for packaging at a lower cost by increasing the volume of purchase, lengthening contracts, or changing from a plant-to-plant contract to a national contract. A supplier that is closer to the plant may be asked to offer better rates, since its freight cost might be considerably lower than that of a more distant supplier. Similarly, a supplier with equipment that is better suited for a manufacturer's need could offer a better price. For example, 40-inch paper bags might be provided more cheaply by a supplier with a 40-inch press than by one with a 60-inch press.

Production

At times a company can realize a substantial savings by producing packaging material itself. An economic analysis must be performed to evaluate this alternative. Increasing the productivity of the packaging process by increasing utilization of machines, obtaining an efficient and balanced assembly line, and minimizing the idle time should reduce packaging cost. Eliminating or reducing defects and maintenance cost and providing better training for workers are also alternatives that may be considered in the effort to control costs.

11.5 DESIGNING A PACKAGING AREA

Packaging is the final task before shipping the product. Like any other job, it must be analyzed to determine an appropriate method of performance. An operation chart and an operation process chart are useful tools in developing the necessary sequence. However, there are a few points that should be noted.

1. Both the means of packaging—for example, strapping, boxing, and stretch wrapping—and the size of the unit load should be established on the basis of the consideration given before.

Designing a packaging area. *Left*: The layout and material flow pattern of the packaging area should be carefully designed. In the manual packaging line shown here, a belt conveyor transports a box from station to station. Operators, based on the order form, fill the required number of products they handle into the box. *Right*: In the packaging work station shown here, the boxed product is sealed, labeled, and stamped, and the date is entered into a computer.

2. If a machine is to be used, the rate of packaging must be determined. It is possible to buy machines with different speeds—for example, one package per second or ten packages per second.

3. The decision as to which machine to buy is also influenced by the variety in packaging the machine will have to perform. Generally, high-speed machines do not adapt as well to changes in specifications as do slower machines.

The layout and floor space required for the packaging area must be given consideration similar to that in designing a work station or a series of work stations for an assembly line. The methods of obtaining a smooth and continuous flow in an assembly line using conveyors were discussed in an earlier section, and the means for calculating area for a work station will be developed in Chapter 13. However, one must keep in mind that packaging most often reflects the product line. It could be an assembly line operation if only one product is involved, a batch-processing operation if there are a number of products manufactured in groups, or in some cases it might even be a job shop arrangement. Because of such possibilities, automation in material handling (e.g., a fixed conveyor or skids and forklift truck) should also be established on the basis of the product mix. However, most packaging works well with conveyors; hence conveyors are normally a dominant feature in packaging areas.

11.6 COMPUTER PROGRAM DESCRIPTION

Two interactive programs for queuing analysis are presented for this chapter. One is for a single server model, and the other is for a multi-server model. The notations used are as follows:

p(I) = Probability of I units being in the system

L = Mean number of units (length) in the system

LQ = Mean length of the queue

W = Mean waiting time in the system

WQ = Mean waiting time in the queue

The programs calculate each of the above system performance measures. Inputs are supplied with answers to the following questions.

1. Single server model M/M/1: *What is the arrival rate? and the service rate?* (Enter the rate of arrival and rate of service.)

2. Multiserver model M/M/C/K (the model is for 'c' servers and a finite maximum queue length):

   a. *What is the arrival rate? and the service rate?* (Enter the total arrival rate and the rate of service per server.)
   b. *How many servers are there?* (Enter the total number of servers.)
   c. *What is the queue limit?* (Enter the value of the maximum number of units that are allowed in the queue.)

SUMMARY

Selecting the right equipment for moving material is a challenging task. One must know what is to be moved, how frequently it is to be moved, what the physical limitations are, and what costs are involved. The chapter illustrates with numerous examples how this can be done. It begins with a simple model showing basic cost calculations and proceeds toward the more complex problem of equipment selection.

Work-volume analysis is one technique that is used. It decides the number of handling units required to perform all necessary material moves if the type of equipment needed for each move is known. The heuristic procedure goes one step further and determines both equipment and their number such that it minimizes the total cost of operation. Selection of AGVs involves slightly different analysis and is also shown by an example.

The next topic of discussion is flow lines, a study of workplace arrangement to obtain a smooth and continuous flow of material within a plant. Conveyors are most often used for moving material between stations, and they are arranged in either a serial, modular, or closed-loop form.

Conveyors can be made to move intermittently or continuously; both modes are commonly used in industry. However, to provide a safety cushion of unfinished goods in between the stations, decoupling of successive stations is desirable and is obtained by providing banking. Various banking procedures are discussed here, and examples illustrate ways to determine the sizes of the banks. Techniques such as queuing analysis and simulation are helpful in this regard.

A closed-loop conveyor system presents a special challenge, and the basic rules for designing such a system are stated. Two illustrative examples also give some flavor of the design analysis.

Automated control and transfer mechanisms can improve productivity greatly. Controls allow one to perform functions such as count size and weight while units are moving on conveyors. Accessories allow units to change direction and orient the position while still on the conveyors. Automated transfer machines and monorails are two of the most commonly used devices for automated transfer. They transport a unit from one station to the next and position it for subsequent operation with very little, if any, human interaction.

Various configurations are possible in designing an assembly line or a flow line within a plant. Straight line, U-shape, Z-shape, and circular are a few of these arrangements. The contributing factors in such decisions are the physical structure of the plant and the required location for receiving and shipping departments. To develop a plant in a multistory building necessitates planning for vertical flows as well. The chapter presents various arrangements for building such a system.

The third topic of discussion is packaging. It is generally the last operation before a product is shipped to a customer. Packaging serves to protect the product from damage in handling and environmental factors. It is important that the final product package has aesthetic appeal if the product is consumer oriented. For industrial use, protection, cost, and convenience are much more important. Styrofoam molds, trays and chips, air cushion mats, and paperboard partitions are some of the materials used for holding units in place. The material used in packaging the product itself is also important in providing protection. Glass bottles, jars, and tumblers are often used for holding liquids and viscous and semisolid or granual substances. Glass keeps foreign elements such as moisture, gases, and odors from contaminating the product. Steel and aluminum are used in the form of cans and drums to hold liquids. Paper products such as cartons and bags are used to hold granual and small substances, while wooden boxes are used to carry heavy and bulky materials.

Additional protection to the product can be given by consolidating many units together. For example, a box may contain twelve individual items. Furthermore, the number of such boxes can be packed together on a pallet; this method of shipping is called palletization. Use of a shrink/stretch film or metal straps is very common in forming such a pallet load or unit load.

Various packaging machines are available to increase the productivity in packaging operations. Automatic adhesive sealers, stitchers, staplers, strappers, wrappers, and palletizers are some of the machines that the chapter describes.

The chapter also presents a discussion on how to reduce packaging costs. Evaluation of design and the material used in packaging as well as the performance of a make-or-buy analysis are some of the ways suggested for lowering the packaging cost.

Designing the packaging area is a problem similar to designing the rest of the plant. One must define the mode of operation and then develop corresponding work stations and material-handling facilities. However, a dominant feature in most packaging areas is the use of conveyors, especially if automatic packing machines are used.

PROBLEMS

11.1 What factors affect the selection of material-handling equipment?

11.2 A $10,000 tractor truck has an expected life of 8 years. The truck operates 280 days a year and travels an average of 15,000 feet per day. If the fuel costs about $9.00 a day and maintenance costs are $0.75 an hour, determine the cost per foot for operating this vehicle. Operator cost is $8.00/hour.

11.3 A plant produces 10,000 telephones per day. The dimensions of each telephone are 8 × 3 × 3 inches. Management wishes to have a one-week (5 days) supply in stock. One hundred telephones are packed in a box that forms a unit that may be stacked five high. Determine the floor space required. Allow aisles of 10 percent.

11.4 Three different material-handling units are available for moving items of material. The costs of these lifts are $3000, $2500, and $3300, respectively. These units must make six moves. Data on the operating cost, W_{ij}, and the operating time, h_{ij}, are given in the accompanying table. Determine which type of equipment should be used.

Move	W_{1j}	h_{1j}	W_{2j}	h_{2j}	W_{3j}	h_{3j}
1	50	0.8	40	0.5	80	0.7
2	60	0.7	70	0.3	50	0.8
3	70	0.4	M	—	30	0.9
4	80	0.5	100	0.3	60	0.2
5	100	0.1	50	0.2	40	0.1
6	40	0.2	M	—	M	—

11.5 Parts needed for production in Departments 1, 2, and 3 are moved from storage by automated guided vehicles. Using the distance from the storage to each department as shown in Figure P11.5, and given the number of trips that will be made per hour to each department and the time needed for loading and unloading given in the accompanying table, determine how many vehicles are needed. Assume an AGV speed of 200 feet/minute and a congestion factor of 0.85.

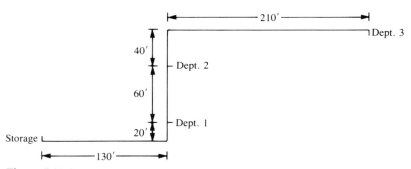

Figure P11.5

Department	Number of Trips per Hour	Loading/ Unloading (minutes)
1	10	2
2	15	4
3	20	3

11.6 Automated guided vehicles are being considered for use in moving supplies from storage to each of four departments. Using Figure P11.6 and given the number of trips made from storage each day and the loading time in the accompanying table, determine how many machines will be needed. As-

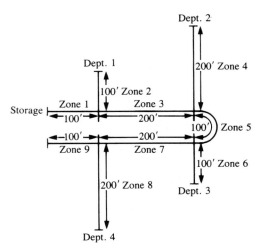

Figure P11.6

sume that we are working with a zone system allowing only one AGV to visit a zone at one time and that the demands are uniformly distributed; for example, for Department 2, demand occurs at 8:00 A.M., 8:20 A.M., 8:40 A.M., and so on. The first demand from each department can be staggered according to number of trips; for example, the first demand from Department 2 may be at 8:00 A.M. or 8:20 A.M., and so on. However, the earliest possible delivery is preferred. (The problem might involve scheduling.) Zone right-of-way rule: first come, first served (FCFS). Storage loading rule: 1 AGV at a time FCFS.

Department	Number of Trips per Hour	Loading/ Unloading (minutes)
1	4	2
2	3	3
3	3	1
4	3	3

11.7 What is banking in material handling? Describe the methods of achieving banking and advantages and disadvantages of each method.

11.8 A rotary table serves as a bank for one work station. If items arrive at the station following a Poisson distribution with a mean of 3 units per minute and an average service time of 0.2 minute per unit, what should be the minimum table size? The probability of not being able to accommodate a unit on the table is to be less than 5 percent. If the operation is to be modified so that the table now will serve as a bank for two operators, each serving the unit with a mean of 0.2 minute, and there are two belts unloading the units on the table, each with a mean of 3 units per minute, what should be the minimum size of the table for keeping the probability of not accommodating a unit to less than 5 percent?

11.9 A product has a work distribution as shown by the precedence relationship diagram in Figure P11.9. The number above a node indicates the time in minutes needed to perform that particular operation. A closed-loop hook conveyor system is to be built that will produce 30 items/hour. Four other products are to be built on the same conveyor system with the dimensions (W × L) shown in the accompanying table. If an operator table is to be limited to no more than 6 feet in length, determine the number of stations needed, the length of each station, the spacing of the hooks on the conveyor, and the length of conveyor if the length is to be at least 1½ times the minimum required length for storing the units and the spacing between stations is to be 2 feet. The input/output rates to the system can be 30 units/hour per person and 20 units/hour per person, respectively, and these stations are in addition to the stations necessary for the assembly jobs. The work on input/output station can be speeded up by increasing the number of persons. For example, 2 workers can unload 60 units/hour.

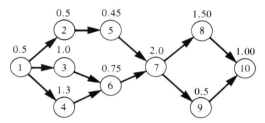

Figure P11.9

Product	1	2	3	4	5
Dimensions	4′ × 2′	1′ × 1′	3′ × 2′	1′ × 2′	3′ × 5′

11.10 In the closed-loop conveyor system shown in Figure P11.10, units are loaded automatically from a box onto the conveyor at the input point. The finished products are collected automatically in a box at the discharge point. The input and output boxes can hold 100 units and 60 units, respectively. The input source can be replenished every 30 minutes by a forklift operator, while the output can be removed every 20 minutes by another forklift operator. Station lengths can, at the most, be 4 feet.

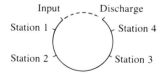

Figure P11.10

a. Given the information in the accompanying table concerning the time needed at each station to perform each job, determine the best speed for the conveyor.

Station	Necessary Time (minutes)
1	1.5
2	2
3	2
4	1

b. What is the minimum capacity of the containers (boxes) needed at the input and output stations for the speed of the conveyor fixed in part a.

c. What adjustment must be made if the required production rate is to be increased to 60 units/hour?

11.11 At Pink Dot Battery an accumulation line is used in conjunction with a roving operator on the acid line. When the flow of empty batteries is too great for one operator to handle, the rover will come and load a batch of batteries (one batch equals 32 batteries) on the accumulation line. After the operator catches up, the rover comes and replaces the batch on the line. Because of the flow arrangement shown in Figure P11.11, if the rover did not divert a batch or two, the batteries could back up into assembly. This would cause a slowdown in the assemblers' work and a decrease in their incentive paȳ (labor problem). Each batch is released by the assembly operator following a uniform distribution with parameters 50 + 8 minutes. The acid line operator takes 3 + 1 minutes to fill each battery. The U-shape within the main conveyor can hold at the most one batch. The rover will take 10 + 3 minutes to load or unload a batch. (The units must be handled in batches for quality control reasons.) The assembly line works 4 hours a day, while the acid line works 8 hours a day. Perform simulation for two days of work and decide how frequently the rover should visit the work station. (For simplicity, assume that the rover can visit the station at intervals of 15 minutes only, for example, 8:00 A.M., 8:15 A.M., 8:30 A.M., and so on.)

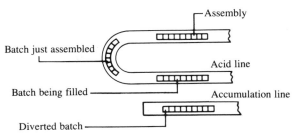

Figure P11.11

11.12 In the above example, assume that the time between releasing batches by the assembly line operator follows an exponential distribution (arrival follows a Poisson distribution) with a mean of 50 minutes, and the acid line operator processes a battery following an exponential distribution with a mean of 3 minutes. Determine the number of acid line operators needed if no accumulation line is provided. Assume that the assembly line will now work 8 hours per day.

11.13 What are the goals in packaging? Describe how each of the following materials is used in packaging.
  a. Glass
  b. Paper carton
  c. Wood
  d. Styrofoam chips
  e. Air cushion mats
  f. Formed styrofoam

11.14 What differences in packaging would be apparent between a household cleaner sold for home use and the same cleaner packaged for industrial use?

11.15 Why is a label of a product important? What information should it display? What are the commonly used materials for labels? Why are these materials chosen?

11.16 What are the four types of bag packaging. Describe and give an example of each.

11.17 Describe the following packaging equipment, specifying their characteristics and how they can be used.
  a. Automatic adhesive sealer
  b. Strapper
  c. Wrapper
  d. Palletizer

SUGGESTED READINGS

Apple, J.M., *Material Handling Systems Design,* John Wiley and Sons, New York, 1972.

Basics of Material Handling, The Material Handling Institute, Inc., 1973.

"Can You Computerize Equipment Selection?", *Modern Materials Handling,* Vol. 21, No. 1, 1966, pp. 46–50.

Hassan, M.M.D.; Hogg, G.L.; and Smith, D.R., "A Construction Algorithm for the Selection and Assignment of Materials Handling Equipment," *International Journal of Production Research,* Vol. 23, No. 2, 1985, pp. 381–392.

Hudson, W.G., *Conveyors and Related Equipment,* John Wiley and Sons, New York, 1954.

Keller, H.C., *Conveyors: Application and Design,* Ronald Press, New York, 1967.

Kwo, T.T., "A Method for Designing Irreversible Overhead Loop Conveyors," *The Journal of Industrial Engineering,* Vol. 11, No. 6, 1960, pp. 459–466.

"Materials Handling—The Trends to Watch," *Modern Materials Handling,* Vol. 30, No. 1, May 1975, pp. 52–59.

Matson, J.O., and White, J.A., "Operational Research and Material Handling," *European Journal of Operational Research,* Vol. 11, 1982, pp. 309–318.

Ravindran, A.; Phillips, D.T.; and Solberg, J.J., *Operations Research—Principles and Practices,* John Wiley and Sons, New York, 1987.

Taha, H., *Operations Research,* 3rd edition, Macmillan, New York, 1982.

Webster, D.B., and Reed, R., Jr., "A Material Handling System Selection Model," *AIEE Transactions,* Vol. 3, No. 1, 1971, pp. 13–21.

CHAPTER

12

Storage and Warehousing

Managers have always sought a method for obtaining a continuous production flow in their plants. Ideally, the raw material coming in should immediately be processed and the final finished products promptly shipped, eliminating any need for storage at either end. This theoretical concept is appropriately called just-in-time (JIT). The idea can be extended within a plant as the product moves from one work station to another. A station should receive the item and the necessary parts just when it is due to process the unit; and upon completing the required task the station should transfer the unit immediately to the next station, which in turn is scheduled to receive the assembly just at that time. In JIT, one makes (produces) only what is needed only when it is required. JIT dramatically reduces the need for storage of raw materials, semifinished assemblies, and finished products. To an appreciable extent, however, the use of JIT is practical only for large, stable manufacturers. The company must be able to predict its raw material requirements in advance so as to coordinate the activities of all its vendors. This requires exact forecasting of the demand, which generally means a firm production schedule and confidence that all the items produced will be sold immediately. The vendors also attempt to implement their own JITs, but whether they achieve JIT in their own plants or not, they still work to deliver the manufacturers' orders on time, the main reason being that the customers will probably take their business elsewhere if the deliveries are not made when promised.

Though one should strive to achieve JIT, perfect implementation of such a system is not possible, and manufacturers will always have some storage requirement, however small. The need to store raw material, partially finished products, and finished goods must then be answered by storage and warehousing. The term storage is generally, but not always, associated with raw material and in-process goods, whereas warehousing refers to the storing of finished goods. A company might have one or more storage and/or warehouse facilities. In some cases, both storage and warehousing share the same building; in others, the storage might be located near the production facilities and the warehouses might be built separately to serve as distribution centers. Within a plant, however, the terms storage and warehousing are frequently used interchangeably to mean either facility.

12.1 WAREHOUSE OWNERSHIP

Once the need for a warehouse has been established, the next step is to decide whether a company-owned warehouse is necessary or a commercial warehouse can better be utilized. Commercial warehouses are built to serve many different customers, and they generally have sufficient personnel, equipment, and storage space to satisfy both the long-term and short-term needs of a customer. Commercial warehouses have two important advantages: felxibility and professional management. The manufacturer is not tied down to a specific warehouse location; as the distribution pattern changes, the company has the option of relocating its distribution centers. Unstable or seasonal demands can make renting the space on an as-needed basis more attractive than building a large warehouse to accommodate the maximum expected demand. If the demand for the company's product(s) is expected to continue for a long time, however, and if its commitment to the present location is strong, a private warehouse might be appropriate.

12.2 STORAGE/WAREHOUSE LOCATION

As in the case of plant location, the selection of storage and warehouse sites is very important. Some of the basic conclusions are obvious. If the warehouse is to contain mainly finished products, it should be close to the customers. If the material stored is to be used in manufacturing, the storage facility should be near the production plant. The location must have sufficient land and good transportation available. The site should not be cut off from suppliers and markets by geographical barriers such as rivers, mountains, and lakes. The site itself must be appropriately zoned by the local authorities and should have good fire and police protection. The company must also be able to obtain the necessary utilities and the required labor for operation of the facility. The site must be of sufficient size to accommodate any future expansion; as a general rule it should be about five times as large as the current needs dictate.

As a first step, a table similar to the one that will be shown for the plant site selection (Chapter 15) could be developed for the warehouse site selection as long as it is feasible to place the warehouse away from the plant. Appropriate weights could be assigned to each factor, and the evaluation would be made in a fashion similar to that shown in the plant site selection procedure. Many of the models developed in Chapters 2, 3, and 4 can be applied in such decisions as the necessary information is generally available to make such determination.

12.3 BUILDING

Factors to be considered in the construction of a warehouse (storage) building include the following:

· Location
· Size of the site

- Building placement
- Approach roads and railroad sidings
- Layout, dock site, and receiving and dispatching areas
- Column patterns and clear height needed for vertical storage
- Aisle layout and width and number, sizes, and arrangement of stacks
- Equipment to be used in material handling
- Lighting, heating, plumbing, and airconditioning

Building and Layout Considerations

A warehouse building within a plant or on a different site should receive the same basic considerations as a manufacturing plant. The location is important, since it determines the cost of transportation, and the plot should be large enough to handle present truck traffic and future expansions; furthermore, it should offer flexibility to change. The building itself should be large enough to handle present requirements and expected demands in the future for a 5- to 10-year period. A square building shape has proven to be very efficient, giving the shortest average distance to be traveled during pickup and distribution operations when the units having about the same demand are stored over the entire floor evenly. However, the plot size most often dictates a rectangular building. Up to some practical limit, the incremental cost of construction of a building decreases as the height increases. It is therefore more economical to construct a taller building (18–20 feet) than one with a broader base for the same volume requirement. A constraint is imposed on usable height, however, by material-handling equipment and its cost, as well as the cost of storage racks. The warehouse can also be multistoried, but the cost of operation in such a building is normally greater than that in a sin-

Warehouse aisles. The aisles here have been made just wide enough for movement of a forklift truck, and the material is stacked on pallets as high as is practical to maximize the use of valuable floor space.

gle-story structure of equal capacity. The structure of warehouse exterior walls is most often a steel frame with concrete blocks or 4-inch-thick brick walls. The floor of the building is usually made of reinforced concrete to ensure strength, rigidity, and uniformity; and it should be able to support the weight of the stored material and equipment.

Aisles are necessary in a warehouse for allowing material-handling equipment to reach different parts of the storage areas. From another viewpoint, however, aisles also constitute wasted space, a space that is not used for the main purpose of the warehouse, which is to store material awaiting use or shipment. To minimize this loss, many warehouses have two types of aisles, main or working aisles and utility or secondary aisles. The main aisles are wide (generally 10–12 feet) to allow a material-handling unit such as a forklift truck to operate.

The utility aisles are used to gain access to racks, offices, elevators, and utility rooms, and they are much narrower—anywhere from 2 to 9 feet—depending upon the size and configuration of the units being processed and the material-handling equipment required for this function. The main aisles connect the receiving and shipping areas, and their placement dictates the material flow. Figure 12.1 presents some of the possible aisle configurations.

The aisle space percentage ratio defined in Chapter 10 is a good measure to identify the efficiency of aisle space allocation. This requirement can be reduced considerably by using the automated storage and retrieval systems to be described in Section 12.8.

The loading docks for the building are described in detail in Section 12.9. Other service facilities that are generally available in the building include both 110 V and 220 V electricity, water, sprinklers, drains, fire extinguishers and fire exits, telephones, compressed air in maintenance areas, battery chargers, and toilets.

Space Determination

It is important that a building has sufficient space to accommodate all the items we intend to store, and yet is not overly large to keep the cost of construction and maintenance down. The following examples illustrate the procedures necessary to determine the space requirement for the building. These methods should be followed for each item in storage.

Example: Floor Space Determination 1

A plant produces 75 units per hour of an item with dimensions of $0.5 \times 0.5 \times 0.1$ foot. Management wishes to store a one-week supply in containers measuring $7 \times 7 \times 4$ feet. A minimum of 3 inches of space is required between adjacent units in each direction for packaging and handling. Determine the number of containers needed. If these containers may be stacked two high, determine the floor space required.

Solution: If there are n items in a row (or column), then $n + 1$ packing spaces must be provided (see Figure 12.2). Thus the number of units that can be stored in each direction (n_1, n_2, and n_3) must satisfy the equation $n_i w_i + (n_i + 1)s_i = d_i$, where w_i is the dimension of the unit, s_i is the packing space required, and d_i is the dimension of the package in the i direction, respectively.

Figure 12.1   Aisle configurations

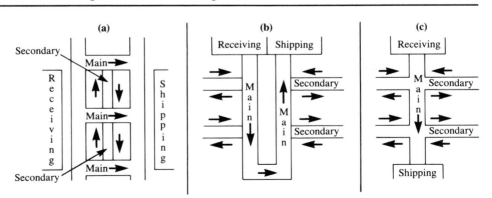

Figure 12.2   Packaging of the units

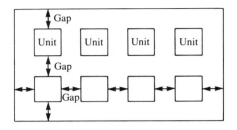

The length of the box is 7 feet, and therefore

$$0.5(n_1) + 0.25(n_1 + 1) = 7$$

Hence the maximum n_1 equals 9.1 ≈ 9.

The width of the box is also 7 feet; therefore

$$0.5n_2 + 0.25(n_2 + 1) = 7$$

Hence n_2 is also equal to 9.

The height of the box is 4 feet; thus

$$0.1 \times n_3 + 0.25(n_3 + 1) = 4$$

Hence $n_3 = 3$.

Accordingly, the total number of units that can be stored in a box is $n_1 \times n_2 \times n_3$ or 243.

Now let us calculate the number of units to be stored in a week and the associated container requirements:

75 units/hour × 40 hour/week
$$= 3000 \text{ units/week}$$

$$\frac{3000 \text{ units}}{243 \text{ units/container}}$$
$$= 12.34, \quad \text{or 13 containers}$$

$$\frac{13 \text{ containers}}{2 \text{ containers/stack}} = 6.5, \quad \text{or 7 stacks}$$

The floor space needed for each stack is

7 × 7,   or 49 square feet

49 square feet × 7 stacks
= 343 square feet of floor space required

Example: Floor Space Determination 2

A plant produces 100 cartons a day to be stored in the warehouse. The company would like to keep a three-day supply of stock in inventory as a safety cushion to fall back on if production is temporarily interrupted. The maximum time between orders is 30 days. Determine the maximum and average amount of stock to be stored. If each unit is 4 × 3 × 2 feet and if they may be stacked six units high, determine the floor space required for storage. (When small loads are being stacked, aisle space of 5–10 percent of the total storage area should be allowed; however, the required space would increase with the size of the load to as much as 30 percent. For this problem, 8 percent will be used.)

Solution:

Maximum units = Inventory for safety
in stock   + Number of units for a
                  single order
        = (3 days)
          × (100 units/day)
          + (30 days)
          × (100 units/day)
        = 300 + 3000
        = 3300 units

Average units = Inventory for safety
in stock   + ½ of the units for a
                  single order
        = 300 + ½(3000)
        = 1800

The maximum storage required is for 3300 units.

$$\frac{3300 \text{ units}}{6 \text{ units/stack}}$$
$$= 550 \text{ stacks (each 4 feet × 3 feet)}$$

550(4 × 3) = 6600 square feet

6600 × 0.08 = 528 square feet for aisles

6600 + 528

    = 7128 square feet of floor space required

12.4 STORAGE/WAREHOUSE FUNCTIONS

In managing a storage and warehouse facility, one must perform many differnt activities related to the processing of raw materials, semifinished products, and finished goods. These tasks range from receiving, inspecting, and storing raw materials to packing, labeling, and shipping orders. The following is a brief description of common activities.

1. *Receiving:* The warehouse receives the material from outside suppliers and accepts responsibility for it. The operation consists of unloading the goods from trucks and/or railroad cars and unpacking the containers.

2. *Identifying and sorting:* Material is identified and then recorded by using tags, codes, or other means. The items are sorted to find any breakage, and shortages are determined by checking receipts versus packing slips. Appropriate action is taken to inform the shippers and vendors of any discrepancies.

3. *Dispatching to storage:* The goods are transferred to appropriate areas for storage.

4. *Storing:* The units are held in inventory until needed.

5. *Picking the order:* Items needed for an order are retrieved from storage. Picking of the items for a particular order can be accomplished by one or more people depending upon the number of items and their locations in the warehouse.

6. *Assembling the order:* All items in a single order are grouped together. Any shortage, breakage, or nonconforming item is recorded, and the item is replaced or the order is modified.

7. *Packaging:* All units in an order are packed together.

8. *Dispatching the shipment:* Appropriate shipping orders and documents are prepared, and the order is sent to the transport vehicles.

9. *Maintaining records:* Records such as the following are kept for each item: the amount received, on-hand inventory, orders received, and orders processed. These records are critical for good inventory management.

12.5 STORAGE AND WAREHOUSE OPERATIONS

Within a plant, management must decide whether to build a centralized warehouse or multiple storage facilities, each near its point of use (e.g., near each assembly line station). The latter approach reduces material handling and halting of production due to delays in delivery from a centralized warehouse. It also allows for tighter inventory control. Many times, such storage facilities can be built to use space that would otherwise not be utilized.

Storage Policies

Within a storage facility, several policies influence its layout, locations of storage cells, and assignment of items to these cells. These policies are briefly described below.

1. *Physical similarity:* Items with similar physical characteristics are grouped together in one area. For example, large items are stored in one area, while small items are located in another. This allows the use of similar material-handling equipment and similar physical care for each area. Also, special environmental controls such as refrigeration, humidity, and fire safety can be concentrated in one area as the needs of the items dictate. (See Figure 12.3.)

Figure 12.3   Storage arrangements (layout pattern based on similarity, either physical or functional)

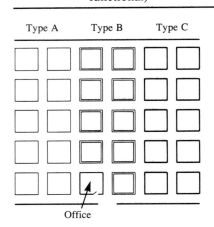

Type A   Type B   Type C

Office

Storage and warehouse operations. In the area storage system, items are stored in a logical way, enabling warehouse workers to easily collect the items required for an order, using equipment such as this high-rise battery-powered forklift truck. This particular piece of equipment also helps to reduce air pollution in a confined space such as a store room.

2. *Functional similarity:* Functionally related items can be stored together. For example, electrically, hydraulically, and mechanically operated items are grouped in segregated storage areas. The system is especially convenient in manually operated storage facilities in which each warehouse worker becomes knowledgeable in a specific functional area. (See Figure 12.3.)

3. *Popularity:* Every warehouse has items that are retrieved more often than others. In this system these fast-moving items are stored close to receiving and shipping areas, and the slow-moving items are assigned to spaces that are farther away. This arrangement minimizes the distance traveled by warehouse workers in picking orders. Actual studies have shown that, on the average, 15 percent of the goods account for 85 percent of the work in a warehouse. (See Figure 12.4.)

4. *Reserve stock separation:* It may be advantageous to separate reserve stocks from working stocks. All working stocks are kept together in a compact area from which picking is relatively easy. Reserve stocks from outlying areas replenish the working stocks as the need arises.

5. *Randomized storage:* With modern information-processing systems (computerized inventory control systems) it is no longer necessary to assign a fixed and unique location to an individual stock item. Changing from dedicated storage to randomized storage might result in considerable savings in the space requirement for the warehouse. The items are stored in spaces that are available when needed

Figure 12.4   General layout pattern based on popularity for one-entrance storage buildings

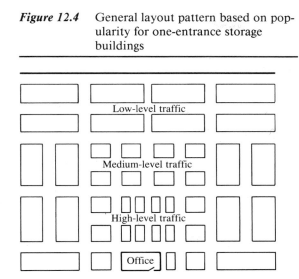

without reserving any space for items that are not currently in stock.

6. *High-security storage:* If there are items that are particularly valuable and subject to significant pilferage (e.g., gold, watches), an area might be needed that is under lock and key and/or other security measures.

Order-Picking Policies

Another important factor affecting the performance and layout of a warehouse is the policy followed in filling an order, called order picking. There are several such policies; we will limit our description to some of the more popular ones.

1. *Area system:* Items are stored in the warehouse in some logical manner. The warehouse personnel circulate through the area, picking the items required for an order until the entire order is filled.

2. *Modified area system:* This system is applicable where reserve stocks are separated

from working stocks. Order picking follows the area system, while secondary personnel are utilized to replenish the working stock from the reserve stock.

3. *Zone system:* The warehouse is divided into zones, and the order is distributed among the order pickers, each picking units from his or her assigned zone.

4. *Sequential zone system:* Each order is divided into zones as in the zone system, but the order is passed from one zone to another as it is assembled. Many orders can be processed simultaneously as each proceeds from one zone to the next.

5. *Multiple-order or schedule system:* A group of orders is collected and analyzed to determine the total items needed from each zone. In a manner similar to the zone system, these items are picked by making one trip through each zone. The orders are assembled in a common area for further dispatching. A slight variation of this operation is scheduling simultaneous arrival of parts from each zone associated with each order, then putting them together for dispatching.

The area system is the simplest of all and is widely used when the average number of items in an order is not large. If this number increases, the order is either picked simultaneously (zone system) or sequentially (sequential zone system). The multiple-order system is beneficial only when there are large numbers of orders, each containing but a few items to be processed.

Example: Storage Arrangement

Four different items are to be stored in the warehouse shown in Figure 12.5. Table 12.1 shows the number of pallets received each week, the number of trips from receiving to storage, the average size of each order shipped, and the number of trips from storage to ship-

Table 12.1   Data on load movement

Item (1)	Pallets Received (2)	Average Receiving Trips per Week (3)	Average Pallets Shipped per Week per Shipment (4)	Shipping Trips per Week (5) = (2)/(4)	Sections Needed (6) = (2)/100
A	275	138	2.7	102	3
B	425	213	2.0	213	5
C	150	75	0.4	375	2
D	550	275	1.2	459	6

ping. Each of the 16 sections of the warehouse stores 100 pallets. The rectilinear distance from section to section is 10 units. Determine the most efficient storage arrangement for the warehouse.

Solution: First determine the ratio of receiving trips to shipping trips as shown in Table 12.2. The items with the higher ratios have more trips from receiving to storage than the reverse. Therefore these items should be located as close to receiving as possible. Items with ratios less than 1 have more trips to shipping and should be located as close to shipping as possible. That is, Item A should be close to receiving, and Items C and D should be located

close to shipping in that order. Item B, with a ratio of 1.0, may be placed in any available space.

This solution is based on an implied understanding of the workings of the warehouse. For instance, for Item A for some order, two pallets might be carried from the warehouse and shipped; in another order, three pallets might be carried and shipped, giving an average of 2.7 per trip. If for any reason only two pallets can be carried at a time, then the average shipping trips should be modified to 275/2 ≈ 138, and a new solution to the problem develops.

Table 12.3 shows the distances calculated for rectilinear travel to each section from the receiving and shipping departments. The assignment of the items now can be made as follows. Item A requires the three sections closest to receiving: 1, 2, and 3. Item C requires the two sections closest to shipping: 15 and 16. Item D requires the six next closest sections to shipping: 8, 10, 11, 12, 13, and 14. Sections 4, 5, 6, 7, and 9 are left for Item B. Figure 12.6 shows the final warehouse assignments.

Figure 12.5   Initial warehouse layout

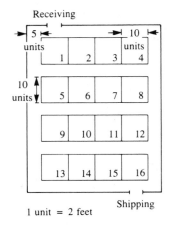

1 unit = 2 feet

Table 12.2   Calculation of ratio of receiving/ shipping for each item

Item	Receiving/Shipping
A	138/102 = 1.35
B	213/213 = 1.00
C	75/375 = 0.20
D	275/459 = 0.60

Table 12.3 Data calculations for each section

Section	Rectilinear Distance to Receiving	Rectilinear Distance to Shipping
1	10	95
2	20	85
3	30	75
4	40	65
5	35	80
6	45	70
7	55	60
8	65	50
9	50	65
10	60	55
11	70	45
12	80	35
13	65	40
14	75	30
15	85	20
16	95	10

Figure 12.6 Final warehouse layout

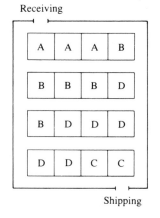

12.6 ACCESSORIES

For storing individual and/or small items, various accessories are available. Almost all storages and warehouses will utilize one or more of these to develop an orderly storage facility.

1. *Bins:* There are many different sizes and shapes of bins that can be used when a great variety of small parts is to be stored.

2. *Shelves:* In most cases, steel or wooden shelving is used for the storage of unpalletized loads or large items. Steel shelving is simply sheet metal that has been fastened to upright posts. The posts usually provide for flexibility in adjusting the height of the shelves and the vertical spaces between shelves.

3. *Racks:* The most commonly used items in storage are racks, which are described in detail in the next subsection.

4. *Stacking:* Unit loads on pallets or in boxes, bags, or sacks can be stacked on top of each other to better utilize vertical space.

5. *Conveyor storage:* Conveyor racks can make an effective storage accessory. These are a series of roller or skate wheel conveyors placed one above the other in adjacent stacks and slanted from input to output. The stored items should be contained in boxes or tote pans. Very compact storage can be achieved by utilizing automatic loaders and eliminating excess aisles.

6. *Yard storage:* High construction costs of enclosed warehouses have made outside storage more desirable for bulk items such as coal, sand, pulp wood, and scrap metal. The availability of containers and protective coatings has also contributed to greater use of yard storage.

Storage Racks

The purpose of storage racks is to facilitate storage and retrieval of loads in the warehouse. The racks are commonly made of steel frames with vertical posts and horizontal bars to support the loads with additional strength provided by diagonal or X-braces. Different types of racks are available in the market, and the decision as

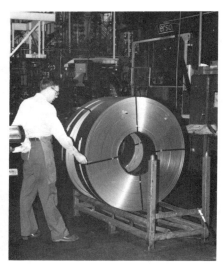

Accessories. *Top left*: A monorail conveyor finds application in a dry-cleaning plant. *Top right*: A selective pallet rack, as shown here, offers versatile storage. Because of its simplicity, this type of rack can be customized to suit the items to be stored such as metal sleeves of different diameters and boxes containing loose parts. *Bottom left*: Steel shelves are commonly used to store unpalletized items. In the steel warehouse storage system shown here, the shelf heights have been adjusted to fit different sizes of items. *Bottom right*: Here, a portable strip and wire rack holds two coils of 8″ metal strips. Portable racks such as this one can be moved with or without their contents for flexible and efficient use of storage space.

to which one to use depends on the type of material that is to be stored. The following is a description of a few typical racks.

Selective Pallet Racks

The most commonly used storage rack is the selective pallet rack. This rack is formed with several pairs of supporting posts, the number depending on the length of the rack, several pairs of longitudinal beams placed to accommodate the height of the utilized pallet loads, and a number of horizontal braces at right angles to the beams in the rack to support the load. The depth of storage can be increased by placing two or more racks adjacent to each other. For example, two adjacent racks give storage room for two deep pallets, three adjacent racks can increase the room to stack pallets three deep, and so on. Owing to the simple design of this type of rack, it can easily be customized to provide maximum efficiency for almost any operation. A few common modifications are back-to-back ties (used to join two adjacent racks), drop-in skid supports, drop-in front-to-rear members, drum supports (used to support barrels), and deck surfaces.

Movable-Shelf Racks

The movable-shelf rack is simply a selective storage pallet rack that has been modified to make it mobile. The modifications include diagonal rear bracing and a permanent top shelf that provides more stability to the rack during movement. This rack can be moved with machinery that is joined to the shelf by lugs that extend from the support beams of the shelf. The arms of a forklift or another vehicle are placed under these lugs and raise the rack. If there is a need to place these racks end-to-end, connectors will have been placed at both ends of each rack. In some cases the rack may take the place of a pallet.

Drive-In and Drive-Through Racks

Drive-in and drive-through racks are systems of racks arranged to form a central aisle. There are also several tunnels that are perpendicular to the central aisle. On each of these sections it is possible to stack a number of pallets. This configuration makes it possible to drive a forklift down the central aisle for loading and unloading. This storage system is useful when a large number of the same items requires storage, for example, beer, tobacco, and frozen foods.

Cantilever Racks

Cantilever racks are used for items that are extremely long (bar stock, for example) and do not conform to other types of storage racks. These racks consist of arms that are fastened to two or more support posts, depending upon the size of the item being stored. In some cases it is convenient to have arms on both sides of the racks, called double-sided cantilever racks.

Stacker Crane Racks

Stacker crane racks, used with automated storage and retrieval systems (described in Section 12.8) and characterized by their extreme height, utilize the available vertical space in a warehouse. These racks usually range in height from 50 feet to 100 feet tall. Because of its extreme height, the basic structure of a stacker crane rack differs from that of most other racks. These racks are built with only one bay opening and the upright frame assemblies. To provide the needed stability, overhead ties and many braces are attached to the racks. The pallets can be supported by rails or arms such as those used on a cantilever rack. Stacker cranes are needed to load and unload the pallets in such a system. Other accessories that are found with these racks include overhead guide rail supports, building attachments, conveyor supports, mezzanine attachments, and loading/unloading stations.

Portable Racks

Portable racks are designed to be mobile, with or without a load. Some of these racks also come equipped with knockdown or nesting features, allowing racks to be disassembled and stored compactly when not in use. The purpose of this is to provide the best possible utilization of the available space. Two of the more common portable racks are portable drum stacking racks and a portable rack for rolled strip or wire.

Rack Buildings

In some cases the racks support the roof and sides of a building. These rack buildings greatly reduce the cost of the storage facility. It is necessary to determine the roof live loads and wind loads to be sure that no building codes are violated.

Fire Prevention

With many wooden pallets supporting cardboard boxes or some other combustible product, storage facilities are in danger of fires. Fire tests have shown that the best fire prevention method is a water sprinkler system. The type and location of the sprinkler head will depend upon the items stored and their configuration.

For further information about storage racks and systems the Rack Manufacturer's Institute may prove valuable. The RMI may be contacted at 1326 Freeport Road, Pittsburgh, Pennsylvania, 15238. See Table 12.4 for typical costs of some types of racks.

12.7 STOCK LOCATION

A system to identify the location of items of stock should be developed to permit quick and easy access to the right unit when needed. The significant location symbol system is one such

Table 12.4   Typical 1986 costs for one linear foot 8′ high × 36″ deep sections with 4000-pound capacity

Type of Rack	*Cost*
Pallet	$60/foot
Drive-in and drive-through	$90/foot
Cantilever	$70/foot
Portable (generally 3′ to 6′)	$85/foot

coding system. It consists of a nine-digit number (e.g., 152012102); the first two numbers identify the building, the next one the floor, the next three the row, the next two the stack number, and the last digit the level as shown below:

Building	Floor	Row	Stack	Level
15	2	012	10	2

The code can be modified to suit the building and its layout. For example, in a building with one major aisle and different spacing between the stacks as shown in Figure 12.7, the areas can be identified by the letters A and B, and a code such as A12153 would identify the location as being in section A, row 12, stack 15, third level.

Figure 12.7   Storage building layout

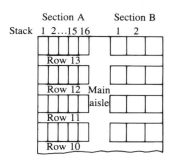

12.8 AUTOMATED STORAGE AND RETRIEVAL

In recent years the automated storage and retrieval (AS/R) system has had a most dramatic impact on storage and warehousing operations. These high-rise storage units are becoming commonplace in companies that deal with numerous items in large volumes, those unable to expand their present warehouses because of the limitation of the available floor space, plants that operate in the environment of high labor cost and/or extended travel and search times in the warehouse, and those for whom delivering all orders accurately is critical. Unlike traditional warehouses, where the record of location and inventory levels may be kept manually, in AS/R such controls are maintained by computers within the system. In addition, the storage capacity of the warehouse is expanded twofold to fivefold by using the same floor space but installing racks that allow high-density storage and by using specialized equipment that can work in narrow aisles. It is quite common to observe racks that are 80–90 feet high being served by computer-controlled machines (stacker cranes) carrying 3000–4000 pound loads. They can travel at speeds of 500 feet per minute in aisles that are only six inches wider than the narrow aisle cranes. Computer control of inventory can account for savings of as much as 20 percent of the inventory cost. Such a fully mechanized system requires very little manpower to operate; a single person can operate a warehouse containing thousands of parts. The system also minimizes the need for material-handling equipment and even material handling itself by reducing the average distance traveled and by being correct every time in identifying the location of an item. Pilferage and breakage, which are generally proportional to the amount of material handled, are also reduced. The AS/R system does, however, require a high initial investment, and a thorough economic analysis must be performed to deter-

Automated storage and retrieval. In an AS/R system, a warehouse's capacity and efficiency can be markedly increased. High-density storage facilities served by narrow aisle cranes maximize the use of space while minimizing pilferage. Also, fewer workers are needed to operate this system, yet a large number of many different items can be handled.

mine the feasibility of using such a system in a particular warehouse.

The AS/R system has four major components. They are S/R machines, the storage structure, conveying devices, and controls.

Storage retrieval cranes form one of the most important parts of the system. These machines can carry heavy loads and can simultaneously move horizontally and vertically to reach the required location. They travel on floor-mounted rails guided by electrical signals and may be equipped to function in the single-command mode, allowing the machines to either store or retrieve in a trip, or they may have double-command mode capability, being able to perform both tasks in one trip.

The storage structure interfaces with S/R machines and, in doing so, requires very close tolerances in its construction. The guide rails within it must allow the S/R cranes to move in and out freely, stopping exactly at the required cubbyhole. The structure itself must be adequately protected against fire. These high-rise structures (heights of 90 feet are not unusual) can be free-standing within a building or can form part of the supporting frame of the building. In the overall design of the structure, as we will see in the next section, many factors play important parts. They include load characteristics such as weight, shape, and size of the items to be stored and environmental factors such as dimensions of the building, activity rate, and limitations imposed by the cost of construction.

Conveying devices are the auxiliary equipment that interface with the S/R machines and various departments within a plant such as shipping, receiving, and manufacturing. They include forklift trucks, various kinds of conveyors, towline and shuttle trolleys, guided vehicles, and other equipment. The items are transported by the conveying devices to and from the department, while they are further handled by the S/R machines in the storage structure. Many of the devices such as conveyors and guided vehicles can be made interactive with the AS/R system's computer to achieve maximum flexibility and rapid real-time response.

Computers and their software and controlling mechanisms that tie the computers to the S/R and auxiliary machines are the key to the control of the AS/R system. They process the information and activate the necessary equipment. Most modern systems use several small computers, each controlling a separate device, rather than one large computer controlling the entire system. These small processors communicate with one large computer, which is in charge of inventory maintenance, cost calculation, and billing information. Such an arrangement is called a distributed system. It is flexible in design and use; despite the failure of one computer, the system can still operate, and the installation and maintenance of small computers are more easily accomplished than would be the case for one large computer. The distributed system also provides a faster response than does a system consisting of a single computer.

Design of AS/R Systems

To design an AS/R system means determining all three dimensions of the physical storage space. A vertical stack or storage, going from floor to ceiling, is referred to as a bay; a series of bays placed side by side are called rows, and the spaces between the rows form the aisles (Figure 12.8). The aisles are used for stacker cranes to move up and down between the rows. Each crane can serve both sides of the aisle and, as mentioned earlier, can be either a single- or dual-cycle crane. A dual-cycle crane can start from a pickup station, store an item, retrieve another item, and return to the pickup station in one command, whereas a single-cycle crane can work only one phase of the cycle at a time. For example, if the command is to store an item, the crane will go from the station to storage and perform the storing operation, but it must wait for a retrieve command before it can fetch another item from the storage and return to the pickup station. Typically one crane can perform 32 single operations per hour or 22 dual cycles per hour. Single operational mode is preferred for speed if the operating policy is to store all items at one time and retrieve all the necessary items at some other time. On the other hand, if storing and retrieving are to occur simultaneously, the dual-cycle crane is

Figure 12.8

AS/R system definitions

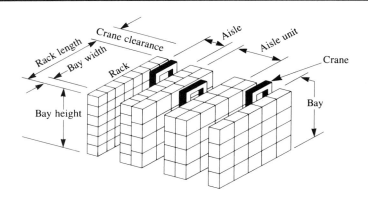

preferred. The efficiency of a crane system depends on several factors that go into making up the system. Generally, 85 percent availability is considered to be the average attainable.

The steps involved in designing AS/R systems are as follows:

1. Determine the dimensions and weight of the load to be stored.
2. Determine the number of units to be stored.
3. Determine the throughput rate per hour.
4. Determine the number of cranes needed.
5. Determine the number of rows required.
6. Determine the building height and load height.
7. Determine the number of bays.
8. Determine the system length.
9. Determine the system width.

In determining the sizes of the loads involved, the storage will typically carry a variety of items in unit loads, many of which could be irregular in shape; therefore the maximum required length, width, and height should be noted. Load orientation also plays an important part; loads are to be stored parallel to the cubbyhole or at some angle and will affect the width and depth of the required storage space.

The weight of the loads influences the design of the structure; the maximum storage weight of an individual unit load should therefore also be noted.

The next step is to estimate how many storage spaces will be needed. In developing this number, one must consider not only the present requirements, but also the expected demands in the foreseeable future for the facility. About two years of lead time are required in building an AS/R system, and therefore the demand for at least two years in the future should be estimated so that the storage does not become inadequate soon after it goes into operation.

The third step is to determine the throughput. This value is the sum of the maximum number of loads in and loads out that will be carried per hour.

In developing the necessary number of cranes, one must first fix the mode of operations that will be most often used: single- or dual-crane systems. The number of cranes then can be determined as follows.

For single-cycle cranes:

$$\text{Number of cranes} = \frac{\text{Throughput}}{32 \text{ (cycles/hour)} \times 0.85 \text{ (efficiency)}}$$

For double-cycle cranes:

Number of cranes

$$= \frac{\text{Throughput}}{22 \text{ (cycles/hour)} \times 0.85 \text{ (efficiency)}}$$

Since one crane can serve two rows, the number of rows in the system is obtained by multiplying the number of cranes by 2.

A note about operational cycles per hour for a crane: P & H, a major manufacturer and system installer of AS/R systems, suggests using 32 and 22 cycles per hour for single-cycle and double-cycle cranes, respectively, for the initial design. The cycles per hour might change slightly depending on the length of the aisle and the length-to-height ratio. It is therefore important to contact the manufacturer after planning the initial design to obtain more accurate estimates.

The next step is to determine the height of the storage. This height may range from 30 to 90 feet; however, the most efficient height is between 50 and 70 feet. We should choose a value within this range and determine how many loads can be stacked in such a structure. Light loads of less than 2500 pounds require a 6-inch clearance for rack support and crane entry, while heavy loads of over 2500 pounds demand a slightly larger clearance of 9 inches. The total number of stacks is obtained by dividing storage height by the sum of the load height and the needed clearance and then subtracting 1 to allow for floor and ceiling clearance.

The number of bays (a bay is one vertical stack of storage from floor to ceiling) needed can now be calculated by applying the following relationship:

Number of bays

$$= \frac{\text{Number of units to be stored}}{\text{Number of rows} \times \text{Number of stacks}}$$

The length of the storage facility depends on the bay width and the number of bays determined thus far. The length of a storage rack is equal to the width of the bay times the number of bays in a row. To this we must add 25 feet for crane runout clearance and an additional allowance for any equipment.

Finally, we must determine the width of the system. To begin, we should establish an aisle and the two adjacent storage racks. A storage cell should be deep enough to accommodate the load, and the aisle should be wide enough to facilitate movement of the load. Thus the aisle unit, which includes 2 times the storage width plus the aisle width plus clearance, should be at least equal to 3 times the depth of a load plus 2 feet for clearance. The width of the storage system is then the aisle unit times number of aisles in the system.

For example, consider the following problem. We wish to store a unit load on $42'' \times 48''$ pallets that is 48 inches high and has a weight of 2000 pounds. Fifty dual-cycle transactions are anticipated per hour, and the total storage requirement is for 10,000 unit loads.

The load characteristics are the following: length, 48 inches; width, 42 inches; height, 48 inches; and load weight, 2000 pounds. Suppose we select the height of the storage building as 60 feet. Then the numbers of stacks that can be accommodated with a 4-foot height load is (taking only the integer portion of the quotient and allowing for 6-inch clearance between stacks)

$$\frac{60}{(4 + 0.5)} - 1 = 13.33 - 1 = 12 \text{ loads}$$

The number of dual-cycle cranes necessary to handle a throughput at 50 transactions per hour are $50/(22 \times 0.85) = 2.67$, or 3 (round off to the next higher integer value of division).

The number of bays necessary would be

$$\frac{10,000 \text{ storage units}}{2 \times 3 \text{ cranes} \times 12 \text{ loads high}}$$
$$= 138.8, \quad \text{or } 139 \text{ bays}$$

For each bay the width is 42 inches for the

load plus 6 inches clearance, giving a total of 4 feet. The length of the storage is then $4 \times 139 + 25$ feet for crane clearance, or 581 feet.

To obtain the width of the system, multiply the aisle unit by the number of aisles (cranes). The depth (length) of a unit $\times 3 + 2$ feet gives the aisle unit, which is $4 \times 3 + 2 = 14$ feet. The width of the system is then $14 \times 3 = 42$ feet. Thus the storage dimensions are $42 \times 581 \times 60$ feet.

It is obvious that by choosing different heights for the building it is possible to obtain different dimensions for the storage building.

Cost Estimate

P & H (1984) uses the cost parameters in Table 12.5, accurate to within 15%, as of 1984.

Changes from that date can be approximated by using the Industrial Commodities Index or Iron Age Steel Index.

For the AS/R system configuration developed in the example, the estimated cost can be calculated as in Table 12.6.

The price will range somewhere between the low value and the high value. For budget purposes, one might consider the cost to be $3.78 million, the average of two values.

Order Picking in an AS/R System

Order picking can be either for a full unit load or for part loads. In a full unit load, two methods may be used: man-ride or out-of-aisle. In the man-ride system the operator rides an S/R machine, picks the goods from storage, and re-

Table 12.5   Estimated costs for AS/R components

Equipment	Cost	
	Low	*High*
Storage spaces (drive-in design), each		
Free-standing racks (building not included)	$100	$160
or rack-supported building (roof, siding, slab, HVAC, sprinklers, lighting not included)	$150	$220
Floor-running, automated crane, each	$210,000	$275,000
Aisle transfer car (if required), each	$140,000	$185,000
Dedicated warehouse computer system		
Basic control and inventory	$80,000	$150,000
Full function	$150,000	$800,000
Aisle hardware for crane or crane transfer car (included floor rail, top rail position sensors, electrical wiring, hydraulic end stops)		
Cost per aisle	$15,000	$20,000
Plus cost per foot	$95	$160
Conveyors (powered chain or roller type)		
Cost per aisle (input and output spurs)	$50,000	$70,000
Plus cost per foot (all other conveyors)	$500	$1000
Project services (includes project management, site supervision, and training)	$50,000	$200,000

Table 12.6 Cost determination for AS/R system

Equipment	Cost Low	Cost High
Storage space		
Rack-supported building 12 loads high × 139 bays × 6 rows = 10,008	$1,501,200	$2,201,760
Cranes, 3 aisles, automated	$630,000	$825,000
Dedicated warehouse computer, full function	$150,000	$800,000
Aisle hardware		
3 aisles	$45,000	$60,000
581 feet/aisle × 3 = 1743 feet	$165,585	$278,880
Conveyors, 3 aisles with I/O spurs	$150,000	$210,000
about 200 feet	$100,000	$200,000
Project services	$50,000	$200,000
Total cost	$2,791,785	$4,775,640

trieves the entire order. The operator is assisted by auxiliary lifting devices and by a computer terminal in the carrying platform, which displays information such as which item to pick, where it is in storage, and what quantity is required.

In out-of-aisle picking, the unit loads (boxes) are picked and brought to the ends of the aisles by machines. The operator then assembles the order, and the remaining items are returned to storage.

12.9 LOADING DOCKS

Let us turn our attention to loading/unloading docks, another important feature in storage operations. Design of loading and unloading docks is a task that depends on the type of transportation used and the amount of material handled. Some warehouses load directly onto planes or trains and require special docks to accommodate these situations. Most loading (and the type concentrated on here) is from warehouse loading dock to truck. The amount of

material handled by weight at the busiest time of day divided by the productivity rate of 7500 pounds per man-hour is used to determine the minimum number of docks needed. For example, suppose that trucks hold an average of 4290 pounds each; then it takes 0.572 man-hour to unload each truck. If 10 minutes are allowed for the truck to leave the dock and the next one to pull up, then 44.3 minutes are needed to serve each truck. Therefore one dock can serve $60/44.3 = 1.35$ trucks per hour. If n trucks are to be served per hour, the number of docks needed would be $[n/1.35] + 1$ where the quantity in brackets stands for the integer portion of the division. For example, to serve three trucks per hour will require $[3.0/1.35] + 1 = 3$ docks.

One might consider the operation of docks as a queuing phenomenon, with trucks arriving randomly (following a Poisson distribution) and the time of service being exponentially distributed or following some other probability distributions. The standard formulas are noted in Appendix C.

In our example, suppose now that the ar-

Receiving and shipping. *Top left*: Loading docks must be appropriately equipped for the types of loads they will handle. Here, a leveling platform is in position for unloading. It makes adjustments for any unevenness between the forklift truck bed and the dock platform, giving a constant gradient surface on which to operate. *Top right*: Size and elevation are important features of loading docks that depend on the type and number of vehicles to be loaded as well as the material being handled. Special docks are required to load directly onto trains or airplanes. Here, a forklift truck loads a rail car. *Bottom right*: The apron depth of a dock must be sufficient for the longest truck that the dock will service. Here, in this exterior view of a loading dock, the dock and apron, which includes several truck bays, are covered by a roof for weather protection.

rival pattern of trucks follows the Poisson distribution (random) with a mean of 3/hour and that the time for service follows an exponential distribution with a mean of 1.35/hour, that is, $\lambda = 3$/hour and $\mu = 1.35$/hour, respectively. Clearly, since the service rate is less than the arrival rate, we need more than one dock.

Assume that we decide to provide three docks. The corresponding characteristic values are (since $\lambda/\mu = 2.22$)

$$P_0 = \left[\frac{(2.22)^0}{0!} + \frac{(2.22)^1}{1!} + \frac{(2.22)^2}{2!} \right.$$
$$\left. + \frac{(2.22)^3}{3!\left(1 - \frac{2.22}{3}\right)} \right]^{-1}$$

$$= 0.0788$$

$$P_1 = \frac{(2.22)^1}{1!} \times 0.0788 = 0.175$$

$$P_2 = \frac{(2.22)^2}{2!} \times 0.0788 = 0.194$$

$$P_3 = \frac{(2.22)^3}{3!} \times 0.0788 = 0.143$$

P_0, P_1, and P_2 represent probabilities of having zero, one, and two trucks in the system, respectively. With three docks in the warehouse, the sum of these probabilities indicates the probability that an arriving truck will find an empty dock and will not have to wait. The average number of trucks in the queue can now be determined:

$$L_q = \frac{(0.0788)(2.22)^3 \left(\frac{2.22}{3}\right)}{3!\left(1 - \frac{2.22}{3}\right)^2}$$

$$= 1.55 \text{ trucks}$$

$$L = \frac{\lambda}{\mu} + L_q = 2.22 + 1.55$$
$$= 3.77 \text{ trucks}$$

The average waiting time for the truck is $W_q = 1.55/3 = 0.5166$ hour, and the average time spent by a truck on the premises is

$$W = W_q + \frac{1}{\mu} = 0.5166 + \frac{1}{1.35}$$
$$= 1.257 \text{ hours}$$

Both quantities can be reduced by increasing the number of docks. We can determine the optimum number of docks by performing an economic analysis similar to ones shown below. Suppose the cost of operating one dock is $12/hour, which includes the cost of an operator, forklift truck, and construction and maintenance cost of the dock spread over an hourly basis. The cost of operation of the truck is $16/hour, including the driver's pay and truck expenses. Since each arriving truck spends 1.257 hours on the premises, it costs $1.257 \times 16 = \$20.112$ per truck. On the average, three trucks arrive per hour, giving an hourly cost of $3 \times 20.112 = \$60.34$ for truck operations. We must add to this the hourly cost of operations of three docks, which is $3 \times 12 = \$36.00$, giving a total hourly cost for the loading/unloading operation of $96.34.

Similar calculations for a different number of docks lead to the following results:

Number of Docks	Cost of Loading/ Unloading per Hour
3	$96.34
4	$88.19
5	$96.66

Therefore on the basis of present data we should have four docks in the plant.

Important aspects of a dock include the length, width, and elevation. The elevation of the dock should be 48 inches for pickup trucks and 52 inches for larger trucks. If a variety of vehicles is used in loading and unloading operations, a scissor dock (table) can be used to vary the elevation. The width and length of the dock depend on the size of the load to be stored on it before being moved into the warehouse or into the truck. The space should be at least 12 feet wide to accommodate a truck. Other factors of importance are the volume of freight, the size of units handled, and the material-handling equipment used.

Suppose a warehouse will handle pallets with a maximum size of 48 square inches. The management expects to have no more than six pallets in storage on the dock at one time. At least 3 inches of clearance between the pallets is needed. To accommodate truck loading and unloading, the dock is already 12 feet wide. If 1 foot is added to the platform, three pallets can be stored across it with 3 inches on the sides and between the pallets; therefore the platform would be made 13 feet wide. If three pallets are placed across the platform, only two rows will have to be placed down the length. That would be 8 feet, 9 inches in length. A forklift truck will be used to handle the pallets. The dock width allows 13 feet for the truck and now another 12 feet must be added to the length for maneuvering space, since the forklift truck chosen is about 8 feet long and another 4 feet of clearance is needed for the truck to safely back up and swing around. The length of the dock is then 20 feet, 9 inches. This means that the overall minimum dimensions should be 20′9″ × 13′.

After determining the number and size of the docks, approaches for the trucks must be designed. These approaches are to handle the traffic coming and going and to allow space for the trucks to turn around and back to the docks. Space should also be allowed for trucks waiting if all docks are occupied. Two points are to be kept in mind: first, the closer the turn from the main road is to 90°, the more difficult it is for the vehicles; and second, large trucks can make left turns more easily than right turns.

Two characteristics of the approach are the apron depth and the bay width. The apron depth is the amount of space needed to accommodate the truck length, and the bay width corresponds to the truck width. The angle of the

Figure 12.9   Apron depth and bay width

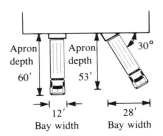

Figure 12.10   Dock arrangements

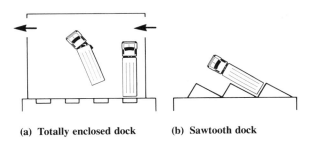

(a) Totally enclosed dock   **(b) Sawtooth dock**

dock can be varied to make the best use of the available space (Figure 12.9). Since truck lengths range from 40 to 70 feet, the apron depth should be sufficient to accommodate the largest vehicle delivering to or picking up from the plant.

The dock could be completely enclosed (Figure 12.10a) or simply covered by a roof, providing protection against the environment. The parking surface for the trucks and trailers immediately adjacent to the dock should be almost level but with a slight incline to provide for adequate drainage. When a sawtooth dock arrangement is available (Figure 12.10b), it helps to reduce maneuvering problems.

12.10 DOCK DOORS

When docks are not in use, their entrances must be secured. The types of protection vary and can be selected according to the company's needs and the advantages of the particular door. The following is a list of door types and their characteristics:

1. *Air curtains:* Used for high traffic situations. High-velocity airstreams are blown down across the door opening. Wind, heat, cold, and insects are kept out by the airflow. Air curtains are not very effective for radical temperature differences or for controlling noise.

2. *Strip doors:* Inexpensive, transparent plastic strips that part easily to accommodate traffic. Strip doors are not a good barrier to heat transfer (in or out) and insects, but they allow quick and easy entrances.

3. *Hinged doors:* Can be made of wood or metal and may have windows. They offer maximum security, but it takes time to open and close the doors each time.

4. *Swinging doors:* Also made of wood or metal with 180° swing radius. They are self-closing and can be secured.

5. *Impact doors:* Swinging doors with impact bumpers so that trucks can push through and bump them open. They can be single or double doors and can be equipped with electronic controls that open the door upon impact.

6. *Sliders:* Single or double doors on track-and-trolley hardware. They can be telescoping if there is not enough wall space to slide the door back. Sliders can be manual or power operated.

7. *Overhead doors:* Similar to garage doors. The doors travel on tracks up to the ceiling. They can be manual or power, but they require time to open. Overhead doors provide the best use of space because the doors do not take up any vertical space.

With the energy conservation and associated savings offered to a company, further sealing of the dock opening might be an important consideration. One way to achieve this is by using conventional insulation or weather stripping, which can be placed around the door. Air seals can also be used that can be inflated and sealed around the truck while it is being loaded or unloaded.

Another consideration in designing dock doors is the choice of controls used to open and close the doors. There are two types to consider, manual and power, and there are some trade-offs in using each type. Power controls open and close the doors more quickly and require little manpower; however, they are more expensive and require more maintenance. Manual controls require more time and manpower, and they expose the building interior to outside temperatures for longer periods. The type of control that is better depends on the volume of activity, which should be analyzed carefully.

SUMMARY

To apply the concept of just-in-time (JIT) in manufacturing organizations requires that raw material and partially finished products are available where and when they are needed. Furthermore, the finished products should be shipped to their final destination immediately so as to avoid any finished goods inventory. Though desirable, such a system is very hard to achieve if one cannot predict with certainty all the demands and sales or is at the mercy of subcontractors who are not reliable in their promised delivery dates or times. Most manufacturers therefore provide for storage and warehouse facilities to absorb variations in production and sales and also to take advantage of economic lot sizes and quantity discounts. Storage is almost always required in retail business, where a variety of customers desires immediate delivery of different products and also where display of the items is so very important to sales.

To select a site for storage and warehousing, one should follow the site selection procedure described in Chapter 15 and apply the models from Chapters 2, 3, and 4. In general, the site should be accessible, appropriately zoned, and protected by the local authorities. Considerations for the building itself involve analysis of factors such as approach roads, size of the site, and inside layout of the building (aisle layout, column pattern, and storage rack arrangement).

The chapter contains two examples that show how the space requirement for each item in storage may be calculated. Such an analysis should be performed on all items stored in a warehouse. Obtaining a good estimate of the total space requirement is important in building a warehouse that is neither too large nor too small.

Many different activities are regularly conducted in storage and warehouse operations. They range from receiving, inspecting, and storing items to packaging, labeling, and shipping them. Physically arranging the warehouse so that these acitivities can take place efficiently is critical in keeping the cost of operation down. The items may be stacked based on physical similarity, functional similarity, and popularity or by separating the reserve stock from the regular stock or storing items randomly but with computer control. Expensive and high-security items may need additional considerations. How the items are to be collected to fill an order also influences the layout of the storage structure. The area system, modified area system, zone system, sequential zone system, and multi-order system are some of the ways in which the order picking may be accomplished. The chapter provides an example to show how a storage arrangement may be developed.

Many accesssories such as bins, shelves, and racks are available for storing small and individual items. Racks are a common feature in most warehouses, and they are available in many different styles and shapes. Some of the common types of racks are the selective pallet rack, movable-shelf rack, drive-in and drive-through racks, cantilever rack, stacker crane rack, and portable rack. Each type has certain properties that make it attractive in a particular operating situation. Storing and then retrieving items from a particular area in the storage structure requires development of a storage location system, and one such coding system is illustrated in the chapter.

Automated storage and retrieval (AS/R) is a nontraditional storage system that uses high-rise storage units, fast-moving stacker cranes, and controls and management systems that are under computer control. It has high initial cost but provides economy in operations by requiring smaller floor space, fewer workers for operation, and less time in order picking (which is also more accurate). It also reduces pilferage. An AS/R system has four major components—namely, S/R machines, the storage structure, conveying devices, and controls. However, to obtain a good estimation of the initial cost, one must develop in detail two major components, S/R machines and the storage structure. The chapter presents the steps involved in the design of an AS/R system and then illustrates them with a numerical example. By using cost data provided by P & H, a major AS/R system manufacturer, the cost estimate for the example problem is obtained.

Loading docks are another important feature in storage operations. The chapter shows how one may use queuing analysis to determine the number of docks necessary to support a certain level of loading and unloading activities. The chapter also discusses dock design, which involves determination of the physical dimensions and layout of docks.

When docks are not in use, their entrances must be secured by dock doors. A number of different types of doors are commercially available, with each having some distinct characteristics. Depending on the activity level and type of activities in a plant, one may use air curtains, swinging doors, impact doors, or overhead doors.

PROBLEMS

12.1 What are the advantages and disadvantages of JIT?

12.2 What factors are to be considered in planning the construction of a warehouse?

12.3 Why are aisles important? What factors are to be considered in planning the aisle width?

12.4 A company has three rooms for storing the units on pallets. (See Figure P12.4.) If 6-foot aisle space is needed, along with clearances of 3 inches between stacks and 1 foot overhead, determine how many $4' \times 5' \times 3'$ (W × L × H) pallets can be stored. The ceilings are 10 feet high. By what percentage will storage capacity increase if the two non-load-bearing (NL) walls are removed and 6-foot aisles can be rearranged either north–south or east–west?

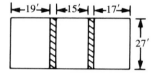

Figure P12.4

12.5 Discuss which method of storage is used in each of the examples below.
a. A grocery store
b. An auto parts store
c. A department store such as Sears
d. A catalog store

12.6 Determine the best storage arrangement for the six items in the accompanying table, along with the number of sections required and the size of the building if it can have only four sections in a row with an arrangement as shown in the sketch in Figure P12.6. Other information is as follows: Items B, E, and F have the same unit size. The unit size of Items A and D is one half the size of Items B, E, and F, while C is one third the size of Items B, E, and F. One hundred units of Item B fill one storage section.

Item	Units Received	Receiving Trips	Average Units Shipped
A	100	50	0.5
B	400	80	3
C	250	10	5
D	200	20	2
E	500	100	5
F	175	15	4

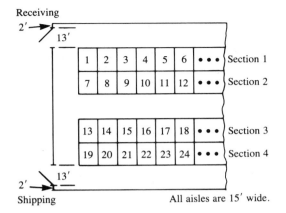

Figure P12.6

12.7 Discuss the advantages of automated storage and retrieval systems.

12.8 A plant produces 75 units per hour of an item with dimensions of $1' \times 1' \times 1'$, storing one week's supplies in containers measuirng $5' \times 5' \times 5'$. A minimum of 3 inches of space is required between adjacent stacks. Determine the number of containers needed.

12.9 A one-week inventory of the boxed telephones of Problem 11.3 are to be stored by an AS/R system.
a. Determine the number of cranes needed if single-cycle cranes are used.
b. Determine the number of cranes if double-cycle cranes are used.
c. Determine the system width and length for part a.
d. Determine the system width and length for part b.
e. Estimate the cost of the system.

12.10 A manufacturing company wishes to store a unit on a $36'' \times 48'' \times 24''$ pallet having a weight of 1400 pounds and 75 dual cycles per hour. The total storage is 18,000 unit loads. The height of the building is 80 feet, but clearances of 2 feet from the ceiling and 6 inches for the rack support are needed.
a. Determine the number of stacks that can be accommodated with the height of the load.
b. Determine the number of dual cranes needed.
c. Determine the number of bays needed.
d. Determine the storage dimensions.
e. Determine the cost of the system.

12.11 A company receives materials delivered by trucks with a maximum weight of 3500 pounds each. If at most ten trucks per hour will be delivering at the plant, determine how many docks they should have, using the standards presented in the chapter. Allow 10 minutes for one truck to leave and another to pull in.

12.12 The docks of Acme Plumbing Manufacturing receive trucks that hold 8000 pounds each. It takes 1.15 man-hours to unload each truck, and 5 minutes is allowed for the truck to leave and another to pull in. If six trucks are to be serviced at the busiest time of the day, how many docks are needed?

12.13 Suppose the arrival pattern of the trucks follows a Poisson distribution with a mean of two trucks per hour, and the service rate follows an exponential distribution with a mean of two trucks per hour. Find the number of docks needed so that a truck does not spend more than ½ hour in the plant.

12.14 If in Problem 12.13 the truck cost is $50/hour and the dock operator is paid $10/hour, what is the most economical dock size? (Ignore the constraint on the time spent by the truck.)

12.15 If in Problem 12.13 it is decided that a truck should not wait more than 10 minutes (excluding time for loading and unloading), how many docks are required?

12.16 Discuss the types of dock doors and their strong and weak points. In one industrial plant in your area, observe the dock door and determine whether the choice was appropriate or whether you would recommend a change.

SUGGESTED READINGS

Storage and Warehousing

Binning, R.L., "New Uses of Traditional Storage Systems Can Minimize Inventory and Work-In-Process," *Industrial Engineering,* Vol. 16, No. 3, March 1984, pp. 81–83.

Schonberger, R.J., "Just-In-Time Production Systems: Replacing Complexity with Simplicity in Manufacturing Management," *Industrial Engineering,* Vol. 16, No. 10, October 1984, pp. 52–63.

Taff, C.A., *Management of Physical Distribution and Transportation,* Richard D. Irwin, Inc., Homewood, Ill., 1972.

Operations Research

Hiller, F.S., and Lieberman, G.J., *Introduction to Operations Research,* Holden-Day, Oakland, Calif., 1980.

Phillips, D.T.; Ravindran, A.; and Solberg, J.J., *Operations Research—Principles and Practice,* John Wiley and Sons, New York, 1976.

Taha, H., *Operations Research—An Introduction,* 3rd edition, Macmillan, New York, 1982.

Automated Storage and Retrieval

Everything You Ever Wanted to Know About System Planning, Harnischfeger P & H, Milwaukee, Wisc., 1978.

9 Simple Steps to Determine the Layout, Design, and Estimated Cost of an Automated Storage/Retrieval System, Harnischfeger P & H, Milwaukee, Wisc., 1984.

Docks

Bolta, H.A. (Ed.), *Materials Handling Handbook,* The Ronald Press Company, New York, 1958.

Goble, E., *Buildings for Industry,* F.W. Dodge Corporation, New York, 1957.

Mallick, R.W., and Gandreau, A.T., *Plant Layout Planning and Practice,* John Wiley and Sons, New York, 1951.

Plant Engineering Directory and Specifications Catalog, Technical Publishing, Chicago, 1985.

Reid, K., *Industrial Buildings,* F.W. Dodge Corporation, New York, 1951.

CHAPTER

13

Plant and Office Layout: Conventional Approach

Since the beginning of organized manufacturing, considerable effort has been expended to make the manufacturing facility as efficient as possible. The locations and arrangements of the departments and work centers contribute in a large measure to the manner in which a facility is operating.

The "right" solution to plant layout problems is important for two reasons. First, material-handling costs comprise anywhere from 30 to 75 percent of total manufacturing costs. Any savings in material handling realized through a better arrangement of the departments is a direct contribution to the improvement of overall efficiency of the operation. Second, plant layout is a long-term, costly proposition, and any modifications or rearrangements of the existing plant represent a large expense and cannot be easily accomplished.

Characteristics of plant layout problems and their data requirements will be discussed in detail in Chapter 14, which also illustrates the use of computers in solving such problems. The basic objective is to achieve an orderly and practical arrangement of departments and work centers to minimize the movement of material and/or personnel while allowing for sufficient working space and perhaps space for future expansion within an area that may be predefined. This goal is kept in mind in every phase of plant development from assignment of overall areas for each of the departments to the generating of a detailed layout of each individual department within its space. These applications are associated with, for example, production, warehousing, offices, toolrooms, food services, and maintenance. We will discuss in this chapter conventional methods of determining plant layout.

13.1 PROCEDURE

In developing a plant layout, such as the sample shown in Figure 13.1, the procedure generally follows the steps indicated below.

First, determine the area required for each work center. Careful analysis must be performed to establish the desired or necessary content of each center and its associated area requirements. For example, an office center might include the offices for the president and

Figure 13.1 Plant layout of Easy Light, Inc.

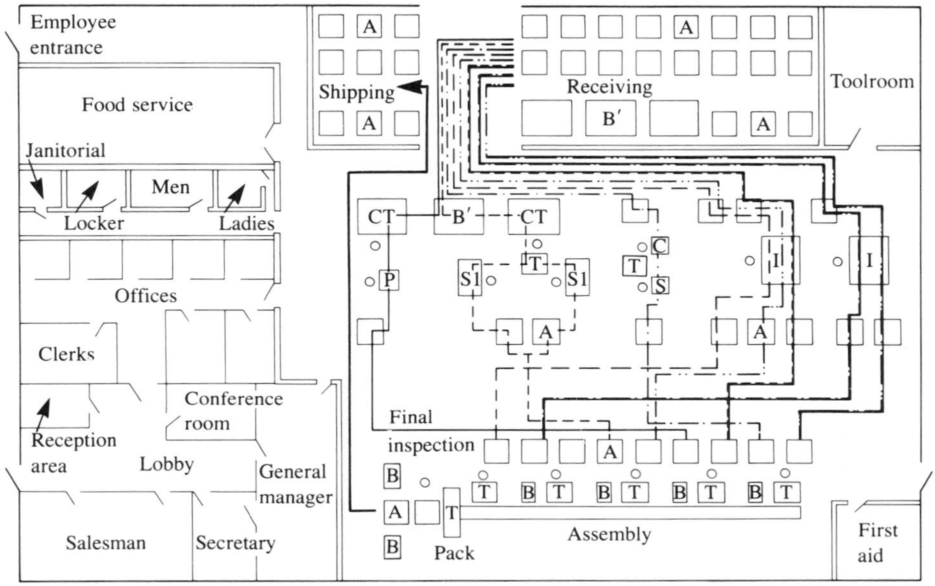

the general manager; areas for sales, personnel, engineering, and accounting departments; conference rooms; reception areas; and space for secretaries. A production area could include an area to locate each machine; room to conduct maintenance on the machine; and space for the operator, in-process inventories, assorted tools, and auxiliary equipment. Also, space must be reserved for quality control, general offices, maintenance benches, and storage areas as well as for water fountains, lavatories, ventilation ducting and equipment, and so forth. It is helpful to develop a simple table listing the contents and their associated area requirements for each work space to obtain an accurate estimate. Table 13.1 and the associated sketch in Figure 13.2 show how this might be done for the production area at a small manufacturing plant. Table 13.2 shows how this might be done for office space. (More about office area development in Section 13.6.)

It should be noted that allowances for aisles in the plant and hallways in offices must be added in their respective area calculations. This allowance is taken to be between 20 and 40 percent of the total area otherwise required.

The second step is to draw a rel-chart (relationship chart) or establish a from-to chart. The rel-chart describes qualitatively the degree of closeness that the analyst feels should exist between different work centers. This proximity might be dictated by the flow between the departments, convenience, necessity of using the same personnel, or facilities being used in two or more departments or need for communication. The standard codes used in describing this closeness, in decending order of priority, are shown in Table 13.3. In addition, there is code X to describe the undesirability of having two activities close together. For example, noisy punch press operation should be isolated from other manned areas, especially offices. X is assigned a negative value such as -1 (or even -10, based on severity) in future evaluations.

Table 13.1 Space requirements (production area)

Machine	No. Needed	Dimensions (feet)	Clearance* $5 \times L + 4 \times W$ (square feet)	Auxiliary Equipment	In-Process Inventory	Single-Unit Area	Total
Lathe	3	3×8	52	$2 \times 4 = 8$	10	94	282
Drill press	2	3×4	32	$2 \times 2 = 4$	15	63	126
⋮	⋮	⋮	⋮	⋮	⋮	⋮	⋮
Quality control room			$20' \times 15'$				300
Lavatory and showers			3 at $3' \times 20'$				90
Lockers and wash basins			30 at $2' \times 6'$				360

*Estimated by the engineer on the project.

Figure 13.2 Sketch of a work station for a lathe

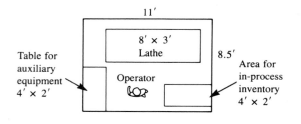

Table 13.2 Office space requirements

Office/Occupants	Area (square feet)
President	250
General manager	200
Sales manager	200
Production manager	200
Accountants (4)	800
Engineers (6)	775
Sales representatives (6)	600
Secretaries (7)	700
Receptionist	150
Conference room	250
Copy room	100
Coffee room	200
Rest Rooms	350
Total	**4775**

An example of a rel-chart is shown in Table 13.4. Only half of the entries are required, since the table is symmetrical about the diagonal.

There is a close correlation between the from-to chart and the rel-chart. The from-to chart describes the estimated trips or movements of unit loads between various departments, for example, 600 trips between production and warehouse per day. Thus it forms a quantitative measurement of needed closeness; for instance, the work centers with the most trips between them should be as close together as possible to minimize travel. In transforming the quantitative measure into a qualitative measure—that is, transferring information from a from-to chart to a rel-chart—the desired degree of closeness between such departments

Table 13.3 Rel-chart priority codes

Code	Priority	Value
A	Absolutely necessary	4
E	Especially important	3
I	Important	2
O	Ordinary	1
U	Unimportant	0
X	Undesirable	−1

is denoted by code A, E, or I, based on the number of trips between them. Similarly, the departments with almost no direct flow among them would be marked with a U or an O for closeness. In essence, the from-to chart supplies information similar to that of a rel-chart and, in fact, can easily be transformed into one.

For example, suppose the company estimates that its flow of goods between departments will be as shown in Table 13.5.

The flow of incoming materials for production is from receiving to warehouse to production, while goods produced flow from production to warehouse to shipping. Sometimes production materials received will go directly to production while produced goods may go directly to shipping. The total flow between the departments is then given as shown in Table 13.6.

It is imperative, considering the large

Table 13.4   Rel-chart

Nodes	PR	WA	OF	TR	FS	MA	LR	SR
(1) Production (PR)	—	A	E	A	E	A	E	E
(2) Warehouse (WA)		—	O	O	U	O	U	A
(3) Office (OF)			—	U	O	O	U	O
(4) Toolroom (TR)				—	O	A	U	U
(5) Food services (FS)					—	U	U	U
(6) Maintenance (MA)						—	U	O
(7) Locker room (LR)							—	U
(8) Shipping/Receiving (SR)								—

Table 13.5   Sample from-to chart, average estimated flow of goods per day

	To		
From	Production	Warehouse	Shipping/ Receiving
Production		600	100
Warehouse	2000		600
Shipping/Receiving	300	2000	

Table 13.6   Total flow between the departments

	To		
From	Production	Warehouse	Shipping/ Receiving
Production	0	2600	400
Warehouse		0	2600
Shipping/Receiving			0

Table 13.7   Rel-chart

	To		
From	Production	Warehouse	Shipping/Receiving
Production	—	A	E
Warehouse		—	A
Shipping/Receiving			—

flow, that production and the warehouse be as close together as possible. The same holds true for the warehouse and shipping/receiving; therefore in the rel-chart shown in Table 13.7 we assign A's between these departments. Since there is some flow between receiving and production, these departments are assigned an E relationship.

Consider another illustration, which shows how to develop a from-to chart in a plant where multiple items are produced. Suppose three products are scheduled for production with a sequence of operations and with known weekly demands as shown in Table 13.8.

A flowchart is developed by noting the flow between departments as the products move from one department to the next in their defined sequence. For example, there is a flow of 500 units between Departments A and C of Product 1 and 1000 units of Product 2 as shown in Table 13.9. The rest of the entries are similarly noted.

The flows between the departments can now be collected to produce a from-to chart, as in Table 13.10.

The entries show the total flow between two departments. The data could be converted to generate a relationship chart by noticing the strong flow between Departments A and C, C and D, and D and F and assigning an A relationship between them. Departments A and B, with a flow of 1000 units, could be given an E relationship, and the departments with a flow of 300 units will be converted to an O relationship. The remaining relationships could be de-

Table 13.9   Flowchart

	To					
From	A	B	C	D	E	F
A			500 1000			300
B	1000		300			
C	300			500 1000		
D						500 1000
E		300				
F						

Table 13.8   Proposed production plan

Product	Sequence of Operations (Departments)	Production per Week
1	A—C—D—F	500
2	B—A—C—D—F	1000
3	E—B—C—A—F	300

Table 13.10   From-to chart

	A	B	C	D	E	F
A		1000	1800			300
B			300	300		
C				1500		
D						1500
E						
F						

fined as unimportant and identified by the letter U. The resultant is as shown in Table 13.11.

The remaining steps of the procedure are now first stated and then illustrated by applying them to an example problem.

The third step is to develop a graphical representation of the rel-chart. Richard Muther (1973) suggests a way of transforming information from the rel-chart to a pictorial or graphical representation. The work centers are represented by nodes, and the number of lines between two nodes represents the closeness between the nodes. (See Figure 13.3 in the next section.) The decoding scheme is as follows. Code A is shown by four lines, E by 3, I by 2, and O by 1. A wiggly line represents X.

The proper arrangement of the nodes and their relationships is critical in plant layout development, since it provides for the starting arrangement of the departments. The objective is to arrange the nodes so that there is a minimum number of areas crossed when going from one department to another with the frequencies indicated by the decoded rel-chart. It might take a number of trials before one obtains a well-defined arrangement.

The procedure starts by conversion or decoding of the rel-chart to what we will denote as a value chart, using the values associated with the codes shown in Table 13.3, reproduced here as Table 13.12 for convenience. The measure of importance of each area, which ascertains the degree of closeness one department has with all other areas, is obtained by adding

the row and column values for that department together. This is necessary because the rel-chart and the corresponding value chart are symmetrical about the diagonal, and only half of the table is normally displayed.

Select the department with the highest total and place it in the center of the nodal diagram. Locate around it any departments that have four-relationships with it. Next, from the other departments in the diagram, select the one with the next highest total and place around it the departments with four-relationships to it. Continue the procedure with each appropriate department in the diagram. After all four-relationship departments are exhausted, if some of the departments are still not in the diagram, continue with three-relationships, following the same sequence of steps as were used with four-relationships. Continue until all the areas are in the diagram, using two- and even one-relationships if necessary.

For the next step, go over the diagram again and adjust the positions of departments, switching them if necessary to satisfy closeness in between. For example, closeness of a department with three-relationships is more important than one with two-relationships. Again, the basic objective is to develop an arrangement in which departmental crossovers are at a minimum when travel between departments is made based on the value chart.

The fourth step is to develop an evaluation chart. The chart provides a measure of effectiveness of the nodal arrangement developed

Table 13.12   Rel-chart priority codes

Code	Priority	Value
A	Absolutely necessary	4
E	Especially necessary	3
I	Important	2
O	Ordinary	1
U	Unimportant	0
X	Undesirable	-1

Table 13.11   Rel-chart

	A	B	C	D	E	F
A		E	A	U	U	O
B			O	U	O	U
C				A	U	U
D					U	A
E						U
F						

in Step 3. Different arrangements can be evaluated by developing a chart for each, and the one with the lowest value is selected as the best arrangement to be used in Step 5.

The evaluation chart is developed by first converting the nodal representation into a semiscaled grid representation. For each department the necessary area is equated to the approximate number of blocks needed, using a convenient scale. (For example, 200 square feet could equal one block.) The total of all the blocks that are needed is determined, and an approximate square or rectangular shape is designed. For example, if a layout needs 50 blocks, the layout could be developed in a 10 × 5 or a 9 × 6 arrangement.

In developing such a grid, it might be beneficial to note the expected column span and accordingly fix one dimension, generally the width. For example, a 20-foot column span can be arranged as a 20′ × 20′ block of 400 square feet or could be divided as a 10′ × 10′ block of 100 square feet.

In the overall grid arrangement, we should make certain that the numbered blocks in one direction (e.g., width) are such that the resultant dimension is an integer multiple of the column span. Though ideally one might wish for a square layout (to minimize distance traveled), most plants are built in a rectangular form to accommodate the shape of the plot of land.

The next step is to place the individual departments within the grid arrangement. The departments are placed in the grid, represented by the necessary blocks for each, by using the arrangement of the nodal diagram in Step 3 as a guide. The closeness measure can now be defined; it is equal to the shortest rectilinear distance between two areas multiplied by the value of the relationship between those two departments. An effectiveness evaluation chart, similar to a value chart, is useful in developing this measure of all departments. The grand total gives the measure of effectiveness of the nodal diagram; the grid chart with minimum sum should be selected for the next step.

The fifth step is to develop templates to represent each area. Spacing of columns in the building—generally 20, 30, or 50 feet—gives a good first dimension. The other dimension is calculated on the basis of the area required for the center. For example, with a 20-foot span in a building, an area of 300 square feet needed for a quality control room is represented by 20′ × 15′ on a template.

The sixth and final step is to arrange the templates in the same fashion as the graphical representation of the rel-chart (Step 3). The sizes and shapes of the templates might have to be changed to correct the resulting odd shape of the building and to minimize wasted space. (Sometimes constraints are imposed on changes in the shapes by the machines within the department.) Again, a few trials might be needed to obtain a desired layout.

The entire procedure is facilitated if some of the basic material to be described in Section 13.3 is used in the development process. Base grid paper, for example, is excellent to use in determining the optimum shape of a department in Step 4. At times a quick sketch might be better for adding clarity and understanding than working with templates.

Example: Layout for a Small Manufacturing Firm

Steps 3–6 are now illustrated by applying them to the example associated with the rel-chart of Table 13.4. Suppose the area needed for each department (including aisle space) is as shown in Table 13.13. To begin the process, the letters in the rel-chart are converted to numbers using the codes given in Table 13.12, and the resulting value chart is shown in Table 13.14. This chart will be used as an aid in developing the nodal representation. To get the total measure of importance of a department, we add its relationship values to all other departments.

Table 13.13 Space requirements

Department	Area (square feet)
(1) Production	4800
(2) Warehouse	3050
(3) Office	2400
(4) Toolroom	1150
(5) Food service	750
(6) Maintenance	1000
(7) Locker room	600
(8) Shipping/Receiving	1900

This is easily done by adding all the numbers in the row and the column for an individual department. For example, the total for Department 4 is obtained by adding the entries in Row 4 $(1 + 4 + 0 + 0)$ and Column 4 $(4 + 1 + 0)$ for the total of 10. Table 13.14 shows these values in the Total column.

Begin the nodal representation by placing the department (area) with highest total, in this case Department 1. Place around it any departments with which it has a four-relationship: 2, 4, and 6. Continue the process by selecting a department with a high total from those that are now in the diagram (i.e., Departments 2, 4, and 6) and place around it all the departments having four-relationships with it. Here we break the tie between Departments 2 and 6 arbitrarily

by selecting Department 2 and placing Department 8 beside it. Continue the process for all departments. Care must be taken to look ahead and notice other departmental relationships. For example, Department 8 has a four-relationship with Department 2, but it also has a three-relationship with Department 1 and therefore must be placed adjacent to both Department 1 and Department 2. After all of these four-relationship departments have been placed, go back to the department with the highest total and begin placing the three-relationships. After all three-relationships are exhausted, adjustments are normally made with the two- and one- relationships, but in this case all the departments are placed. The result is the diagram shown in Figure 13.3.

A good nodal representation has the fewest lines crossing the least number of departments. Depending on how the lines are placed and the number of departments involved in the layout, this can often be a tedious and difficult task to evaluate. Take, for example, the nodal representation of Figure 13.4. Departments 4 and 6 have four-relationships with each other. The nodal representation shows the four lines crossing no departments, but when the departments are actually placed, Department 5 might fall directly between Departments 4 and 6.

Table 13.14 Value chart

	1	2	3	4	5	6	7	8	Total
Department									
1	—	4	3	4	3	4	3	3	24
2		—	1	1	0	1	0	4	11
3			—	0	1	1	0	1	7
4				—	1	4	0	0	10
5					—	0	0	0	5
6						—	0	1	11
7							—	0	3
8								—	9

Figure 13.3 Nodal representation

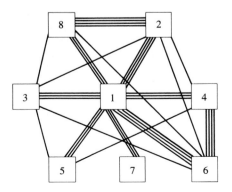

Figure 13.4   Nodal representation with a potential placement problem

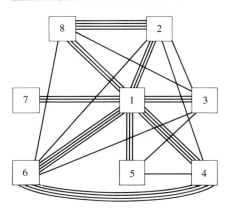

A grid formulation as explained in Step 4 can be quite helpful in evaluating potential layouts. Begin by reducing the area of each department to an approximate number of blocks. For our example, use a scale of 400 square feet equals one block (20′ × 20′). For instance, the toolroom is 1150 square feet, which approximates three blocks. The numbers of blocks needed for each area are indicated in Table 13.15.

With the total blocks needed being 40, let us limit the dimensions of the grid to 5 × 8 blocks. The departments may be placed in the grid according to the nodal representation of Figure 13.3. The shape of the departments should remain as regular as possible, preferably rectangles as shown in Figure 13.5.

Now the grid can be evaluated with respect to distances and departmental crossings. The shortest rectilinear distance, the number of blocks crossed, between departments is multiplied by the value of the relationships between the corresponding departments. With the relationships from Table 13.4 and the rectilinear distances from Figure 13.5 the effectiveness of the departmental placement depicted in the latter is evaluated and the results displayed in Table 13.16. For example, the shortest rectilinear distance between Departments 2 and 6 is three blocks, and these departments have a one-relationship between them, giving 3 × 1 as the entry for Row 2, Column 6 in Table 13.16. The numbers are totaled for rows and then overall to get the effectiveness measure; in this case it is 16.

To evaluate the alternate nodal representation of Figure 13.4, we shall apply the procedure followed with Figure 13.3 and compare the effectiveness measures for the two nodal diagrams. Figure 13.6 converts Figure 13.4 to a grid representation of the layout.

Even though the nodal arrangement of

Table 13.15   Block calculations

Department	Area		Blocks
1	4800		12
2	3050		8
3	2400		6
4	1150		3
5	750		2
6	1000		2
7	600		2
8	1900		5
		Total	40

Figure 13.5   Grid representation

8	8	2	2	2
8	8	2	2	2
3	8	2	2	4
3	1	1	1	4
3	1	1	1	4
3	1	1	1	6
3	1	1	1	6
3	5	5	7	7

Table 13.16 Effectiveness calculation for Figure 13.3

				Department					
	1	*2*	*3*	*4*	*5*	*6*	*7*	*8*	*Row Value*
1	—	0	0	0	0	0	0	0	0
2		—	1×1	0	4×0	3×1	4×0	0	4
3			—	3×0	0	3×1	2×0	0	3
4				—	4×1	0	2×0	2×0	4
5					—	2×0	0	4×0	0
6						—	2×0	5×1	5
7							—	6×1	0
8								—	0
								Total	16

Figure 13.4 appeared to have fewer lines crossing fewer departments, this will in fact not be the case in the actual layout. This observation is the result of comparing the effectiveness measure of Table 13.17 with that of Table 13.16 (25 versus 16).

Small changes can be made in the grid without really affecting the final layout or the measure. In Figure 13.7 we see that all of the departments have been shifted one block clockwise around Department 1. The corresponding total in Table 13.18 has not changed. Essentially, the layouts developed from Figures 13.3 and 13.7 would be equally effective and efficient.

Figure 13.6 Grid representation of Figure 13.4

8	8	2	2	2
8	8	2	2	2
7	8	2	2	3
7	1	1	1	3
6	1	1	1	3
6	1	1	1	3
5	1	1	1	3
5	4	4	4	3

Table 13.17 Effectiveness calculation for Figure 13.4

				Department					
	1	*2*	*3*	*4*	*5*	*6*	*7*	*8*	*Row Value*
1	—	0	0	0	0	0	0	0	0
2		—	0	4×1	5×0	3×1	1×0	0	7
3			—	0	3×1	3×1	3×0	2×1	8
4				—	0	2×4	4×0	4×0	8
5					—	0	2×0	4×0	0
6						—	0	2×1	2
7							—	0	0
8								—	0
								Total	25

Figure 13.7   Modified grid representation

8	8	2	2	2
3	8	2	2	2
3	8	8	2	2
3	1	1	1	4
3	1	1	1	4
3	1	1	1	4
3	1	1	1	6
5	5	7	7	6

Figure 13.8   Elongated grid representation

8	8	2	2
8	8	2	2
3	8	2	2
3	1	1	2
3	1	1	2
3	1	1	4
3	1	1	4
3	1	1	4
5	1	1	6
5	7	7	6

The shape of the grid can be changed without significantly affecting the results as long as the shape is designed with practicality in mind rather than for the purpose of defeating the procedure. Consider the grid in Figure 13.8, in which the design has been elongated, resulting in the total effectiveness measure being increased by 1. But if this new grid is compared to Figure 13.5, it can be seen that there would be only slight difference between the two as far as travel between departments is concerned. The primary difference would be with regard to the shape and the dimensions of the required building.

Table 13.19 evaluates the effectiveness of the grid shown in Figure 13.8. The total is 17, compared with 16 for the best layouts developed in this chapter, indicating a slight increase in travel.

In applying Step 5 of the procedure the grid should be converted to a scale model using templates like those shown in Figure 13.9. The grid representation should be used as a starting point for developing the size and shape of the

Table 13.18   Effectiveness calculation for Figure 13.7

	\ Department								
	1	*2*	*3*	*4*	*5*	*6*	*7*	*8*	*Row Value*
1	—	0	0	0	0	0	0	0	0
2		—	1 × 1	0	6 × 0	3 × 1	4 × 0	0	4
3			—	3 × 0	0	3 × 1	2 × 0	0	3
4				—	4 × 1	0	2 × 0	2 × 0	4
5					—	2 × 0	0	4 × 0	0
6						—	0	5 × 1	5
7							—	4 × 0	0
8								—	0
								Total	16

Table 13.19   Effectiveness calculation for Figure 13.8

	Department								Row Value
	1	2	3	4	5	6	7	8	
1	—	0	0	0	0	0	0	0	0
2		—	1×1	0	6×0	3×1	5×0	0	4
3			—	2×0	0	3×1	2×0	0	3
4				—	3×1	0	2×0	4×0	3
5					—	2×0	0	5×0	0
6						—	0	7×1	7
7							—	6×0	0
8								—	0
								Total	17

templates for the individual departments. Keep in mind that the grid is only an approximation of the required area; hence it will frequently be necessary to enlarge or reduce the grid to meet exact specifications. These templates are laid out in Figure 13.10 according to the grid shown in Figure 13.3. The resulting layout is somewhat irregular and requires modification.

Small adjustments in Departments 2 and 6 produce the smooth layout of Figure 13.11. Modifying the layout is done basically by trial

Figure 13.9   Departmental templates

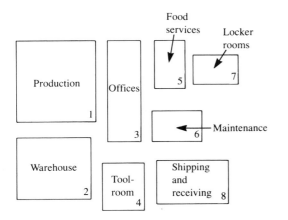

Figure 13.10   Initial layout

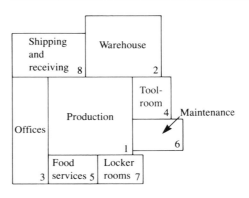

Figure 13.11   Final layout

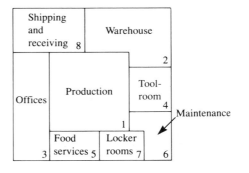

and error; but since the grid has already approximated the layout, the task should be a simple one, requiring only minor adjustments.

The layout planning for any plant can at first appear to be a tedious task. The steps given here are designed to simplify the job and produce an efficient layout.

13.2 DETAILED LAYOUT

Once the positions of major centers in the overall layout have been established, the next step is to develop a detailed layout of each work center. There are three basic ways of representing these layouts: draw sketches, use two-dimensional templates, and use three-dimensional models.

A layout of the production area, for example, must indicate, to scale, the position of each machine and its operator. Adequate room, at least 18 inches of clear space all around the machine and at times more, must be provided to allow maintenance personnel to perform their tasks. The aisles, walls, exits, and columns must be clearly shown on the layout. Each machine should be labeled by name, and arrows should show the material flow—where the material is fed into and cleared from the machine. Space must be provided for in-process inventories, storage, auxiliary equipment, and maintenance benches.

Overhead pipes and ducts are either indicated on the layout or, to avoid clutter, may be presented on an overlay. Outlets for both 240 V and 120 V electrical service, water, drains, compressed air, natural gas, bench lights, and other utilities should be shown and labeled. The flow of material can be indicated by using arrows directly on the layout or on a plastic overlay sheet.

Generally, one begins by selecting a scale to develop the plans; ¼ inch to 1 foot is the most commonly used scale by architects and contractors and therefore should be selected in facility development if at all possible. Draw a plan to show the outside walls; for an existing facility, draw columns, doors, windows, elevators, stairs, and other permanent features. Select the material flow pattern, thus fixing the receiving and shipping departments. Tentatively assign the major departments along the material flow line, making sure there is no interference. If a new building is being planned, the previous steps of placing such items as columns and doors and then defining the material flow pattern could be reversed to give more flexibility in design.

The next step is to define the positions of the aisles, which may or may not be the major material carriers. For example, if forklift trucks are used as the major material handler, then aisles also form the main artery of the material movement; however, even if conveyors are the material movers, aisles are still needed for people, maintenance carts, and truck movement. The aisles are also required for quick evacuation in case of an emergency. Table 13.20 shows the normal aisle widths needed for different types of equipment.

The major aisles are laid out to carry through-traffic associated with all of the departments. Then the departmental aisles branching off the major aisles are established for each department.

Table 13.20   Suggested aisle widths for various flows

Flow	Aisle Width (feet)	Minimum Maneuvering Allowance (feet)
Tractor	12	14
3-ton forklift	11	12
2-ton forklift	10	12
1-ton forklift	9	11
Narrow aisle truck	6	12
Manual platform truck	5	10
Personnel	3	

Plant layout. *Top*: In an assembly line layout, space is assigned in a department for machines, people, and materials once the flow pattern has been determined. The layout, like the one shown here, should be planned as carefully as possible; after all the equipment is in place, changes could prove too costly to be made. *Bottom*: In a job shop layout, flexibility is critical to design since the assemblies, processes, and material flow will vary.

The three characteristics of aisle are length, location, and width. The length of an aisle depends on the area of plant it is to cover and the location of the aisle. Two factors should be considered: All parts of the plant should be quickly accessible, but the more aisles are planned, the larger the plant must be to provide the same production area, increasing the cost of the building. Therefore a minimum number of aisles should be placed to cover a maximum of space.

The aisle's proposed use determines its width. Main aisles throughout the plant should be wide enough, 15–18 feet, to support two-way traffic of both vehicles and people without creating bottlenecks. Smaller aisles that are intended to support only one-way traffic may be 10–12 feet wide or less; and if narrow aisle equipment is being used, aisles may be as narrow as 6 feet. Aisles that are to support only personnel traffic can be as small as 3 feet, depending on the amount of traffic they are to accommodate. The aisles should be located with uniform spacing to increase their effectiveness; they should be straight and not sinuous.

Machines are positioned within a given department once the flow pattern has been designated. Space is assigned for in-process inventory maintenance personnel, benches, operators, and others as was planned in the initial space determination. The arrangements can be adjusted by moving the departmental aisles and material flow, which allows a smooth movement of material and people to each machine. Some modification to the shape of the department might also be necessary during this process.

The next step is to locate personnel and plant services such as washrooms, the cafeteria, vending machines, water fountains, maintenance facilities, and supervisory personnel. They must be placed to achieve the utmost efficiency in operation. For example, the washrooms and the water fountains should ideally be within 60 feet of any person working in the plant. The cafeteria and the main tool crib should be placed so that they are accessible by each department without going through any other department. Checking in and out by personnel should be done at the main entrance gate to the plant building, thus allowing control of who may enter the building. The overall integration of a layout can be checked by following the activities of workers as they follow their daily activities of parking their cars; checking into the plant; working at machines and/or in their work centers; using personal facilities on breaks; requesting maintenance, quality control, and supervisory services; visiting a designated toolroom and tool crib; and finally leaving the plant at the end of the shift. All of these activities should be completed without confusion or too much cross-traffic and congestion in aisles or service areas.

In detailing the plant layout the first alternative of drawing sketches is easy and quickly completed. Usually cross-section or grid paper is used as a base, and the details are drawn in. However, this is not a very flexible process; each change requires some redrawing.

Two-dimensional templates are more popular and relatively inexpensive to construct. Templates made of cardboard, plastic, metal, or some other material (usually to the scale of ¼ inch to 1 foot) are cut to the shapes of various machines. The templates are arranged and rearranged on the same scaled print of the floor layout until a satisfactory plan is obtained.

Three-dimensional models can be constructed and, in a process similar to that for the two-dimensional templates, located so as to achieve the optimum layout. The three-dimensional models are more realistic but are considerably more expensive. If overhead height is of concern, however, perhaps they would be more appropriate, and the models do make selling the plant layout easier.

13.3 MATERIALS USED IN PLANT LAYOUT ILLUSTRATIONS

To make the layout simpler to construct and less difficult to explain and follow, different materials can be used in its construction. Several are explained below.

The base material for the layout must be sturdy and strong, and it should be lightweight in case the display must be moved to different meetings during the course of its final evaluation. Cardboard, plastic, plywood, and metal are some of the more popular materials for the backboard.

The layout should be built or drawn on a background that makes its modification and understanding convenient. Tracing paper for drawing and transparent plastic sheets with grids for scales are good materials on which to develop the layout.

Plastic tape of different colors can be used to show defined lines representing walls, aisles, conveyors, railroad tracks, and material flow. Tape can also be used to show storage areas, pallets, benches, and other miscellaneous items.

Self-adhesive tape is available in any number of different colors and widths. Material flow and aisle boundaries are generally designated by $\frac{1}{16}$-inch tape, conveyors are indicated by $\frac{1}{8}$-inch tape, and walls are shown by ¼- or ⅜-inch tape.

To represent physical objects such as machines, benches, and stools, one can make templates. The material used in their construction should be easy to cut, paste, and form: paper, cardboard, plastic, sheet metal, and wood are popular choices.

Three-dimensional models can be built of materials similar to those used in making templates. Commercially made standard models are also available to represent different machines, office furniture, and other accessories.

All machines, material flows, storage areas, and other miscellaneous items should be labeled for clarity. Though hand-labeling is permitted, a much neater and more professional-looking layout is obtained if a standard lettering set is used.

Clear plastic sheets with tapes representing material flow lines can be used to overlay the basic plant layout to aid in determining the suitability of the plan while avoiding excessive cluttering.

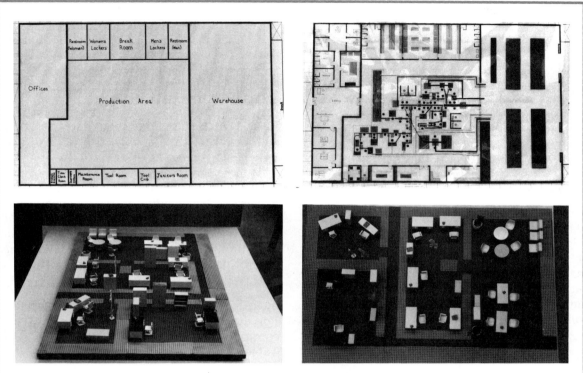

Plant layout illustrations. *Top left*: In a layout base such as the one shown here, the base material should be sturdy and lightweight. The layout itself should be simple to understand and easily altered. *Top right*: Here, transparent plastic overlays are used to show detailed department layouts and material flow paths. *Bottom left*: In the three-dimensional office model shown here, the use of model furniture and equipment helps the audience to visualize the final layout. *Bottom right*: This top view of the three-dimensional model shows how the grid of the plastic base guides in accurate placement of office spaces and furniture.

13.4 DEVELOPING AND ANALYZING PLANT LAYOUTS

Let us now formally define steps that may be helpful in developing and analyzing alternative layouts.

1. Before beginning a layout, the designer should discuss with management the objectives, scope, available finances, and required time schedule for the project.

2. Develop the material flow requirements based on all the products that are scheduled or contemplated for production in the plant.

3. Discuss with management the appropriate layout types that may be used in the plant. Job shop, flow shop, assembly lines, and

flexible manufacturing are some of the prominent types of layouts. The plant might have a combination of two or more of the above types. Everyone concerned should review the data to make certain that all are in agreement with them. If there is disagreement, return to Steps 1 and 2.

4. Develop overall departmental arrangements (layouts). There is seldom a case in which only one arrangement is possible. Provide at least two or three different layouts for discussion.

5. For each overall layout, develop arrangements within each department. These might consist of placement of machines, storage areas, and departmental aisles along with major aisles and material-handling equipment such as conveyers and cranes. Review the alternatives with all departmental supervisory personnel and modify the layouts as appropriate.

6. A tentative solution to any problem should be evaluated with respect to defined objectives; plant layout is no exception. In assembly lines the material and partial assemblies should flow smoothly and continuously from one station to the next in proper sequence following a well-defined flow pattern with no backtracking or congestion in the material flow aisles. The same principles apply in the job shop arrangement except that there might be multiple paths over which the products flow. Sufficient clear space should be available for each machine and its operator, and the layout should offer flexibility for future expansion or changes. The aisles should be wide enough to allow material-handling equipment to travel freely without obstruction. Noisy equipment such as punch presses should be located in an enclosed area. Space should be well utilized overall, and the employees should be satisfied with the layout; if any foreman does not agree with the arrangement, operational problems

may develop. The layout should be organized such that it integrates all the departments and their functions. Many mathematical techniques for plant layout construction and evaluation are now available in computer software. Several of these will be explained in detail in the next chapter.

7. Each alternative should now be evaluated with respect to factors such as capital requirements, operating costs, flexibility in terms of volume and product mix, unit cost of production, future expandability, ease of handling material, safety conditions, ease of supervision, inventory accumulation, maintenance and scheduling feasibility, employee satisfaction, utilization of floor space and volume, and effects of anticipated technological advancements.

8. The above comparisons could be made by using a tabular method of assigning weights. For example, each factor can be subjectively compared and assigned a weight as shown in Table 13.21.

It is obvious that the lower the cost, the higher the weight that is associated with that alternative. In measuring flexibility the higher the flexibility, the more weight the alternative would carry. Each factor has a maximum and minimum weight assigned by the analyst. The minimum weight assures that an alternative that is unsatisfactory in any one factor would be automatically rejected. The maximum weight for each factor reflects what the analyst believes the relative importance of the factor is. In the above example, Alternative 2 does not offer flexibility for future expansion, a factor considered necessary by management, and therefore is not evaluated further. Alternative 4 is selected on the basis of the total point value.

Another method of evaluating the alternatives is to make only financial comparisons. Determine the cash flows for a planning period

Table 13.21　Tabular method for layout evaluation

Factors	Maximum Points	Minimum Points Needed	Alternatives			
			1	2	3	4
(1) Cost						
Initial capital investment	100	50	80	70	90	95
Operating cost	80	50	75	79	73	70
Maintenance cost	80	30	50	60	40	50
Unit cost of production	100	60	85	80	95	100
(2) Flexibility in						
Future expansion	50	30	40	10	45	40
Volume change	50	30	40		30	35
Production mix change	50	20	35		40	40
Technology change	30	10	20		20	20
Material handling ease	60	40	50		55	45
Total			475		488	495

and then calculate the net present worths or return on investments. The alternative with the maximum net present worth (maximum return on investment) would then be selected.

There are some major problems in this method. Fixing the planning horizon is a task that cannot be taken lightly. Too short or too long a period can lead to selection of an alternative that indeed is not the best of those presented. The high costs of initial investment in building, material-handling equipment and systems, and modern processing equipment such as NC machines and robots make it important to be realistic in estimating the planning period. As with any engineering economic analysis, inflation, depreciation, tax considerations, and the minimum attractive rate of interest should be included in determining the net present worth. A thorough discussion with management in establishing the cash flows and other factors goes a long way in selling the ultimate solution to the management.

Selling a facility plan to management and the workers can indeed lead one to include features that are oriented more toward "satisfaction" than toward optimization. Managers are more concerned with long-term planning and with factors such as rate of return, flexibility in design, reliability of the system, and initial investment. Plant workers, on the other hand, are more interested in features such as inventory pileups, noise level, safety, light intensity, colors, heating and air-conditioning, computer support, and other factors that will make them feel comfortable in their work surroundings and will help them to perform their tasks. It is important when making the presentation to realize the makeup of the audience and try to resolve its concerns with one or more plant layout design alternatives.

13.5 PRESENTING THE LAYOUT

There are three phases in selling facility plans. Preparing a neat and organized facility layout and plot plan (Section 13.8) is the first step. A well-written report describing the benefits and high points of the new plan is an important follow-up step. Finally, the oral presentation integrating the written report and plans and answering questions from the audience is critical in selling the overall layout.

The written report should be designed specifically to sell the layout and should avoid any unnecessary sidetrack into detailed technical procedures. Most of the parts of the report are standard: a letter of transmittal, title page, index, introduction, body, conclusion, and appendices. The letter at the beginning of the report states the purpose of the report. The cover page gives the title, tells to whom the report is presented, who is making the presentation, and the date. The body of the report should communicate to the reader the positive aspects of the proposed facility plan. It should be brief and accurate. The detailed justification might be illustrated by tables and charts. The description of the planned implementation should illustrate how the activities in the proposed facility plan could be achieved. The dates for starting and completing activities, along with possible necessary resources, are an important part of this segment. The conclusion should summarize the entire report in one or two pages. The appendices list any data or calculations that were used in formulating the report.

Probably the most important aspect of the project presentation is the oral report. It should be as brief as possible, asking the question of what is to be done and answering exactly how to do it. The information given orally should be directed at the interests of the group to whom it is to be presented. For example, if the vice-president of sales is present, emphasis should be on how this arrangement will help meet his or her forecast sales.

Use visual aids to hold attention and clarify points. Visual aid types are numerous and may be chosen on the basis of cost and the attitude of the person to present the materials. Blueprints, plot plans, and three-dimensional models are popular items used in making the presentation. Other visual aids include flowcharts, cost sheets, templates, and computer graphics.

One should not become immersed in the details of how the plan was derived, but rather should address oneself to the issues of what is to be done, why it is needed, and how it will be achieved. The presentation should last no more than an hour, even less if the ideas can be explained in a shorter time. Rehearse the presentation and do not put too much information on one slide or transparency. Dress appropriately and be confident. Listen carefully and address any questions that are raised. Be communicative without being aggressive. Be open-minded and answer questions truthfully. If you don't know the answer, admit it and ask whether the information may be supplied later; then follow up.

Selling changes to an existing plant presents some special problems. One should realize throughout the plant development and its presentation why people resist new ideas. Inertia, fear of the unknown, conflict of personalities, loss of authority, loss of job or job content, and tendency to defend an existing method of work because of the fear that changes might mean criticism of one's work are some of the major causes of opposition. To work on these, one must stress the positive aspect of the new plan, and the changes should then be introduced gradually over an extended period of time. The planned implementation should illustrate this point. During development of the plan, letting others suggest ideas and/or letting them think that the necessary changes are really their ideas helps tremendously in obtaining their support for the overall plan. Convincing the boss and letting him or her sell the plan to subordinates is also a very effective method of selling. Ultimately, however, for the plan to succeed, all the people in the plant must be satisfied with the changes.

13.6 OFFICE LAYOUT

An office in a manufacturing facility is a communication center that is responsible for such activities as record keeping, accounting, production planning, employee management, inventory control, and sales. People who are responsible for these activities will work from the spaces provided within an office complex, and therefore it is appropriate in the overall plant layout analysis to also consider office layout.

The dimensions of individual offices are a function of the number of people using the office and the amount of furniture and equipment to be placed in the office. A desk might be 30 × 60 inches with a 24-inch chair. Filing cabinets and additional chairs are almost always necessary. In addition, at least 36 inches must be provided for chair clearance, and the room should include 36-inch walking spaces. After all necessary equipment has been decided upon, the dimensions of these items should be added to the necessary clearance and aisle dimensions to calculate the minimum office space required.

Consider an office that is to contain a desk that is 30 × 60 inches, a 24-inch armchair, a 24-inch side chair, and two file cabinets that are each 18 × 24 inches. The armchair requires clearance from the desk, while the side chair can sit against the wall. Figure 13.12 gives an example of a possible layout. The length of the

Figure 13.12 An office layout

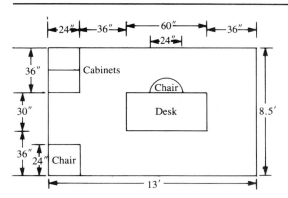

office is 60 inches for the desk, 72 inches for two aisles, and an additional 24 inches for the cabinets and the side chair. Adding these together gives a length of 156 inches, or 13 feet. The width is 30 inches for the desk, 36 inches for chair and clearance, and a 36-inch aisle for 102 inches, or 8.5 feet. Of course, all of these are minimum dimensions and should be adjusted to suit the status or needs of the person occupying the office.

The following guidelines suggest typical average assignments; however, they can be made larger or smaller depending on needs and economy:

· Top executive: 250–500 sq. ft
· Executive: 200–400 sq. ft
· Junior executive: 100–250 sq. ft
· Middle management (engineer, programmer): 80–150 sq. ft
· Clerical: 50–100 sq. ft
· Minimum work station: 50 sq. ft

The steps involved in the office layout process are similar to those in a plant layout analysis. As mentioned before, the first step is to determine the departments and the number of people working and facilities and furniture needed in each. This in turn leads to the estimation of the area needed for each department;

Office layout. In an open office plan such as this one, each person's work space is defined by partitions that can be rearranged according to the department's needs.

Table 13.22 Office space requirements

Office/Occupants	Area (square feet)
President	250
General manager	200
Sales manager	200
Production manager	200
Accountants (4)	800
Engineers (5)	775
Sales representatives (6)	600
Secretaries (7)	700
Receptionist	150
Conference room	250
Copy room	100
Coffee room	200
Rest rooms	350
Total	4775

for example, spaces must be provided for a conference room, a reception area, and the president's office. Table 13.2, reproduced here as Table 13.22, is a listing of space that might be needed and/or people who would be working in the various areas. The figures include a 20 percent allowance for hallways.

The space can be of the "open" type, in which a large number of desks are placed in one area, each person's working space being separated from the others by movable, 6- to 7-foot-high, sound-absorbing partitions. An office can be given more privacy by constructing walls to the ceiling, making it more a permanent structure. In our example, major departments are given open space, although their managers might require private offices. The accounting department, for example, includes space for four accountants and two secretaries, and the accounting manager is provided with a private office. Table 13.23 shows the space distribution that is planned, including a 20 percent hallway allowance.

The next step is to construct a rel-chart between the work areas. The necessary communications and personal contacts between the

departments are of major consideration in developing the rel-chart. Managers should be close to the departments they are supervising, and the secretaries should be close to the departments for which they are responsible to maintain communication and records. The production department generally has more contacts with engineers than with accountants. The

Table 13.23 Modified space allocation

Office	Area
(1) President	250
(2) President's secretary	100
(3) General manager	200
(4) Sales manager	200
(5) Production manager	200
(6) Managers' secretary	100
(7) Accountants (4) and secretaries (2)	900
(8) Engineers (5) and secretary (1)	875
(9) Sales personnel (6) and secretaries (2)	800
(10) Receptionist	150
(11) Conference room	250
(12) Copy room	100
(13) Coffee room	200
(14) Rest room	350
Total	4775

chain of command and the organization chart also play important parts; the president's office is more inclined to have direct communication with departmental managers than with an individual within a department. Table 13.24 shows the closeness relationships.

Table 13.25 converts the letter representation to its value and calculates the total. Again note that the entire table is not needed to calculate the total closeness. The sum of the row and column totals yields the same information. For example, for Office 5 the row total

Table 13.24 Closeness relationships

	Office													
	1	*2*	*3*	*4*	*5*	*6*	*7*	*8*	*9*	*10*	*11*	*12*	*13*	*14*
1	—	A	A	I	I	U	O	U	U	U	U	U	O	O
2		—	U	U	U	U	U	U	U	U	U	I	O	O
3			—	A	A	E	O	U	O	U	U	U	O	O
4				—	U	E	E	O	A	U	U	U	O	O
5					—	E	E	A	U	U	U	I	O	O
6						—	U	U	I	U	U	O	O	O
7							—	U	I	U	U	O	O	O
8								—	U	U	U	O	O	O
9									—	U	U	U	O	O
10										—	U	U	O	O
11											—	U	U	U
12												—	U	U
13													—	E
14														—

Table 13.25 Value table

	Office														
	1	*2*	*3*	*4*	*5*	*6*	*7*	*8*	*9*	*10*	*11*	*12*	*13*	*14*	*Total*
1	—	4	4	2	2	0	1	0	0	0	0	0	1	1	15
2		—	0	0	0	0	0	0	0	0	0	2	1	1	8
3			—	4	4	3	1	0	1	0	0	0	1	1	19
4				—	0	3	3	1	4	0	0	0	1	1	19
5					—	3	3	4	1	0	0	0	1	1	19
6						—	0	0	0	0	0	2	1	1	13
7							—	0	2	0	0	1	1	1	13
8								—	0	0	0	1	1	1	8
9									—	0	0	1	1	1	11
10										—	0	0	1	1	2
11											—	0	0	0	0
12												—	0	0	7
13													—	3	13
14														—	13

is 13 and the column total is 6, giving an overall total of 19.

The closeness total from Table 13.25 is used to develop the nodal representation. The highest total is 19, associated with Departments 3, 4, and 5; therefore choose one of these arbitrarily, say Department 3, and locate it in the center. Place all departments with four-relationships with it—Departments 1, 4, and 5—around it. Choosing the department already in the plot that has the next highest total—Department 4—place all four-relationship departments around it, that is, Department 9. Since there are no four-relationship departments left, go on to the three-relationship departments, again beginning with the highest-total department. Following the procedure leads to the nodal representation shown in Figure 13.13. Departments 10, 11, and 12 can be placed essentially anywhere, and Departments 13 and 14 need only be adjacent to each other.

The nodal representation can now be checked by using an approximate area representation. Using the scale of 100 square feet per block, Figure 13.14 shows the approximate number of squares assigned to each department. The grid can be used further to evaluate the proposed layout.

To evaluate the grid, the shortest number of squares between two departments is multi-

Figure 13.13 Nodal representation for offices

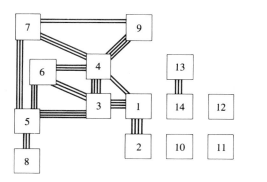

Figure 13.14 Grid representation for offices

7	7	7	11	11	11
7	7	7	9	9	9
7	7	7	9	9	9
5	6	4	4	9	9
5	3	3	1	1	12
8	8	8	2	14	14
8	8	8	13	14	4
8	8	8	13	10	10

plied by the relationship between those departments. All of these numbers are added, and the arrangement with the smallest total should be the best. For the example, evaluating the effectiveness between Departments 9 and 13 requires going straight down from Department 9 to Department 13. Three squares are crossed in the trip; and since there is a one-relationship between the two departments, the effectiveness measure is 3 × 1, or 3. For brevity, only the product is shown in Table 13.26.

Now accurately scaled sets of templates like the ones shown in Figure 13.15 must be drawn. These templates are used to obtain a true picture of the layout. Converting the grid of Figure 13.13 to a definite layout results in the layout presented in Figure 13.16. This layout can be altered to correct the irregular shape of the design. The modified layout shown in Figure 13.17 is accurate and efficient and will be used to build the office.

By now it should be apparent that the office layout problem is a smaller, but no less complex version of the plant layout problem. Space requirements and departmental relationships must be carefully considered to produce the best office design resulting in working conditions that help promote efficiency.

Table 13.26 Effectiveness calculation for Figure 13.13

		Office													
	1	**2**	**3**	**4**	**5**	**6**	**7**	**8**	**9**	**10**	**11**	**12**	**13**	**14**	**Total**
1	—	0	0	0	4	0	2	0	0	0	0	0	1	0	7
2		—	0	0	0	0	0	0	0	0	0	4	0	0	4
3			—	0	0	0	1	0	2	0	0	0	2	2	7
4				—	0	0	0	0	0	0	0	0	2	2	4
5					—	0	0	0	3	0	0	0	4	4	11
6						—	0	0	0	0	0	8	4	4	16
7							—	0	0	0	0	4	4	4	12
8								—	0	0	0	1	0	1	2
9									—	0	0	0	3	1	4
10										—	0	0	0	0	0
11											—	0	0	0	0
12												—	0	0	0
13													—	0	0
14														—	0
														Total	67

Figure 13.15

Office templates

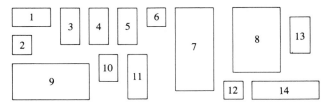

Figure 13.16 Initial layout of offices

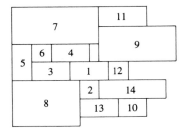

Figure 13.17 Final layout of offices

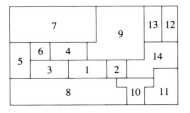

13.7 OFFICE LANDSCAPING

Office landscaping, an alternative to conventional office layouts, is a concept that originated in East Germany in the 1950s. The idea is to make the office an open, airy, attractive, and pleasant place to work, while being flexible enough to change as need dictates. Irregular placement of furniture and the use of partitions are characteristic of office landscaping.

The rows and rows of perfectly lined up desks of the conventional office are eliminated in landscaping. At first glance, a typical landscaped office appears to be haphazardly arranged. Actually, personnel are grouped according to their need for communication. When privacy is required, partitions that are easily moved at any time are utilized. An example of a landscaped layout is shown in Figure 13.18.

The single major problem that must be addressed in landscaping is the level of noise. With many people in one large, open room, noise could become unbearable. But noise can be reduced by reducing the smooth surfaces in the office that reflect sound. Using carpet instead of tiles on the floor and textured walls and ceilings helps to absorb sound. Live or artificial plants and attractive pictures on the wall give the office a homey feeling, which studies have shown to contribute to increased productivity.

13.8 PLOT PLANNING

Once the production and office layout have been determined, the plant and other auxiliary building dimensions have been established, and a piece of land has been purchased or leased for the location of a new plant, complex steps must be taken to decide how the facilities are to be placed and arranged on the piece of land. This is known as plot planning. The basic functions of plot planning are as follows:

· Planning the best use of the space available
· Planning the flow of materials throughout the facilities
· Planning for future needs

Plot planning involves more than just the location of the basic plant on a piece of property. For instance, if external warehouses are planned, they must be located in relation to the

Figure 13.18

A landscaped office

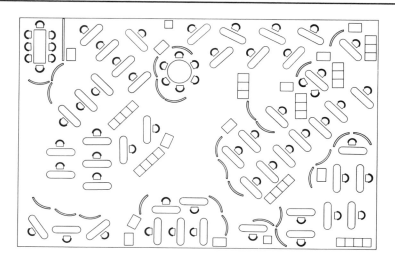

plant. Locations of driveways, entrances, and loading docks, and their relationships to roads and the location of parking relative to entrances must be considered. Other important aspects are locations of sidewalks, steps, and utility lines. Finally, space should be allocated now for future expansions. The plot plan is completed with landscaping and provision for installation of traffic control and other informational signs.

The plot plan always begins with the constant features, which include property lines, railways, waterways, and roadways. The main building is placed; then entrances, exits, roads for loading docks, sidewalks, and parking lots are located in relation to the building and the constant features. Keep in mind that at this point the plant layout has already been de-

signed and is simply being placed in the best position. If the number of factors to be considered is relatively small, the task is not a difficult one. If the complex consists of several buildings and a large piece of land, the task becomes more difficult. In such a case the problem can be approached much like the plant layout problem. Identify all points of consideration and determine adjacency requirements. Then place the buildings, maintaining the necessary adjacencies to the extent possible.

The plot plan should always be drawn to scale with dimensions shown. To improve its attractiveness, a few minor changes can be made in the final drawing as well as adding landscaping. Figure 13.19 is an example of a plot plan.

Figure 13.19

A plot plan

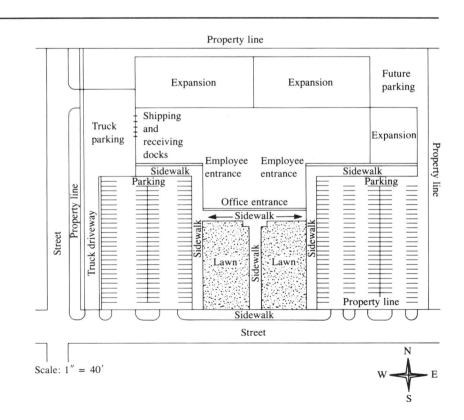

13.9 EVALUATING AND IMPLEMENTING THE LAYOUT

When construction of the plant is complete and production begins, some problems might still develop owing to its layout, product change, or growth in demand. Sometimes the problems are due to inadequate initial analysis. Once such concerns have been identified, they should be clearly defined so that solutions can be developed. The difficulties might include bottlenecks, unsafe conditions, idle or overloaded equipment, poor quality, shipping delays, wasted materials, lack of materials, idle workers, excess handling of materials, and backtracking. Some problems might involve multiple causes, and for this reason each should be carefully analyzed by starting at the problem area and going back through the production processes.

The causes are of course unique to the problem and system in question. Many times they involve not only the plant layout, but often also other factors such as pattern of material flow, environmental conditions, material-handling equipment, poor work methods, obsolete equipment, inadequate storage and warehousing conditions, poor allocation of personnel, or any number of other reasons.

The company can best benefit by seeking out potential problem areas before they become an economic burden. A model of the plant can aid in evaluting the plant and identifying such areas inexpensively. In addition, the model can be used to design and evaluate alternative plans. If the original plan has some flexibility, the necessary changes can be made quickly, easily, and with a minimum of expense.

Other methods of discovering problems include personnel suggestions and quantitative measures. Floor personnel frequently identify difficulties simply through experience. Comments and suggestions from these people should be recorded and investigated. The factors in quantitative methods depend on exactly what is being measured. If the interest is in machine usage, the amount of time the machine is being used can be compared to its total capacity. The same principles can be applied to personnel problems and aisle difficulties that might be causing bottlenecks or a slowdown in the movement of materials.

If there is some degree of flexibility in the plant processes, minor problems can often be handled at low cost by changing sequences, revising the flow pattern, rearranging the workplace, and combining handling with the process. Other problems might be as simple as better housekeeping or following established procedures, but there are times when major expenses in terms of money and time cannot be avoided. A final evaluation in this case would include an economic analysis. If the cost of the changes outweighs the economic benefits, the management would probably decide to live with the difficulty.

13.10 PLANNING ACTIVITIES

Planning of a plant and office building and their construction and start-up activities all occur in real time. The date to purchase or lease the land must be fixed, construction of the buliding must follow some predefined schedule, the equipment must be purchased and installed at a specific time, and workers must be hired and trained so that they will be productive when needed. Each one of these major projects can be further divided into hundreds of subtasks, some of which can be performed simultaneously while others can start only after a number of preceding jobs are completed.

A question arises as to when each activity should be started and completed. Scheduling each task to take place as soon as possible will simply add to the haste and associated costs.

On the other hand, indefinite delays can also add considerably to the costs as well as to the schedule. Circumstances change quite frequently, and any planning done too far in advance is subject to more errors and changes than are jobs planned for immediate actions. Plant layout could be an extended project taking anywhere from 6 months to 3 years. Time and costs estimates of the activities that far in advance could be difficult. Frequent changes in organizations and personnel in a dynamic company present special challenges in adhering to firm schedules and plans.

It might still be possible, however, to develop some reasonable estimates on many tasks by noticing historical and traditional data. For example, it can be observed that delivery of equipment takes 3–6 months from the date of order, and construction of a building takes between 1 and 2 years.

Activities must be scheduled to meet certain target dates. Although these dates are set by the management, they should be tested frequently to make sure they are realistic. Unrealistic target dates set to encourage speed are frequently recognized as meaningless and quite often are ignored. The jobs must start on or very near a prespecified target date and then proceed with realistic time estimates.

Once the tasks have been identified and precedent relationships have been established, it is appropriate to construct a network that can be analyzed by the critical path method (CPM). We will not go into the details of CPM here, since many standard operations research books (Phillips et al., 1987; Taha, 1982) have excellent descriptions of the method. Establishing the critical activities (jobs that must be started and completed on time to achieve the scheduled completion date) and activities with delays does provide guidelines for management in determining the projects that should be accelerated or delayed to meet the desired target dates economically.

13.11 COMPUTER PROGRAM DESCRIPTION

This program constructs a layout for a facility/ plant using certain input parameters. The interactive nature of the program guides the user through the process and constantly checks for the input validity. A description of the sequence of input prompts follows.

The number of departments must be inputted first. The user is then asked to give the interrelationships between the departments. The prompts are as follows:

1. "How many departments (max = 20)?"
2. Input departmental relationships; upper triangular form is assumed. Use A, E, I, O, U (or ⟨Return⟩ key), and X, between department . . . and department . . . where A = absolutely essential, E = essential, I = important, O = ordinary, U = unimportant, and X = undesirable.

If an invalid code is typed, the user can correct the mistake immediately. The user can also give quantitative values to the codes or use the set of default values. A table of the interrelationships is then displayed.

The so-called nodal diagram is generated for the next screen. This diagram shows the relative positions of the departments assuming unit areas for each as described in the chapter. A prompt informs the user of how to get a printout of the diagram: "Press the 'Shift' key and the 'Prt Sc' at the same time to get a printout of the diagram. Otherwise, hit 'Return' to continue."

The areas of the departments may now be inputted either by the actual square foot or by a certain number of blocks representing each area. In the former case, an appropriate number of blocks is calculated for each area automatically. The user is asked: "Do you want to input the number of blocks for each department (Y/N)?"

1. If Y: "Number of blocks for department: ...?"
2. If N: "Please input departmental areas (square feet)?" ... "Department #?" ... "What is the *column span* in your layout or block size of the layout?" *Note:* Only *even* numbers are to be used. The area for a block is calculated by squaring the input to this question.

If the maximum number of blocks exceeds 800, an error message will be printed and the program is terminated.

The next screen displays the final layout with appropriate headings and a score for it at the bottom. The user is then asked if he or she wants to swap any one or a series of blocks. If the answer is yes, the user is given the opportunity to make the necessary changes after which the new layout is displayed with its score. In the case of no swaps, the program is terminated. The prompt is as follows: "Do you wish to interchange any of the blocks; you will as the result see the grid and its score (Y/N)?"

1. If Y: "Please pick your option—1, for interchanging any *two* blocks; 2, for interchanging two groups of blocks each on one row only. Your option?"

   a. If 1: "*Row* and *column* for the first block __, __? *Row* and *column* for the second block __, __?"
   b. If 2: "*Row* and *column* positions for the *first* element of the *first* group (e.g., X5, A4) __, __? *Column* position for the *last* element of the *first* group (e.g., A8) __? *Row* and *column* positions for the *first* element of the *second* group (e.g., Y2, A3) __, __? *Column* position for the *last* element of the *second*

group (e.g., A7) __?" *Note:* The program can exchange only one block at a time (option 1) or one row at a time (option 2).

2. If N: The program is terminated.

SUMMARY

A well-designed plant aids in reducing material-handling costs and contributes to the overall efficiency of operations. There are six basic steps that must be followed in obtaining a good plant layout. The procedure starts by determining the area requirement for each work center and is completed when templates of all the work centers have been prepared to obtain a regularly shaped buliding. Where the templates are to be placed, how large the building should be, and how you know that a good decision has been obtained are the types of questions answered by the remaining steps in the procedure. The procedure is efficient and effective and is illustrated in the chapter by applying it to the development of a layout for a small manufacturing firm. Effectiveness calculations measure the relative value of a design, and the one with the smallest effective value (this value can be negative for an X-type relationship) is generally the most efficient.

Once the positions of the departments are established, the next step is to develop a detailed layout for each department. All items such as aisles, columns, location of machines, the material flows, spaces for in-process inventories, and spaces for maintenance and auxiliary equipment must be clearly shown. Special attention should be given to aisle locations and their widths since they form the connecting paths between the various departments. Commonly used facilities such as washrooms, cafeterias, water fountains, and toolrooms should be placed so as to obtain the utmost efficiency in operations. The chapter gives some guidelines suggesting how this can be achieved.

Various materials may be used to construct a layout. Two- and three-dimensional templates representing physical objects placed on a grid and plastic sheets laid on a sturdy base material are a good start. Plastic tapes of different colors and widths may be used to show aisles, walls, conveyors, material flow, storage areas, and other interesting features and may be further identified by professionally lettering them. Clutter can be reduced by developing different overlay sheets, with each overlay showing a different aspect of the layout.

Analysis of a good solution requires evaluating it both quantitatively and qualitatively. Solutions should offer smooth and continuous material flow, no congestion in aisles, sufficient clearance for machines and operators, and flexibility to change and expand. Quantitatively, a weighted analysis and/or rate-of-return analysis can be performed. A well-written report, a good oral presentation, and a question-and-answer session are the three critical steps in selling a design. At times, features oriented more toward satisfaction than optimization may have to be chosen to obtain consents from both manage-ment and workers, since each may have a different prospective on the design.

Development of an office layout follows the same six-step procedure as that for a plant layout. An example illustrating an office layout is presented in the chapter. Offices can be made more informal, airy, and pleasant to work in by following an office landscaping program. Office landscaping can also reduce the noise level.

Plot planning is the next step in development. It involves placing a newly designed plant with its warehouses, streets, driveways, parking lots, loading docks, and lawns on the plot of land that has been acquired for it. Other features such as sidewalks, utility lines, and security fences must now be considered. The plan should offer a degree of flexibility for change and future expansion.

In time, reevaluation of the existing plant layout may be warranted if inefficiencies develop. It becomes helpful if the model representing the current layout is available. It is therefore advisable to maintain the records and layouts of current development for possible future use. One should also understand that there is a time lapse between the design phase and the construction phase and that situations might arise that demand reanalysis.

PROBLEMS

13.1 Why is the plant layout problem so important?

13.2 Discuss reasons for placing the following departments. Tell what other areas they should be close to and what departments they should be removed from.
  a. Rest rooms
  b. Paint department
  c. Receiving

13.3 Convert the following rel-chart into a nodal representation.

	1	2	3	4	5
1	—	I	A	O	O
2		—	U	O	E
3			—	A	X
4				—	E
5					—

13.4 List the steps for developing a layout.

13.5 Using the rel-chart in Problem 13.3, determine a layout for the five departments. Each department requires 1000 sq. ft, with each block representing 250 sq. ft.

13.6 Using the following information, lay out the six departments. Assume X = −1 and building width = 4 blocks with block size of 20′ × 20′; calculate the effectiveness (the lower the number, the better the arrangement, including negative values). Develop the final layout.

Department	Size
(1) Painting	2100
(2) Welding	1800
(3) Maintenance	700
(4) Machine shop	1500
(5) Office	1000
(6) Storage and shipping	2500

	1	2	3	4	5	6
1	—	E	I	U	X	A
2		—	E	E	U	U
3			—	I	U	U
4				—	U	U
5					—	E
6						—

13.7 The production area of the XYZ Machine Company is a process-type layout. The present layout is as follows:

```
A        C
D        E
B        F
```

In the layout above, A, B, C, D, E, and F are the six processing areas. For safety resons it is undesirable to place areas A and F adjacent to each other. A study of the manufacturing activities of the XYZ Machine Company indicates that four major products accounted for 85 percent of the total manufacturing activity in the above production area last year. The production sequence of these products and the number of units produced per year can be found in the following table.

Product	Production Sequence	Number of Units Produced per Week	Number of Units per Trip
1	B— E— F	8,000	50
2	A— D— E— F	500	25
3	C— B— E— A	5,000	50
4	A— C— B— E— F	15,000	100

Standard 36″ × 36″ unitized pallets and a conventional narrow aisle forklift truck are the material-handling equipment involved in the production of the four products. The average numbers of units carried per trip by the lift truck are also indicated in the above table.

a. Develop an activity relationship chart.
b. Develop a layout of the processing areas.
c. Discuss the present layout versus the proposed layout.
d. If the areas for A, B, and C are twice as large as those for D, E, and F, discuss how the layout will change in terms of relative positions of the departments and shape of the overall layout.

13.8 How does the office layout problem differ from the plant layout problem?

13.9 Using the following information, determine a layout for a plant.

A. Office

Department	Size
(1) Executive office	500
(2) Bookkeeping	800
(3) Sales	800
(4) Engineering	500
(5) Secretarial pool	1000

	1	2	3	4	5
1	—	E	E	E	U
2		—	E	O	A
3			—	U	A
4				—	O
5					—

B. Manufacturing

Department	Size
(1) Machine shop	1500
(2) Assembly line	3000
(3) Welding	1800
(4) Painting	2000
(5) Manufacturing office	250

	1	2	3	4	5
1	—	A	U	U	O
2		—	A	U	O
3			—	A	O
4				—	O
5					—

C. Storage

Department	Size
(1) Receiving raw materials	900
(2) Storage of raw materials	3000
(3) Receiving finished product	900
(4) Storage of finished product	5000
(5) Shipping	900

	1	2	3	4	5
1	—	A	U	U	U
2		—	U	U	U
3			—	A	U
4				—	A
5					—

D. Auxiliary

Department	Size
(1) Rest rooms	400
(2) Lockers	750
(3) Cafeteria	1000
(4) Maintenance	1100

	1	2	3	4
1	—	E	E	O
2		—	U	U
3			—	U
4				—

	A	B	C	D
A	—	O	E	O
B		—	A	A
C			—	O
D				—

Specifically, the bookkeeping department has an E relationship with receiving raw materials and with shipping, painting has an A relationship with receiving finished product, and maintenance has an A relationship with the manufacturing department.

SUGGESTED READINGS

Plant and Office Layout

Conway, H.M., and Liston, L., *Industrial Facilities Planning,* Conway Publishing, Atlanta, 1976.

Francis, R.L., and White, J.A., *Facility Layout and Location—An Analytical Approach,* Prnetice-Hall, Englewood Cliffs, N.J., 1974.

Muther, R., *Systematic Layout Planning,* Cahners Books, Boston, 1973.

Pile, J., *Open Office Planning,* The Architectural Press, New York, 1978.

Tomkins, J.A., and White, J.A., *Facilities Planning,* John Wiley and Sons, New York, 1984.

Operations Research

Phillips, D.T.; Ravindran, A.; and Solberg, J.J., *Operations Research—Principles and Practice,* John Wiley and Sons, New York, 1987.

Taha, H., *Operations Research—An Introduction,* 3rd edition, Macmillan, New York, 1982.

Computer-Aided Plant Layout

Traditionally, the development and evaluation of plant layouts were accomplished subjectively by designers who resorted to graphical techniques and template manipulation. With the emergence of operations research and the use of digital computers, however, more analytical-based procedures were applied in the layout developments. This trend culminated in what is now known as computer-aided plant layout. The approach involves using computer programs that assist in generating layouts quickly and comparing them on an objective basis. It is more suitable for the job shop than for product layouts, which by their very nature are simpler to develop. The term "computer" is used mainly to distinguish between the recent analytical approach and the traditional manual approach. Thus while some of the analytical procedures do not produce block layouts in the final computer output, as others do, they can be included in the discussion of computer-aided plant layout, since the computer is still necessary to perform the required calculations.

Using the computer in solving the layout problem has several advantages over the traditional manual approach. First, the computer can perform the necessary calculations and generate several solutions much more quickly than manual procedures. Second, the computer can solve large-sized problems, which usually involve huge amounts of data. Third, since the computer can develop solutions quickly, it is more economical to use a computer as an aid in the design process than to depend on human planners and designers alone. And fourth, by using the computer, solutions will be developed on the basis of mathematical expressions and operations that can be evaluated objectively. These solutions are often better than those produced solely by subjective judgment.

It should be emphasized that these computer programs are by no means a substitute for human involvement in the design process; they are merely an aid to facilities planners. These programs have

---

*This chapter was contributed by Mohsen M. D. Hassan.

several drawbacks, and the designers must usually modify the solutions obtained from them to guarantee a realistic and useful design. In addition, these programs do not produce a detailed layout of the facility, but rather a block layout of the arrangement of work centers or departments with respect to each other. Hence the output produced by these programs should not be considered the final step in the design process.

14.1 CHARACTERISTICS OF THE PROBLEM

The plant layout problem has some characteristics that make it difficult to formulate and solve completely by analytical means. For one thing, since each location is a candidate for each department, all combinations of departments and locations have to be evaluated. Hence if there are n departments and n locations, the maximum number of combinations that has to be considered is $n!$. Evaluation of all possible combinations could take an inordinate amount of time for medium- to large-sized problems. Even if the numbers of departments and locations are not the same—say, n and m respectively, where $n < m$—the number of combinations that require evaluation is $m!/(m - n)!$, which, although less than $n!$, is still large.

Minimizing the material-handling cost is usually considered the most important objective of the problem. However, several other critical objectives exist such as achieving maximum adjacencies (closeness) between the departments, flexibility of arrangement and operation, and effective utilization of the available space. When considered simultaneously in an analytical model, these objectives complicate the formulation. Furthermore, it is difficult to translate qualitative information into quantitative objectives and constraints.

The layout problem (particularly in industry) involves locating areas, not points. These areas affect the shapes and dimensions of the departments, which in turn affect the distances between these departments. The number of configurations a department can assume is infinite, and the distance between the departments changes accordingly, thereby affecting the material-handling cost. The interrelation between departmental shapes and the distances is difficult to control or predict.

When a new plant is to be constructed, the data required for solving the problem, such as the material flow between departments and the operating cost of material-handling equipment, are usually not available and must be estimated, which is not an easy task.

14.2 DATA REQUIREMENTS

The basic desired input for working with an analytical model varies from method to method. The following list gives the different types of data that might be required. Among these, departmental relationships and areas are the minimum necessary in every computerized method of analysis.

1. The area of each department expressed in square feet or as a number of unit squares.
2. The rectilinear distances between candidate locations or between departments, usually measured between their centers.
3. Departmental relationship measures that can be expressed either quantitatively in a

from-to chart or qualitatively in a relationship chart (rel-chart).

4. A scale for plotting the layout by the computer, required in some of the procedures. In addition, some have several options regarding departmental shapes, location, and the layout outline.

14.3 APPROACHES AND TYPES OF PROCEDURES

In practice, plant layout problems can be solved by any one or more of the following approaches:

· Exact mathematical programming procedures, mainly branch and bound
· Heuristics
· Probabilistic approaches
· The application of graph theory

Notice, however, that whatever the approach used, it attempts to solve an optimization problem. The classification is mainly based on the dominant theoretical basis of the procedure. Thus while the probabilistic approach is completely dependent on heuristics, it is not classified as such, since what distinguishes it from the other heuristics of computer-aided plant layout is the inclusion of probability in its solution procedures. Similarly, graph theory–based procedures use both heuristics and the branch-and-bound techniques in developing layouts; but since they utilize the concepts and ideas of graph theory, they are classified as such.

The objective of these procedures depends on the manner in which departmental relationships are expressed. When qualitative relationships are used, as expressed in a rel-chart, the objective is to maximize the adjacencies (closeness) between departments. On the other hand, when departmental relationships are expressed quantitatively by a from-to chart,

the objective is to minimize the material-handling cost, that is, the product of departmental distances, flow, and unit-handling cost. In many cases, especially for a new plant, the handling system will not be known until the layout has been established. A handling cost of 1 is assumed in such cases, which results in the cost to be minimized being equal to the flow multiplied by the distance only. (Henceforth this cost will be referred to as the movement cost.)

The solution procedures for any of these approaches are either of the construction or the improvement type. Construction procedures, which are used when a layout is developed for the first time, involve three steps. First, a criterion for selecting the departments to enter the layout is established on the basis of departmental relationships. Second, the placement of each department in a suitable location is attempted such that the inherent objective is achieved. Finally, steps toward development of departmental configurations are undertaken. Construction procedures differ among themselves by the rules used to execute each step. The solution obtained from a construction procedure is almost always suboptimal, which is the result of the combined effect of the three steps.

Improvement procedures attempt to reduce the movement cost of an initial layout, which could be an existing layout or one developed by a construction procedure. The improvement is usually achieved by interchanging the locations of departments until no further reduction in movement cost is possible. Thus the solution obtained will most likely be better than that of a construction procedure, but there is still no guarantee of optimality.

14.4 MATHEMATICAL PROGRAMMING

In this section, only two mathematical formulations are illustrated without going into the details of the solution procedures. An interested reader can explore this further by consulting the references, while those who prefer simpler heuristic approaches can go directly to the next section.

The plant layout problem is most often formulated as a quadratic assignment problem (Francis and White, 1974; Pierce and Crowston, 1971), where m locations are available to which n departments are to be assigned. All the assignments must be done simultaneously, and the arrangement that minimizes the movement cost is then selected. If $n \le m$, let

c_{ikjh} = Cost of placing departments i and j at locations k and h, respectively

$X_{ik} = 1$ if department i is located at k
$\quad\;\; = 0$ otherwise

Then we

$$\text{Minimize } Z = \frac{1}{2} \sum_{i=1}^{n} \sum_{k=1}^{m} \sum_{j=1}^{n} \sum_{h=1}^{m} c_{ikjh} X_{ik} X_{jh}$$

subject to

$$\sum_{i=1}^{n} X_{ik} = 1 \qquad k = 1, \ldots, m$$

$$\sum_{i=k}^{m} X_{ik} = 1 \qquad i = 1, \ldots, n$$

$$X_{ik} = \{0, 1\} \qquad \text{for all } i, k$$

Several branch-and-bound algorithms have been developed for solving this structure; however, they are feasible only for small-sized problems. The difficulty increases rapidly as the number of departments exceeds 15 because of the increase in the number of solution combinations. Moreover, unless some constraints are established to prohibit a large number of adjacencies of departments from being achieved or certain departments from being placed in certain locations, every branch in the solution tree is a feasible solution that might have to be evaluated completely. However, branch-and-bound procedures guarantee an optimal solution and can provide alternative layouts in addition to the optimal one.

An alternative mathematical formulation has been suggested (Bazaraa, 1975) as a quadratic set-covering problem and is solved by branch-and-bound procedures. The areas of the layout and the individual departments are expressed as unit squares. Several candidate locations are made available for each department to be located, which prevents each location from being available for every department as in the quadratic assignment problem. Restricting the availability of locations to departments helps in reducing the number of combinations that must be evaluated.

The procedure is capable of producing desirable shapes for the departments as well as for the layout outline by specifying the shapes for these as part of the input data. However, specifying these shapes implies additional constraints to the problem, which in turn might affect the objective function adversely. The model has two drawbacks concerning some of its assumptions. First, candidate locations for each department must be established in advance, which may not be an easy task. Second, the specification of departmental shapes might not be practical, since it is usually not known why a particular shape would be preferred to another or how a good department shape should look. The problem formulation is shown below. Let

$d(k_i, h_j)$ = Rectilinear distance between the centers of the kth and the hth locations for departments i and j

f_{ij} = Flow between departments i and j

F_{ik} = Fixed cost of locating department i at k

$I(i) = I$ = Number of candidate locations for department i

$I(j) = J =$ Number of candidate locations for department j

$J_i(k) =$ Set of squares occupied by department i if it is located at k

$a_{ikt} = 1$ if block $t \in J(k)$

$\quad\ = 0$ otherwise

$X_{ik} = 1$ if department i is located at k

$\quad\ = 0$ otherwise

The problem requires that we

$$\text{Minimize } Z = \sum_{i=1}^{m} \sum_{j=1}^{m} \sum_{k=1}^{l} \sum_{h=1}^{j} f_{ij} X_{ik} X_{jh} d(X_i, h_j)$$
$$+ \sum_{i=1}^{m} \sum_{k=1}^{l} F_{ik} X_{ik}$$

subject to

$$\sum_{k=1}^{l} X_{ik} = 1 \qquad i = 1, 2, \ldots, m$$

$$\sum_{i=1}^{m} \sum_{k=1}^{l} a_{ikt} X_{ik} \qquad \text{all } t$$

$$X_{ik} = \{0, 1\} \qquad k = 1, 2, \ldots, I(i)$$
$$i = 1, 2, \ldots, m$$

A third formulation appeared recently (Hassan, 1983) in which the layout problem is formulated as a generalized assignment problem. However, the solution procedure is heuristic, and therefore optimality is not guaranteed. Practical application of the above procedure is limited owing to its inherent complexity. Therefore little would be gained in pursuing it through a numerical example or further discussion.

14.5 HEURISTICS

As was mentioned previously, applying an exact procedure that produces an optimal solution is feasible only for small problems be-cause of the number of calculations involved in the process; hence the exact procedures are not very widely used. Heuristics, on the other hand, enjoy a large following. The main advantages of heuristics are their ability to produce good sub-optimal solutions and their capacity for han-dling large problems with a reasonable amount of computational effort. The most popular and widely used procedures are now presented in some detail.

CORELAP

Computerized Relationship Layout Planning (CORELAP) (Lee and Moore, 1967) is a con-struction program that uses the relationship chart and attempts to develop a layout whose objective is to achieve maximum adjacencies between the departments. In CORELAP's se-lection method, each element of the relation-ship chart is assigned a numerical value, and departments are ranked in a nonincreasing order of the sum of their relations with one an-other. The first department in the list is selected to enter the layout. Next departments are se-lected by determining which of them has the strongest relation with one of the previously se-lected departments, starting with the first se-lected department, then the second selected, and so on. When ties arise, the department with the larger sum of relations is selected first. CORELAP places the first department in the middle of the layout, and the rest of the depart-ments grow around it. The program attempts to place each department adjacent to the one with which it has the strongest relation, or as close as possible. Departmental shapes are con-structed to be as close to a square as possible. An example of the layout produced by the com-puter for CORELAP is shown in Figure 14.1.

As an example of the selection method of CORELAP, consider the rel-chart shown in Table 14.1. Further, assign the following values to the elements of the chart:

Figure 14.1   Computer output for a layout produced by CORELAP (From Lee and Moore, "CORELAP—Computerized Relationship Layout Planning," *Journal of Industrial Engineering,* Vol. 18, No. 3, 1967. By permission from the Institute of Industrial Engineers.)

```
          0   0   0   0   0   0   0   0   0   0   0   0   0   0   0   0   0   0   0   0   0   0
  0   0   0  25   0   0   0   0   0   0   0   0   0   0   0   0   0   0   0   0   0   0   0
  0   0   0  25  25  25  26   0   0   0   0   0   0   0   0   0   0   0   0   0   0   0   0
  0   0   0  25  25  25  26  26  26  26  26  26   0   0   0   0   0   0   0   0   0   0   0
  0   0   0  25  25  19  19  19  26  26  14  14  14  14  14  14  14  14  14   0   0   0   0
  0   0   0  25  25  19  19  27  27  26  26  14  14  14  14  14  14  14  14   0   0   0   0
  0   0  21  21   0  32  32  33  27  27  24  24  14  14  14  14  14  14  14   0   0   0   0
  0   0  21  21   0  32  32  33  27  27  24  24  24  14  14  14  14  14  14   0   0   0   0
  0  37  37  29  29  29  29  34  34  34  34  34  30  30  14  14  14  14  14   0   0   0   0
  0  37  37  29  29  29  29  34  34  34  34  34  30  30  14  14  14  14  14   0   0   0   0
  0  37  37  37  29  29  34  34  34  34  36  36  36  13  13  14  14  14  14   0   0   0   0
  0  37  37  37  37  31  31  34  34  34  34  36  36  13  13  12  14  14  14   0   0   0   0
  0  37  37  37  37  37  37  37  37  37  36  36  13  13  14  14  14  14  14   0   0   0   0
  0  37  37  37  37  37  37  37  37  37  36  36  36  35  35  14  14  14  14   0   0   0   0
  0  37  37  37  37  37  37  37  37  37  36  36  36  35  14  14  14  14  14  23  22  22   0
  0  37  37  37  37  37  37  37  37  37  11  11  35  35  14  14  18  23  23  23  22  22   0
  0  37  37  37  37  37  37  37  37  37  11  11  35  35  15  15  18  17  17  22  22   0   0
  0  37  37  37  37  37  37  37  37  11   0  35  35  35  35  15  18  18  17  22  22   0   0
  0   0   0   0   0   0   0   0   0   0   0   0  28  28  15  15  15  16  17  17  22  22   0
  0   0   0   0   0   0   0   0   0   0   0   0  28   0   0   0  16  16  16  20   0   0   0
  0   0   0   0   0   0   0   0   0   0   0   0   0   0   0   0   0   0  20  20   0   0   0
  0   0   0   0   0   0   0   0   0   0   0   0   0   0   0   0   0   0   0   0   0   0   0
```

A = 6	O = 3	
E = 5	U = 2	
I = 4	X = 1	

Note that these values are different from those we used in Chapter 13. These values are the measure of closeness between the departments, and the way they are defined is one possible variation when using heuristic methods.

By substituting these values the rel-chart now becomes as shown in Table 14.2.

The sum of the relations between a department and all other departments is obtained by adding entries in corresponding departments, rows, and columns. For example, the relationship value for the third department is obtained by adding the third row and the third column values, that is,

$$5 + 4 + 6 + 5 = 20$$

Similarly, if we perform the calculations for other departments, these sums are 13, 14, 20, 15, and 12. Arranging the departments accord-

Table 14.1   Rel-chart for a CORELAP illustration

Department	1	2	3	4	5
1	—	U	E	I	U
2		—	A	O	O
3			—	E	I
4				—	O
5					—

Table 14.2   Value chart for CORELAP illustration

Department	1	2	3	4	5	Sum of Relations
1	—	2	5	4	2	13
2		—	6	3	3	14
3			—	5	4	20
4				—	3	15
5					—	12

ing to these sums will result in the following ranking: 3, 4, 2, 1, 5.

Department 3 is selected first, since it has the largest sum. In looking for a department with an A relation with Department 3, the program finds and selects Department 2. No other department having an A relationship with Department 3 exists. CORELAP next looks for an A relationship with Department 2 but finds none. Next the program hunts for an E relationship with Department 3, and it finds both Departments 1 and 4. The tie is resolved according to the sum of relationships; since Department 4 has a larger sum of relationships than Department 1, 4 is chosen first and then 1. The final order of selection is 3, 2, 4, 1, 5.

PLANET

Another construction program is Plant Layout Analysis and Evaluation Technique (PLANET) (Apple and Deisenroth, 1972), which uses flow data expressed by the from-to chart or the rel-chart. Three methods are available for selecting departments after the latter are divided into groups according to a user-specified priority. The first method (A) selects from among all the departments the two having the strongest flow relationship with each other. The department having the strongest relationship with any of the previously selected departments is selected next. (In case of a tie, arbitrarily select either.) The second method (B) selects the first two departments in the same way as is done by method A, but the next department selected is the department having the strongest sum of relationships with all the previously selected departments. The third method (C) ranks departments in a decreasing order according to the sums of their flow values with each other and selects them according to this ranking. PLANET's objective is to minimize the total movement cost.

PLANET locates the first selected department in the middle of the layout area, and the rest of the departments cluster around it. The best location for each department is determined by examining several points around the periphery of the partial layout formed by the previously selected and placed departments. These points represent the candidate centers of the entering department. The location that results in minimum movement cost between the entering and placed departments is then selected. This search scheme is repeated for each selected department until the layout is complete. Once a location has been determined for a department, its shape is made as close to a square as is possible. This is accomplished by placing the unit squares that constitute a department side by side in a spiral fashion. An example of the layout produced by PLANET is shown in Figure 14.2.

To demonstrate the selection methods of PLANET, consider the symmetric from-to chart in Table 14.3, and assume that all departments have the same priority.

Figure 14.2   Computer output of a layout by PLANET (The program does not plot the lines representing department borders.)

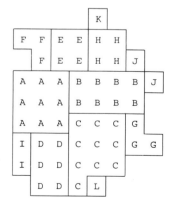

Table 14.3 From-to chart for a PLANET illustration

	A	B	C	D	E
A	—	0	36	18	0
B		—	9	10	0
C			—	12	8
D				—	5
E					—

Method A

Departments A and C are selected first, since they have the strongest relationship, with a flow of 36 between them. The program looks for the department with the next strongest relationship with either A or C, and Department D is found to have the strongest relationship with A, with a flow of 18 between the two. Department D is selected, and the program then looks for the department with the strongest relationship with either A, C, or D. The flow between B and D is 10, which is larger than that between E and any of A, C, or D; hence B enters next and then E, since it is the remaining department. The order of selection is then A and C, followed by D, B, and E.

Method B

A and C are selected first as in method A. Next the sum of the flow between B and both A and C is evaluated as $0 + 9 = 9$, that between D and both A and C as $18 + 12 = 30$, and that between E and both A and C $= 0 + 8 = 8$. Therefore Department D is selected because it has the largest sum with Departments A and C.

Now both B and E are compared by finding the largest sum with the previously selected departments. Department B with A, C, and D yields $0 + 10 + 9 = 19$, and Department E yields $0 + 8 + 5 = 13$. Accordingly, Department B is selected next and then E; and the order of selection is A and C, followed by D, B, and E.

Method C

The sums of the relationships between the indicated departments and all others are:

A: $0 + 36 + 18 + 0 = 54$

B: $0 + 9 + 10 + 0 = 19$

C: $36 + 9 + 12 + 8 = 65$

D: $18 + 10 + 12 + 5 = 45$

E: $0 + 0 + 8 + 5 = 13$

If we arrange the values in decreasing order, the sequence of the departments becomes C, A, D, B, E.

Although all three selection methods produce the same order of selection in this example, this would not necessarily be the case with other data.

To demonstrate the placement procedure of PLANET, assume that all departments have the same area of one square foot. The first two departments (A, C) enter the layout and are placed as shown in Figure 14.3. The movement cost between the two departments is $36 \times 1 = 36$. The candidate location centers for the third entering department are numbered 1 through 6.

The distances are calculated as a rectangular measurement. For example, the distance between Location 4 and the center of C

Figure 14.3 Candidate locations for Department D

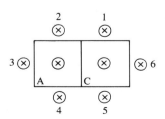

(marked with ⊗) is two units, and the distance between Location 5 and C is one unit.

The movement costs between entering Department D and the placed departments at each of these locations are as follows:

			Location				
Flow		*Dept.*	*1*	*2*	*3*	*4*	*5*	*6*
AD CD								
(18 12) ×		A	2	1	1	1	2	2
		C	1	2	2	2	1	1

= (48, 42, 42, 42, 48, 48)

The minimum movement cost is 42 given by Locations 2, 3, and 4 and any of them can be selected. Arbitrarily select Location 4. The candidate locations for the next selected department, Department B, are shown in Figure 14.4. The movement cost at each location is calculated as follows:

				Location					
			Dept.	*1*	*2*	*3*	*4*	*5*	*6*	*7*
AB	BC	BD								
(0	9	10) ×	A	2	1	1	2	2	2	2
			C	1	2	2	3	3	1	1
			D	3	2	2	1	1	1	3

= (39, 38, 38, 37, 37, 19, 39)

Figure 14.4 Partial layout and candidate location of Department B

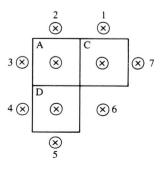

Figure 14.5 Final layout produced by PLANET for the example problem

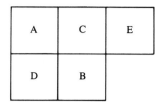

The minimum movement cost is 19 and is given by Location 6.

Similarly, the location of Department E can be determined and is found to have a cost of 23. The final layout is shown in Figure 14.5. The total movement cost of this layout is 36 + 42 + 19 + 23 = 120.

MAT

Another heuristic construction procedure, perhaps less popular than CORELAP and PLANET, is Modular Allocation Technique (MAT) (Edwards et al., 1970), which attempts to solve the quadratic assignment problem. MAT is built on the theory that the sum of the products of two sets of numbers arranged in different order, decreasing and increasing, is minimum. Flow values between pairs of departments are arranged in a decreasing order, while the distances between each pair of the candidate locations are arranged in an increasing order. Each pair in the first set is matched against a pair in the second set, and departments are assigned to locations accordingly. Although in theory this process should lead to the optimal solution, in practice this does not usually occur, owing to the conflicts that arise when departmental distances and flow relations are matched together. For example, in the assignment of a pair of departments to locations, one of the departments might have been assigned previously. In this case the other department of the pair has to be assigned to the remaining location in the corresponding location pair (in which the first department has been assigned a location) if it is vacant or to some other location. Neither location might be the best for that department.

CRAFT

Now let us study a layout method based on improvement procedures, Computerized Relative Allocation of Facilities Technique (CRAFT) (Buffa et al., 1966). It also relies heavily on heuristics because of the magnitude of the computations involved. The program requires an initial layout as an input. The iterative steps within the procedure attempt to interchange the locations of departments that have common borders or equal areas (most often pairwise, but other interchanges are also possible). The interchange that results in the greatest reduction in the movement cost is executed, and the procedure is repeated until no further reduction in the cost is possible. The maximum number of pairwise interchanges that can be performed in an iteration is less than $n(n-1)/2$, where n is the number of departments.

As an example of the calculations of CRAFT, consider the initial layout and the symmetric from-to chart shown in Figure 14.6(a) and Figure 14.6(b), respectively. The distance between the departments in the initial layout are shown in the from-to chart in Figure 14.6(c).

The movement cost of the initial layout is calculated by multiplying the elements of both charts in Figure 14.6(b) and Figure 14.6(c) as follows:

$$\begin{aligned}\text{Movement cost} &= 1(3) + 1(1) + 2(4)\\ &\quad + 2(4) + 1(1) + 1(2) = 23\end{aligned}$$

The possible pairwise interchanges for the initial layout are A–B, A–C, A–D, B–C, B–D, and C–D. If Departments A and B are inter-

Figure 14.6 CRAFT initial layout and data

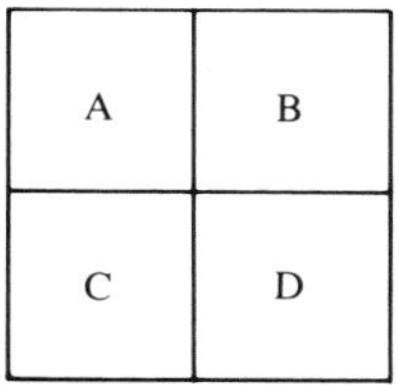

(a) Initial layout used with CRAFT

From/To	A	B	C	D
A	—	3	1	4
B		—	4	1
C			—	2
D				—

(b) Flow data

From/To	A	B	C	D
A	—	1	1	2
B		—	2	1
C			—	1
D				—

(c) Distance between the departments

changed, the distance matrix of the resulting layout will be as shown in Figure 14.7(a). The movement cost of the resulting layout will be 17, which results in a savings of $23 - 17 = 6$. Next, Departments A and C in the initial layout are used in an attempted interchange. The movement cost of the resulting layout is 20, with a cost saving of 3. The other interchanges are attempted, and the cost savings are calculated. The largest saving is obtained by interchanging Departments A and B; in this case the interchange is actually made. The resulting layout at the completion of this iteration is shown in Figure 14.7(a). The distance between the de-

Figure 14.7   Results of interchanging Departments A and B

```
+-----+-----+
|  B  |  A  |
+-----+-----+
|  C  |  D  |
+-----+-----+
```

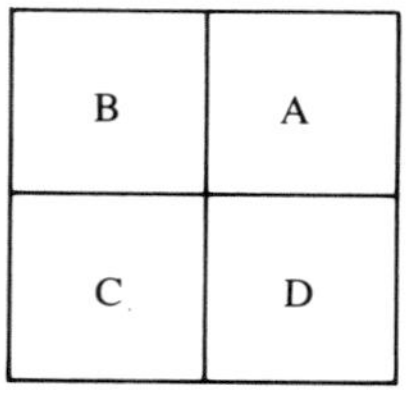

(a) **Current layout after interchanging Departments A and B**

	A	B	C	D
A	—	1	2	1
B		—	1	2
C			—	1
D				—

(b) **Distance matrix if Departments A and B are interchanged**

Figure 14.8   Computer output of initial and final layouts produced by CRAFT (The program does not plot the lines separating the departments.)

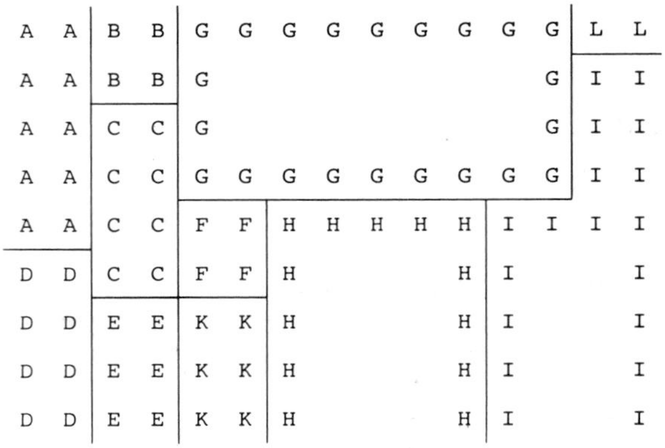

```
A A | B B | G G G G G G G G G | L L
A A | B B | G                 G | I I
A A | C C | G                 G | I I
A A | C C | G G G G G G G G G | I I
A A | C C | F F | H H H H H | I I I I
D D | C C | F F | H         H | I     I
D D | E E | K K | H         H | I     I
D D | E E | K K | H         H | I     I
D D | E E | K K | H         H | I     I
```

(a) **Initial layout**

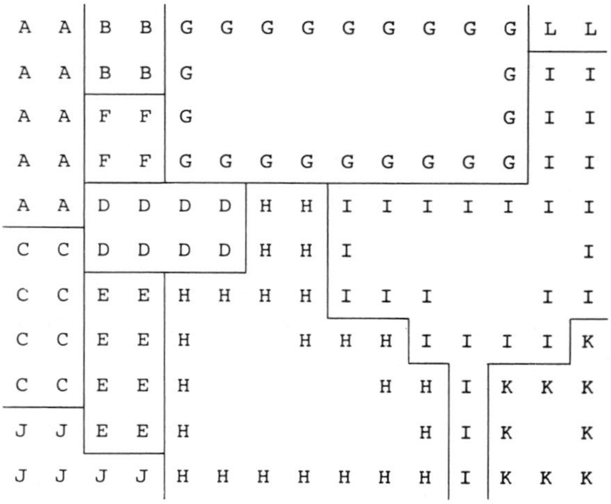

```
A A | B B | G G G G G G G G G | L L
A A | B B | G                 G | I I
A A | F F | G                 G | I I
A A | F F | G G G G G G G G G | I I
A A | D D D D | H H | I I I I I I I
C C | D D D D | H H | I             I
C C | E E | H H H H | I I I     I I
C C | E E | H       H H H | I I I I | K
C C | E E | H           H H | I | K K K
J J | E E | H             H | I | K   K
J J | J J | H H H H H H H | I | K K K
```

(b) **Final layout**

partments is shown in Figure 14.7(b). The movement cost of this current layout is 17.

With the layout of Figure 14.7(a), new pairwise interchanges are attempted between A–C, A–D, B–C, B–D, and C–D. The interchange that results in the most savings in movement cost is performed, and a new layout results. The procedure stops when no cost savings are produced. Notice that the program does not actually interchange the location of departments when the cost savings are compared; only when the cost savings of all interchanges in an iteration have been calculated and the greatest saving has been found does the actual interchange take place. Figure 14.8 shows the computer output of the initial and final layouts when CRAFT is used.

Other improvement procedures differ in the manner in which each carries out the interchange. Some of these procedures consider departmental areas, while others do not. Examples of these procedures are the following.

1. A procedure by Hillier (1963) attempts to move a department to a location to the right of, left of, below, or above its current location. This procedure has been extended to allow exchanging nonadjacent departments.

2. Another procedure (Hillier and Connors, 1966) interchanges at each iteration the two departments having the highest movement cost with other departments.

3. CRAFT-M (Hicks and Cowan, 1976) is an extension of CRAFT that compares the savings in material-handling cost that result from interchanging departments with the cost of reconstructing the departments and decides whether to perform the interchange or not.

4. SPACECRAFT (Johnson, 1982) is an extension of CRAFT that considers multiple floors and nonlinear material-handling costs.

5. Plant Layout Optimization Procedure (PLOP) (Lewis and Block, 1980) examines all the possible exchanges and performs the one that achieves the greatest cost reduction.

14.6 PROBABILISTIC APPROACHES

Probabilistic approaches, like heuristics and exact methods, use known data. However, the probabilistic aspects stem from the manner in which departments are selected to enter the layout in construction procedures or to interchange locations in improvement ones. Three probabilistic approaches exist: random selection, biased sampling, and simulation.

Random Selection

Automated Layout Design Program (ALDEP) (Seehof and Evans, 1967) is the first program to use a probabilistic aspect in its solution methodology. ALDEP is a construction program that attempts to maximize the closeness between departments. The first department to enter the layout is selected randomly. The rest of the departments are selected according to whether their relationship with a placed department is higher than a user-specified value. If no department can be found, or if there are ties, the selection is accomplished randomly. The program places the first selected department in the left-hand side of the layout, starting from the top of the layout outline. The next department is placed adjacent to the last-placed department and begins where that one ended. Departments are constructed as consisting of rectangular strips that are placed beside one another. The width of the strip is user-specified.

Biased Sampling

The biased sampling approach is a probabilistic version of improvement procedures. The procedure depends on the fact that each problem has different solutions with different costs. These solutions are considered to be the population from which some solutions are sampled. The sampling is not performed randomly, but rather in a biased manner to select only good solutions. Thus for a given layout, for each exchange of department locations that reduces the movement cost a probability of executing the exchange is calculated. The value of this probability is a function of the cost reduction anticipated by the exchange. Different samples are taken, and different solutions are obtained, each having an objective function value that is better than the initial one. Notice that in a program such as CRAFT the probability of performing the best cost-saving exchange is equal to 1, while in the biased sampling approach the probability of selecting exchanges is less than 1, and therefore any exchange that reduces cost may be executed, perhaps contributing to savings in computer time.

Simulation

One of the existing procedures of plant layout employs simulation in developing layouts. The procedure depends on considering all feasible layouts that can be produced (using the same set of data) as a statistical population. The simulator generates the sequence by which departments enter the layout in the biased sampling technique to increase the chances of obtaining good solutions. Each run generates a layout, and the simulation is terminated when no considerable improvement can be achieved over previous runs. Simulation usually requires large amounts of computation time compared with other computer-aided plant layout procedures.

Table 14.4 shows a classification of the plant layout procedures reviewed herein according to the approach used and whether the approach is of the construction or the improvement type. Table 14.5 summarizes the most important features of some of these procedures.

Table 14.6 compares the selection procedures, placement, and construction of departments of three construction programs.

Computer Programs for Microcomputers

An increasing number of programs are being developed for use with microcomputers. Micro-Craft (1986), distributed by the Institute of Industrial Engineers, produces a block plan for reducing material-handling costs. BLOC-PLAN (1986), distributed by Moore Productivity Associates, includes random, construction, and improvement algorithms for developing layouts, which improves adjacency. There are many others, and the interested reader is directed to an article by Filley (1985) for the present status of microcomputer programs. (One such program for facility layout planning was illustrated in Chapter 13.) Frequent articles published in *Industrial Engineering* are a good source for obtaining additional current information.

Table 14.4 Classification of plant layout approaches

Approach/Classification	Mathematical Programming	Heuristics	Probabilistics Approaches	Graph Theory
Construction		CORELAP MAT PLANET FATE	Simulation	Foulds (1983) Foulds and Robinson (1976, 1978) Seppanen and Moore (1970, 1975)
Improvement		COFAD CRAFT FRAT Hillier Procedure PLOP Vollman et al. (1968)	Biased sampling	
Both construction and improvement	Bazaraa procedure works in 1975		ALDEP	

Table 14.5 Summary of the features of some plant layout routines

Feature/Name	Objective	Type of Data	Area Consideration	Specific Shape	Outline	Fixing Departments	Multiple Floors
Bazaraa procedure	Minimize cost	Rel-chart or from-to chart	Yes	Any shape	Yes	Yes	No
CORELAP	Maximize closeness	Rel-chart	Yes	Possible	No	Only on the sides	No
MAT	Minimize cost	From-to chart	No	No	No	Possible	No
PLANET	Minimize cost	Rel-chart, from-to chart, or parts list	Yes	Square shapes	No	No	No
CRAFT	Minimize cost	From-to chart	Yes	No	Yes	Yes	No
ALDEP	Maximize closeness	Rel-chart	Yes	No	Yes	No	Yes
Simulation	Minimize cost	From-to chart	Yes	Square shapes	Yes	Yes	Yes
Graph theory	Maximize closeness	Rel-chart	Yes	Yes	Yes	No	Yes*

*Divides the departments into groups corresponding to the floors, then treats each separately as a single-floor problem.

Table 14.6 Comparison of construction programs

Step/Program	Selection	Placement	Construction
ALDEP	Randomly, then according to the elements of the rel-chart	First department starts from the upper left side; the following department starts from where its predecessor ended	Rectangular strips put adjacent to each other
CORELAP	Start with the department having the highest sum of relationships, then according to the elements of the rel-chart	First department in the middle; the other departments around it and as close as possible to the one they have a strong relationship with	Square-shaped departments whenever possible
PLANET	According to user priority and one of the following: 1. Strongest relationship between two departments 2. Strongest relationship with all the placed departments 3. The sum of all relationships	First department in the middle; all the locations around the placed departments are examined for locating an entering department	Square-shaped departments constructed in a spiral fashion

14.7 GRAPH THEORY

During the 1970s, a new approach that utilized graph theory was applied to the plant layout problem. The approach depends heavily on an understanding of graph theory. Since the subject matter is generally not covered in undergraduate classes in an industrial engineering curriculum, nor can it be easily explained in a limited space, we will not go into the details of this approach here. The interested reader may refer to Foulds and Robinson (1976), Seppanen and Moore (1970, 1975), and Hassan (1987) for more information.

14.8 FACILITY DESIGN

The previous approaches concentrated on solving the layout problem independently of another closely related problem, namely, selection of material-handling equipment. Both problems together are known as the facility design problem. In reality, however, the two problems are interdependent, and each one supplies the other with information necessary for its solution. That is, to solve the plant layout problem, the material-handling cost, which depends on the equipment that performs departmental moves, should be known. On the other hand, to select the equipment and assign it to departmental moves, the layout must be known. This classic paradox is very difficult to resolve in the facility design, especially when one considers that each problem also has a multitude of possible solutions.

Computerized Facilities Design (COFAD) (Moore, 1974) attempts to solve this problem and is explained briefly here. The model is a combination of CRAFT and an existing analytical procedure for selecting material-handling equipment. Initially, CRAFT is applied to a given layout in which all the operating costs of the equipment are assumed to be equal to $1. The equipment selection procedure is then applied that assigns the moves to the candidate equipment types. With the operating cost of each move now known, CRAFT is applied again to the last layout developed. A new layout results that is used again for a new selection of the material-handling equipment based on the new layout. COFAD continues in this manner, alternating between CRAFT and the equipment selection procedure, until no further improvement in material-handling cost can be obtained.

Variations of COFAD exist that have different objectives. One of these variations considers safety and is known as COSFAD. It attempts to determine a minimum-cost material-handling system for a layout that has minimum total risk. Another program, known as COFAD-F, develops a facility that can adapt to changing production levels at the smallest increase in material-handling cost.

The way COFAD (and its variations) works by moving back and forth between CRAFT and the equipment selection model seems the only feasible way of solving the facilities design problem. COFAD, however, does not attempt to correct any of the shortcomings of the programs that it uses. Moreover, the calculations performed are extensive and the user cannot obtain intermediate results that might be acceptable until the run of the program is completed. Also, preparing the necessary data for running the program consumes a lot of time and effort, and some of the data, such as the probability distributions required by COSFAD and COFAD-F, might not be readily available.

14.9 COMMENTS

With the increasing use of computers in plant layout, it is important to note some limitations and shortcomings of computer layouts. The second part of this chapter is an assessment of computer-aided plant layout. As was mentioned earlier, analytic approaches are by no means a substitute for human designers, since layout procedures solve a model of the problem but not a real-world situation. Their solutions are merely representations of how things might look, not how things should look.

All the available procedures solve a problem with a single objective, either cost minimization or adjacency maximization, while in reality several other equally important objectives might have to be considered, such as safety and flexibility of arrangement. Considering some of these objectives simultaneously has been attempted by combining both objectives of cost minimization and adjacency maximization. The approach, however, appears erroneous, since the elements of the from-to chart (used in cost minimization) and the rel-chart (used in adjacency maximization) follow different measurement scales, and the layout score produced by assigning values to the elements of the rel-chart and combining them with the elements of the from-to chart has questionable meaning. A better approach has been suggested that considers cost minimization as the sole objective of the problem, subject to some constraints that represent the elements of the rel-chart.

Other drawbacks of these procedures concern the assumptions of linearity of material-handling cost and the treatment of the elements of the from-to chart as deterministic when in fact these data exhibit variations and cannot be determined with certainty. In addition, the distance measured in these procedures is assumed to be rectilinear between departmental centers. Obviously, the movement between two departments in reality is not restricted to the centers and can originate and terminate anywhere in a department. While the assumption of a rectilinear distance is normally a valid one, owing to the existance of perpendicular aisles, it depends to a large extent on the type of material-handling equipment used. For example, when an overhead conveyor is used, a straight line dis-

tance might be appropriate. Notice, however, that in most situations the types of material-handling equipment are not known before the layout is developed.

There might also be some question about the conceptual development of these programs. For example, the elements of rel-charts follow an ordinal scale of measurement, yet they are assigned numerical values (arbitrarily), and mathematical operations are performed with them. Other drawbacks are related to departmental shape and area effect, scale effect, and selection and placement procedures.

14.10 DEPARTMENTAL AND LAYOUT SHAPES

Most of the available plant layout programs often produce unusual shapes for the departments as well as for the outline of the layout, making the resulting layout unrealistic. Most actual departments or work centers need regular shapes such as rectangles and squares. The outputs of these programs are merely an aid to designers, and the resulting layout will usually require manual adjustment. Much time and effort might have to be spent on these adjustments if the resulting layout has too many irregularities. Obviously, this tends to reduce the advantages of using the computer. Moreover, the modified layout might result in an objective function value that is worse than that of the layout that needed modification. This is particularly serious when it is noticed that the majority of the available programs produce only a suboptimal layout, and the optimal solution is usually unknown.

Several factors affect the shapes of the departments and the layout outline. One is the scale needed by some of the programs for plotting the layout. Another is the departmental area required, which affects the shapes produced. The impact of the latter factor can be realized by noticing that a program produces a specific shape for a given area and that any change in the area, as a result of change in configuration of machines, might result in another shape.

The effect of the areas on departmental shapes will also affect the adjacencies achieved and the distances traveled between the departments. For example, consider Figure 14.9(a) and Figure 14.9(b). In each figure, two layouts have the same number of departments, but the areas of the first layout are equal to each other and different from those of the second layout, which results in different shapes. In Figure 14.9(a), only four adjacencies are achieved; in Figure 14.9(b), six adjacencies are obtained. Obviously, the increase in the number of adjacencies is a result of the different areas and the different shapes of the departments.

The available programs also do not substantially consider the distance between the departments and its effect on departmental shapes and vice versa. For example, consider two departments, 1 and 2, which, if constructed by PLANET, will most likely have square shapes as shown in Figure 14.10(a). Each square shape guarantees obtaining the shortest distance between its center and the midpoint of any of its sides. However, this does not necessarily mean that the resulting distance between both department centers is the shortest possible. Figure 14.10(b) shows both departments as rectangles; in Figure 14.10(c), one of them is a rectangle and the other is a square. In the last two figures, the distance between Departments 1 and 2 is less than the distance in Figure 14.10(a), owing to the different shapes assumed by the two departments.

Some of the programs such as ALDEP

Figure 14.9   Area and shape effects on the adjacencies of departments

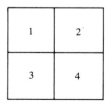

(a) A layout with equal areas and shapes

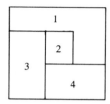

(b) A layout with different areas and shapes

Figure 14.10   Departmental shape effect on distances, $d_2 < d_3 < d_1$

(a)

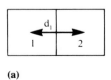

(b)

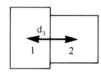

(c)

manage to obtain a regular shape for the outline of the layout by specifying it. Others such as PLANET try to obtain a regular shape for the departments by restricting their shapes to be squares whenever possible. However, it is rare to find a program that produces regular-shaped departments and a rectangular or square outline. For example, ALDEP most often generates irregular-shaped departments, and PLANET seldom produces a regular shape for the outline of the layout.

A few procedures can produce layouts with any shape as specified by the user for the departments and the layout outline, but this is achieved at the expense of the problem objective. Whatever the stipulation in a program about the layout and the departmental shapes, it always represents additional constraints on the problem, and as a result the value of the objective function will most likely be changed adversely.

Apart from obtaining a regular shape for the departments, specifying a particular shape might not be a good idea. For a start, it might affect the objective function value. Furthermore, the designer might not know in advance what a "good" departmental shape should look like or according to what criterion a square can be judged better than a rectangle. Also, as was explained earlier, one must be aware of the effect departmental shapes can have on the distances and the achieved adjacencies. Existing procedures have not addressed this issue properly, with the possible exception of a recently developed procedure in Hassan (1983).

14.11 SCALE EFFECT

Many of the programs require a scale for plotting the layout, but there are no general rules for selecting this scale. Therefore each time a different scale is used with a program (while the rest of the data remain unchanged), a different layout might result. The effect of the scale can

be realized by noticing that the area of each department is usually divided into unit squares or rectangular strips according to the scale value, and the shape of each department changes according to the number of squares or strips into which it is divided. Consequently, when departmental shapes change because of changes in scale, the locations available for each department, the distance between the departments, the number of adjacencies achieved, and the value of the objective function will also change. An example utilizing ALDEP should clarify these points.

ALDEP constructs departments in a way similar to placing vertical strips adjacent to one another. The width of the strip is user-specified and represents the scale required by the program, and its length equals the width of the layout outline as departmental area permits. For example, if a department needs 10 square units, the strip width is specified as 2, and the width of the layout is restricted to 4 (the computer is free to choose the length), then the department will need 1¼ strips, as shown below:

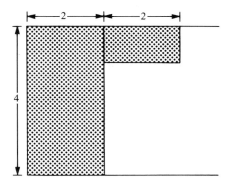

That is,

$$\frac{10}{2 \times 4} = 1.25$$

When an entering department is placed in the layout, it always starts where its predecessor ended. If the width of the strip is small, Departments 1 and 2 will look as shown in Figure 14.11(a). On the other hand, when the scale (i.e., the strip width) is doubled, the two departments will appear as shown in Figure 14.11(b). From both figures, it can be seen that the shapes of the departments, as well as the location of Department 2, are different.

A more serious effect can arise owing to the selected scale (combined with other features of ALDEP, such as its selection and construction procedures). Suppose Department 3 has a strong relationship with Department 2 and an undesirable one with Department 1. Assume that Department 3 is selected to enter the layout according to its relationship with one of the placed departments (Department 2 in this case). In Figure 14.11(a), Department 3 is placed away from Department 1 as it is supposed to be. However, in Figure 14.11(b) the situation changes, and Department 3 is placed adjacent to Department 1 (at the same time being adjacent to Department 2), which is undesirable.

Figure 14.11   Scale effect of the layouts produced by ALDEP

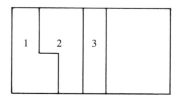

(a) A layout with a small scale

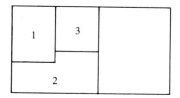

(b) A layout with a large scale

Since a rule concerning the selection of the scale cannot be established, an alternative approach would be to eliminate the scale and consequently its effect. One way of accomplishing this is by specifying a ratio for the dimensions of each department. However, specifying this ratio, which represents a constraint to the problem, might not be a simple task. Besides, there is no guarantee that the resulting layout outline will be regular. Another approach would be to express each area in the input data to the program in unit squares, thus rendering the layout independent of the scale. Even if these areas are originally expressed in square feet, they can be normalized by considering the smallest area as equal to one unit square and dividing the other areas by it.

14.12 SELECTION PROCEDURES

Since most construction programs depend on heuristics, and therefore produce suboptimal solutions, some rules must be established to enable the solution procedure to continually improve the stated objective function. The way in which departments are selected to enter the layout affects whether or not this improvement is achieved. By selecting the departments with strong relationships to enter the layout first, they are guaranteed a favorable place in the layout where they will be surrounded by the other departments, thus moving their flows shorter distances. Although the selection procedures of current programs assist in obtaining good solutions, they suffer from some drawbacks.

ALDEP's selection procedure works easily and quickly, yet at the expense of the value of the objective function and the resulting layout. Apart from the first department, which is selected randomly, the rest of the departments are selected according to whether they have a relationship with any placed department that is equal to or greater than a user-specified relationship. Thus the procedure selects a department according to its relationship with one department only as opposed to its relationship with all the previously selected and placed departments. There is no guarantee that this selection will maximize the closeness of departments, since the selected department is not guaranteed to be adjacent to the one with which its relationship resulted in its being chosen. In fact, it may be placed adjacent to a department with which it has a relationship that is relatively weak or undesirable.

Although CORELAP's selection procedure specifies how to resolve ties when they develop, it does not always accomplish that on the basis of departmental relationships. The first selected department has the highest sum of relationships with all the other departments. If more than one department has this sum, CORELAP resolves the ties by selecting the department having the larger area. An alternative way of resolving this tie could be to select the department having the larger number of strong relationships. In addition, in subsequent iterations when more than one department are candidates for entering the layout, the ties are resolved according to the sum of relationships. However, it the ties still persist, they are again resolved on the basis of departmental sizes. These types of situations can arise, particularly if the relationship chart is "tight," that is, has a high proportion of strong relationships.

Although PLANET has three optional selection methods (A, B, and C), each method suffers from some deficiencies. In PLANET's method A, the two departments having the strongest relationship with each other are selected to enter the layout first. However, there could be more than two departments having equally strong relationships with one another, or there might exist different pairs of departments such that the relationship of each pair is the same as the relationships of the other pairs. PLANET does not indicate how to resolve these ties. It would have been more efficient, for

instance, to select the pair whose elements have the largest sum of flows with all the other departments. Another drawback is the fact that the rest of the departments enter the layout according to their relationships with only one of the placed departments without regard to their relationships with the other placed departments.

The second PLANET selection method, method B, corrects the last drawback of method A by selecting the departments, apart from the first two, according to their relationships with all previously placed departments. It still suffers from the first shortcoming, however, since it selects the first two departments in the same way as does method A.

PLANET's third method, method C, ranks and selects the departments according to the sum of the flow values between a single department and all the others, which then does not consider whether a department has a relationship, especially a strong one, with any of the departments that precede it in the ranking.

14.13 PLACEMENT

The second phase of a construction program attempts to place departments in suitable locations such that the inherent objective is achieved. Although a multitude of factors influence the objective function, in addition to the appropriate placement of departments, the placement procedure of a layout program contributes significantly to the objective function. While each of the available programs attempts to place a department as efficiently as possible, they all have some limitations.

PLANET's placement procedure requires extensive computations, since all the locations around the placed departments are examined to find the best location for entering department. Apart from this limitation, however, the philosophy of PLANET's (and for that matter also CORELAP's) placement procedure is indeed attractive. The first department is placed in the middle of the layout, and the rest of the departments "grow" around it. Thus its placement in the middle of the layout will most likely guarantee its adjacency to as many departments as possible and will ensure shorter distances between it and the rest of the departments.

ALDEP's placement procedure can affect the number of adjacencies achieved. The first department that enters the layout is placed at the left-hand side of the layout, starting from the top. If this department has several strong relationships with other selected departments, it might be adjacent to only one or two of them, thus affecting the value of the objective function.

Another point concerning ALDEP's placement procedure deserves comment. If the first selected department has strong relationships with two other departments that enter the layout after it, and these two departments have an undesirable relationship with each other, then the resulting layout is not very satisfactory, as is shown in Figure 14.12(a). If the first department could have been placed in the middle of the layout, however, each of the other two departments could have been placed at its sides as shown in Figure 14.12(b), thus minimizing the conflicts of this step.

Although CORELAP's placement procedure performs some computations, it is doubtful that the objective function value reflects the true closeness between the departments. In CORELAP, an entering department is placed adjacent, or as close as possible, to the department with which it has the strongest relationship. The procedure does not pay any attention to the relationships between the entering department and the other placed departments, even if these relationships are strong.

Figure 14.12 Effect of ALDEP placement procedure on adjacencies

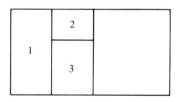

(a) Current placement of departments by ALDEP (Departments 2 and 3 should not be adjacent.)

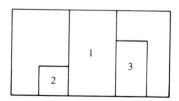

(b) Placement of the first department in the middle of the layout

14.14 CRITICISMS CONCERNING COMPUTER-AIDED PLANT LAYOUT

In addition to the stated shortcomings of the available plant layout programs, some criticism in the literature has been directed to computer-aided plant layout as an approach to solving the problem. One such criticism is that some departments such as receiving and shipping may be located in the middle of the layout, since the computer does not distinguish between departments according to their function or name. For example, in attempting to minimize the movement cost it is quite possible to locate the receiving and shipping departments in the middle of the layout, since they usually interchange large amounts of flow with the other departments. Hence by locating them in the middle they will most likely have short travel distances to all the other departments in the layout. Had any of the plant layout programs been designed to distinguish among the functions of the departments and locate the departments accordingly, the above situation might not have occurred. Several firms have adopted this particular layout configuration, called the "spine concept," quite successfully—particularly when in-process materials are also stored between operations in different departments.

Another limitation that has been mentioned is that the available programs assist in only a small portion of plant layout design processes. It should be pointed out, however, that definite steps for solving plant layout problems already exist (see Chapter 13). Plant layout programs are only one step among several others within this framework. Therefore it is not their purpose to acquire data or provide a detailed layout as some expect. Though it is true that much time and effort is spent on obtaining and preparing the necessary data for running these programs, these data are always required to solve the problem, whether manually or by the computer. Another objection is that some of the available programs do not provide good solutions. As was mentioned before, the plant layout problem has numerous possible solutions that render complete enumeration impossible. Moreover, branch-and-bound procedures are feasible only for small-sized problems. Consequently, heuristics are used as a means of obtaining good solutions quickly at reasonable expense. The key words here are "good," "quickly," and "reasonable expense." A layout designer who wants a better solution than the one obtained by any of these programs must attempt an enumeration scheme such as the branch-and-bound approach, which is impractical, especially for large problems.

14.15 DATA GUIDE TO COMPUTER PROGRAMS

The computer programs discussed in this chapter are available commercially. In Appendix D, a condensed manual describing data and input formats is presented for several popular programs—namely, CRAFT, CORELAP, ALDEP, and PLANET.

SUMMARY

This chapter has presented an overview of the application of analytically based computerized procedures to the plant layout problem. The emphasis in this chapter has been on the conceptual bases of these procedures rather than on how to use the computer to run them. Some methods have been presented in more detail than others, owing to the ease of understanding them and because of their pioneering nature. The capabilities of the available procedures, as well as their similarities and differences, have been discussed in depth in the evaluation of procedures, particularly construction ones, in an attempt to show their hidden limitations.

Despite these limitations and some criticisms that are directed to the approach, its popularity is increasing because of the development of new and more powerful procedures and the increased use of computers. An important point to notice is that the layouts produced by these procedures serve as good starting points for planners, and their use is sure to grow.

Current research in computer-aided plant layout is directed toward achieving one or more of the following enhancements in the procedures.

- Development of detailed layouts.
- Use of the graphic and interactive capabilities of the computers.

· Development of procedures capable of generating multifloor layouts.
· An increase in using graph theory.
· More powerful exact mathematical programming procedures.
· Development of procedures capable of designing layouts for the group technology method of manufacturing.
· Development of procedures suitable for generating flexible layouts, that is, ones that account for future changes in production volumes or process sequence.

PROBLEMS

14.1 What is computer-aided plant layout and what are its advantages over traditional approaches?

14.2 Differentiate between construction and improvement programs.

14.3 What is the relationship of computer-aided plant layout to systematic layout planning?

14.4 Should we implement the layout produced by the computer as it is? Justify your answer.

14.5 How do improvement programs agree with and differ from one another?

14.6 How do construction programs agree with and differ from one another?

14.7 What should be the characteristics of a good plant layout program?

14.8 What are the limitations of available layout programs? What are the most serious ones concerning construction procedures?

14.9 Compare ALDEP, CORELAP, and PLANET regarding the following points:
   a. Speed of running
   b. Quality of output report
   c. Ability to provide different options for the user

14.10 Do we need any one of the standard computer programs to find the optimal layout for the following examples?
   a. Given five locations and the following data:

	A	*B*	*C*	*D*	*E*
A	—	20	20	20	20
B		—	20	20	20
C			—	20	20
D				—	20
E					—

Department	Area
A	5
B	5
C	5
D	5
E	5

b. Given five locations and the following data:

	A	*B*	*C*	*D*	*E*
A	—	25			5
B		—	30		
C			—	20	10
D				—	5
E		—		—	—

14.11 For the data in the accompanying table, run PLANET using scales of 1 unit square equals 10, 20, 50, and 100 feet. Are there any changes in the resulting layouts or the objective function values?

Department	Area (feet)
A	100
B	600
C	200
D	400
E	500
F	300

	A	*B*	*C*	*D*	*E*	*F*
A	—	32	19	48	10	25
B		—	7	53	40	15
C			—	34	23	19
D				—	7	9
E					—	16
F						—

14.12 For the layouts produced by PLANET in Problem 14.11, run CRAFT. Is the improvement significant? Justify your answer.

14.13 For the data in the accompanying table, run CORELAP using scales of 2, 5, and 10 and observe the changes in the resulting layout and objective function value.

Department	Area
A	100
B	600
C	200
D	400
E	500
F	300

	A	*B*	*C*	*D*	*E*	*F*
A	—					
B	A	—				
C	I	E	—			
D	U	U	A	—		
E	X	O	I			

SUGGESTED READINGS

Apple, J.M., and Deisenroth, M.P., "A Computerized Plant Layout Analysis and Evaluation Technique (PLANET)," *AIIE Technical Papers,* Twenty-third Conference, Anaheim, Calif., 1972.

Armour, G.C., and Buffa, E.S., "A Heuristic Algorithm and Simulation Approach to Relative Location of Facilities," *Management Science,* Vol. 9, No. 2, 1963, pp. 215–219.

Bazaraa, M.S., "Computerized Layout Design: A Branch and Bound Approach," *AIIE Transactions,* Vol. 7, No. 4, 1975, pp. 432–437.

Block, T.E., "On the Complexity of Facilities Layout Problems," *Management Science,* Vol. 25, No. 3, 1979, pp. 280–285.

BLOCPLAN, Moore Productivity Software, Blacksburg, Va., 1986.

Brennan, B.M., "Computerized Facilities Planning, Design, and Management Options for Today's Offices," *Industrial Engineering,* Vol. 17, No. 5, May 1985, pp. 70–74.

Buffa, E.S., "Sequence Analysis for Functional Layouts," *J. Industrial Engineering,* Vol. 6, No. 2, 1955, pp. 12–13, 25.

Buffa, E.S.; Armour G.C.; and Vollmann, T.E., "Allocating Facilities with CRAFT," *Harvard Business Review,* Vol. 42, No. 2, March-April 1964, pp. 136–158.

Carrie, A.S., "Computer-Aided Layout Planning—The Way Ahead," *Int. J. Prod. Res.,* Vol. 18, No. 3, 1980, pp. 283–294.

Carrie, A.S.; Moore, J.M.; Roczniak, M.; and Seppanen, J.J., "Graph Theory and Computer-Aided Facilities Design," *OMEGA, The Int. J. Management Sci.,* Vol. 6, No. 4, 1978, pp. 353–361.

Dutta, K.N., and Sahu, S., "A Multigoal Heuristic for Facilities Design Problems: MUGHAL," *Int. J. Prod. Res.,* Vol. 20, No. 2, 1982, pp. 147–154.

Edwards, H.K.; Gilbert, B.E.; and Hale, M.E., "Modular Allocation Technique (MAT)," *Management Science,* Vol. 17, No. 3, 1970, pp. 161–169.

El-Rayah, T.E., and Hollier, R.H., "A Review of Plant Design Techniques," *Int. J. Prod. Res.,* Vol. 8, No. 3, 1970, pp. 263–279.

Filley, R.D., "Three Emerging Computer Technologies Boost Value of, Respect for Facilities Function," *Industrial Engineering,* Vol. 17, No. 5, 1985, pp. 27–29.

Foulds, L.R., "Techniques for Facilities Layout: Deciding Which Pairs of Activities Should Be Adjacent," *Management Science,* Vol. 29, No. 12, 1983, pp. 1414–1426.

Foulds, L.R., and Griffin, J.W., "A Graph Theoretic Heuristic for Multi-Floor Building Layout," *Proceedings of IIE Spring Conference,* Chicago, Ill., 1984.

Foulds, L.R., and Robinson, D.F., "A Strategy for Solving the Plant Layout Problem," *Operational Res. Quart.,* Vol. 27, No. 4, 1976, pp. 845–855.

Foulds, L.R., and Robinson, D.F., "Graph Theoretic Heuristics for the Plant Layout Problem," *Int. J. Prod. Res.,* Vol. 16, No. 1, 1978, pp. 27–37.

Francis, R.L., and White, J.A., *Facility Layout and Location—An Analytical Approach,* Prentice-Hall, Englewood Cliffs, N.J., 1974.

Hassan, M.D., "A Computerized Model for the Selection of Materials Handling Equipment and Area Placement Evaluation (SHAPE)," unpublished Ph.D. dissertation, Texas A&M University, College Station, Texas, 1983.

Hassan, M.D., and Hogg, G., "Review of Graph Theory Applications to the Facilities Layout Problem," *OMEGA, The Int. J. Management Sci.,* Vol. 15, No. 4, 1987, pp. 219–300.

Heider, C.H., "An N-Step, 2-Variable Search Algorithm for the Component Placement Problem," *Naval Res. Logist. Quart.,* Vol. 20, No. 4, 1973, pp. 699–724.

Hicks, P.E., and Cowan, T.E., "CRAFT-M for Layout Rearrangement," *Industrial Engineering,* Vol. 8, No. 5, 1976, pp. 30–35.

Hillier, F.S., "Quantitative Tools for Plant Layout Analysis," *J. Industrial Engineering,* Vol. 14, No. 1, 1963, pp. 33–40.

Hillier, F.S., and Connors, M.M., "Quadratic Assignment Problem Algorithms and the Location of Indivisible Facilities" *Management Science,* Vol. 13, No. 1, 1966, pp. 42–57.

Johnson, R.V., "SPACECRAFT for Multi-Floor Layout Planning," *Management Science,* Vol. 28, No. 4, 1982, pp. 407–417.

Khalil, T.M., "Facilities Relative Allocation Technique (FRAT)," *Int. J. Prod. Res.,* Vol. 2, No. 2, 1973, pp. 174–183.

Krejcirik, M., "Computer-Aided Plant Layout," *Computer-Aided Design,* Vol. 2, No. 1, 1969, pp. 7–19.

Lee, R.C., and Moore J.M., "CORELAP—Computerized Relationship Layout Planning," *J. Industrial Engineering,* Vol. 18, No. 3, 1967, pp. 195–200.

Lewis, W.P., and Block, T.E., "On the Application of Computer Aids to Plant Layout," *Int. J. Prod. Res.,* Vol. 18, No. 1, 1980, pp. 11–20.

Moore, J.M., "Computer-Aided Facilities Design: An International Survey," *Int. J. Prod. Res.,* Vol. 12, No. 1, 1974, pp. 21–40.

Moore, J.M., "The Zone of Compromise for Evaluating Layout Arrangements," *Int. J. Prod. Res.,* Vol. 18, No. 1, 1980, pp. 1–10.

Muther, R., *Systematic Layout Planning,* Cahners, Boston, 1973.

Muther, R., and McPherson, K., "Four Approaches to Computerized Layout Planning," *Industrial Engineering,* Vol. 2, No. 2, 1970, pp. 39–42.

Nugent, C.E.; Vollman, T.E.; and Ruml, J., "An Experimental Comparison of Techniques for the Assignment of Facilities to Locations," *Operations Research,* Vol. 16, No. 1, 1968, pp. 150–173.

Pierce, J.F., and Crowston, W.B., "Tree-Search Algorithm for Quadratic Assignment Problems," *Naval Res. Logist. Quart.,* Vol. 18, No. 1, 1971, pp. 1–36.

Plant Layout, Industrial Engineering and Management Press, Institute of Industrial Engineers, Norcross, Georgia, 1986.

Rosenblatt, M.J., "The Facilities Layout Problem: A Multi-Goal Approach," *Int. J. Prod. Res.,* Vol. 17, No. 4, 1979, pp. 323–332.

Seehof, J.M., and Evans, W.O., "Automated Layout Design Program," *J. Industrial Engineering,* Vol. 18, No. 12, 1967, pp. 690–695.

Seppanen, J., and Moore, J.M., "Facilities Planning with Graph Theory," *Management Science,* Vol. 17, No. 4, 1970, pp. B-242–B-253.

Seppanen, J., and Moore, J.M., "String Processing Algorithms for Plant Layout Problems," *Int. J. Prod. Res.,* Vol. 13, No. 3, 1975, pp. 239–254.

Shore, R.H., and Tompkins, J.A., "Flexible Facilities Design," *AIIE Transactions,* Vol. 12, No. 2, 1980, pp. 200–205.

Tompkins, J.A., *Facilities Design,* North Carolina State University, Raleigh, 1975.

Tompkins, J.A., "Safety and Facilities Design," *Industrial Engineering,* Vol. 8, No. 1, 1976, pp. 38–42.

Tompkins, J.A., and Reed, R., Jr., "Computerized Facilities Design, *AIIE Technical Papers,* Twenty-fifth Conference, Chicago, 1973.

Trybus, T.W., and Hopkins, L.D., "Humans vs. Computer Algorithms for the Plant Layout Problem," *Management Science,* Vol. 26, No. 6, 1980, pp. 570–574.

Volmann, T.E., and Buffa, E.S., "The Facilities Layout Problem in Perspective," *Management Science,* Vol. 13, No. 10, 1966, pp. B-450–B-468.

Vollmann, T.E.; Nugent, C.E.; and Zartler, R.L., "A Computerized Model for Office Layout," *J. Industrial Engineering,* Vol. 19, No. 7, 1968, pp. 321–327.

Vollmann, T.E.; Nugent, C.E.; and Zartler, R.L., "A Computerized Model for Office Layout," *J. Industrial Engineering,* Vol. 19, No. 7, 1968, pp. 321–327.

Zoller, K., and Adendorff, K., "Layout Planning by Computer Simulation," *AIIE Transactions,* Vol. 4, No. 2, 1972, pp. 116–125.

15
Plant Site Selection and Service (Support) Considerations

This chapter briefly discusses topics that relate the financial well-being of a manufacturing firm. In the first section the traditional, nonquantitative method of selecting a plant site is examined. This analysis could serve as a screening process for the quantitative methods described in the location analysis segment, Part I of the book. Section 15.2 deals with utilities. No plant can function without utilities, which can be a major expense. Every effort should be made to reduce the cost, and Section 15.2 examines sources and alternatives for providing utilities. Section 15.3 deals with insurance. To remain in business, almost every firm must have insurance or be self-insured. This section examines types of insurance policies that a manufacturing organization must have and provides examples of total insurance programs. A safe operating environment goes a long way toward reducing insurance cost, and Section 15.4 discusses safety practices and gives an example of a program that encourages safety in a plant. Taxes are a major financial burden that is unavoidable. The projected sales cost of the product must reflect the tax liabilities, and Section 15.5 describes some of the taxes that frequently must be considered. Section 15.6 is on financial statements; it briefly describes how a firm might evaluate total cost and determine the unit cost of production. If the sale price is unknown, it might be possible to develop a balance sheet from the information generated thus far.

15.1 PLANT SITE SELECTION

The geographical location of a new plant often has an important effect on the ultimate profitability of a venture. Frequently, the selection of the site for a company's first plant is made in a less than scientific manner; sometimes the city chosen is simply the hometown of the person beginning the business. The location for an additional plant usually receives more attention because industrial organizations are rather immobile and, once settled in an area, usually remain there. The decision to locate a new plant away from the area of the company's roots is usually made only after completing a thorough analysis. Even the selection of the country in which the plant is to be located is important, as evidenced by the numerous multinational companies that have located their assembly facilities nearer the final product market or have taken advantage of lower labor costs in different countries.

Factors Influencing Site Selection

Several factors are presented below that may be considered in site selection; the detailed listing that is actually used will depend upon the type of business being located.

- Transportation facilities
- Labor supply
- Availability of land
- Nearness to markets
- Availability of suitable utilities
- Proximity to raw materials
- Geographical and weather characteristics
- Taxes and other laws
- Community attitudes
- National security
- Proximity to the company's existing plants

Some of these factors are interrelated, and one could easily produce a list with a slightly different breakdown. The number of alternative sites can usually be reduced by eliminating the locations that do not conform to the minimum requirements as defined by the evaluator in certain critical factors. Owing to the varied nature of different industries, however, these factors can vary from case to case. In each instance, however, the critical criteria should be determined and used in preliminary screening. The other factors will be considered as the site selection process narrows the field of alternative locations.

Transportation Facilities

Suitable transportation facilities must be available to move personnel, equipment, raw materials, and products to and from the plant. Waterways, railroads, and highways are the usual choices for shipping, but air transport might also be appropriate. The volume and types of raw materials and products will often determine the mode of transportation best suited to the plant.

Present transportation systems have made possible the decentralization of U.S. industry. Except in large, congested urban areas, people now travel more by personal cars than by railroads. The interstate highway system has made the suburbs as easily accessible as most cities, and so a plant can draw its work force from an area easily as large as 75 miles in diameter. Although large trucks are hauling increasing amounts of raw materials and finished products, many industries still choose railroads, especially when heavy or bulky material must be transported long distances. An innovation combining the better characteristics of trucks and railroads is the "piggyback" system, in which loaded truck trailers are driven to the rail depot, transported by rail, and then again driven to their ultimate destination. Shipping by barge is a low-cost form of transportation for manufacturing plants that are located on navigable waterways. Companies that provide or require bulk commodities such as coal, petroleum and petroleum products, iron ore, and bulk chemicals find barge transportation especially advantageous. Suitable airports in the vicinity of the plant permit the expeditious transportation of personnel or equipment as required. When the cost of shipping products or raw materials is a significant portion of the finished product cost, transportation can be a major consideration; however, when transportation is only a small part of the final product cost, it may not be as important a factor.

Adequate Labor Supply

Even in the coming age of robotics, no company will be able to operate without employees. The plant location study must ensure that the types and numbers of employees that will be needed will be available, even though most highly trained personnel are very mobile and can be recruited from other areas. The following factors are important: prevailing wages, work week restrictions, existence of competing companies that can cause high turnover or labor unrest, productivity level, labor problems, and the education and experience of available potential employees. Many semiskilled factory positions can be filled by trainable unskilled workers. Some light industries have found that locating plants in smaller communities allows them to use lower-cost, more productive country labor. Other companies have chosen to locate new plants in rural areas to avoid the labor difficulties generally associated with areas that are staunchly union-oriented.

Availability of Land

Communities that are attempting to attract new industry often provide land at a low cost, but the company must make certain that the site does not have problems that are too serious to be overcome. The soil characteristics and topography of the location must be evaluated because they can seriously affect building costs. The site should be almost level, and the soil load-bearing value should be greater than 2000 pounds per square foot. Additional space for future expansion required by unforeseen market conditions should be readily available and accessible.

Nearness to Markets

It is highly desirable that a plant be located in or near the market area for the product, since costs for timely product distribution are a function of location. This is especially true for bulky items for which the cost of shipping is significant in comparison to the cost of materials and labor, as with fertilizer and building materials. For products for which the cost of labor and material is high, such as jewelry, computers, and watches, market proximity is not so critical. Consideration should be given to nearness to the market for any by-products that may result; for instance, sawdust from a lumber mill might be sold to a paper mill or a particle board manufacturer.

Suitable Utilities

Many industrial plants require a great deal of electrical power and steam. Some plants burn fuel (oil, gas, wood, coal) to produce their own power and steam. For these plants the availability of an inexpensive fuel supply is very important. For other plants, many areas of the country have ample supplies of low-cost electricity.

The availability of an adequate water supply is often important. Plants that consume small volumes of water often purchase it from public utilities, but plants that use large quantities might need to locate near a large river or lake or in an area where a deep-water well could be drilled. The state geological survey should be reviewed, and the U.S. Corps of Engineers should be contacted for available information on water tables, seasonal fluctuations of lake and river levels, and future plans for water sources.

The proposed location should have adequate facilities for waste disposal, not only of solids, but perhaps of gases and liquids as well. Care must be taken to ascertain that there are no unreasonable state or municipal regulations that will severely restrict the future plant's ability to dispose of waste economically. The permissible limits for alternative methods of disposal should be determined, and attention should be given to the possibility of providing additional waste disposal facilities.

If the plant site under evaluation is exposed to potential flood damage, the availability of flood protection facilities, such as drainage pumps, levees, canals, and causeways, must be reviewed. The quality and availability of a local fire department should also be determined. The exposure to fire damage because of the proximity of hazardous operations to the potential site should likewise be considered.

Proximity of Raw Materials

The cost of shipping raw materials and fuel to the plant site should be considered along with the cost of transporting the products to market so as to minimize the total transportation cost as much as possible while balancing that cost against other operating expenses. Plants whose raw materials are perishable or bulky tend to locate near their source of raw materials. Plants whose raw materials lose much of their weight during the manufacturing process, such as ore refineries, steel plants, and papermills, also often locate as near their raw material sources as is practical.

Geographical and Weather Considerations

The geographical characteristics of the site can greatly affect the cost of the buildings to be constructed there as well as the plant operating costs. A severely cold climate will necessitate the building of additional shelters for equipment. A very hot climate could require the plant to have additional air-conditioning for personnel comfort and additional cooling towers for process equipment. The geographical and weather factors to be considered are altitude, temperature, humidity, average wind speed, annual rainfall, and terrain.

Taxes and Legal Considerations

Since taxes will be an operating cost, the types, bases, and rates of taxes charged by state and local governments must be considered seriously. Many cities offer tax incentives such as exemptions to encourage companies to locate new plants in their communities. Some taxes to be evaluated are property, income, and sales taxes. Unemployment compensation taxes should also be determined. Local regulations concerning real estate, health and safety codes, truck transportation, roads, acquisition of easements and rights-of-way, zoning, building codes, and labor codes should all be evaluated.

Community Considerations

The community (both local authorities and the people) under consideration should be pleased to have the plant located in its area. The community should be able to provide essential services such as police and fire protection, street maintenance, and trash and garbage disposal. Good community living conditions will be needed to attract and maintain motivated workers. Cultural facilities, churches, libraries, parks, good schools, community theaters, community symphonies, and recreational facilities are indications of a community's character and its growth potential. If something essential is lacking, the company should determine the cost of providing for the need.

National Security

The U.S. government sometimes encourages companies that supply strategic materials to locate such that the sources for the products are widely dispersed. A company can increase its probability of being awarded a government contract if it locates its manufacturing facilities where the government wishes. In other cases the government might want to locate a very security-sensitive manufacturing facility within the limits of a government installation.

Proximity to an Existing Plant

Many companies locate a new satellite facility in the general area of another major plant. Doing so facilitates upper-level management supervision of the new plant. Executives and consultants can minimize travel time between plants if they are near each other. Care must be taken, however, to locate the plants far enough apart that they do not have to compete with each other for labor.

Procedures for Site Selection

From the preceding discussion it is evident that a large amount of information must be accumulated to aid in the decision-making process. Federal and state geological surveys can provide maps and other geographical information. The Department of Labor, the Federal Communications Commission, the Federal Power Commission, and the Federal Trade Commission are other good sources of indispensable information. Local agencies such as industrial commissions and chambers of commerce can provide details more specific to a particular location. Railroads frequently provide information regarding available shipping and the associated costs for any of numerous locations.

Once the necessary information about alternative locations has been gathered, the advantages and disadvantages of one location over another must be evaluated. One location might have the lowest raw material cost, an-

other the lowest utility cost, and yet another the best labor supply. A commonly used aid in selecting from alternative sites is a rating procedure. Each of the major factors is rated from 0 to 100 with regard to its importance. Each individual location is then rated from 0 to the maximum for each factor. The scores for the locations determine the final ranking. An example of this procedure is shown in Table 15.1.

It must be understood that the ranking obtained by this method is, to a large extent, subjective. When the factors are evaluated, they express an analyst's feelings, which are measured in terms of assigned weights. It is quite possible that different analysts might choose varying weights for the same physical conditions, leading to entirely different site selections. Such variations in judgment can be eliminated if all the locations that meet certain minimum requirements are made candidates and the procedures developed in Chapters 2–4 are then applied on the basis of the cost data that can be collected for each site. Since the ultimate objective of any business is to provide goods and/or services at a profit, the site that can minimize the operational cost would be a natural choice. The models developed in earlier

Table 15.1 Sample location rating procedure

Considerations	Maximum Weight	Location X	Location Y	Location Z
Transportation facilities	100	90	80	90
Labor supply	100	70	80	90
Land	100	80	70	70
Markets	100	40	70	80
Utilities	75	70	70	65
Raw materials	75	50	75	60
Geographical/weather	50	40	35	50
Taxes and legal factors	50	30	20	40
Community	40	20	40	35
National security	30	10	5	15
Proximity to existing plant	30	30	10	20
Total	750	530	555	615

chapters would evaluate each site in a quantitative manner.

The minimum requirements can be set in many different ways. For example, for each factor a minimum necessary value could be set, and the site that fails to meet the minimum is then eliminated from any future consideration; or a minimum could be set just for the important factors such as the first four considered in Table 15.1; or an overall minimum could be set, for instance at 550, eliminating location X from further consideration.

Another approach is to allow all the sites to be considered for objective evaluation by first applying the methods from Chapters 2–4 and then checking to see whether the best site selected also meets the minimum requirements of the subjective factors mentioned above. If it does not and an alternative site must be chosen, management will at least understand the "cost" associated with that decision.

Industrial Parks

A large tract of land designed and maintained for the use of several industries is referred to as an industrial park. The industrial park is usually located on a major highway or near an airport. The industries located in the park are usually quite varied and most often have no relationship to each other except for the location.

Industrial parks have become popular because of their several advantages. Successful industrial parks tend to draw more industry to the area. The locations provide access to transportation, are industrially zoned, and already provide utilities. Big buildings are standard, and since the entire area is industry, the neighbors are compatible.

The advantages of industrial parks are not limited to industry; decided advantages for the community also exist. These parks provide for efficient land usage and management, which is quite important in some areas. The centralized location of the area industry also helps to improve the appearance of the community. By drawing and localizing industry, parks help to offer new employment opportunities.

Difficulties certainly do exist that hinder the development of parks. To begin with, a large plot of land is required. This can be difficult to obtain, expecially in large, developed cities. Careful planning is required to ensure a minimum of traffic congestion. Also, building costs can be high when buildings are constructed for adaptability instead of specialization so that if a plant is vacated, it can easily be used by a new company. Thus in considering developing an industrial park, the benefits to the community and the businesses must be weighed against the disadvantages. At the same time an industry that can locate in an industrial park should examine such opportunities in its evaluation for the site.

15.2 UTILITIES SPECIFICATIONS

Specifications for utilities should be carefully developed because they can make up a significant portion of the variable operating cost of a manufacturing facility. Some items that are typically considered are

- Water—drinking, cooling, deionized, and process
- Electricity
- Refrigeration
- Steam
- Compressed air—instrument, control, and power
- Waste disposal and treatment
- Air-conditioning/heating
- Telephone

Water

Potable water might be provided by a public utility or by an in-plant source such as a well or a waterway. The best source is usually the public utility because its water is tested, treated, and certified as being safe for human consumption. If plant water is to be used, some testing and treatment might be required to ensure that it is satisfactory for drinking.

The piping and pumps for cooling water must be large enough to handle the peak demand even if the system has some scale buildup, which can occur as the result of long-term mineral deposits in the pipes effectively reducing their inside diameters. If practical, lines should be large enough to allow for future additions of equipment requiring cooling without necessitating significant piping changes.

Water used in air-conditioning and other equipment cooling can be obtained from the same sources as drinking water; however, since it need not be of the same quality as the drinking water, well water is frequently less expensive to use for this purpose. Cooling water is usually recycled to a central cooling tower, where its temperature is lowered through partial evaporation; water that is lost to the atmosphere must then be replaced.

Provisions must be made in the design of the system to incorporate the necessary pH, algae, and scale control facilities. Instrumentation should be included to measure the total cooling water flow and the individual flow to each piece of equipment. This will allow proper monitoring of the flows and accurate allocation of the cooling water cost if desired.

Process water is that which is used directly in, or is consumed in, the operation of the facility. In some cases the process water need not be potable; in others, such as pharmaceutical manufacturing, the water might have to be mineral- and germ-free. The facilities necessary to treat the water must be carefully specified.

Electricity

The cost of electricity is often a significant operating expense for a manufacturing facility. Most electrical power is provided by a public utility to the plant transformer(s), and distribution of power within the plant is the user's responsibility. Typical contracts for purchasers of electrical power contain provisions for a demand charge, energy usage charge, and fuel adjustment charge.

The maximum power used by a facility during some specified period of time (for example, 30 minutes) within the billing period determines the demand charge portion of the bill. The total energy consumed, in kilowatt-hours, during the billing period (normally a month) is the basis for the usage charge. Typical contracts specify a lower usage charge with increased energy consumption. A facility's usage and demand charges per unit will typically diminish with increased electrical energy requirements. Most contracts are also structured so that more savings are realized by improving the load factor (the ratio of the average use to demand) than by increasing the electrical load. The fuel adjustment charge is used to "pass on" to the user any increase in fuel cost experienced by the utility.

Private electrical power generation (often called "co-generation") is seldom economical. In isolated instances, usually when a plant has a large demand for low-pressure steam, the possibility of producing high-pressure steam to generate electricity should be investigated. The low-pressure steam, the by-product of electrical power generation, could then be utilized in the plant.

Lighting

The amount of lighting that is adequate for a job is highly dependent on the kind of work done. Detailed work, reading, and general office work require much more lighting than tasks requiring only gross skills such as shipping materials out of a warehouse. The Occupational Safety and Health Administration (OSHA) has set certain standards, as shown in Table 15.2, which gives the normal range of illumination levels suggested in different areas of a plant.

The intensity of light depends on the detail requirements of the task. The lower intensity for each task is under normal conditions working with materials that are easy to see.

Lighting should be considered in the plant planning stages. This is important because such things as the materials used and the colors chosen can greatly affect the lighting situation. Rough surfaces absorb more light than do smooth, shiny surfaces, and rooms decorated in dark colors require more lighting than those with bright colors. It is also important to plan the lighting for the specific tasks being performed in the various portions of the plant.

The levels of illumination need to be planned by a professional to avoid overlighting, glare, shadows, and other such problems. An area that is lit too brightly is just as much of a problem as one that is too dim. Eyestrain and fatigue develop quickly, lowering production rates and quality of work. Surface glare and shadows should also be reduced to a minimum to avoid eyestrain.

Compressed Air

Many plants use compressed air for actuating valves and operating controls and instruments. The air for instruments must be thoroughly filtered so that it will be as free of particulates as possible. Compressed air that is used to drive equipment usually does not require such special care, except for moisture removal.

Process Waste Products

It is not uncommon for a by-product of a process that is considered waste to be recyclable.

Table 15.2 OSHA guidelines for illumination in a plant

Illumination Levels	Footcandles
Warehouses and shipping	5
Assembly line	30–500
Office	30–300
Shops	50–500
Inspection and handling	20–500
Extra-fine work	1000

The by-product might be usable in another process in the plant or by another company. With chemical treatment, some substances can be converted into useful products. Flue gases can be used to preheat materials needing oven treatment. The economic advantages of reusing or selling process waste products make the usefulness of such substances worth investigating.

Steam

Steam can be furnished by low-pressure (15 psi or less) boilers for building heat and by power boilers (greater than 15 psi) for steam-driven air compressors, turbine pumps, and many other processes. Generating steam at a high pressure is more efficient than at a low pressure, even if it must be reduced in pressure downstream from the boiler. Initial investment and a large, continuing operating cost make steam an expensive utility; but in most cases there is no substitute when it is to be used in the process.

Ventilation

Often dusts, mists, and chemical vapors are given off in plant processes. These by-products are not only irritants but can be health as well as fire hazards. Proper ventilation is necessary to provide breathable air for employees. The type of ventilation provided depends on the type and concentration of the contaminant and should be planned early in the plant-planning stage.

The type of ventilation used is based on one of two principles: dilution or removal of the contaminant. If the concentration is low or the contaminant is not particularly hazardous, fresh air can be pumped into the room to dilute the concentration. This method requires little maintenance. Particularly hazardous substances must be removed from the air by drawing the contaminated air out, filtering it, and replacing it with decontaminated air. Filters must be constantly maintained. One should be absolutely certain that contaminants are not passed through to other sections through the ventilation system.

Waste Disposal and Treatment

Although waste treatment and disposal are not often thought of as utilities, they can be considered as such. The cost for treatment depends upon the facility served, and the cost factors included in this element of plant operation include those for depreciation, chemical usage, labor, and maintenance. Engineering and cost studies can be conducted to determine the feasibility of disposing of wastes by in-plant treatment, discharging into municipal facilities, incineration, and/or transporting to a landfill.

Air-Conditioning and Heating

To keep interior environments of plants and offices comfortable, the temperature within should be maintained at a fairly constant level: in the range of 68–78°F with a relative humidity of about 50 percent. Depending on the seasonal variation, the ambient air temperature might necessitate the use of heating or air-conditioning systems or both. Such units can be either electric or gas-operated, and their efficiency ratings need to be considered in making purchase recommendations.

Telephone Services

A company's management has many choices in selecting a telephone system. They can contact the local telephone company or any of the private firms that install such systems. The equipment can be bought or rented. Local and state telephone service is provided by the local phone company, while out-of-state service can be obtained from different firms such as MCI, Sprint, GTE, and AT&T. The charges in each case should be examined before contracting for the services; some companies charge a low transmission fee but set monthly fees, some charge no setup cost but have a higher transmission cost, and some companies give discounts for calling during certain time periods, while others do not.

Conservation Measures

The high cost of utilities can be reduced by conducting and following up on careful energy audits and by following conservation measures. A large energy loss occurs if the insulation in the building is poor, but the condition can be corrected by adding or replacing insulation in ceilings and walls. Other ways to conserve energy are making windows airtight, installing self-closing doors, and hanging heavy plastic-strip (2–6 inches wide) curtains on large door openings such as those in receiving and shipping that are seldom closed. Turning off or setting back heating and air-conditioning units when not in use and setting the temperature a couple of degrees lower than standard during winter months and higher than standard during summer also contribute to energy savings. It is desirable that all natural gas-fired equipment such as boilers, hot water heaters, and furnaces be regularly inspected and adjusted if necessary.

In parking lots and warehouses, low-pressure sodium lights, which provide a yellow illumination, are very economical and are quite satisfactory. In the plant and offices, metal halide lights are preferred to fluorescent lights, both giving the same color of light but the latter being more expensive to operate than the former. In a plant with high ceilings, infrared or radiant heaters located close to the floor are more economical than heating the entire area. In some cases, ducts direct heat or ventilation to the immediate vicinity of workers to avoid the added expense of heating or cooling an entire room or building when the space is very big and largely unoccupied. If possible, heating with electric resistance heaters should be avoided because they cost almost 2½ times as much to operate as do gas-fired heaters. Management can investigate the feasibility of using waste heat from manufacturing in heating or preheating. Combustible scrap such as wood, sawdust, and solvents can be used in heating whenever it is possible to safely do so.

Floating balls or wafers of polypropylene in vats used in dipping and painting operations can save as much as 60–70 percent of the heat loss due to radiation. Production processes with waste gases that need to be exhausted should be isolated and provided with their own air inlets so as not to use any treated air (heated or cooled) from the plant. Preventive maintenance programs in most cases eliminate air leaks in compressed air systems.

Management should become familiar with utility charge structures, especially the structure for electricity. If there is significant reduction in cost by having a high power factor, the possibility of using a power factor controller can be investigated. Running electric motors at low loads should be avoided. If there is an item of equipment with a large electrical demand, such as a smelting furnace or a foundry, consideration should be given to using a demand limiter. Whenever equipment such as an electric motor is to be replaced, it is desirable for the management to investigate the feasibility of buying a high-efficiency replacement. Using

special drive belts in pulley-driven devices is more efficient than operating with regular V belts.

Example: Utility Costs for Ace, Inc.

The following is a typical set of utility specifications and associated cost estimates for a small manufacturing plant.

1. *Water.* It is estimated that the plant will require approximately 10 gallons of water per hour, or 1/6 gallons per minute. A 6-inch-diameter water line will be installed. The annual cost for water is estimated at $2000.

2. *Sewage.* The sewage outflow is estimated to be three-fourths of the water intake. The plant will connect onto the city sewerage system at a cost of $744 per year and dispose of waste through a 4-inch-diameter pipe. The annual cost of the water disposed is approximately $1500 of the $2000 listed above.

3. *Telephone.* Three lines will be required to service seven phones. Two lines will be connected to the front desk switchboard, and the third line will be located in the maintenance department. The costs are as follows:

   a. Installation: $175
   b. Monthly charge: $147 (phone rental included)
   c. Total annual telephone cost: $1764

4. *Gas.* Not required.

5. *Electricity requirements:*

   a. Offices: 120 volt, single phase, 50 amp
   b. Heating/air-conditioning: 460 volt, 3 phase, 100 amp
   c. Production machines (each of three lathes): 230 volt, 3 phase, 20 amp
   d. Drill press: 230 volt, 3 phase, 10 amp

   Total annual electricity cost: $5636

   Total annual cost for utilities: $10,144

Example: Sample Calculation for Determining the Amount of Cooling Water Required

Given that a cooling tower must handle a load of 5 million BTUs per hour, that the tower can reduce the water temperature by an average of 75°F during the year, and the following data:

q = 5,000,000 BTU/hour

c_p = 1 BTU/pound °F

t_1 = 115°F

t_2 = 75°F

e_w = 8.33 pounds/gallon

where:

q = Cooling load, BTU/hour

c_p = Heat capacity of cooling water, BTU pound °F

t_1 = Inlet cooling water temperature, °F

t_2 = outlet cooling water temperature, °F

e_w = density of water, pounds/gallon

determine the minimum cooling water flow rate required for the manufacturing plant.

Solution: It is usually not desirable to heat cooling water above 120°F because mineral compounds will begin to precipitate, causing fouling of equipment. Knowing that the maximum allowable cooling water inlet temperature will be 115°F, we have

$$f_w = \frac{q}{c_p(t_1 - t_2)60}$$

where

f_w = Required cooling water flow rate, gallons/minute (gpm)

$$= \frac{5,000,000}{1(115 - 75)(8.33)60}$$

$$\text{in } \frac{\text{BTU}}{\text{hr}} \quad \frac{\text{lb}}{\text{BTU}} \quad \frac{1}{°F} \quad \frac{1}{\text{lb}} \quad \frac{\text{gal hr}}{\text{lb min}}$$

$$= 250 \text{ gpm}$$

Example: Sample Calculation Showing the Profitability of a High Power Factor for Electrical Demand

Given that a manufacturer has an industrial electrical power rate schedule that specifies a demand charge of $3.00 per kilovolt-amp (kVA) per month and that the plant's induction motors operate at an average power factor of 0.63 and draw 55 kilowatts (kW) from the 440 V supply, determine (1) the capacitor rating required to raise the power factor to 0.85 and (2) the annual savings that would result from such an increase in the power factor. The symbols used are as follows:

P = Power (sometimes called real power), measured in kW

S = Apparent power, measured in kVA

Q = Reactive power, measured in kVAR

kW = kilowatts; measure of power

kVA = kilovolt-amps; measure of apparent power

kVAR = kilovolt-amps, reactive; measure of reactive power

P_f = Power factor, cosine of the angle by which apparent power leads or lags real power

Solution: Use the power triangle method. (See Figure 15.1.)

First, for the 0.63 power factor:

arccos 0.63 = α_1 = 50.945

tan 50.945 = 1.2325

Q_1 = 55 × 1.2325 = 67.79 kVAR

$$S_1 = \sqrt{55^2 + 67.8^2} = 87.3 \text{ kVA}$$

For the 0.85 power factor:

arc cos $0.85 = \alpha_2 = 31.788$

tan $31.788 = 0.6197$

$Q_2 = 55 \times 0.6197 = 34.08$ kVAR

$S_2 = \sqrt{55^2 + 34.1^2} = 64.71$ kVA

Reactive kVA at 63% power factor
= 67.8 kVAR

Reactive kVA at 85% power factor
= 34.1 kVAR

The reactive load rating of the capacitor required to raise the power factor from 63 percent to 85 percent is $67.8 - 34.1 = 33.7$ kVAR.

Second, determine the annual savings as follows:

kVA demand at 63% power factor
= 55 kW/0.63 = 87.3 kVA

kVA demand at 85% power factor
= 55 kW/0.85 = 64.7 kVA

The annual demand charge savings are:

Month/year × (kVA at 63% power factor
− kVA at 85% power factor)
× cost/kVA/month
= 12 × (87.3 − 64.7) × 3.00
= $813.60

The initial cost of the capacitor, then, can be economically evaluated against the energy savings.

15.3 INSURANCE

Potential loss to a firm due to fire, other accidents on the premises, or lawsuits resulting from allegedly defective products could be considerable, even catastrophic. Insurance is a contractual system wherein the risk of individual loss is spread over the large number of members who pay a fee and share in both the protection and the risk. The contract is an insurance policy, the fee is a premium, the member is the insured, and the administrator of the system is the insurance company.

The system under which all insurance companies operate is based in large measure on statistical principles. For example, actuarial tables have been developed (and are periodically reviewed and revised) that provide listings of life expectancies for people in a variety of categories. Reference to the appropriate table will advise how many more years the average 34-year-old white male in the United States, for example, may expect to continue living. Many such tables inject other variables such as level of education, principal occupation, and location of residence. It should be anticipated that, all other things being equal, an individual in a category of people expected to die within five years will pay a higher premium for a new life insurance policy than will someone in the group that statistically is expecting to live an additional 25 years.

Types of Insurance

Life insurance is but one of a great many different types of insurance coverage available. Some insurance companies specialize: Some write only life insurance; others limit themselves to insuring carnivals and circuses; still others insure nothing but boilers and machinery.

Figure 15.1 Power factor correction

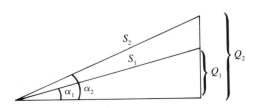

A basic division for the larger, multiline companies is the division between personal lines and commercial lines. The former might include such coverages as private automobiles, life, homeowners, and personal liability, which are written for individuals. The commercial lines portion of the company handles business insurance ranging from that for the mom-and-pop delicatessen on the corner to coverage for the largest oil refinery or automobile assembly plant. Since our objective is to discuss insurance as it relates to business and industry, we will concentrate on the latter.

Commercial Lines

At times the owner or risk manager of a business might feel that he or she is faced with an infinite number of different coverages from which to choose. Certainly no one wishes to be overinsured, but being underinsured in a particular area could prove devastating. Choosing insurance means making business decisions that are no less important than decisions in other facets of the operation.

Many insurance companies split the commercial lines into property and casualty; however, there are sometimes overlaps. Some arbitrary distinction must be made, and we shall consider property to include primarily fire, along with contractors' equipment and boiler and machinery protection. A fire policy is not nearly as limited as the name would suggest; it also includes windstorm, lightning, hail, explosion, and water and smoke damage. By endorsement (amendment) it can also provide coverage against glass breakage, fire sprinkler leakage, riot damage, damage by aircraft and other vehicles, and other perils. When the direct loss is insured, protection against the indirect loss due to business interruption is also available by endorsement.

Casualty insurance normally includes workers' compensation, business automobiles (including trucks), the various kinds of liability,

and burglary. When a package policy is written that contains the elements of both property and casualty—for example, fire and liability—it is usually the property side of the insurance company that issues it.

Essential Coverages

It would be a poor management practice for a business to own its building and not have property insurance or to open its doors to the public without premises liability insurance. In this litigation-prone society a manufacturer without product liability insurance is in danger of bankruptcy.

Practically all states require annual certification that a company's boilers are safe to operate. If the boilers are insured, the insurer is obliged to conduct the inspections necessary to make the determination.

It would be extremely difficult to find many businesses that do not insure their vehicles. Just as with a personal automobile, the company truck can be insured against collision, theft, liability, and comprehensive loss. Some large companies are able to self-insure against everything except liability, but very few have the resources necessary to sustain the large liability loss that can occur at any time. Hence liability insurance is almost universal.

Elements that influence the amounts of the premiums for different lines of coverage are relatively simple in most cases. For example, the installation of an automatic sprinkler system will reduce the fire premium, whereas the use of flammable liquids in the manufacturing process will increase it. Company drivers who have successfully completed a defensive driving course might contribute to a reduction in the automobile premium; however, a number of moving violations and/or accidents by a driver will result in an increase.

Workers' Compensation

Probably the most important type of insurance for business and industry is workers' compensation (WC), owing mainly to the very high premiums and losses associated with this line of coverage. Because the factors affecting the WC premium determination are more complicated than those for other types of insurance, we will limit the remainder of our discussion to workers' compensation.

Every state regulates this protection to some degree because the welfare of so many people is involved; more than three of every four employees in the United States are covered by WC laws. A case could be made for a federal WC law because of the tremendous variance among the laws of the individual states. Some states prohibit self-insurance, and some permit it; a few require that a state insurance fund be used, some simply allow it, and others do not even have such a fund. Private insurance is either required or permitted in almost all states.

Self-insurance for workers' compensation is not as simple as saying that if an employee is injured, he or she will be compensated from the company's operating funds. States that permit self-insurance have strict rules that require approval by the state for each prospective plan. The employees are protected by requiring the employer to post a bond or securities or set up a cash reserve to indemnify each worker who might be injured if the injury resulted from an accident arising from his or her employment and such accident occurred during the course of that employment. (These three elements for a legitimate claim are applicable to all WC coverage, not just self-insurance.)

The biggest difference found in regulations in the various states is in the schedules of payments for injuries. In some states the loss of a thumb will bring the employee higher payments than the loss of an arm at the shoulder will in other states. In some states the loss of a leg will not pay as much as will the loss of a foot in other states. The WC laws in all states require payment of medical benefits; but some do not pay the entire bill or may limit the time for

which the bills are paid, or they may set an upper limit on the total amount that will be paid.

The premium for WC insurance can vary considerably on the basis of several factors. The first of these is the location of the business. Each state regulates the premium paid in that state, but sometimes the insurance company is permitted to reduce the premium below the maximum allowed in order to be competitive.

A second factor is the categories and numbers of workers employed. The ISO (Insurance Services Office) prepares a set of tables that, in effect, tell how hazardous all of the different job categories are. The underwriter finds the appropriate factor in the table, multiplies the payroll for the category by that factor, and thus determines the basic premium for that group of workers. The justification for the factor should be fairly obvious; for example, the likelihood of a compensable injury occurring in an office with only two workers is far less than it is for a company with 25 pulpwood cutters. Chain saws in the woods are more dangerous than typewriters in the office. Payroll enters the picture not only to account for the number of employees, but to consider how expensive an injury might be. Compensation awards are based on percentages of salaries, so the more highly paid the worker, the higher the award. Thus the premium must be higher to statistically cover that predictable (within limits) cost.

A third factor is the loss experience of the particular business. The Bureau of Labor Statistics annually publishes a series of tables that list two important items: the rates of incidence (frequency) of all accidents and, more serious, the lost-time accidents for specific types of businesses (e.g., tire manufacturers) and for broad categories (e.g., all manufacturing). The insured firm (the account) can be compared to its particular category, and the underwriter can then form an opinion concerning the desirability of insuring that company. The loss ratio (value of claims paid divided by the premium) is an indicator of the probable profitability of the account to the insurance company. From this historical (experience) ratio the underwriter can determine by what factor the premium previously determined must be multiplied so that the insurance company, or the state fund, will probably not suffer a net loss for this account for the coming policy period.

A fourth factor can be the size of the company to be insured. A major cost to an insurance company is that for its overhead and administration, usually in the range of 30–40 cents of every dollar of premium collected. The cost, however, to service a WC policy with a premium of $1,000,000 is nowhere near 100 times as high as that to service the $10,000 policy. Because of this and perhaps other favorable factors (instituting training and safety programs and installing newer, safer equipment), premiums may be "deviated" or discounted if the underwriter feels the reduction is justified.

A different system of rating is the retrospective rating plan. The premium as determined above is set as the target, and the account pays this amount at the beginning of the policy period (normally one year). At the end of the period the target is adjusted upward or downward depending upon the loss experience of the account for the period. Thus the employer is rewarded for safe operation and penalized if its employees suffered more (or more costly) injuries than were anticipated.

Insurance Planning

In order for the reader to gain some appreciation of the cost of business insurance, an example is presented that itemizes the costs of insurance for one type of business.

Example: Folding Table Manufacturer

This small company is located in a town of some 10,000 persons. It operates out of a prefabricated metal building of approximately 7900 square feet valued at $435,000. There are seven office workers (salaries totaling $12,500 per month) and 19 hourly production employees (salaries of $30,000 per month). Sales average $2,400,000 per year, and products are shipped by common carrier. The annual insurance premiums for this company are itemized in Table 15.3.

15.4 SAFETY

Accidents can seriously affect the financial resources of a company. Because of the possible high cost in terms of human suffering and lost

Table 15.3 Annual insurance premiums for the folding table manufacturer

Insurance	Premium
Property ($13.79 per $1000 of value)	6,000
Workers' compensation	
Office ($0.24 per $100 of salary)	360
Production ($4.28 per $100 of wages)	15,819
Hospital and medical ($60.00 per month per employee)	39,600*
Premises liability	720
Product liability ($0.05 per $100 of sales)	14,500
Automobile (1 pickup truck)	644
Total	$77,643

*This premium is partially offset by workers' contributions.

production, a business should place particular emphasis on safety measures.

When an accident occurs, either machines, people, or both are involved. If equipment is damaged, the machinery must be repaired or replaced, and the unsafe conditions must be rectified. The cost of an accident in such cases includes the use of emergency equipment, loss of production, and the need for cleanup equipment and crews.

In addition to humanitarian concerns, the loss to the company when an employee is injured can be extreme. A trained employee is at least temporarily displaced, causing a slowdown in production and perhaps a degradation of quality, all of which can contribute to a reduction of income. There is also the additional cost in terms of payment of claims and legal fees, training of a new employee, and an increased insurance premium.

If an accident results in injury or damage that is minor, it should be taken as a warning to be heeded. The employee who gets up rubbing his lower posterior after slipping in spilled oil is lucky; if corrective action is not taken, the next person passing by might not be able to get up at all.

When to Plan

Safety considerations should be deliberated and implemented during the early stages of plant development, preferably when the building is being planned. Appropriate aisles and adequate storage space should be provided. Exits that are easily accessible should be properly located and prominently marked. Stairs, ramps, and ladders should be sturdy and coated with slip-resistant material.

After the building has been designed, the next step in setting up a plant is usually the selection and installation of equipment. The safety standards for the equipment are very important; for instance, there should be a minimum of exposed moving parts that could injure an operator. Also, when the equipment is set up, there should be no obstruction of movement of people and materials through aisles or loading spaces.

When hazardous materials are being handled, extra precautions must be taken to ensure the safety of the employees. These materials should be properly stored and handled, and ventilation should carry away dangerous vapors. Fire equipment should be readily available around combustibles, and eye and body showers should be installed to be used in case of chemical burns. Potential danger to the eyes, as in the case of electric arc welding, can be guarded against with the use of proper goggles or shields.

Once a safe plant has been set up, it is relatively easy to maintain the safe conditions; this job becomes the responsibility of every employee. When all the safety procedures have been defined, they should be followed religiously. Care should be taken never to leave equipment or materials protruding into or obstructing aisles or work spaces. Dangerous areas should never be entered without the proper equipment and instructions. Potentially hazardous situations such as slippery floors should be corrected quickly.

If the equipment in a plant is properly set up and all care has been taken to maintain a safe environment, then most resulting accidents will be caused by improper procedures or human error. This is why it is important to see that employees are properly trained and that appropriate safety methods are constantly being implemented. Employees should be aware of potential hazards such as worn machinery and should report such conditions immediately.

Regulation

Because the safety of all employees is so very important, the federal government has enacted laws to help ensure safe and healthful working conditions for all. The most important law in this regard, passed in 1970, is known as the Occupational Safety and Health Act, or OSHA. (The abbreviation also stands for Occupational Safety and Health Administration.)

One of the purposes of OSHA is encouraging efforts to reduce the number of occupational hazards. OSHA has authorized the Secretary of Labor to set mandatory safety and health standards and to divide the responsibilities between employers and employees. From the accident prevention standpoint the act has provided for the National Institute for Safety and Health (NIOSH) to conduct research in occupational safety and health problems. A company that is found to be in violation of any of the standards set by OSHA may be punished by a fine, or its management may be imprisoned in extreme cases.

Government regulations are not limited to OSHA. Most states have their own safety standards that must be adhered to by industries within their jurisdictions. In many instances, OSHA has provided grants to states to assist in their developing occupational health and safety standards.

Responsibility within a Plant

The responsibility for safety in a given plant, as well as for almost everything else there, rests ultimately with the president of the corporation. The president may be able to assign the responsibility for devising and implementing safety procedures to the plant manager, but that just makes the manager accountable to the president, who remains accountable to the stockholders and the government. If the plant manager should hire a safety manager, the latter would be directly accountable to the plant manager, who in turn still reports to the president, and so on. Of course, if any individual in the chain were to violate the law (OSHA) with regard to safety, that person would be subject to prosecution just as he or she would be for breaking any other law.

Viewed from the positive side, industrial safety is good because it protects the health of workers, reduces injuries, helps prevent deaths, and is cost-effective: it increases productivity. The primary difficulty in implementing a safety program is apathy; very many people are victims of the it-can't-happen-to-me syndrome. Unfortunately it can and does—maybe only once, but once is often enough.

Because constant attention is required to establish and maintain safe conditions and procedures in an industrial setting, the best alternative is generally to have a safety professional (safety manager or safety engineer) on the staff. We know intuitively that a plant with a two-person production department cannot support a staff including a full-time safety manager, but we should also realize that a metal-fabricating plant employing a production force of 1000 people would probably be negligent if it did not employ a safety manager.

In addition to top management and the safety professional, production supervisors bear a great deal of responsibility for the safe operation of a plant. These are the people who are in continual contact with the individual workers, and thus the supervisors are in a position to observe unsafe conditions in their facilities and equipment and unsafe procedures by their crews. The maintenance supervisor is, of course, responsible for ensuring that all items of equipment and machinery are kept in safe operating condition; and the workers, the people with the most to gain or lose, are responsible for following safe procedures and reporting unsafe conditions.

Safety Program

Every business should have a safety program, the details and complexity of which depend upon the size and type of the operation. The plan should always include a policy statement and, in almost every case, safety rules. Most plants need written work procedures that should emphasize safety. Likewise, most will profit from safety meetings. Safety inspections should be held regularly, and the possibility of instituting a system of providing safety incentive awards should be investigated.

Like all policy statements, the safety policy statement should be in general terms, but it must be sincere, and it must be followed. Lip service to a principle is quickly recognized for what it is and can be more damaging than doing nothing. The statement should be signed by the most senior official at the plant or by the president in the home office if a company-wide policy is appropriate. There should be widespread dissemination of the policy statement; each employee should receive a copy when it is first published (in the pay envelope perhaps), each new employee should be handed one upon being hired, and every bulletin board in the plant should display a copy prominently. (We include as an example the safety policy statement for JOROCO Electronics, Inc. See page 374.)

Safety rules vary widely with the needs of the organization, and it can prove dangerous to attempt to implement a "standard" set of rules that are probably not specific enough for a given set of circumstances. As in the case of the safety policy statement, safety rules should reach every employee. The vehicle might be an employee handbook, which would also contain information about such matters as insurance, taxes, vacations, holidays, promotion policies, and retirement. The ever-present bulletin board should supplement the handbook. (The safety rules for the JOROCO Electronics are furnished as an example below.)

Safety meetings can be either an invaluable means of reinforcing the idea of safety in the minds of the workers or a boring, unproductive waste of time. Each meeting must be organized, and everyone must know that it is important. Sometimes the safety meetings are small, brief, informal gatherings held by the supervisor during working hours; sometimes they include a company or departmental dinner after work. In the latter case the plant manager or the president might say a few words about the safety record, and the safety director or an outside consultant might present a formal program on a timely topic. Regardless of the format chosen, the employees should always be afforded the opportunity to ask questions and present opinions.

Incentive awards can be given to remind the employees about the importance of safety and to encourage them to work in a safe manner. There is almost no limit to what the award can be. If the business has several locations, the plant with the best safety record for the year could be awarded a bonus, either a fixed amount or a percentage of salary, as a tangible sign of appreciation by management. Pen-and-pencil desk sets, plaques, playing cards, certificates, calendars, jackets, tee-shirts, sweatshirts, belt buckles, and ball caps are items that, when suitably engraved or imprinted, can be presented during safety meetings in recognition of accident-free periods (months or years) for individuals or groups. Some companies have a standing offer of an all-expense-paid vacation in some attractive locale for employees who complete a lengthy period (perhaps 10 years) without an accident. In general, the awards system should combine relatively frequent presentations of inexpensive gifts for minor achievements with the occasional awarding of more valuable items for significant accomplishments.

The safety manager, or the person with that collateral duty, usually operates in a staff function, reporting directly to the plant manager or the president. The responsibility for a properly functioning safety program is this person's, and all matters relating to safety must be routed to or via his or her desk. The safety manager should conduct inspections in addition to those made by department heads and area supervisors and should maintain records of those inspections and check on efforts to correct deficiencies. One of the safety manager's more important tasks is analyzing accident reports and taking action to make certain that no one repeats the circumstances that caused an earlier accident. He or she is also responsible for the preparation and posting of all OSHA-required reports.

Example: Safety Policy Statement

The primary purpose of this statement is to establish throughout JOROCO Electronics, Inc. the concept that our employees and property are our most important assets. Protection of people from injury and the conservation of property are our top priorities and will receive support and participation by all levels of management.

It is the policy of JOROCO Electronics to ensure that this plant remains a safe place in which to work, one that is free from fire and accidents. I seek your help in accomplishing this goal and will appreciate your efforts toward that end.

The safety of our employees and the prevention of loss are principal responsibilities of management. Efficient production cannot exist without accident prevention. Management and supervisory personnel must keep in mind that the safety and well-being of all employees are their primary responsibility. It can be met only by consistently striving to promote safe work practices and maintaining property and equipment in safe operating order.

The keystone in our safety program is the manager because the manager is continually involved with employees, and safety is people. Neither a foreman, a supervisor, nor an operating head may at any time delegate any portion of his or her responsibility for safety to a staff organization; it is an operating function.

Plant workers must always remember that safe practices are an integral part of all operations. If workers have not followed every safety rule and taken every precaution to protect themselves and their fellow workers, the job has not been completed efficiently. Production and safety are inseparable at JOROCO.

Accidents do not have to happen; they are preventable, and we must not believe otherwise. We all have our part, and it includes acting safely and talking safety at all times. A healthy attitude toward preventing accidents and improving our safety record can be achieved if each of us does his or her part.

Reducing accidents and their related insurance expenses will also enable us to be more competitive in our industry. As we improve as a team in this regard, we enhance our chances as individuals for greater tangible benefits.

/s/ J.R. Coleman
President

Example: Safety Rules

In keeping with the established policies of JOROCO Electronics, Inc. the following rules pertaining to the safety of personnel and property are made effective immediately. Anyone failing to comply with an applicable rule will be subject to disciplinary action including termination if appropriate.

1. Personal protective equipment must be worn as specified.

   a. Hard hats are to be worn in production and loading areas.
   b. Steel-toed safety shoes are to be worn by all employees in production and loading areas.
   c. Goggles are to be worn by all employees operating drills or working in painting or polyurethaning areas.
   d. Earplugs are to be worn in designated high-noise areas.
   e. Respirators are to be used in painting and polyurethane-dispensing areas.

2. Documented procedures, to be followed in an emergency situation, are to be made available to all employees. Evacuation procedures are posted along with emergency phone numbers throughout the plant. Exits are to be clearly marked and kept easily accessible.

3. Sprinklers are installed and fire extinguishers spaced according to regulations set forth by the fire marshal. These are to be tested by authorized personnel only and may not be used except in the event of an emergency.

4. Eyewash and shower equipment are placed in areas where hazardous or flammable materials are handled. The posted instructions regarding their use must be followed to prevent possible serious injury.

5. First aid kits are available in each department and are to be used for minor injuries suffered on the job. Department heads will ensure that the kits are inventoried monthly and replenished as necessary.

6. All accidents and injuries shall be reported to the appropriate supervisor as soon as practicable. Those persons at the scene where serious injury has occurred should take immediate action in accordance with those procedures referenced in paragraph 2 above. The supervisor is to complete an accident report and forward it to the safety manager as soon as possible following any accident.

7. Continuing safety is the responsibility of all plant employees. Each department head will publish safety procedures for operation of all machinery and equipment within his or her jurisdiction. The procedures will become a part of these safety rules and will be enforced as such. No employee is to operate any equipment with which he or she is not familiar.

8. Each area supervisor is required to complete a safety check sheet weekly. He or she will mark appropriate answers—yes, no, or not applicable—for each of the 22 statements and give an explanation for any unsatisfactory conditions. (See Figure 15.2.)

Figure 15.2   Safety check sheet

DEPARTMENT _____ DATE _____	YES	NO	N.A.
1. RECORD OF INJURIES IS UP TO DATE			
2. WORK AREAS ARE CLEAN			
3. AISLES ARE CLEAR			
4. ALL FLOORS ARE CLEAN AND DRY			
5. LOAD LIMITS FOR MACHINERY AND STORAGE AREAS ARE CLEARLY MARKED			
6. ALL PLATFORMS 4 FEET ABOVE FLOOR LEVEL HAVE SECURE RAILS			
7. EXITS ARE CLEARLY MARKED AND EASILY ACCESSIBLE			
8. NOISE EXPOSURE LEVELS HAVE BEEN CHECKED FOR PERMISSIBLE LEVELS			
9. HARD HATS ARE WORN BY WORKERS			
10. STEEL-TOED SHOES ARE WORN BY WORKERS			
11. GOGGLES ARE WORN			
12. RESPIRATORS ARE USED			
13. "NO SMOKING" SIGNS ARE POSTED			
14. FIRST AID KIT IS COMPLETE AND EASILY ACCESSIBLE			
15. FIRE EXTINGUISHERS ARE MOUNTED IN PROMINENT LOCATIONS			
16. FIRE EXTINGUISHERS ARE PROPERLY CHARGED			
17. EVACUATION PLAN IS CLEARLY VISIBLE			
18. EMERGENCY PHONE NUMBERS ARE POSTED			
19. EQUIPMENT HAS BEEN INSPECTED FOR HAZARDS			
20. PREVENTIVE MAINTENANCE HAS BEEN PROPERLY PERFORMED			
21. ALL OVERHEAD LIGHTS ARE IN WORKING ORDER			
22. STEPS HAVE BEEN TAKEN TO MAKE NEEDED REPAIRS AND TO CORRECT ANY UNSAFE CONDITIONS			

EXPLANATIONS AND COMMENTS:

15.5 TAXES

Just like individuals, corporations and other businesses are subject to federal, state, and local taxes. They range from an income tax on earnings to contributions to the unemployment compensation fund for the workers. Tax laws change frequently; one detailed source of information about current applicable laws is the *Prentice-Hall Federal Taxes.* It is a complete listing of tax laws, rates, and information in an 18-volume set published annually by Prentice-Hall. The first volume is an index of tax topics making reference to paragraph numbers in the other volumes.

Following is an overview of the major taxes paid by corporations.

Federal Income Tax

The largest single tax paid by corporations is federal income tax, which the company pays on the profits made each year. The 1986 tax laws, when fully implemented, give the tax rates shown in Table 15.4.

A corporation with a taxable income over $100,000 must pay an additional tax equal to 5 percent of the amount in excess of $100,000 up to a maximum additional tax of $11,750. This extra tax phases out the benefits of the graduated rates for corporations with taxable income between $100,000 and $335,000. A corporation having taxable income of $335,000 or more gets no benefits from the graduated rates and pays, in effect, a flat rate of 34 percent on all taxable income.

Table 15.4   Income tax rates for business

Taxable Income	Tax Rate
0–$50,000	15%
$50,000–$75,000	$7,500 + 25% over $50,000
Over $75,000	$20,000 + 34% over $75,000

Social Security Act

In 1935, Congress passed the Social Security Act, which includes provisions for a system of federal-state unemployment insurance, insurance for old age, old age pensions for the needy, pensions for the blind and for relatives caring for orphans and other destitute children, and an appropriation for the Public Health Service. The entire program is officially named the Old Age, Survivors, and Disability Insurance. These programs are supported by taxes placed on employers and employees based on a percentage of the employee's income.

Payments to individuals qualifying under this act depend on the specific situation. Unemployment benefits, for example, differ from state to state. Old age pensions are based on the amount of taxes paid into the program, the number of years paid, and past and present income. Cost-of-living increases are included for all of these programs.

A program that has been added to the Social Security Act is Medicare, which provides partial payment for health care for those aged 65 and over. It automatically qualifies the recipient for Medicaid, a voluntary insurance program for which the individual pays a set premium. The latter covers additional medical expenses such as prescriptions and a physician's care.

Contributions to Social Security are required for most employees by the Federal Insurance Contributions Act (FICA). To receive retirement benefits upon reaching the required age, a person must have paid FICA taxes for at least 40 quarters. That is, since the calendar year is divided into four quarters (January through March, April through June, and so forth), the individual must have paid taxes during 40 of those periods to be eligible for retirement benefits. When the law was first passed, the minimum was just six quarters so that people who were near retirement age at that time would not be denied the opportunity to receive

Table 15.5 Federal insurance contribution act rates

Year	Contribution to Retirement Benefits (percent)	Contribution to Hospital Benefits (percent)	Total FICA Tax Rate (percent)	Maximum Taxable Income (dollars)
1985	5.7	1.35	7.05	37,800
1986–1987	5.7	1.45	7.15	39,600
1988–1989	6.06	1.45	7.51	
1990 and subsequent	6.20	1.45	7.65	

Social Security benefits. The tax rate has been raised several times and is imposed on both the employer and the employee, based upon the employee's income. The present tax rate structure for FICA is shown in Table 15.5.

Unemployment Compensation

The Federal Unemployment Tax Act (FUTA) imposes a tax on employers alone to provide financial aid for the unemployed. Employers have been paying the federal government 3.5 percent of the first $7000 of annual wages of each employee, with a reduction of up to 2.7 percent of the total for contributions made to state unemployment funds. As of January 1, 1985, the tax rate was increased to 6.2 percent with a credit of up to 5.4 percent for state taxes.

Sales Taxes

A tax imposed on the sale of a broad range of items is known as a sales tax, a major source of revenue for states and localities. These taxes are frequently raised as the needs increase. As might be anticipated, there is a great deal of variance in rates between states; for instance, only a few states allow the establishment of local sales taxes. Items such as alcohol, cigarettes, gasoline, and hotel room rental may be taxed differently, frequently at a higher rate, while food and other necessities may be exempt or have lower rates. Sales taxes are normally paid by consumers; but if a business buys a product or service for its own use, it is usually liable for the sales tax.

Property Taxes

Businesses and corporations are required to pay local taxes, the single largest one being property taxes. Property taxes are levied in every city across the United States, although the actual rate varies considerably from city to city. It can be as low as $0.05/$100 value/year or as high as $3/$100 value/year. Most cities depend on this tax as their single major source of revenue.

Both individuals and businesses are subject to property taxes, which may also be known as real estate taxes. The real estate taxed includes any land owned, houses, and business buildings. In addition to these, businesses may be subject to taxes levied on heavy machinery and inventory.

Comments

The list of tax laws is seemingly endless, and many laws can be imposed for specific situations; there are taxes on inheritances, gifts, and accumulated earnings (intangible property), to name a few. But income, property, sales, FICA, and FUTA taxes constitute the major tax burden for any company and are a major source of income for federal, state, and local governments. For more detailed information, the *Prentice-Hall Federal Taxes* is recommended.

15.6 FINANCIAL STATEMENTS

The objective of financial statements is to clearly state the status of the work center that the report covers. Within a company the reports are generally constructed by using the inverted pyramid approach. Detailed reports are generated for each work center (load center), and summary reports are compiled for each higher level of management for the specific area of responsibility. At the end of each year, a company's financial activities are presented to the stockholders in terms of a balance sheet that states the company's sales, operating expenses, and profit. The company's projected plans are very much dependent on its overall financial standing. Figures 15.3(a) through 15.3(d) show examples of a manufacturing expense statement, a statement of cost of goods made, a balance sheet, and a profit-and-loss statement for a manufacturing firm.

A program written by Kailash Bafna and Shamik Bafna (see Appendix D) is quite useful in performing the detailed financial analysis during the planning stages for a manufacturing plant. It permits the flexibility required to conduct repeated trials with "what if" questions and provides as an output well-formulated reports. The financial analysis is conducted for a five-year planning period, with the following set of calculations made for each year.

1. *Sales revenues.* Annual revenues are computed for varying volumes and price changes.

Figure 15.3(a)   Annual indirect manufacturing expense (IME)

INDIRECT SUPPLIES	$ 1,000
INSURANCE	51,444
SALARIES:	
SUPERVISION	49,820
OFFICE AND CLERICAL	49,964
INDIRECT LABOR	83,040
UTILITIES	198,036
REPAIR AND MAINTENANCE	1,000
DEPRECIATION:	
BUILDING (DDB: 40 YEARS)	21,571
MACHINERY (DDB: 12 YEARS)	41,920
TOTAL	497,795
BURDEN THROUGH DIRECT LABOR COST:	
DIRECT LABOR COST/MONTH	14,240
I.M.E. 497,795/12	41,483

$$\text{OVERHEAD RATE} = \frac{41,483 \times 100}{14,240} = 290\% \text{ OF DIRECT COST}$$

Figure 15.3(b)   Clip Corp. statement of cost of good made, monthly
costs—166,667 units, 9″ × 12″ clipboard

DEPARTMENT, WOODWORKING:	
MATERIAL	$20,625
LABOR	6,240
OVERHEAD	18,096
DEPARTMENT, METALWORKING:	
MATERIAL	12,767
LABOR	1,280
OVERHEAD	3,712
DEPARTMENT, CLIP ASSEMBLY:	
MATERIAL	5,000
LABOR	5,600
OVERHEAD	16,240
DEPARTMENT, FINAL ASSEMBLY:	
MATERIAL	1,217
LABOR	1,120
OVERHEAD	3,248
DEPARTMENT, PACKAGING:	
MATERIAL	2,222
LABOR	2,240
OVERHEAD	6,496
TOTAL STANDARD COST	**$106,103**
PRICE PER UNIT	**$0.67**
SELLING PRICE PER UNIT	**$1.20**

Figure 15.3(c)   Clip Corp. balance sheet for the period

ASSETS:	
CURRENT ASSETS:	
INVENTORIES:	
MATERIALS	**$33,588**
FIXED ASSETS:	
LAND	**6,000**
BUILDING	**435,600**
EQUIPMENT	**254,518**
TOTAL ASSETS	**727,706**
LIABILITIES AND OWNER'S EQUITY:	
LIABILITIES:	
ACCOUNTS PAYABLE	**33,588**
OWNER'S EQUITY:	
CAPITAL STOCK + RETAINED EARNINGS	**694,118**
TOTAL LIABILITIES	**727,706**

Figure 15.3(d)   Clip Corp. profit-and-loss statement (first year forecast)

SALES	$2,400,000
LESS: COST OF GOODS SOLD	1,333,236
LESS: SELLING AND ADMINISTRATIVE EXPENSE	300,000
LESS: INCOME TAXES	260,699
NET PROFIT	506,065

$$\text{PAYBACK PERIOD} = \frac{\text{INVESTMENT}}{\text{UNIFORM ANNUAL CASH BENEFIT}}$$

$$= \frac{\text{INVESTMENT}}{\text{NET PROFIT} + \text{DEPRECIATION CHARGED}}$$

INITIAL INVESTMENT: $727,706

$$\text{PAYBACK PERIOD} = \frac{727,206}{506,065 + 63,441} = 1.277 \text{ YEARS}$$

2. *Operating expenses.* Annual expenses are computed for various items considering proposed volumes and the inflation rate.

3. *Depreciation calculations.* The program computes the annual depreciation for various categories of assets using present tax laws.

4. *Interest calculations.* Given the debt ratio and other relevant information, annual interest payments are computed. Interest on working capital is also determined.

5. *Tax computations.* Using present tax laws, the annual tax liabilities are determined.

6. *Income statement.* This is generated for the five years, showing the annual net incomes.

7. *Earnings on investment.* The after-tax cash flow for the five years is shown. Rate-of-return and payback are also determined.

The data entry form and samples of reports generated by the program are shown in Figures 15.4(a) through 15.4(h).

SUMMARY

The chapter discusses five major topics: plant site selection, utilities, insurance, safety, and taxes. Although the topics seem unrelated, they all have considerable influence on the financial well-being of a manufacturing facility. The financial statements at the end of the chapter illustrate how the financial information may be displayed so that it is easy to understand and interpret.

In evaluating the sites for a new plant, one must consider a multitude of factors. However, it is not necessary that each factor be given equal importance. The need changes from industry to industry, and a factor considered to be very important in one industry may not be as crucial in another. A tabular method is suggested in the chapter for selecting one site from among the possibilities. However, selecting a site based on this method may be quite subjective. One may be more objective by combining the tabular method with the location methods discussed in Chapters 2 through 4. Locating a plant in an industrial park has some definite advantages, and should an opportunity arise to do

Figure 15.4(a)   Financial analysis worksheet

1. **NAME OF YOUR COMPANY:** WARD-BAND, INC.
2. **HOW MANY PRODUCTS DO YOU HAVE:** 3

INFORMATION NEEDED	PRODUCT #1	PRODUCT #2	PRODUCT #3
PLANT CAPACITY (IN PCS.)	100,000	30,000	60,000
FIRST YEAR SELLING PRICE (IN $)	15.00	38.50	22.25

	YEAR 1	YEAR 2	YEAR 3	YEAR 4	YEAR 5
PERCENT OF DESIGN CAPACITY	75	100	110	120	125

AVERAGE RATE OF SELLING PRICE INCREASE (% PER YEAR): 6

3. **COSTS FOR THE FIRST YEAR FOR THE FOLLOWING:**

RAW MATERIALS	$784,200
PURCHASED PARTS	$570,500
HOURLY EMPLOYEES	$727,400
SALARIED EMPLOYEES	$403,700
INSURANCE	$ 12,000
MAINTENANCE	$ 48,000
UTILITIES	$ 24,000
MISCELLANEOUS	$ 60,000

AVERAGE RATE OF COST INCREASE (% PER YEAR): 8

4.

ITEM	INITIAL COST	SALVAGE
PRODUCTION EQUIPMENT	350,000	10,000
TOOL ROOM EQUIPMENT	50,000	5,000
MECHANICAL HANDLING EQUIPMENT	75,000	3,000
AUXILIARY HANDLING EQUIPMENT	150,000	5,000
OFFICE EQUIPMENT	80,000	4,000
BUILDING COST	550,000	75,000
LAND	120,000	–

5. **WHAT PERCENT OF CAPITAL IS BORROWED?**   60
   WHAT IS THE INTEREST RATE OF BORROWED CAPITAL (%)?   14
   IN HOW MANY YEARS IS THE LOAN TO BE PAID OFF?   10
   WHAT PERCENT OF THE ANNUAL REVENUES IS THE WORKING CAPITAL?   17
   WHAT IS THE INTEREST RATE OF WORKING CAPITAL (%)?   16

Figure 15.4(b)

Sample sales revenues

```
                                    WARD-BAND, INC.
                                    ---------------


                                     Sales Revenues
                                     --------------

Item                    Year 1     Year 2     Year 3     Year 4     Year 5
--------------------------------------------------------------------------------

Percent of Capacity         75        100        110        120        125

Product 1:
----------
Volume (units)           75000     100000     110000     120000     125000

Selling Price ($)           15       15.9      16.85      17.87      18.94

Total Sales ($)        1125000    1590000    1853940    2143829    2367144


Product 2:
----------
Volume (units)           22500      30000      33000      36000      37500

Selling Price ($)         38.5      40.81      43.26      45.85      48.61

Total Sales ($)         866250    1224300    1427534    1650748    1822701


Product 3:
----------
Volume (units)           45000      60000      66000      72000      75000

Selling Price ($)        22.25      23.59         25       26.5      28.09

Total Sales ($)        1001250    1415100    1650007    1908008    2106758


Total Revenue ($)      2992500    4229400    4931481    5702585    6296604
```

Note: 1.   An annual selling price increase of 6% is assumed.
        2.   Installed capacity for Product 1 equals 100000 units
             per year.
        3.   Installed capacity for Product 2 equals 30000 units
             per year.
        4.   Installed capacity for Product 3 equals 60000 units
             per year.

Figure 15.4(c)

Sample operating expenses

```
                        WARD-BAND, INC.
                        ---------------


                        Operating Expenses
                        ------------------

Item                 Year 1    Year 2    Year 3    Year 4    Year 5
-----------------------------------------------------------------------

* Raw Materials      784200    1129248   1341547   1580586   1778159

* Purchased Parts    570500     821520    975966   1149865   1293598

* Hourly Employees   727400    1047456   1244378   1466103   1649366

- Salaried Employees 403700     435996    470876    508546    549229

- Insurance           12000      12960     13997     15117     16326

* Maintenance         48000      69120     82115     96746    108839

* Utilities           24000      34560     41057     48373     54420

- Miscellaneous       60000      64800     69984     75583     81629

                     -------   -------   -------   -------   -------

Total Expenses      2629800    3615660   4239919   4940919   5531568
```

Note: 1. An annual cost increase of 8% is assumed.
       2. Asterisks (*) indicate directly variable costs.
           Dashes (-) indicate fixed costs.

Figure 15.4(d)

Sample depreciation
calculations

```
                        WARD-BAND, INC.
                        ---------------


                        Depreciation Calculations
                        -------------------------

                        First     Salvage   Tax     Annual    Book Value
Item                    Cost      Value     Life    Deprec.   Sixth Year
-----------------------------------------------------------------------

Production Equip.       350000    10000     7       48571     107143

Tool Room Equip.         50000     5000     7        6429      17857

Mech. Handling Equip.    75000     3000     7       10286      23571

Aux. Handling Equip.    150000     5000    12       12083      89583

Office Equip.            80000     4000    10        7600      42000

Building                550000    75000    30       15833     470833

Land                    120000                          0

Total                  1375000                     100802     870988
```

Note: Straight-line depreciation is assumed in all cases.

Figure 15.4(e)

Sample interest calculations

WARD-BAND, INC.

Interest Calculations

Item	Year 1	Year 2	Year 3	Year 4	Year 5
Borrowed Capital:					
Loan Balance	825000	782336	733700	678254	615046
Annual Payment	158164	158164	158164	158164	158164
Principal Paid	42664	48637	55446	63208	72057
Interest on B. C.	115500	109527	102718	94956	86106
Working Capital:					
Amount of W. C.	508725	718998	838352	969439	1070423
Interest on W. C.	81396	115040	134136	155110	171268
Total Interest	196896	224567	236854	250066	257374

Note: 1. Loan is 60% of total capital.
      2. Capital is borrowed at 14% per year.
      3. Loan is paid off in equal annual installments in 10 years.
      4. Working capital is assumed to be 17% of annual revenues.
      5. Interest on working capital is charged at 16% per year.

Figure 15.4(f)

Sample tax computation

WARD-BAND, INC.

Tax Computation

Item	Year 1	Year 2	Year 3	Year 4	Year 5
Gross Tax Income	65002	288371	353905	410797	406860
Loss Carry Forward	0	0	0	0	0
Net Tax Income	65002	288371	353905	410797	406860
Gross Tax	16601	92546	120328	139671	138332
Net Tax	16601	92546	120328	139671	138332

Figure 15.4(g)

Sample income statement

WARD-BAND, INC.

Income Statement

Item	Year 1	Year 2	Year 3	Year 4	Year 5
Sales Revenues	2992500	4229400	4931481	5702585	6296604
Operating Expenses	2629800	3615660	4239919	4940919	5531568
Gross Profit	362700	613740	691562	761666	765036
Depreciation	100802	100802	100802	100802	100802
Interest	196896	224567	236854	250066	257374
Gross Tax Income	65002	288371	353905	410797	406860
Net Tax	16601	92546	120328	139671	138332
Net Income	48401	195825	233577	271126	268527

Figure 15.4(h)

Sample earnings on
investment

WARD-BAND, INC.

Earnings on Investment

Item	Year 1	Year 2	Year 3	Year 4	Year 5
Net Income	48401	195825	233577	271126	268527
Add Depreciation	100802	100802	100802	100802	100802
Net Cash Flow	149203	296627	334380	371929	369330

Return on investment over a 5-year period = 16.01%

Payback on investment = 4.6 years

so, it may be beneficial to investigate it further.

Utilities are the lifelines of industry—without them, an industry cannot function. The expense of utilities can be controlled by conservation and planning. Typically, we include water, electricity, refrigeration, steam, compressed air, water disposal and treatment, air-conditioning and heating, and telephone under the category of utilities. For each, the chapter discusses what the requirements in a plant could be and proposes a few ways of increasing efficiency in utilization. A chart illustrating typical utility specifications for a small manufacturing firm shows how the total cost for utilities may be estimated.

Making contributions to insurance is an expense that cannot be avoided if one intends to stay in business for long. Various types of insurances are available such as fire, workers' compensation (WC), business, automobile, product and general liability, burglary, and medical and dental care. Some coverages are optional but strongly recommended, while others such as workers' compensation are required by law. The benefits provided under WC are quite different in each state, and therefore associated premiums also vary from state to state. Similarly, one might choose different levels of protection for other types of insurances, thereby reducing or increasing the total cost.

Safety is extremely important in a place of business. Unsafe conditions can lead to an increase in the number of accidents, which results in either higher premiums for insurance or cancellation of the policy. In addition, accidents cause human suffering and loss of production. Safety requires planning—from the time a plant is built to every instance in its operation. OSHA, a regulatory arm of the U.S. government, issues regulations that must be followed. Within a plant, the major responsibility to develop a good safety program lies with the management, and an example of one such safety program is given in the chapter.

Taxes are unavoidable and should be taken into account in planning. A major tax for any business is the federal income tax. However, state tax, Social Security, property taxes, and sales tax also contribute to the total tax bill. Tax laws and rates change frequently, and it is a good idea to refer to current publications for up-to-date information on the subject.

A financial statement or chart shows the financial state of a control unit such as a department within an organization or of the whole organization, depending on the level at which the statement is developed. The details involved in the statement are level dependent; lower levels in the organization receive very detailed reports, while only summaries are presented to the top level of management. The chapter presents samples of various financial reports, along with information about a computer program that can be used to evaluate "what if" questions that occur in the overall planning of a manufacturing plant.

PROBLEMS

15.1 Discuss the factors in selecting a site for:
  a. A hospital
  b. A fresh-produce processing plant
  c. A clothing manufacturer
  d. An electronics corporation

15.2 Discuss specific circumstances when it would be best to ship by:
  a. Air transport
  b. Highway
  c. Rail

15.3 What are the advantages of company insurance? What types of insurance should be purchased?

15.4 Discuss the influences on workers' compensation insurance premiums.

15.5 Discuss the safety equipment that should be used for the following jobs:
  a. Moving pallets of materials in shipping
  b. Operating a drill press
  c. Working at a nuclear power plant

15.6 What major impact has OSHA had on plant and employee safety?

15.7 Go to a business in your community and prepare the safety check sheet in reference to the company's operation.

15.8 An electronics corporation had a net profit before taxes of $150,000 last year. If the state tax is 10% of the federal tax, what is the overall income tax for the company?

15.9 What benefits does the Social Security Act of 1935 provide for?

15.10 A company wishes to build a second cooling tower to accommodate additions to its plant. Its present tower handles 5 million BTUs per hour but has the capacity to handle 280 gallons per minute. The outlet cooling water temperature is 78°F. The system needs to handle 6,500,000 BTUs per hour. Determine the cooling water flow rate of the second tower.

15.11 A company wants to reduce its power expenses by $1000 per year. If the electrical power rate schedule specifies a demand charge of $2.87 per kVa per month, and the average power factor is 0.58, drawing 55 kW from the 440 V supply, determine how much the power factor would need to be increased to save this money.

15.12 What means of conserving energy are available for manufacturing plants? (Exclude reduction of operations.)

SUGGESTED READINGS

Industrial Site

Hamilton, F.E. Ian, *Spatial Perspectives on Industrial Organizations and Decision-Making,* John Wiley and Sons, New York, 1974.

Lochmoeller, D.C.; Muncy, D.A.; Thorme, O.J.; and Viets, M.A., *Industrial Development Handbook,* The Urban Land Institute, Washington, D.C., 1975.

Richardson, H.W., *Regional Economics,* Praeger Publishers, New York, 1971.

Smith, D.M., *Industrial Location,* John Wiley and Sons, Inc., 1971.

Utilities Specifications

Bonbright, J.C., *Principles of Public Utility Rates,* Columbia University Press, New York, 1961.

Lewis, B.T., and Morrow, J.P., *Facilities,* McGraw-Hill, New York, 1973.

Lighting Design and Application, Illumination Engineering Society of North America, New York, 1985.

Lowther, J., *Industrial Energy Conservation,* Louisiana Tech University, Ruston, La., 1985.

Insurance

Greene, M.R., and Swander, P., *Insurance Insights,* Southwestern Publishing Company, Cincinnati, 1974.

Mehr, R.I., and Cammack, E., *Principles of Insurance,* Richard D. Irwin, Homewood, Ill., 1972.

Pfaffle, A.E., and Nicosia, S., *Risk Analysis Guide to Insurance and Employee Benefits,* AMACOM, New York, 1977.

Taxes

Corporate Taxes, A World Wide Summary, Price Waterhouse, New York, 1985.

Corporation and Partnership Tax Return Guide, Research Institute of America, Inc., New York, 1985.

Prentice-Hall Federal Taxes, Prentice-Hall, Englewood Cliffs, N.J., 1985.

Social Security Administration, U.S. Department of Health and Human Services, Washington, D.C.

Safety

Hawdley, W., *Industrial Safety Handbook,* McGraw-Hill, London, 1977.

Lewis, B.T., and Marron, J.D., *Facilities,* McGraw-Hill, New York, 1973.

Marshall, G., *Safety Engineering,* PWS Publishers, Boston, 1982.

OSHA Safety and Health Standards, U.S. Department of Labor, Washington, D.C., 1983.

CHAPTER 16

Computer-Integrated Manufacturing Systems

There has been a dramatic increase in the use of computers in manufacturing-related activities. In the 1950s, these applications were restricted to a few well-defined problems in administration and finance, in which they produced straightforward solutions along with good documentation, mainly handling data sequentially. With the development of high-speed processing and low-cost memory, storing and manipulating large sets of nonsequential data have been greatly facilitated, and new applications in manufacturing have been developed. Inventory control, production scheduling, production requirements, and capacity planning are some of the well-known areas in which computers were successfully used in 1970s. With further advancements in both hardware and software, it is possible to further unite all phases of manufacturing.

Computer-integrated manufacturing systems (CIMS) is a concept for integrating all components involved in the production of an item. It starts with the initial stages of planning and design and encompasses the final stages of manufacturing, packaging, and shipping. It is the combining of all existing technologies to manage and control the entire business that has led to the term "automated factory" to describe the ultimately successful application of CIMS. CIMS includes components such as computer-aided design, computer-aided manufacturing (i.e., the use of CNC and DNC machines), artificial intelligence, computer process planning, database technology and management, expert systems, information flow, just-in-time concepts, material requirement planning, automated inspection methods, process and adaptive control, and robots. All of these elements work together using a common database. Data are entered from the operations of the entire plant: from each level of production on the shop floor, by receiving and shipping, marketing, engineering design, and designated individuals within other departments who are related to the operation of the plant. The information is continuously processed, updated, and reported to people on an as-needed basis in an appropriate format to assist in making decisions. The same data set is used by the NC machines to manufacture products and by automated material-handling systems to move them from one work station to another. This integration of mechanical, electrical, and informational systems is the backbone of CIMS.

16.1 SYSTEMS AND FILES

The system is designed to offer great flexibility in production by using NC machines and increased speed and flexibility in material-handling systems by using computers and automated guided vehicles. It can handle a production mix consisting of up to 800 different parts with a production volume anywhere from a few units to 10,000–12,000 units. The terms "special system," "flexible manufacturing,"

and "manufacturing cell" are often used to describe CIMS with a different combination of part variety and production volume. Table 16.1 shows a popular classification guideline suggesting the features of a system to be utilized.

A special system offers less production flexibility and more material-handling automation, including fast-moving computerized conveyor systems, while a manufacturing cell offers more flexibility in manufacturing with a

Table 16.1   CIMS classification characteristics

System Classification	Number of Different Parts to Produce	Number of Parts Produced per Hour	Features
Transfer line	1 to 5	More than 15 in hundreds	High utilization of machines and plant space, minimum labor cost, mainly single-purpose machining with sequences controlled by programmable controller; no flexibility
Special system	3 to 10	1 to 15	Produces parts in batches, considerable time spent in changeover; may contain one or more machines dedicated to the product
Flexible manufacturing (FMS)	10 to 50	Up to 10	Series of flexible machines with mainly numerically control, equally flexible but highly automated material handling; allows random processing
Manufacturing cell or stand-alone system	More than 20	About 3 or 4	FMS with independent control of each machine and can process units in batch mode

small production volume on the average for each part. Here material handling can be obtained more independently by using automated guided vehicle systems.

As is true of any manufacturing system, the purpose of a CIMS is to produce salable goods at a minimum cost in the shortest possible time. Unlike traditional manufacturing, however, CIMS employs the philosophy that management must work to optimize the whole business operation rather than individual functions or work stations.

CIMS requires a change in management style. It allows management personnel to make decisions on the basis of up-to-the-minute information that is made readily available to them through the use of the computer, which is essential for maintaining, manipulating, storing, and recalling the database. Such a database may consist of information on machine utilization in the plant; the scheduling of parts based on machine availability, promised delivery dates, and the possibility of creating final assemblies without building large in-process inventories; personnel assignments as machine coupling becomes more common and nonproductive time decreases; and/or tools that are to be selected and loaded into the NC machine from the tool drum, based on the information supplied to the machines by the computer concerning parts and operations. The CIMS database might also consist of:

· A file for parts programs to be used in NC machines.
· A file on routing describing the work stations through which the part must be processed.
· A file on parts production indicating for each part the production rate for each machine, in-process stores, quality control requirements, etc.
· A file for each work station indicating the

work scheduled, the tools needed, and the operator assigned.
· A file on all tools indicating tool use, its life, maintenance needed, and the cost for each of those items.

The changes in management concepts are evident if one considers the cooperation the system demands from the various departments. No longer can each department within a plant, such as design, manufacturing, marketing, shipping, receiving, personnel, or purchasing, consider itself a closed cell in its operations. All employees, including even operators on the shop floor, must work jointly using the same manufacturing database.

Each person is responsible for keeping the database current and accurate and for using the information from it when needed. This type of system requires total commitment from people in top management. In particular, they must (1) train everyone in the plant to work with CIMS; (2) employ people who will take responsibility within the sphere of their influence instead of waiting to be told what to do (they must also develop the reward system that will encourage such behavior); and (3) develop a matrix or similar type of organization that allows and encourages different departments to communicate with each other.

It should be clearly understood by all who are working in the plant that there is a person in charge of the CIMS database who is responsible for maintaining and protecting it; however, this database is not assigned exclusively to any single person. Everyone can use it when needed, and everyone is responsible for its upkeep.

16.2 COMPONENTS OF CIMS

Most of the components have already been described in other chapters specifically related to the activities discussed in those chapters. To make the description of CIMS more nearly complete, we will briefly review those functions here.

Computer-Aided Design

A manufacturer must begin with the design, drawings, and part details for a product, and computer-aided design (CAD) helps in every phase of these operations. CAD is especially useful because, on the average, even a small change in one drawing typically affects about five other drawings. Using such a system improves productivity three to four times and also improves the quality of design and drawing details. CAD in design analysis is sometimes referred to as computer-aided engineering (CAE). CAE can, for example, simulate how a tool will perform on a workpiece; how a setup of fixtures will function; how a mold will operate under the effects of friction, heat, and mold pressure; and how these factors will affect the raw material and ultimately the final product. CAD helps to form an interaction between designers and manufacturing engineers during the design stage of the product. Manufacturing engineers can develop production processes, tools, and dies for the proposed design and analyze its feasibility during the initial stages. If changes are needed, they can be made before the product goes to the production phase.

Vendors can also be made part of the team to evaluate supply requirements, materials, and tools that would be needed for the job. Again, initial suggestions by suppliers can be very useful in modifying the product design and lowering the unit cost. (More can be found on CAD in Chapter 5.)

Group Technology

A large quantity of detailed design data can be put in order by group technology (GT), which systematically codes the data and design analysis. Coding can be done by using design attributes such as the size, shape, weight, and material of the part, tools, and equipment needed in parts production. A computer can group similarly coded items together and store this information in its memory. Generally, in an existing production facility a new product design may be able to use anywhere from 20–60 percent of the existing drawings with little or no change required. Without a computer and group technology, however, a designer would most likely design and draw the parts again rather than go through thousands of detailed drawings to find the parts that will fit into the new design. GT also facilitates the formulation of machine cells for flexible manufacturing centers, thus expanding and expediting the work of CAD by reducing product design time as well as by reducing setup and waiting times.

However, GT is time consuming. Even though computer software can allocate the GT codes to a part, each item of design data must be entered manually into the computer. (See also Chapter 7.)

Computer-Aided Manufacturing

Further integration of CAD data to production is achieved through computer-aided manufacturing (CAM). It has three major components: machine tools, production machines, and electronic controls and relays with which to operate production machines.

Numerically controlled machines (NC, CNC, DNC) form the foundation of CAM. NC machines are driven by a program on punched paper tape. CNC machines are NC machines controlled by their own computer. Large numbers of tool programs can be stored in the computer memory and recalled as the need arises, thereby avoiding the necessity of feeding a tape for every program. A DNC (direct numerically controlled) machine is a number of CNC machines connected to one large computer. The loading, unloading, and modification or programs in DNC machines can be accomplished more quickly and conveniently than in CNC machines.

NC machines have higher spindle speeds and more torque on tool tips than do conventional machines, increasing cutting speeds. They also have sensors mounted on the tool spindles and work surface for real-time adaptive tool control.

CNC and DNC machines can be programmed off-line, increasing productivity. The controls for machines are provided by a programmable logic controller (PLC), which replaces thousands of relays and contains active memory for input/output functions. A PLC is a solid state device that is able to withstand high temperatures and dusty environments, the conditions almost always present in production facilities. (See Chapter 6.) If the design and production quantities of a workpiece require increased specialization, process-specialized machine tools can also be incorporated into a system. For example, these specialized machine tools could be used to perform critical tolerance and/or multiple-spindle drilling, tapping, or boring operations and, in more sophisticated systems, washing, assembly, and even inspection operations.

Robots

A robot is a "worker" within the CIMS. It is a unit consisting of sensors, computers, and controls, the use of which is obtained in terms of its arm movements (and sometimes leg or wheel movements). A robot by itself is not very useful unless it is integrated in automated material handling, adaptive processes, and/or multitask functions.

A robot can be used as a part feeder to a single machine or a group of machines, a material handler in a complex assembly, or a visual sensor in an inspection process. Robots are also used in manufacturing involving such tedious tasks as welding and spray-painting.

More and more robots are installed in CIMS as they become affordable. Flexibility and quality of work are two major traits that encourage their use in production. Since robots do not have human feelings, nor do they perspire from heat, shed hair, or breathe fresh air, they are especially suited for working on monotonous and hazardous jobs (e.g., handling toxic chemicals) or where a dirt-free, clean operation is absolutely necessary (e.g., making computer chips). Robots can be attached to the ceiling, freeing the floor area for other activities. With adjustable speeds and motions, robots work precisely and repeatedly, integrating themselves completely with the rhythm of the other equipment within the system.

A robot is programmed either by being physically led through the steps it is required to perform or by use of a specific programming language. Research is continuing to standardize programming languages and to make vision robots more affordable for any size of manufacturing firm. There are also attempts to increase the flexibility of robots by programming on-line, that is, without removing them from production.

Material Handling

The purpose of the material-handling system in CIMS is to automatically deliver and remove workpieces from the machine tools. This eliminates the manual handling of the workpieces, which in turn lowers the work-in-process time. The material-handling system under computer control also maintains a constant supply of workpieces to the machine tools.

With an increased use of computers, servomechanisms, and electronic controls, material handling has been transformed into automated material handling (AMH). For example, a modern forklift truck has a CRT terminal, computer keyboard, radio communications, and bar code scanners to work effectively with different data files. The just-in-time (JIT) philosophy reduces material in inventory and thus the need for material handling. The material requirement planning (MRP) system maintains accurate data on what parts are needed and how many are available at what time on the basis of the production schedule, again reducing material handling. The automated storage and retrieval system (AS/RS) automatically stores and dispenses items from storage. Automated guided vehicles (AGVs) collect and deliver products on flexible paths defined by under-floor wires or an optical guidance system. They are provided with safety bumpers at both ends, which stop the vehicle upon contact with any object. Though controlled by computer, they also allow an operator to override the system and drive the vehicle away from its guided path if needed. For high-volume transportation, conveyors still play a major part, but even they are now automated. For example, powered roller conveyors are used to transport heavy workpieces. Standard machines are linked by a waist-high conveyor loop, allowing it to carry workpieces through stations for different machining jobs. Towline carts are run by an under-the-floor towline system. Chains that are located in slots in the floor pull the carts along permissible paths. A cart is towed to the appropriate machine under computer control; and when it arrives, a plate rises from the floor and disconnects the cart's chain drive. A piston pushes the cart against a hard stop for accurate positioning before loading or unloading the pallet.

In many instances, conveyors provide movement over a fixed path within the work cells, even acting as mobile assembly platforms, while AGVs provide interdepartmental transport.

All of the units within AMH are becoming more flexible, increasing nonsynchronized handling; that is, each unit is able to move independently of the others. There are many additional paths over which AGVs can travel while making decisions automatically: picking up speed and slowing down when needed, posi-

tioning themselves precisely under a work station, and initiating loading and unloading actions by conveyors and by robots. The productivity of modern forklift trucks is increaed tremendously by computers and other equipment. In all, AMH eliminates material handling when possible and performs other necessary material handling efficiently to increase productivity.

Computer Systems

In most systems, machine tools, material handling, and control activities are under computer direction. The magnitude of computer control is based on the system's level of sophistication; the more complex the CIM system, the greater the computer involvement.

Such a computer requires both hardware and software. The computer hardware depends on considerations such as the number of NC machines in the system, the number of parts involved, the amount of data stored, and the physical locations of the data terminals. The following is an example of some of the hardware requirements:

- Mainframe computer for overall control of the system.
- Equipment for control of the material-handling system.
- At least two disk drives, one on-line containing system data files and programs and the other providing backup.
- Keyboard/printer for master control of the system.
- Video display terminals for entry of system commands and display of reports and messages.
- Shop terminals used for semiautomatic entry of commands.

Control Software

Large amounts of data generated when a product requires many different items have created

Figure 16.1 Flow of information and products in a CIM system (redrawn courtesy of Reza Ziai and Sinan Kayaligil)

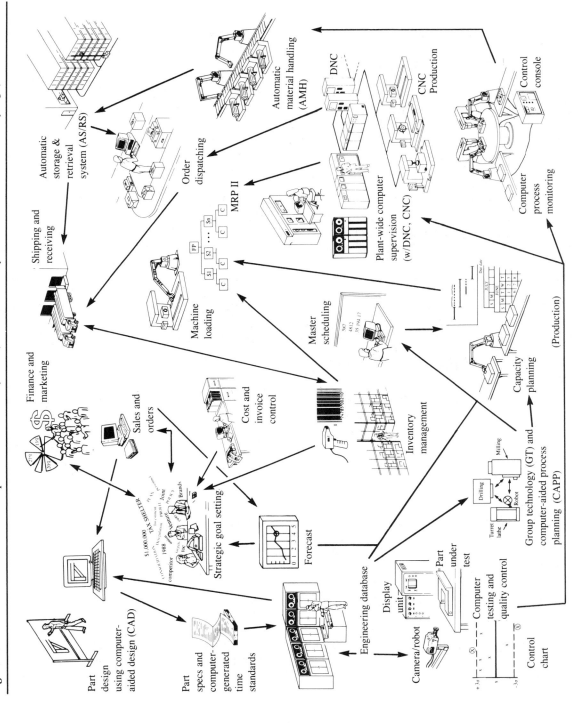

a need for software to efficiently control the system. Typically, a medium-sized firm has four to five million pieces of information that require storage and manipulation. The computer software consists of two major databases—manufacturing and product. The manufacturing database collects data from the process computer and the process controller. The data are used to report shop activities and to plan and control future activities. The product database is used by programmable equipment to perform manufacturing activities such as machining, inspecting, and counting. The data banks might consist of bills of materials, customer orders and promised dates of deliveries, inventories, production schedules, personnel schedules, and tool and equipment requirements among others. We will briefly describe some of the available software in the following paragraphs.

Manufacturing planning and control (MPCS) and its components, manufacturing resource planning (MPC II) and material requirement planning (MRP), are the key software packages in the market today. The MRP system answers pertinent questions such as which products are scheduled to be made, the parts and quantities that are needed to make these products, how much stock of each part the plant has, and how much it must purchase and when. MRP II, which is a simulation program, schedules the operations of each work station, each machine, and indeed the entire plant. Additionally, MRP II provides the status of each machine and its productivity. It makes sure that the required parts arrive at the correct station at the right time in the necessary quantities. MPCS integrates MRP and MRP II with CAD, CAM, and other software.

Those other computer programs include one on statistical quality control (SQC), which provides information on whether a process is in control, and a package for testing the accuracy and reliability of tools and equipment. Simulation packages are also available that allow "what if" analysis of sales, finances, tooling,

and overall operation of the plant. It is certain that more and better software will be developed to increase automation in plants as time passes.

Computer software is an important element of CIMS. It is expensive and yet the only truly flexible part of the system. As was mentioned above, there are some standard packages such as MRP II; but whenever possible, modularity is essential so as to not tie down the entire system to a single program. Many tool programs, for example, are available from various vendors, and a manufacturer should be able to select the appropriate ones and integrate them into its own CIM system. As the products change over the years, such flexibility becomes critical in reducing software costs.

Artificial intelligence (AI), or the expert system, is a recent addition to computer software. It consists of two phases—software to deduce new information, called the "inference engine," and software to help modify decision-making rules based on the latest available information, called a "knowledge base." Observed AI applications in manufacturing are found in the industrial environment (production planning, maintenance, CAD use), the service environment (training, customer services, route selections in shipping), and the management environment (strategic planning, market planning, financial planning).

The flow of information and products in a computer-integrated manufacturing system is illustrated in Figure 16.1 (page 395).

16.3 BENEFITS AND DEFICIENCIES OF CIMS

CIMS is receptive to changes. In a CNC system consisting of ten NC machines, it is possible to produce either ten parts of the same kind or, at the other extreme, ten different parts. Changes in designs and demands can easily be accommodated by CAD systems. Once the software has been developed, it is equally convenient to produce one part or a thousand parts. A manufacturer can therefore respond rapidly to changes in product design, demand, or product mix. The manufacturer can effectively serve both small customers asking for only a few parts and major customers looking for large quantities of many different parts. Each can be made with consistent quality with a minimum of waste. Inventories and setup times are reduced considerably in comparison to those of traditional manufacturing, giving short turnaround times and lower costs. CIMS has the ability to handle parts with complex designs and to produce them continuously with a minimum of unscheduled downtimes. Each worker becomes a "micromanager," increasing the level of his or her contribution to the overall working of the plant. Increased productivity and improved quality of work are the results of a job-satisfied worker.

CIMS also has other benefits. According to a study sponsored by the American Society of Manufacturing Engineers, the use of CIMS has:

- Reduced engineering design cost by 15 percent.
- Reduced overall lead time by 30–60 percent.
- Increased production by 40–70 percent.
- Reduced work-in-process inventory by 30–60 percent.
- Reduced personnel costs by 5–20 percent.

There are also some difficulties associated with implementation of CIMS. Total CIM systems are expensive, costing anywhere from $6,000,000 to $500,000,000, depending on the size of the plant. Very few companies can afford the capital investment to make the entire existing plant compatible with CIMS. Furthermore, there is a great deal of fear and hostility toward computerization within a typical firm when people see their jobs being changed dramatically or even disappear as a result of CIMS.

Most manufacturing plants currently have a combination of both batch and flow-through processing, which makes changeover and control difficult. The technologically skilled personnel needed in CIMS remain in short supply, and the costs of database development and maintenance are considerable.

However, these difficulties are surmountable with research and development initiated by industries, universities, and government. As a result, the use of CIMS is expected to grow.

16.4 PLANNING FOR CIMS

Transferring a manufacturing system from the traditional surrounding to CIMS is indeed a major task. Two prime issues must be addressed before such implementation is possible. They are (1) what planning and design considerations are necessary for a given industrial application and (2) how the large investment normally required for CIMS can be justified.

The outline below might prove to be beneficial in formulating and analyzing the requirements in such a changeover.

1. *Define objectives.* Analyze the engineering and manufacturing capacities and needs and envision how automation might be accomplished. Consider what traditional competitors have done in their efforts to modify and manufacture the product. Give special attention to leading users of factory automation. Such analysis might provide the company with a set of goals to achieve and might also indicate where the competitors are heading and what must be done to remain viable.

2. *Formulate a project team.* The team evaluates the alternatives and performs the necessary engineering and financial analysis. It is not uncommon to have a time span of three years or more before changeover to CIMS, from concepts to production, is accomplished.

There must be continuity during this time, and the project team provides it. The team might consist of one senior manager and a representative from each functional area that would be affected by the changeover, such as design, production, manufacturing, and finance. It might also be beneficial to include workers or a union representation on the team; this inclusion can facilitate acceptance of the changeover by the shop floor personnel.

3. *Keep up-to-date with the technology.* CIMS might be a new field to many of the people on the team. To understand the technology, members can attend seminars and conferences and read technical articles. The team might ask vendors to arrange for a field trip to a plant where their product has been successfully implemented. The educational process should also be extended within the company. For example, before implementing a complete CIM system, it might be beneficial to obtain experience with some of the individual components of CIMS such as NC machines and robots as they are used in producing specific components for the company. Such knowledge is valuable in developing the useful integration of efforts.

4. *Evaluate the concepts.* The company should conduct an analysis to determine suitability of CIMS based on the present and anticipated future product mixes. Group technology should be applied wherever appropriate. The analysis should include current data on unit cost, lead time, quality, manpower, material handling, utilities, and any other appropriate items. A number of different manufacturing systems should be analyzed—for example, stand-alone cell versus integrated manufacturing. If CIMS is appropriate, several detailed possible configurations should be developed with different types of equipment.

5. *Justify the investment.* In making an economic analysis, both benefits and cost must

be estimated. It is relatively easy to determine the investment needed in the purchase and installation of the hardware and software for CIMS. Less obvious is the cost of training and new personnel requirements for implementation of the system. On the benefit side, certain improvements such as quality of the product and reduction of personnel can be easily quantified, but difficulty might arise in trying to estimate the benefit of better material handling and reduction of lead times. To develop a correct financial appraisal, one must consider:

a. the total CIMS picture to include both direct and indirect costs as well as benefits to the company in its future developments,
b. a long time span of five years or more to account for long-term impact of the investment, and
c. probability assignment in the evaluation to estimate uncertainties in the future.

We may ask questions such as:

a. What will be the result if we do not invest in CIMS?
b. Are there other alternatives that will keep the company competitive?
c. What is the competition doing worldwide and how will this affect the future?
d. Are there any new products that can be manufactured using CIMS that might open up new markets?

If the analysis leads to the decision to invest in CIMS, the next step is to ask for specific proposals from vendors.

6. *Evaluate and implement the system.* Information collected so far will lead one to develop a sound proposal to obtain firm bids from different vendors. Many discussions are required for vendors to understand and develop a system for any unique application that

may be associated with the company's product. Proposals should be evaluated not only on the cost basis but also according to factors such as experience of vendors, quality of their products and reliability and availability of technical assistance, should it be needed. Once the vendors have been selected, close cooperation between different suppliers and an understanding of the company's objective are essential to obtain an integrated system.

16.5 SYSTEM PROVIDERS

Currently, no single manufacturer will develop an entire system. "Off-the-shelf" concepts do not fit all the peculiar requirements that are generally associated with a firm. Many manufacturers, such as Allen-Bradley, Sperry, IBM, Tandy, Litton, Digital Equipment, and Easton, can develop an appropriate group among themselves, each performing the task for which it has the greatest expertise, to develop the entire system. For example, Digital computers might manage a flexible manufacturing system (FMS) consisting of Ingersoll CNC machines using Allen-Bradley controllers for individual CNC tools. The computer systems might be developed by Digital, Hewlett-Packard, or IBM. Machine tools might come from Cincinnati Milacron. Material handling might be designed by Litton or Eaton-Kenway, while the robots might be supplied by GCA and GMF.

With experience and higher productivity in the computer industry, the prices of CIMS are sure to come down, making them more affordable to small and medium-sized firms. Then again, one might build the system in modular fashion, reducing the initial expenditure.

A partial listing of CIMS suppliers is given in Table 16.2.

Table 16.2   CIMS suppliers

Supplier	Address
Harnischfeger Engineers, Inc.	P.O. Box 554 Milwaukee, WI 53204
Xerox Computer Services	5310 Beethoven Street P.O. Box 66924 Los Angeles, CA 90086
Schlumberger	4251 Plymouth Road P.O. Box 986 Ann Arbor, MI 48106-0986
Scientific Systems Services	2000 Commerce Drive Melbourne, FL 32901
Accu-Sort Systems	511 School House Road Telford, PA 18969
White Storage and Retrieval Systems	30 Boright Avenue Kenilworth, NJ 07033
Advanced Systems, Inc.	155E Algonquin Road Arlington Heights, IL 60005
Allen-Bradley	1201 South & Second Street Milwaukee, WI 53206

continues

Table 16.2   Continued

Supplier	Address
Sperry Corporation	P.O. Box 500 Blue Bell, PA 19426-0024
Tandem Computers	19191 Vallco Parkway Cupertino, CA 95014
Acco Babcock, Material Handling	P.O. Box 460 Bailies Lane Frederick, MD 21701
Digital Equipment	129 Parker Street Maynard, MA 01754
Eaton-Kenway	515 East 100 South Salt Lake City, UT 84102
Insta Read	2400 Diversified Way Orlando, FL 32804
Interlake Material Handling/Storage Products	100 Tower Drive Burr Ridge, IL 60521
GMF Robotics Corporation	5600 New King Street Troy, MI 48098
Litton UHS	7900 Tanners Gate Florence, KY 41042

PROBLEMS

16.1 Discuss the philosophy behind CIMS.

16.2 Are there any additional subject areas that a person must learn to work in a CIMS environment?

16.3 Define the following:
  a. CAD
  b. GT
  c. CAM
  d. AMH

16.4 Distinguish between NC, CNC, and DNC.

16.5 What is a robot? Give examples of its use in industry.

16.6 What are some of the main advantages in using robots? What are the disadvantages?

16.7 Identify three methods of instituting automated material handling in place of material handling.

16.8 How is computer software utilized in industry today?

16.9 Savings in what areas will contribute toward justification of expenditures in CIMS?

16.10 Differentiate between two types of computer software available for control in industry.

16.11 Give two specific examples of the use of artificial intelligence in industry.

16.12 Discuss the advantages and disadvantages of CIMS.

16.13 Visit a local manufacturing facility. Identify specific ways in which CIMS could contribute to their processes.

SUGGESTED READINGS

"A Look at NC Today," *Manufacturing Engineering,* Oct. 1975, pp. 24–28.

Asfahl, C.R., *Robots and Manufacturing Automation,* John Wiley and Sons, New York, 1985.

Fisher, E.L., "Expert Systems Can Aid Intelligent CIM Decision Making," *Industrial Engineering,* Vol. 17, No. 3, March 1985, pp. 78–83.

Goldhar, J.D., and Jelinek, M., "Computer-Integrated Flexible Manufacturing—Organizational, Economic, and Strategic Implications," *Inter Faces,* Vol. 15, No. 3, May–June, 1985, pp. 94–105.

Gould, L.S., "Computer-Integrated Manufacturing Systems," *Modern Material Handling,* May 1985, pp. 91–144.

Halevi, G., *The Role of Computers in Manufacturing Processes,* John Wiley and Sons, New York, 1980.

Houtzeel, A., "The Many Faces of Group Technology," *American Machinist,* Jan. 1979, pp. 115–120.

"Measuring the Effect of NC," *American Machinist,* Nov. 1983, pp. 113–144.

Ogorek, M., "CNC Standard Formats," *Manufacturing Engineering,* Jan. 1985, pp. 43–45.

APPENDIX

A

Engineering Economy Formulas

All of us realize the time value of money. One thousand dollars in hand today is worth more than a promised $1000 one year from now; the present sum could be invested, for example, at an interest rate of 10% to earn $100, making a total year-end capital of $1100. When we buy an automobile and agree to pay monthly installments, the sum of the installment payments is more than the amount owed on the car, again because of the interest payment on the owed capital, or the time value of money.

It is possible to determine the effect of the time value of money by using mathematical expressions. We will not go into details of derivations but only state the final results. Most standard books on engineering economy also tabulate the values resulting from using the formulas for specific interest rates and time periods; it might be easier for the reader to use these tabulated values. The following notations are used in Table A.1.

i = Interest rate per interest period in decimals (e.g., 10% per year interest is 0.10; if the compounding is done semiannually, $i = 0.1 \times \frac{1}{2} = 0.05$)

n = Number of interest periods (e.g., if compounding is done semiannually, $n = \text{years} \times 2$)

P = Present value of the money

F = Future value of the money; that is, the value of P, n periods from now at an interest rate of i (e.g., $1100 at the end of one year, $P = 1000$, and $i = 10\%$)

A = Annuity, or equal payments (receipts) at the end of each period, for n periods, the present worth of which is P at an interest rate of $i\%$ (e.g., monthly car payments)

G = A uniform arithmetic gradient in payments, starting with zero and increasing at a constant amount G, from period to period for n periods

Table A.1　Time/money relationships

Given	To Find	Multiply the Given by the Factor	Name of the Factor	Mathematical Expression of the Factor	Time/Money Display
P	F	$(F/P, i\%, n)$	Compound amount	$(1 + i)^n$	
F	P	$(P/F, i\%, n)$	Present worth	$(1 + i)^{-n}$	
P	A	$(A/P, i\%, n)$	Capital recovery	$\dfrac{i(1 + i)^n}{(1 + i)^n - 1}$	
A	P	$(P/A, i\%, n)$	Series present worth	$\dfrac{(1 + i)^n - 1}{i(1 + i)^n}$	
F	A	$(A/F, i\%, n)$	Sinking Fund	$\dfrac{i}{(1 + i)^n - 1}$	
A	F	$(F/A, i\%, n)$	Series compound amount	$\dfrac{(1 + i)^n - 1}{i}$	
G	P	$(P/G, i\%, n)$	Gradient present worth	$\dfrac{1}{i}\left[\dfrac{(1 + i)^n - 1}{i} - n\right]\left[\dfrac{1}{(1 + i)^n}\right]$	

APPENDIX B

Quality Assurance Manual, Spring Controls, Inc.

CONTENTS

1.0 SCOPE

The purpose of this manual is to establish methods and procedures to be followed at this facility to assure compliance with quality standards required by contractual agreements with customers, particularly those stipulating military and government specification (see MIL-I-45208).

Methods and procedures specified in this manual are applicable to the inspection of the end product and products in process of manufacture or components thereof.

This manual also is to convey to suppliers of Spring Controls, Inc. their responsibility in the control of the quality of their products. All suppliers shall be provided with drawings, documents, or specifications which shall be the acceptable standards.

This manual is reproduced courtesy of John Cochran.

1.1 REVISION CONTROL

Date	Section	Page	Revised Data
1/2/..	3.0	1	Added "per MIL-STD-105D"
1/2/..	6.0	1	Added "per MIL-STD-105D"
7/5/..	7.0	3	Added list of Process Vendors

1.2 QUALITY CONTROL ORGANIZATION CHART

2.0 BLUEPRINT CHANGE CONTROL

1. Blueprint changes are controlled by the Production Control Department.

2. When a blueprint change is received, the master file is to be updated with the new changes.

3. Obsolete blueprints are to be marked void and filed in the master file for reference.

3.0 INCOMING MATERIAL INSPECTION

1. All raw materials, parts, and assemblies purchased for use in connection with the production of orders for the government or its subcontractors are procured against the supplier's certifications. All such items shall be subjected to inspection after receipt per MIL-STD-105D to assure conformance to purchase order requirements.

2. Certified test reports covering all raw materials and their specifications are to be retained at our plant subject to inspection upon request.

3. Spring wire and flat spring stock are to be checked for dimensional accuracy. Material that does not conform to ordered specifications is to be rejected and returned to suppliers accompanied by a rejection report. A corrective action report is required from the suppliers. Raw stocks that are accepted are to be segregated and tagged upon receipt and stocked in their respective places provided in our warehouse.

4.0 MATERIAL CONTROL

1. Raw material in stock must be carefully controlled at all times. All raw stock shall be tagged and stored according to material, type, size, and weight.

2. Raw stock is to be checked out of Material Stores to Production as required for work and process by a shop traveler. The shop traveler is released by Inventory Control for Production only after verification that the material meets all requirements.

3. After completion by Production of a specific requirement, all raw material not used must be returned to stock and replaced in its designated place, properly tagged. Appropriate notations shall be made on the shop traveler and returned to Inventory Control for the purpose of controlling the expended amount of material.

5.0 FIRST ARTICLE INSPECTION

1. Tooling shall be checked for accuracy in conformance with the latest change of the appropriate drawing by the production supervisor prior to setup.

2. A number of the first parts produced shall be inspected, and production shall not begin until these parts are approved by the inspector. Parts not acceptable must be scrapped, and new samples must be submitted until accepted by the Inspection Department.

3. On the first production article, a comprehensive inspection and tests shall be performed of that article to assure its conformance with all blueprints and specification requirements. Approval shall be stamped on the shop traveler by the inspector.

6.0 FINAL INSPECTION

1. Final inspection will be made by the quality control inspector to establish conformance with the customer's purchase order and drawing requirements.

2. Final inspection includes checking every blueprint requirement. Inspection is both visual and mechanical and covers such items as physical dimensions, load and torsional testing.

3. Inspection shall include sampling in accordance with approved statistical sampling methods per MIL-STD-105D or as specified by contract.

6.1 INSPECTION REPORT

1. Customer:
2. Date:
3. Part Number:
4. Purchase Order Number:
5. Applicable Specifications:

6. Blueprint Dimensions:

7. Actual Dimensions:

8. Heat Treatment:
9. Finish:
10. Inspected by:
11. Stamp
12. Accept or Reject:
13. Approved by:

6.2 INSTRUCTIONS FOR INSPECTION REPORT

In filling out the inspection report, it should be noticed that each title has a number. These instructions "key" on that number, and begin with the items at the top of the form.

Number	*Instruction*
1. Customer	Enter name of customer
2. Date	Enter date of inspection
3. Part Number	Enter part number
4. Purchase Order Number	Enter purchase order number
5. Applicable Specifications	Enter specifications for material finish, heat treat, etc.

7.0 OUTSIDE VENDOR QUALITY AND EVALUATION CONTROL

1. A Quality Control Evaluation Report will be made on each vendor doing any special process or subcontract work for Spring Controls, Inc. This report is intended to furnish data relative to the capability of the vendor to control the quality and conformance of applicable supplies, parts, and services to our requirements.

2. This report shall be prepared by the Quality Control Department at least once each year and as otherwise deemed necessary. The report will be required for any new vendor prior to release of work to that vendor and must be approved by Quality Control.

3. The Quality Control Department shall maintain a list of approved sources for all processes, subcontractors, and other facilities as may be necessary to control outside sources.

4. All outside processing of parts requiring government specifications shall be certified by the processor that the processing complies with these specifications. In addition, parts shall be inspected by the Inspection Department in accordance with approved statistical sampling methods or as specified by contract.

5. Copies of outside certifications shall be kept on file in the office of the Quality Control Department and shall be available for review as required.

6. All processes to be completed by outside vendors shall be controlled in a manner pertaining to the specification. All outside vendors must be approved by the prime customer and/or the government agency which may have control over process approval.

7. All outside vendors, in addition to approval by the prime customer, must be approved by the Quality Control Department of Spring Controls, Inc. and must maintain such records as requested in addition to specification records.

7.1 LISTING OF APPROVED PROCESSORS

Vendor	Process
	All types electroplating
	Glass bead peen
	Passivation
	Phosphate
	Parkerize and oil
	All types electroplating
	All types heat treating

8.0 INSPECTION MEASURING AND TESTING EQUIPMENT

1. All inspection equipment including personally owned calipers and micrometers shall be calibrated quarterly or more frequently as deemed necessary.

2. All inspection equipment shall be listed in a handbook and held in the Inspection Department to indicate when each item should be checked and recorded. This list shall be reviewed by the chief inspector on the first of such month for the purpose of calling in each test piece of equipment for inspection and review. All tests shall be completed within thirty (30) days prior to the expiration date.

3. The inspection equipment used in the Inspection Department shall receive periodic comparisons as outlined below and shall be recorded reflecting the date, type of equipment inspected, and any remarks. This is performed by a certified outside source, and certification shall be maintained in the Quality Control Department.

4. Spring testers shall be tested and recorded for accuracy every three (3) months and any other time as deemed necessary. All testers shall be tagged showing the actual date of test and when the next test is due.

5. All micrometers and calipers used by the Inspection Department shall be checked with gage blocks traceable to the National Bureau of Standards and recorded twice monthly. They shall be labeled and recorded in the Inspection and Testing Book showing the date and identification of the person performing the test.

6. All furnaces and control panels shall be calibrated for uniform heating by an outside certified furnace inspection service every six (6) months. Certifications will be required and shall be maintained in the Quality Control Department.

7. Gage blocks shall be tested once each year, and a certification of calibration source will contain a statement to the effect that the calibration of equipment is traceable to the National Bureau of Standards.

8. A label as designated by the Quality Control Department will be established to use on small gauges and other equipment to indicate the period of time said equipment was checked and recorded.

9.0 SHIPPING AND PACKAGING

1. All packaging and shipping is to be in accordance with the customer's requirements. Our normal procedures ensure that the packing list and identification tags are inside the package. The package is further identified by placing the customer's name, part number, and purchase order number on the outside of the package.

2. Any deviation from the above procedure must be in accordance with the customer's specification.

3. The packing list further states, "Test reports covering all the material itemized as requiring tests are on file subject to examination and indicate conformance with applicable specification requirements."

4. Conformance certifications and/or physical and chemical test reports when required will be attached to the packing list. All certifications will show the customer order number, part number, and quantity shipped.

5. Adequate packaging is to be employed to ensure that all parts receive protection from damage and corrosion during shipment and handling.

6. Final packaging shall be approved by the Inspection Department.

7. The above paragraphs are to be posted in the packaging and shipping area.

10.0 INSPECTION STAMPS AND TAGS

1. All inspectors shall have a stamp that will be numbered to identify them on all articles inspected and stamped. These stamps shall not be used by anyone other than the designated personnel assigned a numbered stamp. The inspector should use diligence in controlling his assigned stamp.

2. Inspection Department stamps shall be used on all first article production parts and final inspections.

3. A list of inspection stamp numbers will be maintained in the Inspection Department with the name of the inspector having control and the date of issuance. New stamps shall be issued only with the approval of Quality Control.

11.0 DISCREPANCY REPORT FORM

1. Customer:
2. Part Number:
3. Purchase Order Number:
4. Date Received:
5. Quantity Received:
6. Discrepancies:

7. Responsible Department:
8. Inspector:
9. Preventive Action:

10. Final Disposition: Rework   Scrap
11. Approval:
12. Quality Control Manager:

11.1 PROCEDURE FOR REPORTING DISCREPANCIES

This procedure is to be used to report discrepancies on rejected parts. In filling out the discrepancy report, it should be noticed that each title has a number. These instructions "key" on that number, and begin at the top of the form. All entries must be typed or printed.

Number	*Instructions*
1. Customer	Enter name of customer
2. Part Number	Enter part number
3. Purchase Order Number	Enter purchase order number
4. Date Received	Enter date parts were received
5. Quantity Received	Enter quantity received
6. Discrepancies	Indicate what is wrong with the part; be as detailed as necessary to show the location of the discrepancy on the part; quote drawings, specification, and condition of the part, etc.
7. Responsible Department	Enter name of department responsible for the discrepancy
8. Inspector	Enter name and number of inspector who checked parts
9. Preventive Action	Enter a statement of the action to prevent recurrence. Include the cause of the discrepancy and the positive corrective action taken.
10. Final Disposition	Indicate whether parts are to be reworked or scrapped
11. Approval	Signature of a company official approving final disposition
12. Quality Control Manager	Signature of the Quality Control Manager

12.0 QUALITY CONTROL

1. It shall be the responsibility of the Quality Control Department to see that articles are produced in accordance with the requirements of the contract and/or purchase order and the amendments or supplements thereto.

2. The responsibility of producing and maintaining all specifications required by purchase orders, drawing, and contracts shall rest within the Production Control Department.

3. The quality control requirements of Spring Controls, Inc. are believed to be consistent with the requirements established by MIL-I-45028.

4. All changes in the Quality Assurance Manual will be incorporated into all existing manuals that are in the hands of prime customers or subcontractors. It shall be the Quality Control Department's responsibility that all outstanding manuals be maintained to the latest revision.

13.0 SHOP TRAVELER

Shop Travelers are required to include the following data:

1. Customer Name
2. Purchase Order Number
3. Part Number and Revision
4. Date of Order Entry
5. Date Order is Due
6. Order Quantity
7. Type and Size of Material
8. Manufacturing Procedures
9. Finish and Heat Treatment Specifications
10. Name of Vendor on Outside Process

14.0 TYPES OF RECORDS AND NUMBER OF YEARS MAINTAINED

Item	Period (Years)
(1) Purchase Orders from Customer	5
(2) Raw Material Purchase Orders	5
(3) Material Certifications	5
(4) Special Process Certifications: Heat Treatment, Plating, etc.	5
(5) Inspection Reports	5

15.0 SPECIAL PROCESS CONTROL

1. Special processing such as heat treating, Magnaflux, plating, etc. is to be performed by outside vendors.

2. All special processing must be certified to specifications by the vendor performing the process. In addition, parts must be inspected by our Inspection Department in accordance with approved statistical sampling methods or as specified by the customer.

APPENDIX
C

Queuing Results

Notation:

λ = Arrival rate in the system

μ = Service rate of each server

n = Number of customers in queuing system

P_n = Probability that exactly n customers are in queuing system

L = Expected number of customers in queuing system

L_q = Expected number of customers in queue

W = Expected waiting time in the system (including service time)

   = L/λ

W_q = Expected waiting time in queue

   = L_q/λ

ρ = Utilization ratio

λ_e = Effective arrival rate (finite queue models) = λP_n if there are only n spaces in the system.

Single server (Poisson arrival, exponential service, M/M/1):

$$\rho = \frac{\lambda}{\mu}$$

$$P_0 = 1 - \rho$$

$$P_n = \rho^n P_0$$

$$L = \frac{\rho}{1 - \rho}$$

$$L_q = \frac{\lambda^2}{\mu(\mu - \lambda)}$$

Multi-server model (Poisson arrival, exponential service):

1. Unlimited queue (M/M/C):

$$P_0 = \frac{1}{\left[\sum_{n=0}^{C-1} \frac{(\lambda/\mu)^n}{n!} + \frac{(\lambda/\mu)^C}{C!} \cdot \frac{1}{1 - (\lambda/C\mu)} \right]}$$

$$P_n = \frac{\left(\frac{\lambda}{\mu}\right)^n P_0}{n!} \quad \text{if } 0 \le n \le c$$

$$\frac{\left(\frac{\lambda}{\mu}\right)^n P_0}{C!C^{n-c}} \quad \text{if } n \ge c$$

$$L_q = \frac{P_0 (\lambda/\mu)^C \rho}{C!(1 - \rho)^2} \quad \text{where } \rho = \frac{\lambda}{C\mu} < 1$$

$$L = L_q + \frac{\lambda}{\mu}$$

2. Queue size limited to K (M/M/C/K):

$$P_0 = \frac{1}{\left[\sum_{n=0}^{C-1} \frac{(\lambda/\mu)^n}{n!} + \frac{(\lambda/\mu)^C}{C!} \cdot \left(\frac{1 - \rho^{K-C+1}}{1 - \rho} \right) \right]} \quad \text{where } \rho = \lambda/C\mu \le 1$$

$$P_n = \begin{cases} \dfrac{\left(\frac{\lambda}{\mu}\right)^n}{n!} P_0 & \text{if } 0 \le n \le c \\ \dfrac{(\lambda/\mu)^n}{C!C^{n-c}} P_0 & \text{if } n > c \end{cases}$$

$$L_q = \frac{(\lambda\mu)^C \rho P_0}{C!(1 - \rho)^2} \{1 - [K - C)(1 - \rho) + 1]\rho^{K-C}\}$$

$$L = L_q + \frac{\lambda}{\mu}(1 - P_K)$$

APPENDIX
D

Computer Programs

All the programs listed in Appendix D are available on a disk. These programs can be run on IBM-PC® or compatible machines (with 128K RAM memory and a printer), except for two programs—2 and 7, which are in FORTRAN WATFIV and should be uploaded and run on an IBM mainframe system. The programs listed in Table D.1 can be loaded by either their given names, e.g., HEUR#1, or by their corresponding number, e.g., 1.

To load a program, please follow one of the appropriate procedures listed below.

For a single drive system,

1. Insert the DOS disk in drive A.
2. Turn on the computer.
3. After receiving the system prompt, insert the BASIC disk in the drive.
4. Type GW-BASIC for MS-DOS or BASICA for PC-DOS.
5. After the BASIC is loaded, remove the BASIC from the drive and replace it with the program disk.
6. Load the desired program by typing RUN "MENU".

For a two-disk drive system,

1. Insert the DOS disk in drive A.
2. Turn on the computer. Remove the DOS disk from drive A after the system has been loaded.
3. Insert the program disk in drive A.
4. Insert the BASIC disk in drive B.
5. Type B;GWBASIC for MS-DOS or B;BASICA for PC-DOS.
6. To load the menu program after the BASIC has been loaded, type RUN "MENU".

IBM-PC is a registered trademark of International Business Machines Corporation.

GW-BASIC is a registered trademark of Microsoft, Inc.

BetterBASIC is a registered trademark of Summit Software Technology, Inc.

Table D.1 Available computer programs

Name of Program Disk	Language	Name of Program in Appendix
1. HEUR#1	GW-BASIC®	Heuristic approach, basic facility location, Chapter 2
2. HEUR#2	FORTRAN WATFIV	Heuristic method for solving problems with fixed cost, Chapter 3
3. SNGL-RDC	GW-BASIC	Single facility, rectilinear distance model, Chapter 4
4. SNGL-QAD	GW-BASIC	Single facility, quadratic cost model, Chapter 4
5. MULT-FAC	GW-BASIC	Multiple facility, quadratic cost model, Chapter 4
6. RATE-IRR	GW-BASIC	Rate of return, Chapter 5
7. GROUPING	FORTRAN WATFIV	Job grouping, Chapter 5
8. ASMB-LCR	GW-BASIC	Assembly line, balancing, largest candidate rule, Chapter 7
9. ASMB-RPW	GW-BASIC	Assembly line, balancing, ranked proportional weighting method, Chapter 7
10. Q-SINGLE	GW-BASIC	Queueing: SINGLE-server M/M/1, Chapter 11
11. Q-MULTI	GW-BASIC	Queueing: MULTI-server M/M/C/K, Chapter 11
12. COMLAD*	BetterBASIC®	COMputer LAyout, Design, Chapter 13
13. DATGUIDE	GW-BASIC	DATa GUIDE to computer programs, Chapter 14
14. FIN ANAL	GW-BASIC	FINancial ANALysis, Chapter 15

*The program on the disk is an EXECUTABLE (EXE) file. It should be run after loading the program disk by typing COMLAD, following the system prompt.

```
10 '
20 '
30 '**********************************************************
35 '*                                                        *
40 '*                    HEURISTIC APPROACH                  *
50 '*            Basic Facility Location Problem             *
60 '*                      Chapter 2                         *
70 '*                                                        *
80 '**********************************************************
90 '
100 CLS
110 '
120 OPTION BASE 1
130 DIM SAVEIT(100,75),SUM(75),DEMAND(100),INDEX(10),COST(50,50)
140 DIM ADEMAND(100)
150 PRINT "REFER TO SECTION 2.6 OF THE BOOK TO GET A DESCRIPTION"
160 PRINT "OF THE PROGRAM AND HOW TO USE THE DATA STATEMENTS. TO"
170 PRINT "GET TO THE FIRST DATA STATEMENT, TYPE 'LIST 1260'
AFTER"
180 PRINT "THIS PROGRAM HAS FINISHED RUNNING. HIT RETURN TO
PROCEED."
190 INPUT Y$
200 READ IROW,ICOL,IMC
210 PRINT "////////// INPUT VALUES \\\\\\\\\\\\\\\":PRINT
220 PRINT "ROWS = ";IROW;" COLUMNS = ";ICOL;" FACILITIES = ";IMC
230 PRINT : PRINT : PRINT "DEMAND FOR EACH LOCATION IS:"
240 PRINT "  LOCATION           DEMAND"
250 FOR I = 1 TO IROW
260 READ ADEMAND(I)
270 PRINT "      ";I;"               ";ADEMAND(I)
280 NEXT I
290 PRINT : PRINT : PRINT "  COST TABLES"
300 FOR I=1 TO IROW
310 FOR J= 1 TO ICOL
320 READ COST(I,J)
330 PRINT COST(I,J);
340 NEXT J
350 PRINT
360 NEXT I
370 PRINT : PRINT
380 FOR I = 1 TO IROW
390 FOR J = 1 TO ICOL
400 COST(I,J) = COST(I,J) * ADEMAND(I)
410 NEXT J
420 NEXT I
430 '
440 K=1
450 FOR J=1 TO ICOL
460 SUM(J)=0
470 NEXT J
480 '
490 FOR J= 1 TO ICOL
500 FOR I=1 TO IROW
510 SUM(J)=SUM(J)+COST(I,J)
520 NEXT I
530 NEXT J
540 '
550 MIN=9999
560 FOR J=1 TO ICOL
570 IF SUM(J)>MIN GOTO 600
```

```
580 MIN=SUM(J)
590 INDEX(K) = J
600 NEXT J
610 '
620 FOR I=1 TO IROW
630 DEMAND(I)=INDEX(K)
640 NEXT I
650 '
660 IF K => IMC GOTO 1120
670 FOR I= 1 TO IROW
680 FOR J= 1 TO ICOL
690 SAVEIT(I,J)=0
700 NEXT J
710 NEXT I
720 '
730 FOR I= 1 TO IROW
740 FOR J= 1 TO ICOL
750 FOR JJ=1 TO K
760 IF J=INDEX(JJ) GOTO 820
770 NEXT JJ
780 II= DEMAND(I)
790 DIF=COST(I,J)-COST(I,II)
800 IF DIF>0 GOTO 820
810 SAVEIT(I,J)=-DIF
820 NEXT J
830 NEXT I
840 '
850 '
860 FOR J=1 TO ICOL
870 SUM(J)=0
880 NEXT J
890 '
900 FOR J=1 TO ICOL
910 FOR I=1 TO IROW
920 SUM(J)=SUM(J)+SAVEIT(I,J)
930 NEXT I
940 NEXT J
950 '
960 MAX=0
970 '
980 FOR J=1 TO ICOL
990 IF SUM(J) < MAX GOTO 1020
1000 INDEX(K)=J
1010 MAX=SUM(J)
1020 NEXT J
1030 '
1040 IF MAX=0 GOTO 1120
1050 FOR I=1 TO IROW
1060 J=INDEX(K)
1070 IF SAVEIT(I,J)=0 GOTO 1090
1080 DEMAND(I)=J
1090 NEXT I
1100 K = K + 1
1110 GOTO 660
1120 TCOST=0
1130 K = K - 1
1140 FOR I=1 TO IROW
1150 J=DEMAND(I)
1160 TCOST=TCOST+COST(I,J)
```

```
1170 NEXT I
1180 '
1190 INPUT"HIT ENTER TO SEE THE RESULTS";A$
1200 CLS
1210 PRINT:PRINT"*************** T H E   F I N A L  R E S U L T
**
1220 PRINT: PRINT
1230 PRINT "DEMAND ASSIGNMENTS ARE:"
1240 PRINT "DEMAND FROM------ASSIGNED TO"
1250 FOR I=1 TO IROW
1260 PRINT "   ";I;"                  ";DEMAND(I)
1270 NEXT I
1280 PRINT : PRINT
1290 PRINT "COST OF THIS POLICY IS ";TCOST;"."
1300 '
1310 DATA 5,5,3
1320 DATA 100,50,150,200,300
1330 DATA 5,3,2,8,5
1340 DATA 3,5,2,6,7
1350 DATA 5,2,0,1,0
1360 DATA 2,1,8,2,3
1370 DATA 3,2,4,0,4
```

```
     // EXEC WATFIV
     //GO.SYSIN DD *
     $JOB
     C***************************************************************
     C                                                             *
     C                                                             *
     C          Chapter 3:   HEURISTIC METHOD FOR SOLVING PROBLEMS *
     C          ************************************************** *
     C                       WITH FIXED COST                       *
     C                       ****************                      *
     C     (This FORTRAN program was written for WATFIV compiler)  *
     C                                                             *
     C***************************************************************
           INTEGER NOCUST,NOFAC,NOLOC,PC,COUNT,IDMND(60)
           INTEGER MSAV(50),EXPLIF(50),N
           REAL FINMAT(60,50),LOCST(60,50),SAVMAT(60,50),SAVNET(50)
           REAL
     FXCM(50),FXCMB(50),FINFC(50),RATE,ASSIGN(60),MAXSAV,TOT(60)
           REAL
     ASSIGB(60),TOTSAV(50),DEMCOS(50),TOTCOS(50),VARCM(60,50)
        1    FORMAT(4(I2))
        2    FORMAT(F7.2)
        3    FORMAT(50(F7.2))
        4    FORMAT(60(I4))
        5    FORMAT(50(I2))
           READ1,NOCUST,NOLOC,NOFAC,PC
           READ4,(IDMND(I),I = 1,NOCUST)
           DO 12 I = 1, NOCUST
              READ3,(LOCST(I,J),J=1,NOLOC)
       12  CONTINUE
              READ3,(FXCMB(J),J=1,NOLOC)
              READ5,(EXPLIF(J),J=1,NOLOC)
           READ2,RATE
           CALL
     OUT1(NOCUST,NOLOC,NOFAC,PC,IDMND,FXCM,EXPLIF,LOCST,FXCMB,
        1          RATE)
           CALL
     STEP1(RATE,EXPLIF,NOLOC,NOCUST,SAVMAT,LOCST,IDMND,FINMAT,PC,
        1          FXCMB,FXCM)
           IF(NOFAC.EQ.1) GOTO 16
           CALL
     STEP2(NOLOC,SAVMAT,FINMAT,MSAV,NOCUST,SAVNET,IDMND,PC,
        1              COUNT)
           CALL
     STEP3(SAVNET,FXCM,NOCUST,NOLOC,FINMAT,MSAV,PC,COUNT,
        1          IDMND)
           CALL
     STEP4(SAVNET,NOLOC,IDMND,MSAV,FINMAT,NOCUST,PC,COUNT,
        1              SAVMAT,VARCM,FXCM,TOTSAV)
           CALL
     STEP5(FINMAT,IDMND,NOLOC,NOCUST,SAVMAT,VARCM,ASSIGN,
        1              FXCM,MSAV,TOTSAV,SAVNET,COUNT,MAXSAV,PC)
       20    IF(MAXSAV.LE.0.OR.COUNT.EQ.NOFAC) GO TO 17
           CALL
     STEP4(SAVNET,NOLOC,IDMND,MSAV,FINMAT,NOCUST,PC,COUNT,
        1              SAVMAT,VARCM,FXCM,TOTSAV)
           CALL
     STEP5(FINMAT,IDMND,NOLOC,NOCUST,SAVMAT,VARCM,ASSIGN,
        1              FXCM,MSAV,TOTSAV,SAVNET,COUNT,MAXSAV,PC)
```

```
                      GOTO 20
      17    CALL OUT6(FINMAT,NOLOC,NOCUST,FXCM,SAVMAT,TOTSAV,
         1             MSAV,COUNT,DEMCOS,ASSIGN,TOT)
            N     = COUNT
            COUNT = COUNT - 1
            DO 19 I = 1 , COUNT
                CALL STEP6(SAVMAT,NOLOC,IDMND,MSAV,FINMAT,NOCUST,PC,
         C                 TOTSAV,FXCM,N,SAVNET)
                N = N - 1
      19    CONTINUE
            GO TO 18
      16   CALL HEUR(FINMAT,NOLOC,NOCUST,FXCM,SAVNET,SAVMAT,TOTSAV,
         CMSAV,COUNT,DEMCOS,IDMND)
           GO TO 40
      18   CALL OUT7(NOLOC,NOCUST,TOT,ASSIGN)
      40   STOP
           END
           FUNCTION UAC(I,K,N)
           INTEGER N
           REAL I,K
           UAC = -1 *((I *(1 + I)**N) / ((1+I)**N - 1)) * K
           RETURN
           END
           SUBROUTINE
HEUR(FINMAT,NOLOC,NOCUST,FXCM,SAVNET,SAVMAT,TOTSAV,
        CMSAV,COUNT,DEMCOS,IDMND)
           INTEGER NOFAC,NOLOC,NOCUST,MSAV(50),ID(50),COUNT,IDMND(60)
           REAL    DEMCOS(50),SAVMAT(60,50),MINSAV
           REAL    FINMAT(60,50),FXCM(50),SAVNET(50),TOTSAV(50)
           DO 1 J = 1,NOLOC
              SUM = 0
              DO 2 I = 1 ,NOCUST
                 SUM = SUM + SAVMAT(I,J)
      2       CONTINUE
              TOTSAV(J) = SUM
      1    CONTINUE
           MINSAV = TOTSAV(1)
           MSAV(1) = 1
           DO 3 J = 2 , NOLOC
              IF(TOTSAV(J).GE.MINSAV) GO TO 3
              MSAV(1) = J
              MINSAV = TOTSAV(J)
      3    CONTINUE
           DO 4 J = 1,NOLOC
              DO 5 I = 1 ,NOCUST
                 IF(J.NE.MSAV(1)) GO TO 8
                    FINMAT(I,J) = IDMND(I)
                    GO TO 5
      8          FINMAT(I,J) = 0
      5       CONTINUE
      4    CONTINUE
           CALL OUT8(FINMAT,MSAV,NOLOC,NOCUST,IDMND,TOTSAV)
           RETURN
           END
C******************************************************************
C                                                                *
C                        STEP 1                                  *
C                        ******                                  *
C******************************************************************
```

```
        SUBROUTINE
STEP1(RATE,EXPLIF,NOLOC,NOCUST,SAVMAT,LOCST,IDMND,FINM
    CAT,PC,FXCMB,FXCM)
        INTEGER NOCUST,NOLOC,PC,EXPLIF(50),IDMND(60),N
        REAL    FINMAT(60,50),LOCST(60,50),SAVMAT(60,50)
        REAL FXCM(50),FXCMB(50),MINSAV,RATE,K,MAXF,MAX1,MAX2
        DO 1 I = 1,NOCUST
            DO 2 J = 1 ,NOLOC
                SAVMAT(I,J) = LOCST(I,J) * IDMND(I)
    2       CONTINUE
    1   CONTINUE
        DO 3 J = 1 , NOLOC
            N = EXPLIF(J)
            K = FXCMB(J)
            FXCM(J) = UAC(RATE,K,N)
    3   CONTINUE
        MAXF = 0
        MAX1 = 0
        MAX2 = 0
        DO 100 J =1 ,NOLOC
            IF(FXCM(J).LE.MAXF) GO TO 102
            MAXF = FXCM(J)
   102      DO 101 I = 1 ,NOCUST
                IF(SAVMAT(I,J).LE.MAX1) GO TO 101
                MAX2 = MAX1
                MAX1 = SAVMAT(I,J)
   101      CONTINUE
   100  CONTINUE
        TOT = MAXF + MAX1 + MAX2
        DO 21 I = 1, NOCUST
            DO 22 J =1 ,NOLOC
                IF(SAVMAT(I,J).GE.0) GO TO 22
                SAVMAT(I,J) = TOT
    22      CONTINUE
    21  CONTINUE
        IF(PC.NE.1) GO TO 4
            CALL OUT2(SAVMAT,NOLOC,NOCUST)
    4   RETURN
        END
C*************************************************************
C                   STEP 2
C                   ******
C*************************************************************
        SUBROUTINE
STEP2(NOLOC,SAVMAT,FINMAT,MSAV,NOCUST,SAVNET,IDMND,PC,
    CCOUNT)
        INTEGER IDMND(60),PC,MSAV(50),COUNT,MSAV1(50)
        REAL FINMAT(60,50),SAVMAT(60,50),SAVNET(50)
        DO 1 I = 1 , NOCUST
            MINSAV = SAVMAT(I,1)
            MSAV(I) = 1
            DO 2 J = 2, NOLOC
                IF(SAVMAT(I,J).GE.MINSAV) GO TO 2
                MSAV(I) = J
                MINSAV = SAVMAT(I,J)
    2       CONTINUE
    1   CONTINUE
```

```
          DO 3 I = 1 , NOCUST
             MINSAV = SAVMAT(I,1)
             N = MSAV(I)
             MSAV1(I) = 1
             DO 4 J = 1, NOLOC
                IF(J.EQ.N.OR.SAVMAT(I,J).GE.MINSAV) GO TO 4
                   MSAV1(I) = J
                   MINSAV = SAVMAT(I,J)
    4        FINMAT(I,J) = 0
             N1 = MSAV1(I)
             FINMAT(I,N) = SAVMAT(I,N1) - SAVMAT(I,N)
    3     CONTINUE
          RETURN
          END
C****************************************************************
C                    STEP 3                                     *
C                    ******                                     *
C                                                               *
C****************************************************************
          SUBROUTINE
STEP3(SAVNET,FXCM,NOCUST,NOLOC,FINMAT,MSAV,PC,COUNT,
       CIDMND)
          INTEGER NOCUST,NOLOC,PC,COUNT,MSAV(50),IDMND(60)
          REAL FINMAT(60,50),SAVNET(50)
          REAL FXCM(50),MINSAV
          DO 1 J = 1 ,NOLOC
             SAVNET(J) = 0
             DO 2 I = 1 ,NOCUST
                N = MSAV(I)
                IF(J.NE.N) GOTO 2
                   SAVNET(J) = SAVNET(J) + FINMAT(I,J)
    2        CONTINUE
             SAVNET(J) = SAVNET(J) + FXCM(J)
    1     CONTINUE
          IF(PC.EQ.0) GO TO 10
             CALL OUT3(FINMAT,NOLOC,NOCUST,FXCM,SAVNET)
   10     DO 4 I = 1, NOCUST
             MSAV(I) = 0
    4     CONTINUE
          MINSAV = SAVNET(1)
          DO 5 J = 2 , NOLOC
             IF(SAVNET(J).LE.MINSAV) GO TO 5
                MSAV(1) = J
                MINSAV = SAVNET(J)
    5     CONTINUE
          COUNT = 1
          N = MSAV(1)
          DO 6 J = 1, NOLOC
             DO 7 I = 1, NOCUST
             FINMAT(I,N) = IDMND(I)
             IF(J.EQ.N) GO TO 7
                FINMAT(I,J) = 0
    7        CONTINUE
    6     CONTINUE
          IF(PC.EQ.0) GO TO 3
             CALL OUT4(FINMAT,NOLOC,NOCUST,FXCM,SAVNET)
    3     RETURN
          END
```

```
C**************************************************************
C                                                            *
C                    STEP  4                                 *
C                    ******                                  *
C                                                            *
C**************************************************************
      SUBROUTINE
STEP4(SAVNET,NOLOC,IDMND,MSAV,FINMAT,NOCUST,PC,COUNT,
     CSAVMAT,VARCM,FXCM,TOTSAV)
      INTEGER IDMND(60),NOLOC,NOCUST,COUNT,MSAV(50),PC,MSAV1(50)
      REAL    FINMAT(60,50),SAVNET(50),SAVMAT(60,50),VARCM(60,50)
      REAL    FXCM(50),TOTSAV(50)
      DO 1 I = 1, NOCUST
         DO 2 J = 1, NOLOC
            IF(FINMAT(I,J).EQ.0) GO TO 2
               MSAV1(I) = J
  2      CONTINUE
  1   CONTINUE
      DO 5 I = 1, NOCUST
         N = MSAV1(I)
         DO 6 J = 1, NOLOC
            IF(SAVMAT(I,J).GT.SAVMAT(I,N)) GO TO 8
               VARCM(I,J)  = SAVMAT(I,N) - SAVMAT(I,J)
               GOTO 6
  8            VARCM(I,J) = -1
  6      CONTINUE
         VARCM(I,N) = 0
  5   CONTINUE
      DO 3 J = 1 , NOLOC
         DO 7 K = 1, COUNT
            N = MSAV(K)
            IF (J.NE.N) GO TO 7
               TOTSAV(J) = 0
               SAVNET(J) = 0
               GO TO 3
  7      CONTINUE
         TOTSAV(J) = 0
         SAVNET(J) = 0
         DO 4 I = 1 ,NOCUST
            IF(VARCM(I,J).LE.0) GO TO 4
               TOTSAV(J) = TOTSAV(J) + VARCM(I,J)
  4      CONTINUE
         SAVNET(J) = FXCM(J) + TOTSAV(J)
  3   CONTINUE
      IF(PC.NE.1) GO TO 9
         CALL OUT5(VARCM,NOLOC,NOCUST,FXCM,SAVNET,TOTSAV)
  9      RETURN
         END
C**************************************************************
C
C                    STEP  5                                 *
C                    ******                                  *
C**************************************************************
      SUBROUTINE
STEP5(FINMAT,IDMND,NOLOC,NOCUST,SAVMAT,VARCM,ASSIGN,
     C FXCM,MSAV,TOTSAV,SAVNET,COUNT,MAXSAV,PC)
      INTEGER NOCUST,NOLOC,PC,MSAV(50),IDMND(60),COUNT
      REAL FINMAT(60,50),SAVMAT(60,50),SAVNET(50),TOTSAV(50)
      REAL FXCM(50),MAXSAV,ASSIGN(60),VARCM(60,50)
      MAXSAV = SAVNET(1)
```

```
              N = MSAV(COUNT)
              COUNT = COUNT + 1
              MSAV(COUNT) = 1
              DO 5 J = 2 , NOLOC
                  IF(J.EQ.N.OR.SAVNET(J).LE.MAXSAV) GO TO 5
                      MAXSAV = SAVNET(J)
                      MSAV(COUNT) = J
        5     CONTINUE
              N = MSAV(COUNT)
              L1 = COUNT - 1
              L = MSAV(L1)
              DO 6 I = 1 , NOCUST
                  IF (VARCM(I,N).LT.0)  GO TO 6
                      FINMAT(I,N) = IDMND(I)
                  IF(FINMAT(I,L).LE.0) GO TO 6
                      FINMAT(I,L) = 0
        6     CONTINUE
              IF(PC.NE.1) GO TO 9
                  CALL OUT4(FINMAT,NOLOC,NOCUST,FXCM,SAVNET)
        9         RETURN
                  END
C*****************************************************************
C                                                                *
C                   STEP 6                                        *
C                   ******                                       *
C                                                                *
C*****************************************************************
          SUBROUTINE STEP6(SAVMAT,NOLOC,IDMND,MSAV,FINMAT,NOCUST,PC,
        C TOTSAV,FXCM,N,SAVNET)
          INTEGER IDMND(60),NOLOC,NOCUST,MSAV(50),N,PC
          REAL FINMAT(60,50),SAVMAT(60,50),SAVNET(50)
          REAL FXCM(50),MINSAV,TOTSAV(50)
          K = MSAV(N)
          L1 = N - 1
          IF (L1.EQ.0.) GO TO 9
          L = MSAV(L1)
          TOTSAV(K) = 0
          TOTSAV(L) = 0
          DO 1 I = 1, NOCUST
              IF(FINMAT(I,L).LT.0) GO TO 1
                  TOTSAV(K) = TOTSAV(K) + SAVMAT(I,K) - SAVMAT(I,L)
        1     CONTINUE
              I1 = TOTSAV(K) + FXCM(L)
          IF (I1.LE.0) GO TO 2
              DO 3 I = 1 , NOCUST
                  IF(FINMAT(I,L).LT.0) GO TO 4
                      FINMAT(I,K) = IDMND(I)
                      GO TO 3
        4         FINMAT(I,J) = 0
        3     CONTINUE
              MSAV(N) = 0
        2 TOTSAV(K) = 0
          TOTSAV(L) = 0
          DO 5 I = 1, NOCUST
              IF(FINMAT(I,K).LT.0) GO TO 5
                  TOTSAV(L) = TOTSAV(L) + SAVMAT(I,L) - SAVMAT(I,K)
        5     CONTINUE
              I1 = TOTSAV(L) + FXCM(K)
          IF (I1.LE.0) GO TO 6
              DO 7 I = 1 , NOCUST
```

```
              IF(FINMAT(I,K).LT.0) GO TO 8
                  FINMAT(I,L) = IDMND(I)
                  GO TO 7
      8           FINMAT(I,J) = 0
      7       CONTINUE
      6    IF(PC.NE.1) GO TO 9
             CALL OUT5(FINMAT,NOLOC,NOCUST,FXCM,SAVNET,MSAV)
      9    RETURN
           END
C****************************************************************
C                                                              *
C                    OUTPUT 1                                  *
C                    ********                                  *
C                                                              *
C****************************************************************
        SUBROUTINE
     OUT1(NOCUST,NOLOC,NOFAC,PC,IDMND,FXCM,EXPLIF,LOCST,
        CFXCMB,RATE)
           INTEGER NOCUST,NOLOC,NOFAC,PC,IDMND(60),EXPLIF(50)
           INTEGER ID(50)
           REAL FXCMB(50),LOCST(60,50),FXCM(50),RATE
           PRINT14,'THIS IS INPUT DATA '
           PRINT13,NOCUST,NOLOC,NOFAC,PC
             PRINT16,(IDMND(I),I=1,NOCUST)
           DO 21 I = 1, NOCUST
             PRINT15,(LOCST(I,J),J=1,NOLOC)
      21   CONTINUE
               PRINT15,(FXCMB(J),J=1,NOLOC)
             PRINT17,(EXPLIF(J),J=1,NOLOC)
           PRINT15,RATE
           DO 1 I = 1, 50
             ID(I) = I
      1    CONTINUE
           PRINT13,NOCUST,NOLOC,NOFAC,PC
           PRINT10,'LOCATION',(ID(I),I=1,NOLOC)
           PRINT12,(EXPLIF(J),J=1,NOLOC)
           PRINT8,'TRANSPORTATION COST AND DEMAND'
           PRINT81,'=============================='
           PRINT9,'CUSTOMER','LOCATION','DEMAND'
           PRINT7,(ID(I),I=1,NOLOC)
           DO 2 I = 1 ,NOCUST
             PRINT6,I,(LOCST(I,J),J=1,NOLOC),IDMND(I)
      2    CONTINUE
           PRINT11,(FXCMB(J),J=1,NOLOC)
C
C                               |  THE INTEGER NO. WILL CHANGE TO THE
SAME AS
C                               U  NUMBER OF LOCATION.
      6    FORMAT('-',30X,I3,5X,5(F7.2),I4)
      7    FORMAT('-',38X,10(I3,4X))
      8    FORMAT('-',40X,A30)
     81    FORMAT(' ',40X,A30)
      9    FORMAT('-',22X,A16,20X,A8,26X,A6)
     10    FORMAT('-',30X,A14,10(I3,4X))
     11    FORMAT('-',23X,'FIXED COST   ',5X,10(F7.2))
     12    FORMAT('-',35X,'EXPECTED LIFE',10(I3,4X),//)
     13    FORMAT('1',//,23X,'CUSTOMER NUMBERS = ',I3,5X,'LOCATION
NUMBERS
     C= ',I3,5X,'FACILITY NUMBERS = ',I3,5X,'PRINT NUMBER = ',I2)
```

```
   14   FORMAT('1',//,20X,A30)
   15   FORMAT('-',//,30X,50(F7.2))
   16   FORMAT('-',//,30X,60(I4))
   17   FORMAT('-',//,30X,50(I2))
        RETURN
        END
C****************************************************************
C
C                        OUTPUT 2                              *
C                        *******                               *
C                                                              *
C****************************************************************
        SUBROUTINE OUT2(SAVMAT,NOLOC,NOCUST)
        INTEGER NOLOC,NOCUST,ID(50)
        REAL SAVMAT(60,50)
        PRINT8,'DEMAND COST    '
        PRINT81,'=========== '
        PRINT9,'CUSTOMER','LOCATION'
        DO 1 I = 1 ,NOLOC
           ID(I) = I
   1    CONTINUE
        PRINT10,(ID(I),I= 1,NOLOC)
        DO 2 I = 1, NOCUST
   .       PRINT11,I,(SAVMAT(I,J),J = 1, NOLOC)
   2    CONTINUE
   8    FORMAT('1',//,41X,A20)
   81   FORMAT(' ',40X,A20)
   9    FORMAT('-',17X,A16,20X,A8,26X,A6)
   10   FORMAT('-',38X,10(I3,4X))
C
C                        | THE INTEGER NO. WILL CHANGE TO THE
SAME AS
C                        U  NUMBER OF LOCATION.
   11   FORMAT('-',30X,I3,6X,5(F7.2),I4)
        RETURN
        END
C****************************************************************
C        OUTPUT 3
C        ********
C                                                              *
C****************************************************************
        SUBROUTINE OUT3(FINMAT,NOLOC,NOCUST,FXCM,SAVNET)
        INTEGER NOFAC,NOLOC,NOCUST,ID(50)
        REAL    FINMAT(60,50),FXCM(50),SAVNET(50)
        DO 1 I = 1 ,NOLOC
           ID(I) = I
   1    CONTINUE
        PRINT8,'MINIMUM SAVINGS '
        PRINT81,'=============== '
        PRINT9,'CUSTOMER','LOCATION'
        PRINT10,(ID(I),I= 1,NOLOC)
        DO 2 I = 1, NOCUST
           PRINT11,I,(FINMAT(I,J),J = 1, NOLOC)
   2    CONTINUE
        PRINT7,'FIXED COST',(FXCM(J),J=1,NOLOC)
        PRINT7,'NET SAVINGS',(SAVNET(J),J=1,NOLOC)
   7    FORMAT('-',17X,A16,5X,10(F7.2))
   8    FORMAT('1',//,41X,A20)
   81   FORMAT(' ',40X,A20)
```

```
      9   FORMAT(' ',17X,A16,20X,A8,26X,A6)
     10   FORMAT('-',38X,10(I3,4X))
C
C                              | THE INTEGER NO. WILL CHANGE TO THE
SAME AS
C                              U  NUMBER OF LOCATION.
     11   FORMAT('-',30X,I3,5X,5(F7.2),I4)
          RETURN
          END
C****************************************************************
C                                                              *
C                    OUTPUT 4                                  *
C                    ********                                  *
C                                                              *
C****************************************************************
          SUBROUTINE OUT4(FINMAT,NOLOC,NOCUST,FXCM,SAVNET)
          INTEGER NOLOC,NOCUST,ID(50)
          REAL    FINMAT(60,50),FXCM(50),SAVNET(50)
          DO 1 I = 1 ,NOLOC
             ID(I) = I
      1   CONTINUE
          PRINT8,'DEMAND ASSIGNMENT'
          PRINT81,'================='
          PRINT9,'CUSTOMER','LOCATION'
          PRINT10,(ID(I),I= 1,NOLOC)
          DO 2 I = 1, NOCUST
             PRINT11,I,(FINMAT(I,J),J = 1, NOLOC)
      2   CONTINUE
      8   FORMAT('1',//,41X,A20)
     81   FORMAT(' ',40X,A20)
      9   FORMAT('-',17X,A16,20X,A8,26X,A6)
     10   FORMAT('-',38X,10(I3,4X))
C
C                              | THE INTEGER NO. WILL CHANGE TO THE
SAME AS
C                              U  NUMBER OF LOCATION.
     11   FORMAT('-',30X,I3,5X,5(F7.2),I4)
          RETURN
          END
C****************************************************************
C                    OUTPUT 5
C                    ********                                  *
C                                                              *
C****************************************************************
          SUBROUTINE OUT5(VARCM,NOLOC,NOCUST,FXCM,SAVNET,TOTSAV)
          INTEGER NOFAC,NOLOC,NOCUST,ID(50)
          REAL    VARCM(60,50),SAVNET(50),TOTSAV(50),FXCM(50)
          DO 1 I = 1 ,NOLOC
             ID(I) = I
      1   CONTINUE
          PRINT8,'SAVINGS TABLE'
          PRINT81,'============'
          PRINT9,'CUSTOMER','LOCATION'
          PRINT10,(ID(I),I= 1,NOLOC)
          DO 3 I = 1, NOCUST
             PRINT11,I,(VARCM(I,J),J = 1, NOLOC)
      3   CONTINUE
          PRINT12,(TOTSAV(J),J=1,NOLOC)
          PRINT13,(FXCM(J),J=1,NOLOC)
          PRINT14,(SAVNET(J),J=1,NOLOC)
```

```
 8   FORMAT('1',//,41X,A20)
81   FORMAT(' ',40X,A20)
 9   FORMAT('-',17X,A16,20X,A8,26X,A6)
10   FORMAT('-',38X,10(I3,4X))
C
C                              | THE INTEGER NO. WILL CHANGE TO THE
SAME AS
C                                U  NUMBER OF LOCATION.
11   FORMAT('-',30X,I3,5X,10(F7.2),I4)
C
C    THE INTEGER NO. IDICATED HERE       |
C    HAVE TO CHANGE TO NO. OF LOCATION.  U
12   FORMAT('-',17X,'   TOTAL SAVINGS',5X,5(F7.2),I4)
13   FORMAT('-',17X,'      FIXED COST',5X,5(F7.2),I4)
14   FORMAT('-',17X,'     NET SAVINGS',5X,5(F9.2),I4)
     RETURN
     END
C*****************************************************************
C                                                               *
C                    OUTPUT 6                                   *
C                    ********                                   *
C                                                               *
C*****************************************************************
     SUBROUTINE OUT6(FINMAT,NOLOC,NOCUST,FXCM,SAVMAT,TOTSAV,
   CMSAV,COUNT,DEMCOS,ASSIGN,TOT)
     INTEGER NOFAC,NOLOC,NOCUST,MSAV(50),ID(50),COUNT
     REAL     DEMCOS(50),SAVMAT(60,50),TOT(60)
     REAL
FINMAT(60,50),FXCM(50),SAVNET(50),ASSIGN(60),TOTSAV(50)
     DO 21 J = 1 ,NOLOC
         DEMCOS(J) = 0
         TOTSAV(J) = 0
         DO 22 I = 1, NOCUST
            IF(FINMAT(I,J).EQ.0) GO TO 22
               FINMAT(I,J) = SAVMAT(I,J)
               DEMCOS(J) = DEMCOS(J) + FINMAT(I,J)
               ASSIGN(I) = J
22       CONTINUE
         IF(DEMCOS(J).EQ.0) GOTO 23
            TOTSAV(J) = DEMCOS(J) - FXCM(J)
            GO TO 21
23       FXCM(J) = 0
21   CONTINUE
     DO 1 I = 1 ,NOLOC
         ID(I) = I
         FXCM(I) = FXCM(I) * (-1)
1    CONTINUE
     DO 4 I = 1, NOCUST
         J = ASSIGN(I)
         TOT(I) = SAVMAT(I,J)
4    CONTINUE
     L  = NOCUST + 1
     L1 = L + 1
     L2 = L1 + 1
     TOT(L) = 0
     TOT(L1) = 0
     TOT(L2) = 0
     N = MSAV(COUNT)
     DO 5 J = 1, NOLOC
```

```
            IF(DEMCOS(J).EQ.0) GO TO 5
                TOT(L) = TOT(L) + DEMCOS(J)
                TOT(L1) = TOT(L1) + FXCM(J)
                TOT(L2) = TOT(L2) + TOTSAV(J)
      5     CONTINUE
            PRINT8,'TOTAL COST      '
            PRINT81,'=========      '
            PRINT9,'CUSTOMER','LOCATION','TOTAL'
            PRINT10,(ID(I),I= 1,NOLOC)
            DO 3 I = 1, NOCUST
                PRINT11,I,(FINMAT(I,J),J = 1, NOLOC),TOT(I)
      3     CONTINUE
            PRINT7,'DEMAND COST',(DEMCOS(J),J=1,NOLOC),TOT(L)
            PRINT7,'FIXED COST',(FXCM(J),J=1,NOLOC),TOT(L1)
            PRINT7,'TOTAL COST ',(TOTSAV(J),J=1,NOLOC),TOT(L2)
C
C                              | THE INTEGER NO. WILL CHANGE TO THE
SAME AS
C                              U  NUMBER OF LOCATION.
      7     FORMAT('-',17X,A16,5X,5(F7.2),F9.2)
      8     FORMAT('1',//,41X,A20)
     81     FORMAT(' ',40X,A20)
      9     FORMAT('-',22X,A16,20X,A8,40X,A6)
     10     FORMAT('-',38X,10(I3,4X))
     11     FORMAT('-',30X,I3,5X,10(F7.2),F7.2)
            RETURN
            END
C***************************************************************
C                                                             *
C                      OUTPUT 7                               *
C                      *******                                *
C                                                             *
C***************************************************************
            SUBROUTINE OUT7(NOLOC,NOCUST,TOT,ASSIGN)
            INTEGER NOLOC,NOCUST,ID(60),FINASS(60)
            REAL    ASSIGN(60),TOT(60)
            DO 1 I = 1 ,NOLOC
                ID(I) = I
      1     CONTINUE
            DO 2 I = 1,NOCUST
                J = ASSIGN(I)
                FINASS(I) = J
      2     CONTINUE
            PRINT8,'FINAL ASSIGNMENT '
            PRINT81,'================ '
            PRINT9,'CUSTOMER       ',(ID(I),I= 1,NOCUST)
            PRINT9,'LOCATION       ',(FINASS(I),I=1,NOCUST)
            PRINT7,'       COST OF ASSIGNMENT ',(TOT(I),I=1,NOCUST)
            L = NOCUST + 1
            L1=L+1
            TOT(L)=TOT(L)+TOT(L1)
            PRINT7,'TOTAL COST OF ASSIGNMENT = ',TOT(L)
            PRINT8,'                '
      7     FORMAT('-',6X,A27,4X,10(F7.2))
      8     FORMAT('1',//,41X,A20)
     81     FORMAT(' ',40X,A20)
      9     FORMAT('-',17X,A16,5X,10(I3,4X))
            RETURN
```

```
                  END
                  SUBROUTINE OUT8(FINMAT,MSAV,NOLOC,NOCUST,IDMND,TOTSAV)
                  INTEGER NOCUST,NOLOC,IDMND(60),ID(50),MSAV(50)
                  REAL FINMAT(60,50),TOTSAV(50)
                  DO 1 I = 1, 50
                     ID(I) = I
         1        CONTINUE
                  PRINT10,'LOCATION',(ID(I),I=1,NOLOC)
                  PRINT8,'TRANSPORTATION COST AND DEMAND'
                  PRINT81,'=============================='
                  PRINT9,'CUSTOMER','LOCATION','DEMAND'
                  PRINT7,(ID(I),I=1,NOLOC)
                  DO 2 I = 1 ,NOCUST
                     PRINT6,I,(FINMAT(I,J),J=1,NOLOC),IDMND(I)
         2        CONTINUE
                  PRINT11,(TOTSAV(J),J=1,NOLOC)
                  PRINT14,'THE FINAL LOCATION ASSIGNMENT ',MSAV(1)
C
C                                 | THE INTEGER NO. WILL CHANGE TO THE
SAME AS
C                                 U  NUMBER OF LOCATION.
     6   FORMAT('-',30X,I3,5X,5(F7.2),I4)
     7   FORMAT('-',38X,10(I3,4X))
     8   FORMAT('-',40X,A30)
    81   FORMAT(' ',40X,A30)
     9   FORMAT('-',22X,A16,20X,A8,26X,A6)
    10   FORMAT('-',30X,A14,10(I3,4X))
    11   FORMAT('-',23X,'FIXED COST   ',5X,10(F7.2))
    14   FORMAT('-',//,30X,A30,I3)
                  RETURN
                  END
$ENTRY
 5 5 3 1
 100   50 150 200 300
      500     300     200     800     500
      300     500     200     600     700
      500     200     000     100     000
      200     100     800     200     300
      300     200     400     000     400
5000.004000.004500.006000.004200.00
 8 71010 9
      10
//
```

```
10 '*********************************************************
20 '*              SINGLE FACILITY                         *
30 '*        Rectilinear distance cost model               *
40 '*                Chapter 4                              *
50 '*    INPUT DATA:                                        *
60 '*              # of machines                            *
70 '*              x coordinate, y coordinate, # of trips   *
80 '*              repeat for each machine                  *
90 '*********************************************************
100 DIM IA(50),IO(50),NT(50),ORDERX(50),ORDERY(50)
110 DIM MCHX(50),MCHY(50)
120 DEFSNG A-H,O-Z
130 DEFINT I-N
140 PRINT "how many machines are there?"
150 INPUT N
160 FOR J=1 TO N
170 PRINT "x coordinate for machine";J;", y coordinate , # of trips"
180 INPUT IA(J),IO(J),NT(J)
190 TOTTR=TOTTR+NT(J)
200 NEXT J
210 LEAST=IA(1)
220 ISAVE=1
230 KNT=KNT+1
240 FOR J=2 TO N
250 IF LEAST<IA(J) GOTO 280
260 LEAST=IA(J)
270 ISAVE=J
280 NEXT J
290 ORDERX(KNT)=LEAST
300 MCHX(KNT)=ISAVE
310 IA(ISAVE)=IA(ISAVE)+1000
320 IF KNT<N GOTO 210
330 OPTTR=TOTTR/2
340 FOR J=1 TO N
350 FIND=FIND+NT(MCHX(J))
360 IF OPTTR<FIND GOTO 390
370 IF OPTTR=FIND GOTO 390
380 NEXT J
390 XAXIS=ORDERX(J)
400 LEST=IO(1)
410 JSAVE=1
420 CNT=CNT+1
430 FOR J=2 TO N
440 IF LEST<IO(J) GOTO 470
450 LEST=IO(J)
460 JSAVE=J
470 NEXT J
480 ORDERY(CNT)=LEST
490 MCHY(CNT)=JSAVE
500 IO(JSAVE)=IO(JSAVE)+1000
510 IF CNT<N GOTO 400
520 FOR J= 1 TO N
530 ABC=ABC+NT(MCHY(J))
540 IF OPTTR<ABC GOTO 570
550 IF ABC=OPTTR GOTO 570
560 NEXT J
570 YAXIS=ORDERY(J)
580 FOR IUJ=1 TO 5
590 PRINT
```

```
600 NEXT IUJ
610 CLS: PRINT
"****************************************************"
620 PRINT "
****************************************************":PRINT:PRINT
630 PRINT "    The location for the new facility is:
(";XAXIS;",";YAXIS;")"
640 PRINT: PRINT:PRINT"
****************************************************"
650 PRINT"****************************************************"
660 DATA 6
670 DATA 20,46,20
680 DATA 15,28,15
690 DATA 26,35,30
700 DATA 50,20,18
710 DATA 45,15,20
720 DATA 1,6,15
```

```
10 '************************************************
15 '*              SINGLE FACILITY                  *
20 '*         Quadratic cost model                  *
25 '*             Chapter 4                          *
30 '*   input data is the same as the RDC model     *
40 '************************************************
50 DIM IA(50),IO(50),NT(50)
60 DEFSNG A-H,O-Z
70 DEFINT I-N
80 CLS
90 PRINT "how many machines are there?"
100 INPUT N
110 FOR I= 1 TO N
120 PRINT "x coordinate for machine";I;" , y coordinate , and #
of trips"
130 INPUT IA(I),IO(I),NT(I)
140 XMULT=IA(I)*NT(I)
150 YMULT=IO(I)*NT(I)
160 TOTX=TOTX+XMULT
170 TOTY=TOTY+YMULT
180 TTRIP=TTRIP+NT(I)
190 NEXT I
200 XAXIS=TOTX/TTRIP
210 YAXIS=TOTY/TTRIP
220
CLS:PRINT"***************************************************"
230 PRINT"
***************************************************":PRINT:PRINT
240 PRINT "   The location for the new facility is:
(";XAXIS;",";YAXIS;")"
250 PRINT :PRINT
260 PRINT"    ***************************************************"
270 PRINT"***************************************************"
280 DATA 6
290 DATA 20,46,20
300 DATA 15,28,15
310 DATA 26,35,30
320 DATA 50,20,18
330 DATA 45,15,20
340 DATA 1,6,15
```

```
10 '**********************************************************
20 '*          MULTIPLE FACILITIES OF THE SAME TYPE          *
30 '*                    Chapter 4                           *
40 '**********************************************************
50 DEFINT J-N
60 DEFSNG A-I,L,O-Z
70 DIM IA(20),IO(20),D(20),SD(20),DIST(20,20),RAD(20)
80 DIM LO(20,20),KEEP(20)
90 CLS
100 PRINT "how many facilities need placement?"
110 INPUT NF
120 LEST=10000 : MKOUNT=0 : L=0 : K=0 : FLAG=0 : IFLAG=0 : Z=N
130 PRINT "how many customers are there?"
140 INPUT N
150 FOR I=1 TO N
160 PRINT"x coordinate of customer";I;" , Y coordinate , AND
DEMAND"
170 INPUT IA(I),IO(I),D(I)
180 IF D(I)>LEST GOTO 200
190 LEST=D(I)
200 NEXT I
210 FOR I=1 TO N
220 SD(I)=D(I)/LEST
230 NEXT I
240 CLS
250 PRINT "FACILITY";SPC(14);"COORDINATES";SPC(14);"ASSIGNED
CUSTOMERS"
260 PRINT"------------------------------------------------------
----------"
270 IF N>NF THEN GOTO 340
280 KEEP(1)=1 :J1=NF :K=1
290 FOR I=1 TO NF
300    LO(I,1)=I
310    KEEP(I)=I
320 NEXT I
330 GOTO 920
340 SM=10000
350 MKOUNT=MKOUNT+1
360 FOR I=1 TO N
370 IF SD(I)=-1 GOTO 460
380 FOR J=1 TO N
390 IF SD(J)=-1 GOTO  450
400 IF I=J GOTO 450
410 DIST(I,J)=(((IA(I)-IA(J))^2)+((IO(I)-IO(J))^2))^.5
420 CON(I)=DIST(I,J)*SD(I)
430 IF CON(I)>SM GOTO 450
440 SM=CON(I)
450 NEXT J
460 NEXT I
470 FOR I=1 TO N
480 IF SD(I)=-1 GOTO 500
490 RAD(I)=SM/SD(I)
500 NEXT I
510 FOR I=1 TO N
520 IF SD(I)=-1 GOTO 680
530 FOR J=1 TO N
540 IF SD(J)=-1 GOTO 670
550 IF I=J GOTO 670
560 YTO=RAD(I)-DIST(I,J)+1
570 IF YTO<0 GOTO 670
```

```
580 ISA=I : JSA=J
590 IF IFLAG=0  THEN K=2 :LO(1,1)=ISA : LO(1,2)=JSA  :GOTO 670
600 LL=0 : ITEMP=ISA
610    FOR J2=1 TO K
620      IF LO(1,J2)=ITEMP    THEN FLAG=1
630    NEXT J2
640 LL=LL+1
650 IF FLAG=0 THEN K=K+1 : LO(1,K)=ITEMP
660 IF LL < 2 THEN ITEMP=JSA : FLAG=0 : GOTO 610
670 NEXT J
680 NEXT I
690 TIA(ISA)=((SD(ISA)*IA(ISA))+(SD(JSA)*IA(JSA)))
700 DIA(ISA)=SD(ISA)+SD(JSA)
710 IA(ISA)=TIA(ISA)/DIA(ISA)
720 TIO(ISA)=((SD(ISA)*IO(ISA))+(SD(JSA)*IO(JSA)))
730 IO(ISA)=TIO(ISA)/DIA(ISA)
740 SD(ISA)=DIA(ISA)
750 'PRINT "facility";MKOUNT;"has coordinates
(";IA(ISA);",";IO(ISA);")"
760 'PRINT "scaled demand for facility";MKOUNT;"is";SD(ISA)
770 SD(JSA)=-1
780 IFLAG=99    'signal the end of the FIRST iteration
790 Z=Z-1   'two demand pts are combined and a new fac. formed
800 KEEP(1)=ISA
810 IF Z>NF GOTO 340
820 IF NF=1 THEN J1=1 :GOTO 920
830 'Assign facilities to the remaining demand points
840 J1=1
850 FOR II=1 TO N
860    FOR J=1 TO K
870       IF LO(1,J) =  II THEN ITEMP=0 :GOTO 910  ELSE ITEMP=II
880    NEXT J
890   J1=J1+1
900   LO(J1,1)=ITEMP : KEEP(J1)=ITEMP
910 NEXT II
920 FOR I=1 TO J1
930   PRINT SPC(2);I ;SPC(16);:PRINT USING
"###.##";IA(KEEP(I));:PRINT",";:PRINT USING "###.##";IO(KEEP(I));
940   PRINT SPC(12);
950 FOR J=1 TO K
960   IF LO(I,J) <> 0 THEN PRINT USING"##";LO(I,J);
970   IF LO(I,J+1)<>0 THEN PRINT ",";
980 NEXT J
990 PRINT
1000 NEXT I
1010 END
```

```
10  '************************************************
15  '*                                              *
20  '*              RATE OF RETURN                  *
30  '*               Chapter 5                      *
35  '*                                              *
40  '************************************************
50  DEFINT I-N
60  PRINT "what is the initial investment?"
70  INPUT SUM3
80  DEFSNG A-H,O-Z
90  PRINT "how many years will the project operate ?"
100 INPUT N
110 FOR J=1 TO N
120 PRINT "what is the income for year";J;"?"
130 INPUT AM(J)
140 NEXT J
150 X1=SUM3-100
160 X2=SUM3+100
170 R=.01
180 R=R+.001
190 SUM2=0
200 FOR I=1 TO N
210 SUM2=(AM(I)*(1+R)^-I)+SUM2
220 NEXT I
230   IF SUM2<X2 GOTO 250
240 GOTO 260
250   IF SUM2>X1 GOTO 300
260   GOTO 180
270 'FOR IUJ=1 TO 100
280 'PRINT "      "
290 'NEXT IUJ
300 FOR IUJ=1 TO 100
310 PRINT "      "
320 NEXT IUJ
330 PRINT "THE INTERNAL RATE OF RETURN FOR THE PROJECT IS=";R
```

```
      ****************************************************
      *                                                  *
      *                   JOB  GROUPING                  *
      *                                                  *
      *                   Chapter 7                      *
      *                                                  *
      ****************************************************
      $JOB
 1            INTEGER RN(40),RN1(40),GN(40)
 2            INTEGER RC(20),WN(20)
 3            CHARACTER*2 MA(40,20)/800*'  '/,MATRIX(40,20),RM(40,20)/800*'  '/
 4            CHARACTER*2 JOB*2(20)
 5            CHARACTER*1 IL(100)/100*'_'/
 6            READ 1001,JB,MC
 7     1001   FORMAT(2I3)
 8            DO 222 I=1,JB
 9      222   READ6,(MATRIX(I,J),J =1,MC)
10       99   FORMAT(' ',5X,'|',20(1X,A2),1X,'|')
11        6   FORMAT(20A1)
12        1   FORMAT('1',5X,100(A1))
13        2   FORMAT(' ',5X,'|',3X,'|',' A  B  C  D  E  F  G  H  I  J  K  L  M'
             1'  N  O  P  Q  R  S','|')
**WARNING**   EXPECTING COMMA BETWEEN FORMAT ITEMS NEAR M' '
14        3   FORMAT(' ',5X,'|',I2,1X,'|',20(1X,A1,1X),1X,'|')
15        4   FORMAT('0',5X,100(A1))
16        9   FORMAT(' ',5X,'|',I2,1X,'|',20(1X,A2),1X,'|')
17            DO 100 I=1,MC
18            WN(I) = 1
19            RC(I)=I
20      100   CONTINUE
21            K = 1
22            DO 14 L = 1,MC
23               DO 11 I = 1, JB
24                  IF(MATRIX(I,L).NE.'X') GO TO 11
*EXTENSION*   A CHARACTER VARIABLE IS USED WITH A RELATIONAL OPERATOR
25                  DO 12 J = L, MC
26                     IF(MATRIX(I,J).NE.'X') GO TO 12
*EXTENSION*   A CHARACTER VARIABLE IS USED WITH A RELATIONAL OPERATOR
27                     RN(K)        = I
28                     WN(J)        = K
29                     RM(K,J)      = 'X '
30                     MA(I,J)      = 'X '
31                     MATRIX(I,J)  = '  '
32       12          CONTINUE
33                  RN(K) = I
34                  K       = K + 1
35       11       CONTINUE
36       14   CONTINUE
37            PRINT 999
38      999   FORMAT('1')
39            PRINT1,IL
40            PRINT 2
41            DO 15 I=1,JB
42            PRINT9,I,(MA(I,J),J=1,MC)
43       15   CONTINUE
44            PRINT4,IL
45            CALL FRC(WN,RM,RC,JB,MC)
46            CALL FGN(RC,RM,GN,JB,MC)
47            CALL FMA(RM,GN,MA,RN1,RN,JB,MC)
48            CALL FLC(MA,RC,JB,MC)
49            PRINT1,IL
50            PRINT2
51            DO 10 I=1,JB
```

```
52                    PRINT9,RN1(I),(MA(I,J),J=1,MC)
53          10  CONTINUE
54              PRINT4,IL
55              STOP
56              END
**WARNING**     FORMAT STATEMENT      99 IS UNREFERENCED
**WARNING**     FORMAT STATEMENT       3 IS UNREFERENCED

57              SUBROUTINE FRC(WN,RM,RC,JB,MC)
58              INTEGER WN(20),RC(20)
59              CHARACTER*2 RM(40,20)
60              RC(1) = 1
61              K     = 2
62              I1    = 1
63              I2    = 2
64              I3    = WN(1)
65              MC1=MC-1
66              DO 1 J=1,MC1
67                 DO 2 L = I2, MC
68                    RC(K) = L
69                    IS = 0
70                    DO 3 I = I1,I3
71                       IF(L.GT.MC.OR.I.GT.JB) GO TO 5
72                       IF(RM(I,L).NE.'X ') GO TO 3
*EXTENSION*   A CHARACTER VARIABLE IS USED WITH A RELATIONAL OPERATOR
73                       IS = IS + 1
74           3  CONTINUE
75              IF(IS.GE.2) GO TO 4
76                 IF(K.GT.MC) GO TO 5
77                    RC(K) = L
78                    K     = K + 1
79                    I1    = I3 + 1
80                    I3    = WN(L)
81                 GO TO 1
82           4  IF(K.GT.MC) GO TO 5
83                 RC(K)    =RC(K-1)
84                 K        = K + 1
85           2  CONTINUE
86           1  I2 = L + 1
87           5  RETURN
88              END

89              SUBROUTINE FGN(RC,RM,GN,JB,MC)
90              INTEGER SUM(40),RC(20),COUNT(20),GN(40)
91              CHARACTER*2 RM(40,20)
92              DO 1 I =1,JB
93              GN(I)=0
94                 K     = 1
95                 I2    = 1
96                 COUNT(1) = RC(1)
97                 J1    = I2 + 1
98                 DO 2 I1 = 1,MC
99                    SUM(I1) = 0
100          2  CONTINUE
101             DO 3 J = 1,MC
102                IF(J1.GT.MC.OR.I2.GT.JB) GO TO 6
103                   IF(RC(I2).EQ.RC(J1)) GO TO 3
104                      COUNT(K+1) = RC(J1)
105          6  J2 = J1 - 1
106             DO 4 L = I2, J2
107                IF(RM(I,L).NE.'X ') GO TO 4
*EXTENSION*   A CHARACTER VARIABLE IS USED WITH A RELATIONAL OPERATOR
108                      SUM(K)=SUM(K)+1
109          4           CONTINUE
```

```
110                       K = K + 1
111                       I2 = J2 + 1
112       3       J1 = J1 + 1
113               M = 0
114               DO 5 L1=1,MC
115                   IF(M.GE.SUM(L1)) GO TO 5
116                   M       = SUM(L1)
117                   GN(I) = COUNT(L1)
118       5       CONTINUE
119       1   CONTINUE
120           RETURN
121           END

122           SUBROUTINE FMA(RM,GN,MA,RN1,RN,JB,MC)
123           INTEGER GN(40),RN(40),RN1(40)
124           CHARACTER*2 RM(40,20),MA(40,20)
125           K = 1
126           DO 1 J = 1,MC
127               DO 2 I = 1,JB
128                   IF(GN(I).NE.J.OR.K.GT.JB) GO TO 2
       C            IF(K.GT.JB) GO TO 2
129                   RN1(K) = RN(I)
130                   DO 3 L = 1,MC
131                       MA(K,L) = RM(I,L)
132       3           CONTINUE
133                   K = K + 1
134       2       CONTINUE
135       1   CONTINUE
136           RETURN
137           END

138           SUBROUTINE FLC(MA,RC,JB,MC)
139           INTEGER LC(20),RC(20),LR(20)
140           CHARACTER*2 MA(40,20)
141           DO 1 I = 1,MC
142               LR(I) = 0
143               LC(I) = 0
144       1   CONTINUE
145           K = 1
146           LC(1) = RC(1)
147           KM    = RC(1)
148           DO 2 J = 2,MC
149               IF(KM.EQ.RC(J)) GO TO 2
150               K       = K + 1
151               LC(K)     = RC(J)
152               KM        = RC(J)
153       2   CONTINUE
154           K = 1
155           LR(1) = 1
156           DO 3 J = 2,MC
157               L1 = LC(J)
158               JB1=JB-1
159               DO 4 I = 1,JB1
160                   IF(LC(J).EQ.0) GO TO 3
161                   IF(MA(I,L1).NE.'X '.OR.MA(I+1,L1).EQ.'  ') GO TO 4
```

EXTENSION   A CHARACTER VARIABLE IS USED WITH A RELATIONAL OPERATOR
EXTENSION   A CHARACTER VARIABLE IS USED WITH A RELATIONAL OPERATOR

```
162                       K       = K + 1
163                       LR(K)     = I
164                       GO TO 3
```

```
165      4         CONTINUE
166      3      CONTINUE
167             DO 11 J = 1,MC
168                IF(LR(J).NE.0) GO TO 11
169                   LR(J)  = JB+1
170                   LC(J)  = MC+1
171                   GO TO 12
172     11      CONTINUE
173     12      MC1=MC-1
174             DO 5 J =2,MC1
175                L = LC(J) - 1
176                IF(LR(J+1).LE.0) GO TO 5
177                   I1 = LR(J-1)
178                   I2 = LR(J+1) - 1
179                   IF(I1.EQ.0) GO TO 5
180                   IF(I2.EQ.0) GO TO 5
181                   DO 6 I = I1,I2
182                      IF(MA(I,L).NE.'X ') GO TO 7
*EXTENSION*  A CHARACTER VARIABLE IS USED WITH A RELATIONAL OPERATOR
183                         MA(I,L) = 'X|'
184                         GO TO 6
185      7                MA(I,L) =    '|'
186      6             CONTINUE
187      5      CONTINUE
188             MC1=MC-1
189             DO 8 J = 2,MC1
190                IF(LC(J+1).EQ.0) GO TO 8
191                   I1 = LC(J-1)
192                   I2 = LC(J+1) - 1
193                   L  = LR(J) -1
194                IF(I1.EQ.0) GO TO 8
195                IF(I2.EQ.0) GO TO 8
196                DO 9 I = I1,I2
197                      IF(L.EQ.0) GO TO 9
198                      IF(I.EQ.0) GO TO 9
199                      IF(I.GT.MC) GO TO 9
200                      IF(L.LT.1) GO TO 9
201                      IF(MA(L,I).NE.'X ') GO TO 10
*EXTENSION*  A CHARACTER VARIABLE IS USED WITH A RELATIONAL OPERATOR
202                         MA(L,I) = 'X_'
203                         GO TO 9
204     10                IF(MA(L,I).NE.' ') GO TO 9
*EXTENSION*  A CHARACTER VARIABLE IS USED WITH A RELATIONAL OPERATOR
205                         MA(L,I) = '__'
206      9             CONTINUE
207      8      CONTINUE
208             RETURN
209             END

         $ENTRY
```

	A	B	C	D	E	F	G	H	I	J	K	L	M	N	O	P	Q	R	S
1	X	X				X													
2							X		X										
3							X	X	X										
4	X		X			X	X	X											
5							X	X	X										
6							X	X						X	X	X	X	X	
7	X	X	X		X	X													
8		X	X																
9				X	X	X													
10	X	X	X																
11				X		X													
12	X	X						X	X										
13			X	X				X	X										
14			X	X				X	X										
15	X	X	X		X	X				X		L							
16	X	X	X		X	X												X	
17										X	X	X	X			X	X		
18				X	X	X				X	X								
19																X	X		
20	X	X																	
21	X	X	X																
22				X		X	X	X											
23	X	X	X																
24				X	X	X													
25				X	X														
26	X							X			X								
27							X	X	X										
28		X		X		X													
29	X	X				X	X												
30	X	X	X				X	X											
31					X	X	X												
32	X	X				X	X												
33	X				X														
34	X	X	X					X	X										
35	X	X						X			K				O				
36			X	X	X		X	X											
37	X						X			X									
38			X	X			X												
39	X						X	X		X									
40			X				X	X											

```
  |   | A  B  C  D  E  F  G  H  I  J  K  L  M  N  O  P  Q  R  S |
  |  1|  X  X     |        X
  |  7|  X  X  X|     X  X
  | 10|  X  X  X|
  | 12|  X  X     |              X  X
  | 15|  X  X  X|     X  X           X        X
  | 16|  X  X  X|     X  X                             X
  | 20|  X  X     |
  | 21|  X  X  X|
  | 23|  X  X  X|
  | 26|  X        |              X           X
  | 29|  X  X     |        X  X
  | 30|  X  X  X|           X  X
  | 32|  X  X     |        X  X
  | 33|  X        |     X
  | 34|  X  X  X|           X  X
  | 35|  X  X     |_ _ _ _ _ _ _ X_ _ X_        X
  | 37|  X        |              X  X  |
  | 39|  X        |              X  X  |
  |  8|     X  X  |
  |  4|  X_ _ X_ _ _ _ _ X_ X_ X_ _ _|
  | 28|     X     |  X     X          |
  | 13|        X  X|           X  X|    |
  | 14|        X  X|           X  X|    |
  | 36|        X  X  X        X  X|    |
  | 40|        X  |           X  X|    |
  |  9|        X  X  X        |    |
  | 11|        X  X  |        |    |
  | 22|        X     X  X  X  |    |
  | 24|        X  X  X        |    |
  | 25|        X  X  |        |    |
  | 38|        X  X  |     X|    |
  | 18|        X  X  X  |  X  X|
  | 31|        X  X  X  |     |
  |  2|           X     X|    |
  |  3|           X  X  X|    |
  |  5|           X  X  X|    |
  | 27|           X  X  X| _ _|  _ _ _ _ X  X  X  X  X _ _
  |  6|           X  X  | _ _|
  | 17|           X  X| X  X        X  X
  | 19|              |_ _        X  X
```

CORE USAGE   OBJECT CODE=   8480 BYTES,ARRAY AREA=   5980 BYTES,TOTAL AREA AVAILABLE= 380928 BYTES

DIAGNOSTICS   NUMBER OF ERRORS=   0, NUMBER OF WARNINGS=   3, NUMBER OF EXTENSIONS=   9

COMPILE TIME=   0.20 SEC,EXECUTION TIME=   0.48 SEC, WATFIV - JUL 1973 V1L4   14.32.03

$STOP

```
1       //SPGRBA JOB (SPGRB),'LTUVMB SPGRB',                        JOB   183
        //              PASSWORD=,
        //              MSGLEVEL=(1,1),
        //              TIME=15,
        //              CLASS=Z
        ***JOBPARM LINES=50
        ***JOBPARM ROOM=GRBX
2       // EXEC WATFIV
3       XXWATFIV     PROC
        ***
4       XXGO          EXEC PGM=WATFIV
5       XXSTEPLIB    DD    DSN=WATFIV.JOBLIB,DISP=SHR
6       XXWATLIB     DD    DSN=WATFIV.FUNLIB,
        XX                 DISP=(SHR,KEEP),DCB=(RECFM=FB,LRECL=80,BLKSIZE=800)
7       XXSYSPRINT DD    SYSOUT=*,SPACE=(TRK,(10,10),RLSE)
8       XXFT05F001 DD    DDNAME=SYSIN
9       XXFT06F001 DD    SYSOUT=*,DCB=(RECFM=FBA,BLKSIZE=532),             *
        XX                 SPACE=(TRK,(10,10),RLSE)
10      //GO.SYSIN DD *
        //
```

```
IEF236I ALLOC. FOR SPGRBA GO
IEF237I 148   ALLOCATED TO STEPLIB
IEF237I 148   ALLOCATED TO SYS00384
IEF237I 148   ALLOCATED TO WATLIB
IEF237I JES2 ALLOCATED TO SYSPRINT
IEF237I JES2 ALLOCATED TO FT05F001
IEF237I JES2 ALLOCATED TO FT06F001
IEF142I SPGRBA GO - STEP WAS EXECUTED - COND CODE 0002
IEF285I    WATFIV.JOBLIB                              KEPT
IEF285I    VOL SER NOS= VS1STG.
IEF285I    SYSCTLG.VVS1STG                            KEPT
IEF285I    VOL SER NOS= VS1STG.
IEF285I    WATFIV.FUNLIB                              KEPT
IEF285I    VOL SER NOS= VS1STG.
IEF285I    JES2.JOB00183.S00102                       SYSOUT
IEF285I    JES2.JOB00183.SI0101                       SYSIN
IEF285I    JES2.JOB00183.S00103                       SYSOUT
IEF373I STEP /GO     / START 87226.1432
IEF374I STEP /GO     / STOP  87226.1432 CPU  0MIN 01.00SEC SRB  0MIN 00.11SEC VIRT  512K SYS  216K
IEF375I  JOB /SPGRBA / START 87226.1432
IEF376I  JOB /SPGRBA / STOP  87226.1432 CPU  0MIN 01.00SEC SRB  0MIN 00.11SEC
```

```
1 DEFSNG A-H,O-Z
2 DEFINT I-N
3 REM ***********************************************************
4 REM *                ASSEMBLY LINE BALANCING                  *
5 REM *                 largest candidate rule                  *
6 REM *                  ---->  Chapter 7  <----                *
7 REM *         INPUT DATA:  cycle time                         *
8 REM *                      number of tasks                    *
9 REM *                      time (task 1),# of predecessors (task 1)*
10 REM*                      time (task 2),# of predecessors (task 2)*
11 REM*                              continue                    *
12 REM*                    predecessor # 1 (task 1)             *
13 REM*                    predecessor # 2 (task 1)             *
14 REM*                              continue                    *
15 REM*                    predecessor # 1 (task 2)             *
16 REM*                              continue                    *
17 REM***********************************************************
20 DIM   TTT(15),KSAVE(15),MM(15),LSAVE(15),TTTT(15)
25 PRINT "what is the required cycle time?"
30 INPUT CTC
31 ETME=0!
32 IHLP=0
40 IP=0
50 LL=0
60 SCCT=0!
70 IPEE=0
75 PRINT "how many tasks are to be assigned?"
80 INPUT N
81 LPRINT "the number of tasks is";N
82 LPRINT "   "
83 LPRINT "   "
120 FOR I = 1 TO N
125 PRINT "time for task";I;",  number of predecessors for
task";I
130 INPUT T(I),M(I)
140 LPRINT "time for task";I;"is";T(I)
150 LPRINT "task";I;"has";M(I);"predessors"
151 IHLP=IHLP+M(I)
152 ETME=ETME+T(I)
160 LPRINT "    "
170 NEXT I
180 FOR IRT =1 TO N
190 MM(IRT)=M(IRT)
200 TTT(IRT)=T(IRT)
210 TTTT(IRT)=T(IRT)
220 NEXT IRT
230 LPRINT "    "
240 FOR J = 1 TO N
250 IZ=M(J)
260 IF IZ=0 GOTO 330
270 LPRINT "task";J;"must be preceded by the following tasks"
280 FOR K = 1 TO IZ
285 PRINT "give the #";K;"predecessor for task";J
290 INPUT L(J,K)
300 LPRINT "task";L(J,K)
310 NEXT K
320 LPRINT "   "
330 NEXT J
340 JJ=0
350 JJ=JJ+1
```

```
351 IF JJ>N GOTO 680
370 TEMP=0!
380 NA=JJ-1
390 FOR II = 1 TO N
400 IF NA=0 GOTO 440
410 FOR IJ = 1 TO NA
420 IF ISAVE(IJ)=II GOTO 470
430 NEXT IJ
440 IF T(II)<TEMP GOTO 470
441 IF T(II)=TEMP GOTO 470
450 TEMP=T(II)
460 ITI=II
470 NEXT II
480 ISAVE(JJ)=ITI
490 IZZ=M(ISAVE(JJ))
500 IF IZZ=0 GOTO 630
510 IF IP=0 GOTO 350
520 IJK=0
530 LL=0
540 LL=LL+1
550 IF LL>IP GOTO 350
560 FOR KK = 1 TO IZZ
570 IF L(ISAVE(JJ),KK)<>JSAVE(LL) GOTO 600
580 IJK=IJK+1
590 IF IJK=IZZ GOTO 630
600 IF IZZ=KK GOTO 540
610 IF LL>IP GOTO 350
620 NEXT KK
630 IP=IP+1
640 JSAVE(IP)=ISAVE(JJ)
650 TT(JSAVE(IP))=T(JSAVE(IP))
660 T(JSAVE(IP))=0!
670 IF IP<N GOTO 340
680 KKOUNT=0
690 CT=0!
700 KOUNT=0
710 KKY=1
720 KOUNT=KKY-1
730 KOUNT=KOUNT+1
740 IF KOUNT=N GOTO 820
750 IF KOUNT<N GOTO 820
760 FOR IIY= 1 TO N
770 IF TT(JSAVE(IIY))=0! GOTO 810
780 KKY=IIY
790 CT=0!
800 GOTO 720
810 NEXT IIY
818 IF KOUNT>N GOTO 1090
820 IF TT(JSAVE(KOUNT))=0 GOTO 730
830 CT=CT+TT(JSAVE(KOUNT))
840 IF CT=CTC GOTO 880
850 IF CT<CTC GOTO 880
860 CT=CT-TT(JSAVE(KOUNT))
870 GOTO 730
880 IF MM(JSAVE(KOUNT))<>0 GOTO 720
890 IF TT(JSAVE(KOUNT))=0! GOTO 720
900 KSAVE(KOUNT)=JSAVE(KOUNT)
910 IPEE=IPEE+1
920 IBUK=KSAVE(KOUNT)
```

```
930 LSAVE(IPEE)=IBUK
940 FOR JFK = 1 TO N
950 NEXT JFK
960 GOTO 970
970 TT(JSAVE(KOUNT))=0!
980 FOR JM = 1 TO N
990 IZT=M(JM)
1000 IF IZT=0 GOTO 1040
1010 FOR JA = 1 TO IZT
1020 IF L(JM,JA)=KSAVE(KOUNT) THEN MM(JM)=MM(JM)-1
1030 NEXT JA
1040 NEXT JM
1041 LSAVE(N)=N
1050 KKOUNT=KKOUNT+1
1060 IF KKOUNT=N GOTO 1090
1070 IF KOUNT<>N GOTO 730
1080 IF KOUNT=N GOTO 690
1090 KLPI=0
1100 KKKNT=0
1110 CCTT=0!
1120 KLPI=KLPI+1
1130 KKKNT=KKKNT+1
1140 IF KKKNT>N GOTO 1180
1150 CCTT=CCTT+TTTT(LSAVE(KKKNT))
1160 IF CCTT>CTC GOTO 1180
1170 GOTO 1130
1180 LPRINT "station";KLPI;"contains"
1190 KLPI2=KKKNT-1
1200 FOR LMNO = 1 TO KLPI2
1210 IF TTT(LMNO)=0! GOTO 1240
1220 LPRINT "task";LSAVE(LMNO);"time";TTTT(LSAVE(LMNO))
1230 TTT(LMNO)=0!
1240 NEXT LMNO
1250 IF KLPI2=N GOTO 1270
1260 CCTT=CCTT-TTTT(LSAVE(KKKNT))
1270 LPRINT "the station time is";CCTT
1271 SCCTV=SCCT-1
1290 IF CCTT<SCCT GOTO 1310
1300 SCCT=CCTT
1310 LPRINT "    "
1320 LPRINT "    "
1330 IF KKKNT >N GOTO 1360
1340 KKKNT=KKKNT-1
1350 GOTO 1110
1360 LPRINT "the cycle time is";SCCT
1370 NCYCLS=KLPI
1375 EFFIC=(1-((SCCT*KLPI)-ETME)/(KLPI*SCCT))*100
1380 LPRINT "efficiency =";EFFIC
```

```
1 DEFSNG A-H,O-Z
2 DEFINT I-N
3 REM ****************************************************************
4 REM *                    ASSEMBLY LINE BALANCING                  *
5 REM *                    largest candidate rule                   *
6 REM *                    ----> Chapter 7  <----                   *
7 REM *         INPUT DATA:  cycle time                             *
8 REM *                      number of tasks                        *
9 REM *                      time (task 1),# of predecessors (task 1)*
10 REM*                      time (task 2),# of predecessors (task 2)*
11 REM*                                continue                     *
12 REM*                      predecessor # 1 (task 1)               *
13 REM*                      predecessor # 2 (task 1)               *
14 REM*                                continue                     *
15 REM*                      predecessor # 1 (task 2)               *
16 REM*                                continue                     *
17 REM****************************************************************
20 DIM  TTT(15),KSAVE(15),MM(15),LSAVE(15),TTTT(15)
25 PRINT "what is the required cycle time?"
30 INPUT CTC
31 ETME=0!
32 IHLP=0
40 IP=0
50 LL=0
60 SCCT=0!
70 IPEE=0
75 PRINT "how many tasks are to be assigned?"
80 INPUT N
81 LPRINT "the number of tasks is";N
82 LPRINT "   "
83 LPRINT "   "
120 FOR I = 1 TO N
125 PRINT "time for task";I;",  number of predecessors for
task";I
130 INPUT T(I),M(I)
140 LPRINT "time for task";I;"is";T(I)
150 LPRINT "task";I;"has";M(I);"predessors"
151 IHLP=IHLP+M(I)
152 ETME=ETME+T(I)
160 LPRINT "    "
170 NEXT I
180 FOR IRT =1 TO N
190 MM(IRT)=M(IRT)
200 TTT(IRT)=T(IRT)
210 TTTT(IRT)=T(IRT)
220 NEXT IRT
230 LPRINT "   "
240 FOR J = 1 TO N
250 IZ=M(J)
260 IF IZ=0 GOTO 330
270 LPRINT "task";J;"must be preceded by the following tasks"
280 FOR K = 1 TO IZ
285 PRINT "give the #";K;"predecessor for task";J
290 INPUT L(J,K)
310 NEXT K
320 LPRINT "    "
330 NEXT J
340 JJ=0
350 JJ=JJ+1
351 IF JJ>N GOTO 680
```

```
370 TEMP=0!
380 NA=JJ-1
390 FOR II = 1 TO N
400 IF NA=0 GOTO 440
410 FOR IJ = 1 TO NA
420 IF ISAVE(IJ)=II GOTO 470
430 NEXT IJ
440 IF T(II)<TEMP GOTO 470
441 IF T(II)=TEMP GOTO 470
450 TEMP=T(II)
460 ITI=II
470 NEXT II
480 ISAVE(JJ)=ITI
490 IZZ=M(ISAVE(JJ))
500 IF IZZ=0 GOTO 630
510 IF IP=0 GOTO 350
520 IJK=0
530 LL=0
540 LL=LL+1
550 IF LL>IP GOTO 350
560 FOR KK = 1 TO IZZ
570 IF L(ISAVE(JJ),KK)<>JSAVE(LL) GOTO 600
580 IJK=IJK+1
590 IF IJK=IZZ GOTO 630
600 IF IZZ=KK GOTO 540
610 IF LL>IP GOTO 350
620 NEXT KK
630 IP=IP+1
640 JSAVE(IP)=ISAVE(JJ)
650 TT(JSAVE(IP))=T(JSAVE(IP))
660 T(JSAVE(IP))=0!
670 IF IP<N GOTO 340
680 KKOUNT=0
690 CT=0!
700 KOUNT=0
710 KKY=1
720 KOUNT=KKY-1
730 KOUNT=KOUNT+1
740 IF KOUNT=N GOTO 820
750 IF KOUNT<N GOTO 820
760 FOR IIY= 1 TO N
770 IF TT(JSAVE(IIY))=0! GOTO 810
780 KKY=IIY
790 CT=0!
800 GOTO 720
810 NEXT IIY
818 IF KOUNT>N GOTO 1090
820 IF TT(JSAVE(KOUNT))=0 GOTO 730
830 CT=CT+TT(JSAVE(KOUNT))
840 IF CT=CTC GOTO 880
850 IF CT<CTC GOTO 880
860 CT=CT-TT(JSAVE(KOUNT))
870 GOTO 730
880 IF MM(JSAVE(KOUNT))<>0 GOTO 720
890 IF TT(JSAVE(KOUNT))=0! GOTO 720
900 KSAVE(KOUNT)=JSAVE(KOUNT)
910 IPEE=IPEE+1
920 IBUK=KSAVE(KOUNT)
930 LSAVE(IPEE)=IBUK
940 FOR JFK = 1 TO N
```

```
950 NEXT JFK
960 GOTO 970
970 TT(JSAVE(KOUNT))=0!
980 FOR JM = 1 TO N
990 IZT=M(JM)
1000 IF IZT=0 GOTO 1040
1010 FOR JA = 1 TO IZT
1020 IF L(JM,JA)=KSAVE(KOUNT) THEN MM(JM)=MM(JM)-1
1030 NEXT JA
1040 NEXT JM
1041 LSAVE(N)=N
1050 KKOUNT=KKOUNT+1
1060 IF KKOUNT=N GOTO 1090
1070 IF KOUNT<>N GOTO 730
1080 IF KOUNT=N GOTO 690
1090 KLPI=0
1100 KKKNT=0
1110 CCTT=0!
1120 KLPI=KLPI+1
1130 KKKNT=KKKNT+1
1140 IF KKKNT>N GOTO 1180
1150 CCTT=CCTT+TTTT(LSAVE(KKKNT))
1160 IF CCTT>CTC GOTO 1180
1170 GOTO 1130
1180 LPRINT "station";KLPI;"contains"
1190 KLPI2=KKKNT-1
1200 FOR LMNO = 1 TO KLPI2
1210 IF TTT(LMNO)=0! GOTO 1240
1220 LPRINT "task";LSAVE(LMNO);"time";TTTT(LSAVE(LMNO))
1230 TTT(LMNO)=0!
1240 NEXT LMNO
1250 IF KLPI2=N GOTO 1270
1260 CCTT=CCTT-TTTT(LSAVE(KKKNT))
1270 LPRINT "the station time is";CCTT
1271 SCCTV=SCCT-1
1290 IF CCTT<SCCT GOTO 1310
1300 SCCT=CCTT
1310 LPRINT "      "
1320 LPRINT "      "
1330 IF KKKNT >N GOTO 1360
1340 KKKNT=KKKNT-1
1350 GOTO 1110
1360 LPRINT "the cycle time is";SCCT
1370 NCYCLS=KLPI
1375 EFFIC=(1-((SCCT*KLPI)-ETME)/(KLPI*SCCT))*100
1380 LPRINT "efficiency =";EFFIC
```

```
10  '************************************************************
15  '*                                                          *
20  '*                 ASSEMBLY LINE BALANCING                  *
25  '*          Ranked Proportional Weighting Method            *
30  '*                      Chapter 7                           *
35  '*                                                          *
40  '************************************************************
50  PRINT "if you want interactive mode enter the number "1""
60  PRINT "if you want data mode enter the number "0""
70  INPUT ONT
80  IF ONT=1 GOTO 160
85  PRINT "data starts with statement # 2540    "
90  PRINT "enter the data in the following manner:"
100 PRINT "   required cycle time"
110 PRINT "   # of tasks"
120 PRINT "   time for each task, # of predecessors for each
task"
130 PRINT "   1st predecessor for the 2nd task"
140 PRINT "   2nd predecessor "
150 PRINT "   ect."
160 DEFSNG A-H,O-Z
170 DEFINT I-N
180 DIM T(15),M(11),L(11,11),ISAVE(12),JSAVE(12),TT(12),TTT(12)
190 DIM KSAVE(12),MM(12),LSAVE(12),TTTT(12),ID(12),IRPW(12,12)
200 DIM TOTAL(12),TOTAL1(12),LW(12,59),M2(12),TP(12,12),TTTTT(12)
210 PRINT "imd";ONT
220 IF ONT<>1 GOTO 260
230 PRINT "what is the required cycle time?"
240 INPUT CTC
250 GOTO 270
260 READ CTC
270 IP=0
280 SCCT=0!
290 IPEE=0
300 ETME=0!
310 IHLP=0
320 IF ONT=0 GOTO 360
330 PRINT "how many tasks are there?"
340 INPUT N
350 GOTO 370
360 READ N
370 PRINT "the # of tasks is";N
380 PRINT "   "
390 PRINT "   "
400 FOR I = 1 TO N
410 IF ONT=0 GOTO 450
420 PRINT "time for task";I;"# of predecessors for task";I
430 INPUT T(I),M(I)
440 GOTO 460
450 READ T(I),M(I)
460 IHLP=IHLP+M(I)
470 ETME=ETME+T(I)
480 PRINT "time for task";I;" is";T(I)
490 PRINT "task";I;" has";M(I);" predessors"
500 PRINT "   "
510 NEXT I
520 FOR IRT = 1 TO N
530 MM(IRT)=M(IRT)
540 TT(IRT)=T(IRT)
550 TTT(IRT)=T(IRT)
```

```
560 TTTT(IRT)=T(IRT)
570 TTTTT(IRT)=T(IRT)
580 NEXT IRT
590 PRINT "   "
600 FOR J = 1 TO N
610 IZ = M(J)
620 IF IZ=0 GOTO 730
630 PRINT "task";J;"must be preceded by the following tasks"
640 FOR K = 1 TO IZ
650 IF ONT=0 GOTO 690
660 PRINT "what is predecessor #";K;"for task";J
670 INPUT L(J,K)
680 GOTO 700
690 READ L(J,K)
700 PRINT "     task";L(J,K)
710 NEXT K
720 PRINT "  "
730 NEXT J
740 ICNT=1
750 ICNT1=0
760 FOR J = 1 TO N
770 IZ=M(J)
780 IF IZ=0 GOTO 850
790 FOR K = 1 TO IZ
800 IF L(J,K)=ICNT GOTO 840
810 ICNT1=ICNT1+1
820 IF ICNT1<IHLP GOTO 840
830 ICHK=ICNT
840 NEXT K
850 NEXT J
860 ICNT=ICNT+1
870 IF ICNT<N GOTO 760
880 IF ICNT=N GOTO 760
890 FOR I = 1 TO N
900 M2(I)=0
910 NEXT I
920 HOLD=N
930 KNT=0
940 FOR J = 1 TO N
950 IZ=M(J)
960 IF IZ=0 GOTO 1030
970 FOR K = 1 TO IZ
980 IF L(J,K)<>HOLD GOTO 1020
990 KNT=KNT+1
1000 LW(HOLD,KNT)=J
1010 M2(HOLD)=KNT
1020 NEXT K
1030 NEXT J
1040 HOLD=HOLD-1
1050 IF HOLD>0 GOTO 930
1060 HLD=N
1070 FOR J = 1 TO N
1080 IZ=M2(J)
1090 IF IZ=0 GOTO 1220
1100 IPIX=0
1110 FOR K = 1 TO IZ
1120 IF LW(J,K)<>HLD GOTO 1210
1130 IPOT=M2(J)+1
1140 IPOT2=M2(HLD)+M2(J)
1150 FOR ITID = IPOT TO IPOT2
```

```
1160 IPIX=IPIX+1
1170 IF ICHK=HLD GOTO 1200
1180 LW(J,ITID)=LW(HLD,IPIX)
1190 M2(J)=ITID
1200 NEXT ITID
1210 NEXT K
1220 NEXT J
1230 IF HLD=0 GOTO 1260
1240 IF M2(HLD)=0 GOTO 1260
1250 IF LW(HLD,M2(HLD))<>N GOTO 1070
1260 HLD=HLD-1
1270 IF HLD>0 GOTO 1070
1280 FOR IS = 1 TO N
1290 FOR IJK = 1 TO N
1300 TP(IS,IJK)=T(IS)
1310 NEXT IJK
1320 NEXT IS
1330 IOUNTR=0
1340 IOUNTR=IOUNTR+1
1350 Z=0!
1360 IZ=M2(IOUNTR)
1370 IF IZ=0 GOTO 1430
1380 FOR J = 1 TO IZ
1390 Z=TP(LW(IOUNTR,J),IOUNTR)+Z
1400 TP(LW(IOUNTR,J),IOUNTR)=0!
1410 TOTAL(IOUNTR)=Z+TTTTT(IOUNTR)
1420 NEXT J
1430 IF IZ=0 THEN TOTAL(IOUNTR)=0!
1440 IF IOUNTR<N GOTO 1340
1450 JJ=0
1460 JJ=JJ+1
1470 IF JJ>N GOTO 1800
1480 TEMP=0!
1490 NA=JJ-1
1500 FOR II = 1 TO N
1510 IF NA=0 GOTO 1550
1520 FOR IJ = 1 TO NA
1530 IF ISAVE(IJ)=II GOTO 1590
1540 NEXT IJ
1550 IF TOTAL(II)<TEMP GOTO 1590
1560 IF TOTAL(II)=TEMP GOTO 1590
1570 TEMP=TOTAL(II)
1580 ITI=II
1590 NEXT II
1600 ISAVE(JJ)=ITI
1610 IZZ=M(ISAVE(JJ))
1620 IF IZZ=0 GOTO 1750
1630 IF IP=0 GOTO 1460
1640 IJK=0
1650 LL=0
1660 LL=LL+1
1670 IF LL>IP GOTO 1460
1680 FOR KK = 1 TO IZZ
1690 IF L(ISAVE(JJ),KK)<>JSAVE(LL) GOTO 1720
1700 IJK=IJK+1
1710 IF IJK=IZZ GOTO 1750
1720 IF IZZ=KK GOTO 1660
1730 IF LL>IP GOTO 1460
1740 NEXT KK
1750 IP=IP+1
```

```
1760 JSAVE(IP)=ISAVE(JJ)
1770 TOTAL1(JSAVE(IP))=TOTAL(JSAVE(IP))
1780 TOTAL(JSAVE(IP))=0!
1790 IF IP<N GOTO 1450
1800 KKOUNT=0
1810 CT=0!
1820 KOUNT=0
1830 KKY=1
1840 KOUNT=KKY-1
1850 KOUNT=KOUNT+1
1860 IF KOUNT=N GOTO 1950
1870 IF KOUNT<N GOTO 1950
1880 FOR IIY = 1 TO N
1890 IF TT(JSAVE(IIY))=0! GOTO 1930
1900 KKY=IIY
1910 CT=0!
1920 GOTO 1840
1930 NEXT IIY
1940 IF KOUNT>N GOTO 2230
1950 IF TT(JSAVE(KOUNT))=0 GOTO 1850
1960 CT=CT+TT(JSAVE(KOUNT))
1970 IF CT=CTC GOTO 2010
1980 IF CT<CTC GOTO 2010
1990 CT=CT-TT(JSAVE(KOUNT))
2000 GOTO 1850
2010 IF MM(JSAVE(KOUNT))<>0 GOTO 1840
2020 IF TT(JSAVE(KOUNT))=0! GOTO 1840
2030 KSAVE(KOUNT)=JSAVE(KOUNT)
2040 IPEE=IPEE+1
2050 IBUK=KSAVE(KOUNT)
2060 LSAVE(IPEE)=IBUK
2070 FOR JFK = 1 TO N
2080 NEXT JFK
2090 GOTO 2100
2100 TT(JSAVE(KOUNT))=0!
2110 FOR JM = 1 TO N
2120 IZT=M(JM)
2130 IF IZT=0 GOTO 2170
2140 FOR JA = 1 TO IZT
2150 IF L(JM,JA)=KSAVE(KOUNT) THEN MM(JM)=MM(JM)-1
2160 NEXT JA
2170 NEXT JM
2180 LSAVE(N)=N
2190 KKOUNT=KKOUNT+1
2200 IF KKOUNT=N GOTO 2230
2210 IF KOUNT<>N GOTO 1850
2220 IF KOUNT=N GOTO 1810
2230 KLPI=0
2240 KKKNT=0
2250 CCTT=0!
2260 KLPI=KLPI+1
2270 KKKNT=KKKNT+1
2280 IF KKKNT>N GOTO 2320
2290 CCTT=CCTT+TTTT(LSAVE(KKKNT))
2300 IF CCTT>CTC GOTO 2320
2310 GOTO 2270
2320 LPRINT "station";KLPI;"contains"
2330 KLPI2=KKKNT-1
2340 FOR LMNO = 1 TO KLPI2
2350 IF TTT(LMNO)=0! GOTO 2380
```

```
2360 LPRINT "task";LSAVE(LMNO);"time";TTTT(LSAVE(LMNO))
2370 TTT(LMNO)=0!
2380 NEXT LMNO
2390 IF KLPI2=N GOTO 2410
2400 CCTT=CCTT-TTTT(LSAVE(KKKNT))
2410 LPRINT "the station time is";CCTT
2420 SCCTV=SCCT-1
2430 IF CCTT<SCCT GOTO 2450
2440 SCCT=CCTT
2450 LPRINT "  "
2460 LPRINT "  "
2470 IF KKKNT>N GOTO 2500
2480 KKKNT=KKKNT-1
2490 GOTO 2250
2500 LPRINT "the cycle time is";SCCT
2510 NCYCLS=KLPI
2520 EFFIC=(1-((SCCT*KLPI)-ETME)/(KLPI*SCCT))*100
2530 LPRINT "efficiency   =";EFFIC
2540 DATA 1.0
2550 DATA 11
2560 DATA .22,0
2570 DATA .7,1
2580 DATA .42,1
2590 DATA .38,2
2600 DATA .48,2
2610 DATA .12,1
2620 DATA .63,2
2630 DATA .52,1
2640 DATA .34,2
2650 DATA .8,2
2660 DATA .7,2
2670 DATA 1
2680 DATA 1
2690 DATA 1
2700 DATA 3
2710 DATA 2
2720 DATA 3
2730 DATA 5
2740 DATA 3
2750 DATA 5
2760 DATA 4
2770 DATA 6
2780 DATA 7
2790 DATA 7
2800 DATA 8
2810 DATA 9
2820 DATA 10
```

```
10 REM *************************************************
20 REM *    QUEUING:   SINGLE SERVER, M/M/1 MODEL      *
30 REM *               Chapter 11                      *
40 REM *    r = arrival rate                           *
50 REM *    t = service rate                           *
60 REM *    INPUT DATA:arrival rate,service rate       *
70 REM *************************************************
80 PRINT "PLEASE TURN YOUR PRINTER ON!" : PRINT
90 PRINT "what is the arrival rate? , and the service rate?"
100 INPUT R,T
110 DIM YP(50)
120 YP(O)=1-(R/T)
130 LPRINT " p(o)=";YP(O)
140 I=0
150 Z=YP(O)
160 I=I+1
170 YP(I)=((R/T)^I)*YP(O)
180 LPRINT " p(";I;")=";YP(I)
190 Z=Z+YP(I)
200 IF YP(I)>.0001 GOTO 160
210 XL=(R/T)/(1-(R/T))
220 XLQ=((R/T)^2)/(1-(R/T))
230 LPRINT
240 LPRINT "L=";XL
250 LPRINT "LQ=";XLQ
260 W=XL/R
270 WQ=XLQ/R
280 LPRINT "w=";W
290 LPRINT "wq=";WQ
```

```
10 REM ********************************************************
20 REM *    QUEUING:   MULTIPLE SERVERS, M/M/C/K MODEL     *
25 REM *                  Chapter 11                       *
30 REM *    r= arrival rate, t= service rate               *
40 REM *    nn= limit on the queue,c= # of servers         *
50 REM *     INPUT DATA:                                   *
60 REM *                   arrival rate,service rate       *
70 REM *                   # of servers                    *
80 REM *                   limit on queue                  *
90 REM*********************************************************
100 PRINT "what is the arrival rate? , and the service rate?
110 INPUT R,T
120 S=R/T
130 PRINT "how many servers are there?"
140 INPUT C
150 DEFSNG A-H,O-Z
160 DEFINT I-N
170 PRINT "what is the queue limit?"
180 INPUT NN
190 N=-1
200 N=N+1
210 X1=N
220 X2=X1
230 X2=X2-1
240 X1=X1*X2
250 IF X1=0,THEN X1=1
260 IF X2>1 GOTO 230
270 C1=C
280 C2=C1
290 C2=C2-1
300 C1=C1*C2
310 IF C1=0, THEN C1=1
320 IF C2>1 GOTO 290
330 ZZ=((S^C*((1-(S/C)^(NN-C+1))))/(C1*(1-S/C)))
340 Z=(S^N)/X1
350 A=C-1
360 Z1=Z1+Z
370 IF N<A GOTO 200
380 YO=1!/(ZZ+Z1)
390 LPRINT "p 0=";YO
400 FOR N = 1 TO C
410 YY1=N
420 IF N=1 GOTO 480
430 YY2=YY1
440 YY2=YY2-1
450 YY1=YY1*YY2
460 IF YY2>1 GOTO 440
470 'LPRINT "fact",YY1
480 YP(N)=((S^N)/YY1)*YO
490 LPRINT "p";N;"=";YP(N)
500 NEXT N
510 K=C+1
520 FOR N=K TO NN
530 YP(N)=((S^N)*YO)/(C1*(C^(N-C)))
540 LPRINT "p";N;"="YP(N)
550 NEXT N
560 TOT=YO
570 FOR I = 1 TO NN
580 TOT=TOT+YP(I)
590 NEXT I
```

```
600 FOR I=C TO NN
610   XLQ=XLQ+(I-C)*YP(I)
620 NEXT I
630 LPRINT "lq";XLQ
640 XL=XLQ+S*(1-YP(NN))
650 LPRINT "L = ";XL
660 WQ=XLQ/R*(1-YP(NN))
670 LPRINT "wq = ";WQ
680 W=XLQ+(1/T)
690 LPRINT "W = ";W
```

```
*************************************************
*                                               *
*           COMPUTER LAYOUT DESIGN              *
*                                               *
*                Chapter 13                     *
*                                               *
*************************************************
                              -- Chapter 13 --
SOURCE
PRECISION= 7
AUTODEF=ON
OPTION BASE=0
ERL=OFF
ERRORMODE=LOCAL
RESUME=LINE
FORMODE=BB
PRINTMODE=BB
SCOPE=ON
STACK=2000
PROCS=4
INTEGER ARRAY(11,11): G
INTEGER ARRAY(31,31): REL
INTEGER ARRAY(31): TOT
INTEGER: ND,I,IJ,C,R,J,CPOS,LS
INTEGER: II,JJ,I1,Maxx,Dpt
INTEGER ARRAY(30): HOLDDPT
INTEGER: POINTER,IR,J1,HII
INTEGER: HJ,VALUE1,NUM,IP,JP,VALUE2
INTEGER ARRAY(31): A
INTEGER: IG
INTEGER ARRAY(31): Tvalue,Hold
INTEGER: Relflag,Yflag
INTEGER: L,Ik,Ic,K,Ndpt,N,NHFLAG,Mtotflag
INTEGER: Hflag,Ix
INTEGER ARRAY(26): Dptx
STRING: A$[?]
INTEGER ARRAY(26): NG
INTEGER ARRAY(30): BLKS,Place
INTEGER ARRAY(101): Hi,Hj,IMNG,Iplace
INTEGER: BL,TL,PN
INTEGER: SUM,RL1,RL,GD,FLAGMID,NBLK,COUNT
INTEGER ARRAY(25,51): GRID
INTEGER: FR,FC
INTEGER: CFLAG,PNFLAG,ONE,LR,LC,JK,NGCOUNT,KEEPI
INTEGER: CNT,FLAG1,Flag1p,Flag2,Flag2p,Flag3,Flag3p,Flag4
INTEGER: Flag4p,Span,M,SCALE,Minn,KI,KJ,KEEPL
INTEGER: KEEPR,RII,CJJ
INTEGER ARRAY(31): FRD,COORD
INTEGER ARRAY(31): FCD
INTEGER ARRAY(51): KDIF
INTEGER: LL,LL1,FLAG,U,DIST
INTEGER ARRAY(51): CL
INTEGER: KD,SWAP_OK,R1,C1,IGD,M1,N1,IJSAVE
INTEGER: R2,C2,GDTEMP,J1SAVE,LL2
INTEGER ARRAY(61,61): ROW,COL,ROW2,COL2
INTEGER: MN,FLAGNH
INTEGER ARRAY(31,31): MinDIST
INTEGER: NBLK2,LDIST,SCORE,I11,J11,JJ1,II1
INTEGER: GD1,NBLK1,I3,J3,I4,J4,L4,NBR
INTEGER: MM,L1,L2,II1P1,NDD,NDM1
```

```
INTEGER ARRAY(11,11): GSAVE
INTEGER: J2,DIST1,O,DIFF,IFlag2p,IFLAG,Li
STRING ARRAY(31,31)[16]: SREL
INTEGER: VA,VE
INTEGER: VI,VO,VU,VX,NBLKS,STACK,Realj
REAL: AJ,B
STRING: STR[40],Srow[2],Scol[2],Sub1[1],Sub2[1],RCC[2],R$[16]
INTEGER: Falg,Rcc1

INTEGER FUNCTION: GETSQR
   INTEGER ARG: ND
END FUNCTION

PROCEDURE: PUSH
   INTEGER ARG: GETDPT
END PROCEDURE

PROCEDURE: PULL
   INTEGER ARG: GETDPT/VAR
END PROCEDURE

INTEGER FUNCTION: DISTANCE
   INTEGER ARG: I1,J1,I2,J2
END FUNCTION

INTEGER FUNCTION: GETSQR
   INTEGER: IDIM
      100 '
      110 FOR IDIM=1 TO ND
      120   IF(IDIM*IDIM >= ND) EXIT
      130 NEXT IDIM
      140 RESULT=IDIM
END FUNCTION

PROCEDURE: PUSH
   EXTERNAL: HOLDDPT( ),POINTER
      10 '
      20 'Put in a stack the dept's that are placed in the nodal
         diagram
      30 '
      100 HOLDDPT(POINTER)=GETDPT
      110 POINTER=POINTER+1
END PROCEDURE

PROCEDURE: PULL
   EXTERNAL: HOLDDPT( ),POINTER
      10 '
      20 'Pull from a stack the dept's placed in the nodal
         diagram
      30 '
      100 POINTER=POINTER-1
      110 GETDPT=HOLDDPT(POINTER)
END PROCEDURE
INTEGER FUNCTION: DISTANCE
      10 'This function calculates the RECTILINEAR DISTANCE
         between two
      20 ' depatrments in a given grid (matrix)
      30 DEFINT I,J
      40 RESULT=(ABS(I1-I2)+ABS(J1-J2)-1)
END FUNCTION
```

```
'MAIN Program:

10  '****************************************************
15  '*                                                  *
20  '*        Microcomputer Based Layout Design         *
25  '*        (A construction type algorithm)           *
27  '*           --->   Chapter 13  <---                *
30  '*                                                  *
40  '*        COMputer LAyout Design  (COMLAD)          *
45  '*                      by                          *
50  '*                                                  *
52  '*                M. Reza Ziai                      *
55  '*                                                  *
57  '*   Mechanical and Industrial Engineering Dept.    *
59  '*           Louisiana Tech University              *
60  '*                 Ruston, La.                      *
62  '*                                                  *
65  '****************************************************
70  DEFINT I-N,C,F,G,R
80  CLS
100 '
102 'GOSUB 10260     'Read input values for the two example
    problems
104 'IFlag2p=2 :GOTO 127        'Skip user input subroutine
106 GOSUB 2000 '
110 GOSUB 10000          'Print the title and get the input for
    the program
120 GOSUB 10500          'Establish the coordinates for THE first
    dept.
125 GOSUB 10800          'Correct the input mistakes
127 INPUT"HIT ENTER TO SEE THE TOTALS";HII
130 GOSUB 11000          'Compute the total value for each
    department
133 IF IFlag2p=2 GOSUB 10500 'get coordinates
140 Ndpt=ND
170 '
180 FOR IR=6 TO 0 STEP -1     'Start with any six possible
    relationships
181    IF IR=6 THEN I4=VA ELSE IF IR=5 I4=VE ELSE IF IR=4 I4=VI
    ELSE IF IR=3 I4=VO ELSE IF IR=2 I4=VU ELSE I4=VX
182    FOR I=1 TO 25 :NG(I)=0: NEXT I
183    FOR N=1 TO ND : Tvalue(N)=TOT(N) : NEXT N
185    GOSUB 12000          'Determine the MAXimum TOTal
187    IF Mtotflag=0 THEN G(II,JJ)=Dpt :Ic=Ic+1 :Hold(Ic)=Dpt
    :Mtotflag=1
190    GOSUB 13000          'Get the dept's with the prevailing
                           relationship for the current highest total
210    GOSUB 14000          'Place the chosen (above) dept's in the
                           nodal diagram where most appropriate.
230    GOSUB 15000          'Find the neighbors for the placed
                           dept's, locate them in the diagram for
                           the current rel'p (IR),and in turn
250    '                    their neighbors and ...
255    IF Ic=>ND THEN EXIT   'If all the depts are already
                           placed print result
260    GOSUB 16000          'Place in the diagram those depts with
                           this IR not already placed.
280 NEXT IR
290 GOSUB 17000          'Output the nodal diagram
```

```
 300 GOSUB 18000 'Get the # of blocks for each dept and set the
                  limits of grid
 310 GOSUB 19000 'Apply the CENTER approach to get a preliminary
                  GRID
 360 GOSUB 27400  'Let user interchange blocks and produce a new
                  GRID with its score!
 370 STOP
1000 END
2000 '=======================================
2010 'Draw the frames for the initial screen
2020 '=======================================
2030 FRAME 4,6,28,18,52
2040 FRAME 4,4,22,20,58
2050 FRAME 4,2,12,22,68
2060 LOCATE 8,31:PRINT "COMputer LAyout Design"
2070 LOCATE 10,39:PRINT"COMLAD"
2080 LOCATE 12,41:PRINT"by"
2085 LOCATE 14,36:PRINT"M. Reza Ziai"
2110 FRAME 6,5,26,19,54
2120 FRAME 6,3,19,21,62
2130 FRAME 6,1,6,23,74
2140 LOCATE 18,30:INPUT"HIT Ret. to continue";HII
2150 CLS:RETURN
10000 '[[[[[[[[[[[[[[]]]]]]]]]]]]]]]]]]]]]]]]]]]]]]]]]]]]]]]]]]]]
10005 'Print out the title and get all the input for the program!
10010 FRAME 9,3,3,7,40
10020 LOCATE 6,5
10026 INPUT " HOW MANY DEPARTMENTS (MAX=20)";ND
10027 IF(ND < 1 OR ND> 20) THEN GOTO 10026
10028 '
10029 '
10072 CLS
10074 PRINT"\\\\\\ Input departmental relationships //////////"
10075 PRINT"////// Upper triangular form is assumed \\\\\\\\"
10076 PRINT"        Use A,E,I,O,U(or <Return> key, and X"
10077 FOR I=1 TO ND-1
10078   FOR J=I+1 TO ND
10080     PRINT"Between department";I;"and department";J;:INPUT
            SREL(I,J)
10082     GOSUB 28000    'check validity of input CODE
10083     IF LR=1 THEN GOTO 10080
10084     SREL(J,I)=SREL(I,J)
10085   NEXT J
10086   PRINT
10087 NEXT I
10090 FOR L=1 TO ND:SREL(L,L)="S": NEXT L  'put S's along the
            diagonal
10092 GOSUB 28100   'ask the user for input values of the CODES
10105 CLS
10107 PRINT SPC(3);
10110 FOR I=1 TO ND
10120   PRINT USING "###";I;
10130 NEXT I
10135 PRINT:PRINT
10140 FOR IJ=1 TO ND
10150   PRINT IJ : IF(IJ=1) THEN C=POS(0) : R=CSRLIN
10160 NEXT IJ
10170 CPOS=C+3 : R=R-1
10180 '
```

```
10185 LOCATE R,CPOS,1,0,12
10190 FOR I=1 TO ND
10200    FOR J=1 TO ND
10210      PRINT " ";SREL(I,J);
10215    NEXT J
10217    PRINT
10220    R=R+1: LOCATE R,CPOS
10230 NEXT I
10255 RETURN
10260 '+++++++++++++++++++++++++++++++++++++++++++++++++
10265 '  READ INPUT VALUES FOR THE TWO SAMPLE PROBLEMS
10270 '+++++++++++++++++++++++++++++++++++++++++++++++++
10275 INPUT "Which problem do you want to run 1 (depts=14) or 2
      (depts=10)";Li
10280 IF Li=2 THEN GOTO 10340
10285 IF Li<> 1 THEN GOTO 10275
10290 READ ND
10295 FOR I=1 TO ND-1
10300    FOR J=I+1 TO ND
10305      READ REL(I,J)
10307      REL(J,I)=REL(I,J)
10310    NEXT J
10315 NEXT I
10320 FOR I=1 TO ND
10325    READ BLKS(I)
10330 NEXT I
10335 IF Li=1 THEN GOTO 10350
10337 IF K=1 THEN GOTO 10350
10340 RESTORE,10375
10342 K=1
10345 GOTO 10290
10350 K=0: RETURN
10355 '------------
10360 DATA 14,4,4,2,2,0,1,0,0,0,0,-1,1,-1,3,2,0,0,0,0,0,0,0,1,
      1,-1,3,3,2,2,0,1,0,0,0,1,-1,0,3,2,1,4,0,0,0,1,-1,2,3,4,1,0,
      0,0,1,1,0,0,0,0,1,2,1,0,0,2,1,1,1,1,0,1,1,0,1,1,0,0,0,1,0,0,
      0,0,0,0,0,0,1,0,0,1
10365 DATA 20,8,7,16,6,8,10,13,10,10,9,9,5,25
10370 'The second problem with 10 departments
10375 DATA 10,4,3,0,0,0,0,0,0,0,2,0,3,3,0,2,0,2,2,0,0,0,0,0,0,
      3,2,2,2,0,0,4,3,0,0,0,4,0,0,0,0,0,0,0,3,0
10380 DATA 13,5,13,11,10,10,6,8,8,5
10385 '
10500 '
10510 '~~~~~~~~~~~~~~~~~~~~~~~~~~~~~~~~~~~~~~~~~~~~~~~~~~~~~~~~~~~~~~~
10520 'Establish the coordinates for the first dept. to enter
        the nodal diagram (this is the dept. with THE highest
        total)
10540 '
10550 '~~~~~~~~~~~~~~~~~~~~~~~~~~~~~~~~~~~~~~~~~~~~~~~~~~~~~~~~~~~~~~~
10560 '
10570 LS=GETSQR(ND)
10580 DO IF (LS < 4)
10590    II=2 : JJ=2
10600 ELSE
10610    II=3 : JJ=3
10620 END DO
10700 RETURN
10800 '~~~~~~~~~~~~~~~~~~~~~~~~~~~~~~~~~~~~~~~~~~~~~~~~~~~~~~~~~~~~~~~}
```

```
10805 ' Correct any mistakes made in the input of the
      relationships
10810 '~~~~~~~~~~~~~~~~~~~~~~~~~~~~~~~~~~~~~~~~~~~~~~~~~~~~~~~~
10815 '
10820 INPUT " Are all the relationships inputted correctly
      (Y/N)";A$
10825 IF ((A$="Y" OR A$="y") AND HII=1) THEN GOTO 10847
10827 IF A$="Y" OR A$="y" THEN GOTO 10850
10828 IF A$="N" OR A$="n" THEN GOTO 10830
10829 IF A$<>"N" OR A$<>"n" THEN PRINT " Type Y  or  N ":GOTO
      10820
10830 INPUT "Correct the relationship between departments _ ,
      _";I,J
10840 INPUT "What is the relationship";SREL(I,J)
      :SREL(J,I)=SREL(I,J)
10842 GOSUB 28000: GOSUB 28240:HII=1 : GOTO 10820
10845 GOTO 10820
10847 GOSUB 10105         'Print the new relationship table
10850 RETURN
10855 '
11000 '~~~~~~~~~~~~~~~~~~~~~~~~~~~~~~~~~~~~~~~~~~~~~~~~~~~~~~~~}
11010 'Compute the total value for each department         }
11020 '                                                     }
11030 '~~~~~~~~~~~~~~~~~~~~~~~~~~~~~~~~~~~~~~~~~~~~~~~~~~~~~~~
11040 '
11050 '
11070 '    Compute totals for each dept. rowwise
11080 FOR I=1 TO ND-1
11090    FOR J=I+1 TO ND
11100       TOT(I)=TOT(I)+REL(I,J)
11110    NEXT I
11120 NEXT J
11130 '    Compute totals for each dept. columnwise
11140 FOR J=2 TO ND
11150    FOR I=1 TO J-1
11160       TOT(J)=TOT(J)+REL(I,J)
11170    NEXT I
11180 NEXT J
11190 '
11195 FOR I=1 TO ND:PRINT "TOTAL FOR DEPT";I;"=";TOT(I) : NEXT I
11200 RETURN
12000 '~~~~~~~~~~~~~~~~~~~~~~~~~~~~~~~~~~~~~~~~~~~~~~~~~~~~~~~~}
12010 '                                                     }
12020 '    Determine the MAXimum TOTal                      }
12030 '                                                     }
12040 '~~~~~~~~~~~~~~~~~~~~~~~~~~~~~~~~~~~~~~~~~~~~~~~~~~~~~~~
12050 Maxx=-9999
12060 FOR I=1 TO Ndpt
12065    IF Hflag = 1 THEN I1=Hold(I) ELSE I1=I
12070    DO IF (Tvalue(I1) >  Maxx)
12080       Maxx=Tvalue(I1)
12090       Dpt=I1
12100    END DO
12110 NEXT I
12112 Ix=Ix+1   'counter for the depts w/ the maximum total,
      found so far
12115 'PRINT "Dpt max=";Dpt;"tot=";Maxx
12120 '
12130 RETURN
```

```
12140 '
12150 '
13000 '~~~~~~~~~~~~~~~~~~~~~~~~~~~~~~~~~~~~~~~~~~~~~~~~~~~~~~~~~~~~~~}
13010 '                                                             }
13020 '   Getting the neighbor departments with the prevailing
          relationship for the current highest total
13040 '~~~~~~~~~~~~~~~~~~~~~~~~~~~~~~~~~~~~~~~~~~~~~~~~~~~~~~~~~~~~
13060 ' Check the neighbor dept's for their relationships
13070 J1=0 : Relflag=0
13075 IF Dpt=ND THEN GOTO 13110
13080 FOR I=Dpt+1 TO ND
13090    IF (REL(Dpt,I) = I4) THEN J1=J1+1 : NG(J1)=I : Relflag=1
13100 NEXT I
13105 '
13110 IF Dpt=1 THEN GOTO 13165
13120 FOR J=1 TO Dpt-1
13130    IF(REL(J,Dpt)=I4) THEN J1=J1+1 : NG(J1)=J : Relflag=1
13140 NEXT J
13160 '
13165 'IF Relflag=1:PRINT "dpt=";Dpt;:FOR I=1 TO J1:PRINT
          "ng:";NG(I);:NEXT I:PRINT
13170 NUM=J1
13400 RETURN
14000 '~~~~~~~~~~~~~~~~~~~~~~~~~~~~~~~~~~~~~~~~~~~~~~~~~~~~~~~~~~~~~}
14030 ' Dept's are placed in the nodal diagram as close to the
          dept's with which they have the strongest relationships
          as possible.~~~~~~~~~~~~~~~~~~~~~~~~~~~~~~~~~~~~~~~~~~~~~
14060 '~~~~~~~~~~~~~~~~~~~~~~~~~~~~~~~~~~~~~~~~~~~~~~~~~~~~~~~~~~~~
14085 '
14100 FOR IG=1 TO NUM            'NUM is the NUMber of NG's for each
                                 relationship
14105    VALUE1=9999    : NHFLAG=0
14110    DO IF NG(IG) > 0      'for a given dept.
14112      FOR N=1 TO Ic
14114        IF NG(IG) = Hold(N) THEN NHFLAG=1 : EXIT
14116      NEXT N
14118      IF NHFLAG=1 THEN EXIT
14120      FOR IP=1 TO LS
14130        FOR JP=1 TO LS
14140          DO IF (G(IP,JP)=0)      'if the position is vacant;
14150            GOSUB 14500                'determine its candidacy
                                                value
14160          DO IF (VALUE2 <= VALUE1)
14170            R=IP : C=JP
14173            'PRINT ")))))))))))) R,C &
                    VALUE1,VALUE2";R;C,VALUE1;VALUE2
14180            VALUE1=VALUE2
14190          END DO
14200          END DO
14210        NEXT JP
14220      NEXT IP
14222      G(R,C)=NG(IG)
14225      Ic=Ic+1 : Hold(Ic)=NG(IG)      'Hold the placed dept in
             memory
14227      'PRINT R;C;"G(IP,JP=";G(R,C)
14230    END DO
14235    Hflag=1               'Set the flag for the max total
                                 subroutine
14240 NEXT IG
```

```
14250 '
14260 '
14270 RETURN
14280 '
14290 '
14500 '~~~~~~~~~~~~~~~~~~~~~~~~~~~~~~~~~~~~~~~~~~~~~~~~~~~~~~~~~~~~~~~}
14510 '                                                           }
14520 ' Determine the candidacy value of the departments          }
14530 '                                                           }
14540 '~~~~~~~~~~~~~~~~~~~~~~~~~~~~~~~~~~~~~~~~~~~~~~~~~~~~~~~~~~~~~~
14545 VALUE2=0
14550 FOR I=1 TO LS
14560   FOR J=1 TO LS
14565     IF G(I,J)=0 THEN GOTO 14580    'If no dept here, no need
                                              to compute
14570       VALUE2=VALUE2+REL(NG(IG) , G(I,J))*((ABS(IP-I)+ABS(JP-
            J)-1))
14580   NEXT J
14590 NEXT I
14600 '
14610 '
14620 RETURN
15000 '
15010 '~~~~~~~~~~~~~~~~~~~~~~~~~~~~~~~~~~~~~~~~~~~~~~~~~~~~~~~~~~~~~~}
15020 '  Get the neighbor departments for the already placed ones
        (in the decreasing order of their total value (TOT), and
        place them in the diagram. Continue this process till all
        the dept's are placed.
15070 '~~~~~~~~~~~~~~~~~~~~~~~~~~~~~~~~~~~~~~~~~~~~~~~~~~~~~~~~~~~~~~
15080 '
15100 FOR K=2 TO Ic        'Ic: # of placed dept's
15120   FOR N=2 TO Ic : Tvalue(Hold(N))=TOT(Hold(N)) : NEXT N
15130   Ndpt=Ic
15134   ' Ix is the index of all the Maximum total depts found so
          far
15135   IF Ix>22 Ix=20
15136   Dptx(Ix)=Dpt
15138   ' Exclude the following depts in the search for the MAX
          TOTal; get the
15139   ' next highest (total) dept.
15140   FOR N=1 TO Ix :Tvalue(Dptx(N))=-999 : NEXT N
15150   GOSUB 12000           'Get the MAX TOTal for the current
                                    dept's.
15160   GOSUB 13000           'Find the neighbors for this MAX TOTal
15170   GOSUB 14000           'Place them in the diagram
15180   ' Tvalue(Dpt)=-999    'Skip this dept in the next search
                                    for the MAX TOTal
15190 NEXT K
15200 '
15499 RETURN
16000 '
16010 '~~~~~~~~~~~~~~~~~~~~~~~~~~~~~~~~~~~~~~~~~~~~~~~~~~~~~~~~~~~~~~}
16020 '  Find any dept(s) with the current relationship not
        already placed in the diagram and locate it (them)
        there.
16070 '~~~~~~~~~~~~~~~~~~~~~~~~~~~~~~~~~~~~~~~~~~~~~~~~~~~~~~~~~~~~~~
16080 '
16090 FOR Ik=1 TO ND
16110   FOR L=1 TO Ic
16115     IF (Ik = Hold(L)) THEN Yflag=1 :EXIT
```

```
16130    NEXT L
16140    DO IF(Yflag = 0)
16143      'PRINT "!!!!!!!!! YOU GOT IN ; Ik=";Ik
16145      Dpt=Ik
16150      GOSUB 13000  'Find any neighbors
16180      IF Relflag=1 THEN NG(J1+1)=NG(1): NG(J1)=Dpt :NUM=2
           :GOSUB 14000
16190    END DO
16195    Yflag=0
16210 NEXT Ik
16220 '
16230 Hflag=0            'Reset the max tot. flag for the next IR
16499 RETURN
17000 '
17010 '~~~~~~~~~~~~~~~~~~~~~~~~~~~~~~~~~~~~~~~~~~~~~~~~~~~~~~~~~~~~~~~}
17020 '              OUTPUT THE NODAL DIAGRAM                       }
17030 '                                                            }
17040 '~~~~~~~~~~~~~~~~~~~~~~~~~~~~~~~~~~~~~~~~~~~~~~~~~~~~~~~~~~~~~~
17090 PRINT:PRINT:INPUT "*** Hit  'Return'  to see the NODAL
                             diagram ** ";A$
17095 CLS
17100 FOR I=1 TO LS
17110    FOR J=1 TO LS
17115      K=G(I,J)
17120      IF K=16 COLOR 1 ELSE IF K=17 COLOR 2   ELSE IF K=18
           COLOR 3 ELSE IF K=19 COLOR 4 ELSE IF K=20 COLOR 5 ELSE
           COLOR K
17130      IF G(I,J)>0 LOCATE I*3-1,J*5-1 :PRINT USING
           "###";G(I,J);
17132      GSAVE(I,J)=G(I,J)    'save the NODAL DIAGRAM for future
           use
17140    NEXT J
17160 NEXT I
17165 COLOR 7: PRINT:PRINT:PRINT
17240 INPUT"Press the 'Shift' key and the 'Prt Sc' key at the
      same time to get a print out  of the NODAL diagram above,
      otherwise, hit 'Return' to continue";A$
17499 RETURN
18000 'MMMMMMMMMMMMMMMMMMMMMMMMMMMMMMMMMMMMMMMMMMMMMMMMMMMMMM
18010 ' Get the number of blocks for each department
18012 '-------------------------------------------------
18018 IF IFlag2p=2 GOTO 18330
18020 PRINT:PRINT
18024 PRINT "Do you want to input the number of blocks for each
      department"
18026 INPUT  "(Y/N)";A$
18028 IF A$="N" OR A$="n" THEN GOTO 18034
18030 IF A$="Y" OR A$="y" THEN FLAG=5 :GOTO 18037
18032 IF A$ <> "Y" THEN GOTO 18026
18034 IF FLAG=5 THEN GOTO 18066
18035 GOTO 18068
18037 FOR L=1 TO ND
18040    IF FLAG=5  PRINT "Number of blocks for department: ";L
         :INPUT BLKS(L)
18045 NEXT L
18047 INPUT "Any MISTAKES ( Y or N)";A$
18050 IF A$="N" OR A$="n" THEN GOTO 18060
18052 IF A$="Y" OR A$="y" THEN GOTO 18055
18053 IF A$<>"Y" OR A$<>"y" THEN INPUT"TYPE  Y  or  N";A$: GOTO
18050
```

```
18055 INPUT "Input   'DEPARTMENT NUMBER , NUMBER OF BLOCKS'--,--
      ";I,BLKS(I)
18057 INPUT "Any more corrections ( Y  or  N )";A$
18060 SUM=0 : FOR J=1 TO ND: SUM=SUM+BLKS(J):NEXT J
18066 IF  FLAG=5 OR IFlag2p=2  THEN GOTO 18330
18068 PRINT:PRINT"Please input departmental areas (square feet)"
18070 FOR I=1 TO ND
18080   PRINT"DEPARTMENT #";I;: INPUT A(I)
18090   IF (A(I) < 0) THEN GOTO 18080
18100 NEXT I
18110 INPUT "Are all the AREAS correct  ( Y  or  N )";A$
18120 IF A$="Y" OR A$="y" THEN GOTO 18220
18125 IF A$="N" OR A$="n" THEN GOTO 18140
18130 IF A$<>"N" OR A$<>"n" THEN PRINT"Type Y or N":GOTO 18110
18140 INPUT "Input the 'DEPARTMENT NUMBER , AREA'";I,A(I)
18145 IF I>Ndpt OR I<1 THEN GOTO 18140
18150 INPUT "All corrections made ( Y or N )";A$ :GOTO 18120
18160 '
18220 CLS:PRINT:PRINT"********** The AREAS as typed by YOU
      ***********" :PRINT
18230 FOR I=1 TO ND :PRINT"The area for the department(";I;")
      =";A(I):NEXT I
18240 PRINT:PRINT "What is the COLUMN SPAN in your layout; "
18245 PRINT"OR, the BLOCK SIZE of the layout (EVEN numbers only)"
18247 INPUT "(AREA of a BLOCK is calculated by squaring the input
      value)";Span
18248 IF Span=0 THEN PRINT:PRINT"Value " 0 " not
      valid!":BEEP:GOTO 18240
18250 M=Span/2 : IF (M-Span/2) <> 0 THEN GOTO 18240
18260 SCALE=Span*Span      'Set the block size
18270 '
18280 FOR I=1 TO ND
18290   BLKS(I)=A(I)/SCALE+.3
18300   SUM=SUM+BLKS(I)
18310 NEXT I
18320 IF SUM >= 608 :PRINT"*** ERROR: #OF BLOCKS HAVE EXCEEDED
      THE MAXIMUM (<=608) ***":STOP
18330 RETURN
19000 '
19010 '---------------------------------------------------------
19020 '     Apply the center approach to get a preliminary GRID
19030 '_____
19035 '
19040 LL=1 : TL=1 : RL=38 : BL=16
19045 KEEPL=LL : KEEPR=RL   :COUNT=0
19050 GOSUB 19200           'Place the highest total dept in the
                            "center"
19055 LL=KEEPL : RL=KEEPR
19057 CNT=1 : POINTER=1
19060 GOSUB 19800           'Find the imm. neighbors and their
                            locations
19065 FR=RII : FC=CJJ
19070 GOSUB 20300           'Place the imm. neighbors in the grid
19080 POINTER=POINTER+1     'Point to the next possible dept to be
                            placed in grid
19090 GD=Place(POINTER)     'Get the dept
19100 GOSUB 19800           'Dept GD has already been placed, find
                            its immediate neighbors.
19107 FR=FRD(GD) : FC=FCD(GD)
```

```
19108 IF J1=0 THEN GOTO 19110
19109 'PRINT: PRINT "GD=";GD;"FR=";FR;"FC=";FC;:FOR L=1 TO
      J1:PRINT "IMNG(";L;")=";IMNG(L):NEXT L
19110 GOSUB 20300          'Place the neighbors in the grid
19120 IF COUNT < ND GOTO 19080
19123 CLS:GOTO 24200   ' PRINT THE G R I D!]
19130 'RETURN
19200 '''''''''''''''''''''''''''''''''''''''''''''''''''''''''''''
19210 ' Place the dept. with the highest total in a relatively
         central position
19220 '''''''''''''''''''''''''''''''''''''''''''''''''''''''''''''
19230 GD=G(II,JJ)
19240 Place(1)=GD             'Keep count of the Placed depts in the
                               grid
19260 R=10 : C=18       'Start placing the first dept here!
19290 TL=1 : LL=1           'Reset Top and Left limits
19294 RII=R : CJJ=C
19295 '******************************************
19300 COUNT=COUNT+1
19310 GD=Place(COUNT)
19320 GRID(R,C)=GD
19325 NBLK=BLKS(GD)
19360 NBLK=NBLK-1
19365 IF NBLK=0  :GOTO 19530
19375 ONE=1      'Search the grid SPIRALLY to place a dept. there
                  (clockwise)
19377 '           to shape the depts as close to a square as
                  possible
19380 FOR I=1 TO 120
19390    CFLAG=0         'Toggle switch for the row and column
19410    DO 2 TIMES
19420      FOR J=1 TO I
19440         IF CFLAG=0 THEN C=C+ONE ELSE R=R+ONE
19441         '     PRINT ">>>> GD=";GD;"  R=";R;"   C=";C
19442         IF R <= 0  OR  C <= 0   THEN GOTO 19490
19445         'If the limits on four sides are not exceeded and the
              grid in
19447         'position R,C is empty,place the dept.there;else
              continue searching
19450         IF (C>RL    OR   R>BL ) GOTO 19485
19455         DO IF ( GRID(R-1,C)=GD OR GRID(R,C-1)=GD OR
              GRID(R,C+1)=GD OR   GRID(R+1,C)=GD   )
19460            IF GRID(R,C)=0    GRID(R,C)=GD : NBLK=NBLK-1
19475            '     PRINT "BLKS=";NBLK
19480         END DO
19485         IF NBLK=0 THEN LR=R : LC=C :EXIT 3 LEVELS
19490      NEXT J
19500      CFLAG=1
19510    REPEAT
19515    ONE=-ONE
19520 NEXT I
19530 RETURN
19550 '
19800 '------------------------------------------------------------
19810 'Find the position of the current dept in the nodal diagram
19820 '& its IMMediate NeiGhbors (i.e. dist=0), and their
         respective locations (1,2,3, or 4); then rank them based
         on the decreasing order of their TOTals.
```

```
19850 '_____
19860 '
19870 ' Find the dept in the nodal diagram
19880 FOR I=1 TO LS
19890   FOR J=1 TO LS
19900     IF G(I,J)=GD THEN Ik=I : JK=J : EXIT 2 LEVELS
19910   NEXT J
19920 NEXT I
19930 G(Ik,JK)=0     'Exclude this dept. from any future search
19940 ' Find the immediate neighbors & set their coordinates
19950 J1=0  : FLAG=0
19958 K=G(Ik-1,JK)
19960 DO IF K <>0 :FLAG=0
19962   FOR L=1 TO CNT
19964     IF K=Place(L) FLAG=1
19966   NEXT L
19968   IF FLAG=0 OR CNT<6 THEN J1=J1+1: IMNG(J1)=K:
            COORD(IMNG(J1))=1
19969 END DO
19970 '
19972 K=G(Ik,JK-1)
19974 DO IF K <>0 :FLAG=0
19976   FOR L=1 TO CNT
19977     IF K=Place(L) FLAG=1
19978   NEXT L
19979   IF FLAG=0 OR CNT<6 THEN J1=J1+1: IMNG(J1)=K:
            COORD(IMNG(J1))=2
19980 END DO
19981 '
19982 K=G(Ik,JK+1)
19983 DO IF K <>0 : FLAG=0
19985   FOR L=1 TO CNT
19987     IF K=Place(L) FLAG=1
19988   NEXT L
19989   IF FLAG=0 OR CNT<6 THEN J1=J1+1:IMNG(J1)=K:
            COORD(IMNG(J1))=3
19991 END DO
19993 '
19995 K=G(Ik+1,JK)
19997 DO IF K <>0 :FLAG=0
19999   FOR L=1 TO CNT
20002     IF K=Place(L) FLAG=1
20004   NEXT L
20006   IF FLAG=0 OR CNT<6 THEN J1=J1+1: IMNG(J1)=K:
            COORD(IMNG(J1))=4
20008 END DO
20070 '
20080 ' Rank the neigbors in the decreasing order of their TOTals
20090 IF J1=0 GOTO 20210
20100 FOR J=1 TO J1
20110   VALUE1=-9999
20120   FOR I=J TO J1
20130     K=TOT(IMNG(I))
20140     IF K > VALUE1 THEN VALUE1=K : KEEPI=I
20150   NEXT I
20160   CNT=CNT+1
20170   Place(CNT)=IMNG(KEEPI)  'Keep the neigh's in the waiting
                                        list to be placed in the grid
20180   SWAP IMNG(KEEPI),IMNG(J)
```

```
20190 NEXT J
20200 '
20205 'FOR L=1 TO J1:PRINT "NG'S IN
        ORDER=";IMNG(L);"COORD=";COORD(IMNG(L));"TOT=";TOT(IMNG(L))
        :NEXT L
20210 RETURN
20300 '-------------------------------------------------------
20310 ' Based on the location of the immediate neighbor (i.e.
        1,2,3 or 4) start the search in that direction to place
        the neighbor
20330 '
20340 IF J1=0 THEN GOTO 20470  'No imm. neighbors found for this
                                dept or they are all already
                                placed
20350 HII=1       'Simulated FOR/NEXT loop
20360 KEEPI=COORD(IMNG(HII))  'Get the position of this imm.
                                neigh. (HII) from storage
20370 DO IF KEEPI=1
20380    GOSUB 20500   'Place this imm. ng. in the grid
20390 ELSE IF KEEPI=2
20400    GOSUB 20750
20410 ELSE IF KEEPI=3
20420    GOSUB 21000
20430 ELSE
20440    GOSUB 21250
20450 END DO
20460 HII=HII+1
20462 IF HII <= J1 GOTO 20360
20470 RETURN
20500 '-------------------------------------------------
20510 ' Search the entire upper sector
20520 '-------------------------------------------------
20530 '
20540 '   1st position
20550 '
20560 FLAG1=1 : Flag1p=1
20570 FOR K=FC TO RL            'Start the search from the upper right
                                sector
20580    FOR L=FR TO TL STEP -1
20590       IF GRID(L,K)=0 THEN R=L:FRD(IMNG(HII))=L: C=K
            :FCD(IMNG(HII))=K:FLAG1=0 : EXIT 2 LEVELS
20600    NEXT L
20610 NEXT K
20620 IF FLAG1=0 GOSUB 19300: GOTO 20720
20630 FOR K=FC-1 TO LL STEP -1    ' Search the upper left sector
20650    FOR L=FR TO TL STEP -1
20660       IF GRID(L,K)=0 THEN R=L : FRD(IMNG(HII))=L: C=K :
            FCD(IMNG(HII))=K: Flag1p=0 : EXIT 2 LEVELS
20670    NEXT L
20680 NEXT K
20690 IF Flag1p=0 THEN GOSUB 19300: GOTO 20720
20700 GOTO 21250      'Search the entire lower sector for any
                       vacancy
20710 '
20720 RETURN
20750 '-------------------------------------------------
20760 ' Search the entire left sector
20770 '-------------------------------------------------
20780 '   2nd position
```

```
20790 Flag2=1 : Flag2p=1
20800 FOR K=FR TO BL          'Search the lower left sector
20810    FOR L=FC TO LL STEP -1
20820       IF GRID(K,L)=0 THEN R=K : FRD(IMNG(HII))=K: C=L :
             FCD(IMNG(HII))=L: Flag2=0 : EXIT 2 LEVELS
20830    NEXT L
20840 NEXT K
20850 IF Flag2=0 GOSUB 19300: GOTO 20950
20860 FOR K=FR-1 TO TL STEP -1   'Now search the lower left
                                  sector
20880    FOR L=FC TO LL STEP -1
20890       IF GRID(K,L)=0 THEN R=K : FRD(IMNG(HII))=K: C=L :
             FCD(IMNG(HII))=L: Flag2p=0 : EXIT 2 LEVELS
20900    NEXT L
20910 NEXT K
20920 IF Flag2p=0 GOSUB 19300: GOTO 20950
20930 GOTO 21000      'Search the entire right sector for any
                       vacancy
20940 '
20950 RETURN
21000 '---------------------------------------------
21010 ' Search the entire right sector
21020 '---------------------------------------------
21030 '    3rd position
21040 Flag3=1 : Flag3p=1
21050 FOR K=FR TO BL       ' Start the search from the lower right
                            sector
21060    FOR L=FC TO RL
21070       IF GRID(K,L)=0 THEN R=K : FRD(IMNG(HII))=K:  C=L :
             FCD(IMNG(HII))=L:Flag3=0 : EXIT 2 LEVELS
21080    NEXT L
21090 NEXT K
21100 IF Flag3=0 THEN GOSUB 19300: GOTO 21200
21110 FOR K=FR-1 TO TL STEP -1 'Continue the search thru the
                                upper right sector
21130    FOR L=FC TO RL
21140       IF GRID(K,L)=0 THEN R=K : FRD(IMNG(HII))=K:  C=L :
             FCD(IMNG(HII))=L: Flag3p=0 : EXIT 2 LEVELS
21150    NEXT L
21160 NEXT K
21170 IF Flag3p=0 THEN GOSUB 19300: GOTO 21200
21180 GOTO 20750     'Search the entire left sector for vacancy
21190 '
21200 RETURN
21250 '-------------------------------------------------
21260 ' Search the entire lower sector
21270 '-------------------------------------------------
21280 '   4th position
21290 Flag4=1 : Flag4p=1
21300 FOR L=FC TO RL      'Start the search from the lower right
                           sector
21310    FOR K=FR TO BL
21320       IF GRID(K,L)=0 THEN R=K : FRD(IMNG(HII))=K:  C=L :
             FCD(IMNG(HII))=L:Flag4=0 : EXIT 2 LEVELS
21330    NEXT K
21340 NEXT L
21350 IF Flag4=0 THEN GOSUB 19300: GOTO 21440
21360 FOR L=FC-1 TO LL STEP -1 'Continue the search thru the
                                lower left sector
21380    FOR K=FR TO BL
```

```
21390       IF GRID(K,L)=0 THEN R=K : FRD(IMNG(HII))=K:   C=L :
            FCD(IMNG(HII))=L: Flag4p=0 : EXIT 2 LEVELS
21400    NEXT K
21410  NEXT L
21420  IF Flag4p=0 THEN GOSUB 19300: GOTO 21440
21430  GOTO 20500          'Search out the entire upper sector for
                            vacancy
21440  RETURN
24000  '!!!!!!!!!!!!!!!!!!!!!!!!!!!!!!!!!!!!!!!!!!!!!!!!!!!!!!!!!!!!!!!!!
24010  '  Print out the NODAL diagram and the GRID
24020  '-------------------------------------------------------------
24030  '
24040  CLS
24045  LOCATE 5,40 :PRINT "         "; : : R=6 :C=44
24050  FOR I=1 TO LS
24060     PRINT SPC(4);I;
24070  NEXT I
24080  FOR J=1 TO LS
24085     R=R+3
24090     LOCATE R,C
24100     PRINT J,: PRINT "    ";:FOR I=1 TO LS:PRINT
           USING"###";GSAVE(J,I);:PRINT SPC(3);:NEXT I
24110  NEXT J
24115  GOTO 24210
24120  C=C+2:LOCATE R-3,C
24130  FOR I=1 TO LS
24140     FOR J=1 TO LS
24150        PRINT SPC(2);G(I,J);
24160     NEXT J
24170     PRINT
24180  NEXT I
24190  ' PRINT THE GRID
24200  FLAG=999
24210  GOSUB 29000   'print out the headings for the GRID
24215  IF FLAG=999 :GOSUB 28700    'Adjust the RL & BL to exclude
           the 0's outside them! : FLAG=0
24217  LOCATE 4,5
24220  FOR I1=1 TO BL
24230     FOR JJ=1 TO RL
24231        I2=GRID(I1,JJ)
24232        IF I2=16 COLOR 3 ELSE IF I2=17 COLOR 4 ELSE IF I2=18
             COLOR 5 ELSE IF I2=19 COLOR 6 ELSE IF I2=20 COLOR 7
             ELSE COLOR I2
24240        PRINT USING "##";GRID(I1,JJ);
24250     NEXT JJ
24260     LOCATE 4+I1,5
24270  NEXT I1
24280  COLOR 7,0
24283  IF Flag4p=8888 RETURN
24285  LOCATE 22,1:PRINT"      The SCORE is being computed, please
           WAIT . . ."
24290  GOSUB 24500        'Compute the score for this configuration
           of GRID
24300  LOCATE 20,1:PRINT " The  S C O R E  for the above LAYOUT
           =";SCORE
24310  PRINT:INPUT"PRESS THE 'Shift' KEY AND THE 'Prt Sc' AT THE
           SAME TIME TO GET A PRINT OUT OF     THE GRID, OTHERWISE, HIT
           'Return' TO CONTINUE";A$: CLS
24400  RETURN
24500  '-------------------------------------------------------------
```

```
24510 ' Score the given GRID; i.e. get the summation of the
          product of the shortest distance for one dept w/ all the
          rest  and  its relationship value with them
24540 '-----------------------------------------------------------
24545 ONE=1 : NDM1=ND-1  : NDD=ND
24550 FOR II1=ONE TO NDM1
24560   IF CFLAG=-999 II1P1=II1+1 ELSE II1P1=ONE+1
24570   FOR JJ1=II1P1 TO NDD
24575     NBLK1=BLKS(II1)    'Get the # of blocks for dept II
24580     DIST1=999    'Initialize the distance
24590     FOR I11=1 TO BL
24600       FOR J11=1 TO RL
24610         GD1=GRID(I11,J11)
24620         DO IF GD1=II1  'Start getting the shortest dist of
                                 dept II with JJ thru ND
24630           NBLK1=NBLK1-1    'Decrement the # of blocks of
                                 dept II by one
24640           GOSUB 24900    'Get the shortest dist of the given
                                 block of II with dept JJ
24650           IF NBLK1=0 THEN MinDIST(II1,JJ1)=DIST1 :
                MinDIST(JJ1,II1)=DIST1 :EXIT 3 LEVELS 'THE
                min-dist found!
24660         END DO
24670       NEXT J11
24680     NEXT I11
24690   NEXT JJ1
24700 NEXT II1
24705 IF CFLAG=-999 GOTO 24720
24710 GOSUB 25100   'Compute the sum of the sums of the products
                        of the MinDIST'swith the   values of the
                        Relationship chart; I.E. THE SCORE!
24720 CFLAG=0 :RETURN
24900 '-------------------------------------------------
24910 'Compute the shortest rectilinear distance of the given
          block of dept II with dept JJ
24930 '-------------------------------------------------
24940 NBLK2=BLKS(JJ1)
24950 FOR I3=1 TO BL
24960   FOR J3=1 TO RL
24970     DO IF GRID(I3,J3)=JJ1
24980       LDIST=(ABS(I11-I3)+ABS(J11-J3)-1)
24990       IF LDIST <= DIST1  DIST1=LDIST
25000       NBLK2=NBLK2-1    'Decrement # of blocks of dept JJ by
                                 one
25010       IF NBLK2=0  EXIT 3 LEVELS   'All the distances of the
                blocks of dept JJ with the given block of dept II
                have been calculated, get next blk of II,if any
25020     END DO
25030   NEXT J3
25040 NEXT I3
25050 '
25060 RETURN
25100 '-----------------------------------------------------------
25110 'Compute the sum of the sums of the products of the
          mindist's with the values of the Relationship Chart
25130 '-----------------------------------------------------------
25140 SUM=0
25150 FOR I4=1 TO ND-1
25160   FOR J4=I4+1 TO ND
```

```
25170      SUM=SUM+REL(I4,J4)*MinDIST(I4,J4)
25180    NEXT J4
25190  NEXT I4
25200  SCORE=SUM       'Score calculated
25210  '
25220  RETURN
27400  '********************************************************
27410  ' User may at this point interchange any of the blocks by
         specifying their coordinates
27430  '----------------------------------------------------------
27435  CFLAG=0 : Flag4p=8888 : GOSUB 24210:Flag4p=0
27440  PRINT:LOCATE 19,2 :INPUT"DO YOU WISH TO INTERCHANGE ANY OF
       THE BLOCKS; YOU WILL AS THE RESULT SEE THE GRID
       AND ITS SCORE (Y/N)";A$:R=CSRLIN:C=POS(0)
27460  IF A$="Y" OR A$="y" THEN GOTO 27485
27470  IF A$= "N" OR A$="n" THEN GOTO 27900
27475  IF A$<>"N" OR A$<>"n" THEN PRINT" Type Y or N ":GOTO 27440
27480  GOTO 27900   'Go to the main routine and end the progrm
27485  STR="                                                    "
27490  LOCATE R-2,C : PRINT "PLEASE PICK YOUR OPTION:";STR+STR
27500  PRINT "    1    for interchanging any TWO blocks                "
27510  PRINT "    2    for interchanging two groups of blocks, EACH
                 ON ONE ROW ONLY"
27520  INPUT "YOUR CHOICE ";MN
27530  IF MN=1 THEN GOTO 27770
27540  IF MN <> 2 THEN :GOTO 27490
27545  LOCATE R-2,C: PRINT
       STR+STR+STR+STR+STR+STR+STR+STR+STR+STR+STR+STR+STR+STR:
       Flag1p=0
27550  LOCATE 20,1 :INPUT"ROW and COLUMN for the first element of
       the FIRST group --,--(e.g. X0,B1)";Srow,Scol : FLAG1=0 :
       FLAG=1 :RCC=Srow :GOSUB 29300: FLAG=0:I1=LL2: RCC=Scol
       :Flag1p=1 :GOSUB 29300: Flag1p=0:J1=LL2
27570  IF ((I1<1 OR I1>BL) OR (J1<1 OR J1>RL)) OR (FLAG1=9)  THEN
       LOCATE 23,C:PRINT"CHECK YOUR COORDINATES!":N=1 :GOTO 27550
27580  IF N=1 THEN LOCATE 23,C:PRINT STR+STR+STR :N=0
27590  LOCATE 21,1: INPUT"COLUMN position for the last element of
       the FIRST group -- (e.g. B1)"; Scol: II1=I1: RCC=Scol
       :Flag1p=1 :GOSUB 29300: Flag1p=0 :JJ1=LL2: IP=JJ1-J1
27600  IF(II1<1 OR II1>BL) OR (JJ1<1 OR JJ1>RL) OR FLAG1=9 OR IP
       <0 THEN LOCATE 23,C: PRINT"Check your coordinates!":N=1 :
       FLAG1=0:GOTO 27590
27605  IF N=1 THEN LOCATE 23,C:PRINT STR+STR+STR: N=0
27610  LOCATE 22,1:   INPUT"ROW and COLUMN for the first element
       of the SECOND group --,--(e.g.X3,A5)";Srow,Scol :RCC=Srow:
       FLAG=1 :GOSUB 29300:FLAG=0:I2=LL2: RCC=Scol :Flag1p=1:
       GOSUB 29300:Flag1p=0: J2=LL2
27620  IF(I2<1 OR I2>BL) OR (J2<1 OR J2>RL) OR FLAG1=9 THEN LOCATE
       23,C: PRINT"CHECK YOUR COORDINATES!": N=1 :FLAG1=0 :GOTO
       27610
27625  IF N=1 THEN LOCATE 23,1 :PRINT STR+STR+STR: N=0
27630  LOCATE 23,C  : INPUT"COLUMN position for the last element
       of the SECOND group -- (e.g. A5)";Scol: II2=I2 :RCC=Scol:
       Flag1p=1 :GOSUB 29300: Flag1p=0:  JJ2=LL2:JP=JJ2-J2
27640  IF(II2<1 OR II2>BL) OR (JJ2<1 OR JJ2>RL) OR FLAG1=9 OR JP<0
       THEN : FLAG1=0 :GOTO 27630
27645  IF JP <> IP THEN LOCATE 20,1 :PRINT"# of COLUMNS MUST be
       the same as the FIRST group's": N=1:GOTO 27630
27650  N=0
```

```
27680 FOR J=J1 TO JJ1
27690    FC=J2+N
27700    SWAP GRID(I1,J),GRID(I2,FC)
27710    N=N+1
27720 NEXT J
27750 GOTO 27820 'Print out the new GRID, its score and end the
               program
27770 LOCATE R-2,C:PRINT STR+STR+STR+STR+STR+STR+STR+STR+STR+
      STR+STR+STR+STR+STR
27774 LOCATE 20,1 :INPUT"ROW and COLUMN for the FIRST block --,--
      (e.g. X4,A2)"; Srow,Scol
27775 RCC=Srow : FLAG=1 :GOSUB 29300:FLAG=0: R1=LL2 : RCC=Scol:
      Flag1p=1 : GOSUB 29300: C1=LL2 :Flag1p=0
27780 IF(R1<1 OR R1>BL) OR (C1<1 OR C1>RL) OR FLAG1=9 THEN LOCATE
      23,1: PRINT"Check your coordinates!": M=1 :FLAG1=0: GOTO
      27774
27785 IF M=1 THEN LOCATE 23,1:PRINT STR+STR+STR :M=0
27790 LOCATE 21,1 :INPUT "ROW and COLUMN for the SECOND block --
      ,--(e.g. Y2,A7)";Srow,Scol
27795 RCC=Srow : FLAG=1 :GOSUB 29300: FLAG=0 :R2=LL2 : RCC=Scol :
      Flag1p=1 : GOSUB 29300:C2=LL2: Flag1p=0
27800 IF(R2<1 OR R2>BL) OR (C2<1 OR C2>RL) OR FLAG1=9 THEN LOCATE
      23,1 :PRINT"Check your coordinates!":FLAG1=0: GOTO 27790
27810 SWAP GRID(R1,C1),GRID(R2,C2)
27820 CLS  :   GOSUB 24210   'Print out the new GRID and its score
27830 Flag4p=8888 :GOSUB 24210:Flag4p=0
27835 LOCATE 20,1 :INPUT"Any more interchange (Y/N)";A$
27840 IF A$="Y" OR A$="y" :GOTO 27485
27845 IF A$="N" OR A$="n" THEN GOTO 27900
27850 IF A$<>"N" OR A$<>"n" THEN PRINT" Type  Y or N " :GOTO
      27835
27900 RETURN
28000 '??????????????????????????????????????????????????????
28005 '  Set all the lower case letters to capital letters
28010 '  & check if SREL( , ) values are inputted correctly?
28020 '------------------------------------------------------
28022 R$=SREL(I,J) : LR=0
28024 IF R$="a" SREL(I,J)="A": R$="A" : GOTO 28050
28026 IF R$="e" SREL(I,J)="E": R$="E" : GOTO 28050
28030 IF R$="i" SREL(I,J)="I": R$="I" : GOTO 28050
28035 IF R$="o" SREL(I,J)="O": R$="O" : GOTO 28050
28040 IF R$="u" OR R$="" SREL(I,J)="U": R$="U" : GOTO 28050
28043 IF R$="x" SREL(I,J)="X": R$="X" : GOTO 28050
28045 '--Check for input validity--
28050 IF R$="A" OR R$="E" OR R$="I" GOTO 28090
28060 IF R$="O" OR R$="U" OR R$="X" GOTO 28090
28070 LR=1     ' set the flag for incorrect relp.
28080 PRINT " ** INVALID CODE **"
28090 RETURN
28100 'I I I I I I I I I I I I I I I I I I I I I I I I I
28110 'User is asked to input values for the CODES
28120 '-----------------------------------------------------
28130 PRINT
28140 PRINT "Do you wish to use the default values for the
               relationships"
28150 INPUT " codes (A=4,E=3,I=2,O=1,U=0,X=-1) (Y/N)";A$
28160 IF A$="y" OR A$="Y" THEN RESTORE,28350: READ
      VA,VE,VI,VO,VU,VX:GOTO 28240
28170 DO IF A$="N" OR A$="n"
```

```
28180     INPUT"Input your values for the codes A,E,I,O,U,X
          respectively, separated by a comma (values must be within
          the range  -20 & 20 ";VA,VE,VI,VO,VU,VX
28190     L=-20 : N=20
28200     IF (VA<L OR VA>N)OR(VE<L OR VE>N)OR(VI<L OR VI>N)OR(VO<L
          OR  VO>N) OR (VU<L OR VU>N) OR (VX<L OR VX>N) THEN GOTO
          28180
28210     EXIT TO,28240
28220 END DO
28230 PRINT "** Type  Y or N **": GOTO 28130
28240 FOR I=1 TO ND-1
28250     FOR J=I+1 TO ND
28260       R$=SREL(I,J)
28270       IF R$="A" THEN REL(I,J)=VA : GOTO 28330
28280       IF R$="E" THEN REL(I,J)=VE : GOTO 28330
28290       IF R$="I" THEN REL(I,J)=VI : GOTO 28330
28300       IF R$="O" THEN REL(I,J)=VO : GOTO 28330
28310       IF R$="U" OR SREL(I,J)=" " THEN REL(I,J)=VU : GOTO
28330
28320       REL(I,J)=VX
28330       REL(J,I)=REL(I,J)
28335     NEXT J
28340 NEXT I
28350 DATA 4,3,2,1,0,-1
28360 RETURN
28700 '-----------------------------------------------------------
28710 ' RL & BL are adjusted so that no 0 outside the four limits
      is printed
28720 '-----------------------------------------------------------
28730 FLAG=0
28750 FOR I=1 TO 21
28760     FLAG1=0
28770     FOR J=1 TO 40
28780       IF GRID(I,J) <>0 FLAG1=99
28790       IF GRID(I,J) <>0 AND FLAG=0 TL=I : Ik=I-1 : FLAG=99
28800     NEXT J
28810     IF FLAG1=99 BL=I
28820 NEXT I
28830 BL=(BL-TL)+1 : TL=1 : FLAG=0   'Top Limit & Bottom Limit are
      adjusted
28840 '
28850 FOR J=1 TO 40
28860     FLAG1=0
28870     FOR I=1 TO 21
28880       IF GRID(I,J) <>0 FLAG1=99
28890       IF GRID(I,J) <>0 AND FLAG=0 LL=J : Ic=J-1 : FLAG=99
28900     NEXT I
28910     IF FLAG1=99   RL=J
28920 NEXT J
28930 RL=(RL-LL)+1 : LL=1 : FLAG=0   'Right Limit & Left Limit are
      adjusted
28940 FOR I=1 TO BL
28950     FOR J=1 TO RL
28960       GRID(I,J)=GRID(I+Ik,J+Ic)
28970     NEXT J
28980 NEXT I
28990 RETURN
29000 '------------------------------------
29010 ' Print out the headings for the GRID
29020 '------------------------------------
```

```
29030 PRINT SPC(12);"A";SPC(19);"B";SPC(19);"C";SPC(19);"D"
29040 STR="\1 2 3 4 5 6 7 8 9 0"
29050 PRINT "      ";STR+STR+STR+"\1 2 3 4 5 6 7 8"
29060 LOCATE 8,1:PRINT"X" : LOCATE 14,1: PRINT ">" :LOCATE
      18,1:PRINT "Y"
29065 LL1=0
29070 FOR L2=1 TO 16
29080    LL1=LL1+1
29090    LOCATE L2+3,2 : PRINT LL1
29100    IF LL1=9 LL1=-1
29110 NEXT L2
29120 RETURN
29300 '------------------------------------------------------------
29310 'convert the STRING position of the coordinate into numeric
      value
29320 '------------------------------------------------------------
29330 Sub1=LEFT$(RCC,1)
29340 Sub2=RIGHT$(RCC,1)
29342 IF LEN(RCC) > 2 THEN FLAG1=9
29344 DO IF Flag1p=1: IF (Sub1="X" OR Sub1="x" OR Sub1="Y" OR
      Sub1="y") THEN FLAG1=9:END DO
29345 DO IF FLAG=1: IF  (Sub1="A" OR Sub1="a" OR Sub1="B" OR
      Sub1="b" OR Sub1="C" OR Sub1="c") THEN  FLAG1=9 :END DO
29346 IF Sub1="X" OR Sub1="x" Sub1="A"
29347 IF Sub1="Y" OR Sub1="y" Sub1="B"
29350 IF Sub1="A" OR Sub1="a" L1=0 ELSE IF Sub1="B" OR Sub1="b"
L1=10  ELSE IF Sub1="C" OR Sub1="c" L1=20 ELSE  L1=30
29360 L2=VAL(Sub2) : IF L2=0 THEN L2=10
29370 LL2=L1+L2
29390 RETURN

ENDFILE
```

EXAMPLE PROBLEM FOR THE COMPUTER PROGRAM

	OFFICE	AREA (Sq. Feet)
1)	PRESIDENT	250
2)	VICE PRESIDENT	200
3)	PLANT MANAGER	200
4)	SALES MANAGER	200
5)	ENGINEERS	500
6)	PERSONNEL	250
7)	SUPERVISORS	250
8)	SALES PERSONS	300
9)	SECRETARIES	400
10)	RECEPTIONIST	200
11)	MEETING ROOM	300
12)	UTILITY ROOM	150
13)	CAFETERIA	250
14)	REST ROOMS	300

RELATIONSHIP CHART

	1	2	3	4	5	6	7	8	9	10	11	12	13	14
1	-	A	E	E	I	O	I	O	A	E	U	X	U	U
2		-	O	I	I	A	E	E	U	U	O	U	U	X
3			-	E	I	U	U	O	I	O	U	U	U	E
4				-	E	I	O	A	E	U	U	U	U	U
5					-	U	O	I	O	I	I	U	O	O
6						-	O	E	I	O	U	U	U	X
7							-	A	A	O	U	U	O	X
8								-	O	O	I	U	U	X
9									-	O	I	O	I	O
10										-	I	O	I	X
11											-	X	E	E
12												-	I	O
13													-	O
14														-

NOTE: In the following pages clarifications for the program are
      illustrated by "NOTES", they do not form part of the
      program.

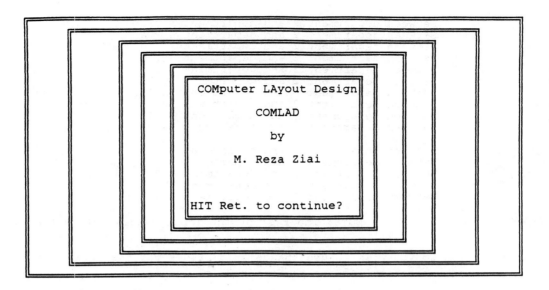

```
COMputer LAyout Design

COMLAD

by

M. Reza Ziai


HIT Ret. to continue?
```

```
HOW MANY DEPARTMENTS (MAX=20)? 14
```

```
\\\\\\\\\\\\ Input departmental relationships ////////////////
//////////// Upper triangular form is assumed \\\\\\\\\\\\\\\\
           Use A,E,I,O,U(or <Return> key, and X
Between department 1 and department 2 ? a
Between department 1 and department 3 ? e
Between department 1 and department 4 ? e
Between department 1 and department 5 ? i
Between department 1 and department 6 ? o
Between department 1 and department 7 ? i
Between department 1 and department 8 ? o
Between department 1 and department 9 ? a
Between department 1 and department 10 ? e
Between department 1 and department 11 ?
Between department 1 and department 12 ? x
Between department 1 and department 13 ?
Between department 1 and department 14 ? y
  ** INVALID CODE **
Between department 1 and department 14 ? u

Between department 2 and department 3 ? o
Between department 2 and department 4 ? i
```

```
Between department 2 and department 5 ? i
Between department 2 and department 6 ? a
Between department 2 and department 7 ? e
Between department 2 and department 8 ? e
Between department 2 and department 9 ?
Between department 2 and department 10 ?
Between department 2 and department 11 ? o
Between department 2 and department 12 ?
Between department 2 and department 13 ?
Between department 2 and department 14 ? x

Between department 3 and department 4 ? e
Between department 3 and department 5 ? i
Between department 3 and department 6 ?
Between department 3 and department 7 ?
Between department 3 and department 8 ? o
```

NOTE: The rest of the input values follow the above format

Do you wish to use the default values for the relationships
 codes (A=4,E=3,I=2,O=1,U=0,X=-1) (Y/N)? y

NOTE: Display of input relationship chart

```
       1  2  3  4  5  6  7  8  9 10 11 12 13 14

   1   S  A  E  E  I  O  I  O  A  E  U  X  U  U
   2   A  S  O  I  I  A  E  E  U  U  O  U  U  X
   3   E  O  S  E  I  U  U  O  I  O  U  U  U  E
   4   E  I  E  S  E  I  O  A  E  U  U  U  U  U
   5   I  I  I  E  S  U  O  I  O  I  I  U  O  O
   6   O  A  U  I  U  S  O  E  I  O  U  U  U  X
   7   I  E  U  O  O  O  S  A  A  O  U  U  O  X
   8   O  E  O  A  I  E  A  S  O  O  I  U  U  X
   9   A  U  I  E  O  I  A  O  S  O  I  O  I  O
  10   E  U  O  U  I  O  O  O  O  S  I  O  I  X
  11   U  O  U  U  I  U  U  I  I  I  S  X  E  E
  12   X  U  U  U  U  U  U  U  O  O  X  S  I  O
  13   U  U  U  U  O  U  O  U  I  I  E  I  S  O
  14   U  X  E  U  O  X  X  X  O  X  E  O  O  S
```
     Are all the relationships inputted correctly (Y/N)? y

HIT ENTER TO SEE THE TOTALS?

```
TOTAL FOR DEPT 1 = 22
TOTAL FOR DEPT 2 = 19
TOTAL FOR DEPT 3 = 16
TOTAL FOR DEPT 4 = 21
TOTAL FOR DEPT 5 = 19
TOTAL FOR DEPT 6 = 13
TOTAL FOR DEPT 7 = 17
TOTAL FOR DEPT 8 = 21
TOTAL FOR DEPT 9 = 24
TOTAL FOR DEPT 10 = 14
TOTAL FOR DEPT 11 = 14
TOTAL FOR DEPT 12 = 3
TOTAL FOR DEPT 13 = 12
TOTAL FOR DEPT 14 = 5
```

```
*** Hit  'Return'  to see the NODAL diagram ** ?


   12   13   11   14


    5   10    4    3
                                        NOTE: NODAL diagram
            8    9    7


            6    1    2
```

Press the 'Shift' key and the 'Prt Sc' key at the same time to get a print out
of the NODAL diagram above, otherwise, hit 'Return' to continue?

Do you want to input the number of blocks for each department
(Y/N)? y

```
            Number of blocks for department:  1
            ? 7
            Number of blocks for department:  2
            ? 6
            Number of blocks for department:  3
            ? 6
            Number of blocks for department:  4
            ? 6
            Number of blocks for department:  5
            ? 14
            Number of blocks for department:  6
            ? 7
            Number of blocks for department:  7
            ? 7
            Number of blocks for department:  8
            ? 9
            Number of blocks for department:  9
            ? 11
            Number of blocks for department:  10
            ? 6
            Number of blocks for department:  11
            ? 9
            Number of blocks for department:  12
            ? 4
            Number of blocks for department:  13
            ? 7
            Number of blocks for department:  14
            ? 9
            Any MISTAKES ( Y or N)? n
```


                NOTE: OR, if the AREAS are to be inputted

```
Do you want to input the number of blocks for each department
(Y/N)? n

Please input departmental areas (square feet)
DEPARTMENT # 1 ? 250
DEPARTMENT # 2 ? 200
DEPARTMENT # 3 ? 200
DEPARTMENT # 4 ? 200
DEPARTMENT # 5 ? 500
DEPARTMENT # 6 ? 250
DEPARTMENT # 7 ? 250
DEPARTMENT # 8 ? 300
DEPARTMENT # 9 ? 400
DEPARTMENT # 10 ? 200
DEPARTMENT # 11 ? 300
DEPARTMENT # 12 ? 150
DEPARTMENT # 13 ? 250
DEPARTMENT # 14 ? 250
Are all the AREAS correct  ( Y  or  N )? n
Input the 'DEPARTMENT NUMBER , AREA'? 14,300
All corrections made ( Y or N )? y


*********** The AREAS as typed by YOU ************

The area for the department( 1 ) = 250
The area for the department( 2 ) = 200
The area for the department( 3 ) = 200
The area for the department( 4 ) = 200
The area for the department( 5 ) = 500
The area for the department( 6 ) = 250
The area for the department( 7 ) = 250
The area for the department( 8 ) = 300
The area for the department( 9 ) = 400
The area for the department( 10 ) = 200
The area for the department( 11 ) = 300
The area for the department( 12 ) = 150
The area for the department( 13 ) = 250
The area for the department( 14 ) = 300

What is the COLUMN SPAN in your layout;
OR, the BLOCK SIZE of the layout (EVEN numbers only)
(AREA of a BLOCK is calculated by squaring the input value)? 6


NOTE: The output layout will be seen on the next screen


NOTE: Depending on the number of departments and the number of
      blocks for each, it may take several minutes before the
      SCORE is computed, so please be patient!
```

```
              A                    B                    C                    D
        \1 2 3 4 5 6 7 8 9 0\1 2 3 4 5 6 7 8 9 0\1 2 3 4 5 6 7 8 9 0\1 2 3 4 5 6 7 8

   1    0 0 0 0 0 0 0 014141414 0
   2    0 0 0 013131313314141414 0
   3    0 012131313111111114 0 0
   4    5 512121211111111111 0 0 0
X  5    5 5 5101010 4 4 4 3 3 3 0
   6    5 5 5101010 4 4 4 3 3 3 0
   7    5 5 5 0 8 8 9 9 9 9 7 7 0
   8    5 5 5 0 8 8 9 9 9 9 7 7 0
   9    0 6 6 6 8 8 9 9 9 7 7 7 0
   0    0 6 6 8 8 8 1 1 1 1 2 2 2
>  1    0 6 6 0 0 0 1 1 1 2 2 2 0
   2
   3
   4
Y  5
   6
```

NOTE: "0" indicates there is no

department in that position

NOTE: All the following prompts are displayed at the bottom of
        the screen

The SCORE is being computed, please WAIT . . .

The S C O R E for the above LAYOUT = 219

PRESS THE 'Shift' KEY AND THE 'Prt Sc' AT THE SAME TIME TO GET A PRINT OUT OF
THE GRID, OTHERWISE, HIT 'Return' TO CONTINUE?

NOTE: Once 'Return' key is HIT, the question below is asked

DO YOU WISH TO INTERCHANGE ANY OF THE BLOCKS; YOU WILL AS THE
RESULT SEE THE GRID AND ITS SCORE (Y/N)? Y

NOTE: In response to the 'Y', two choices are presented to the
        user

PLEASE PICK YOUR OPTION:
    1   for interchanging any TWO blocks
    2   for interchanging two groups of blocks, EACH ON ONE ROW ONLY
YOUR CHOICE ? 1

NOTE: User inputs the coordinates for the TWO blocks of his/her
        choice

NOTE: X and Y coordinates of the blocks are identified by
      associated label headings; for example, A6,X7 indicates one
      block of department 8
ROW and COLUMN for the FIRST block --,--(e.g. X4,A2)? a0,a0
Check your coordinates!
ROW and COLUMN for the FIRST block --,--(e.g. X4,A2)? x0,a0

ROW and COLUMN for the SECOND block --,--(e.g. Y2,A7)? y1,b6
Check your coordinates!
ROW and COLUMN for the SECOND block --,--(e.g. Y2,A7)? y1,a6

      NOTE: The screen below reflects the exchanged blocks and the
            SCORE

```
            A                     B                     C                     D
        \1 2 3 4 5 6 7 8 9 0\1 2 3 4 5 6 7 8 9 0\1 2 3 4 5 6 7 8 9 0\1 2 3 4 5 6 7 8

    1    0 0 0 0 0 0 0 014141414 0
    2    0 0 0 01313131314141414 0
    3    0 0121313131111111114 0 0
    4    5 51212121111111111 0 0 0
X   5    5 5 5101010 4 4 4 3 3 3 0
    6    5 5 5101010 4 4 4 3 3 3 0
    7    5 5 5 0 8 8 9 9 9 9 7 7 0
    8    5 5 5 0 8 8 9 9 9 9 7 7 0
    9    0 6 6 6 8 8 9 9 9 7 7 7 0
    0    0 6 6 8 8 8 1 1 1 0 2 2 2
>   1    0 6 6 0 0 1 1 1 1 2 2 2 0
    2
    3
    4
Y   5
    6
```

      The SCORE is being computed, please WAIT . . .

The S C O R E for the above LAYOUT = 223

PRESS THE 'Shift' KEY AND THE 'Prt Sc' AT THE SAME TIME TO GET A PRINT OUT OF
THE GRID, OTHERWISE, HIT 'Return' TO CONTINUE?

      NOTE: After the 'Return' key is HIT, the user is asked a series
            of questions similar to the previous case

Any more interchange (Y/N) ? y

PLEASE PICK YOUR OPTION:
    1   for interchanging any TWO blocks
    2   for interchanging two groups of blocks, EACH ON ONE ROW ONLY
YOUR CHOICE ? 2

ROW and COLUMN for the first element of the FIRST group --,--(e.g. X0,B1)? x0,b1

COLUMN position for the last element of the FIRST group -- (e.g. B1)? b3
ROW and COLUMN for the first element of the SECOND group --,--(e.g.X3,A5)? x0,a0

COLUMN position for the last element of the SECOND group -- (e.g. A5)? b2

```
            NOTE: The screen below displays the layout reflecting the
                  exchanged blocks and the new SCORE

            A                    B                    C                    D
       \1 2 3 4 5 6 7 8 9 0\1 2 3 4 5 6 7 8 9 0\1 2 3 4 5 6 7 8 9 0\1 2 3 4 5 6 7 8

    1   0 0 0 0 0 0 0 014141414 0
    2   0 0 0 01313131314141414 0
    3   0 0121313131111111114 0 0
    4   5 51212121111111111 0 0 0
X   5   5 5 5101010 4 4 4 3 3 3 0
    6   5 5 5101010 4 4 4 3 3 3 0
    7   5 5 5 0 8 8 9 9 9 9 7 7 0
    8   5 5 5 0 8 8 9 9 9 9 7 7 0
    9   0 6 6 6 8 8 9 9 9 7 7 7 0
    0   0 6 6 8 8 8 1 1 1 2 2 2 0
>   1   0 6 6 0 0 1 1 1 1 2 2 2 0
    2
    3
    4
Y   5
    6
```

The SCORE is being computed, please WAIT . . .

The S C O R E for the above LAYOUT = 215

PRESS THE 'Shift' KEY AND THE 'Prt Sc' AT THE SAME TIME TO GET A PRINT OUT OF
THE GRID, OTHERWISE, HIT 'Return' TO CONTINUE?

```
            NOTE: Three blocks of department 2 are shifted to left by one
                  position to reduce the size of the layout by one column.  To
                  accomplish this, one block of department 1 was moved to a
                  different row to make room for the above shift.  This example
                  shows that only interchange of one block or one row at a time is
                  permissible in the program
```

Any more interchange (Y/N)? n

Ok

```
            NOTE: Since no more interchange is desired, the program is
                  terminated
```

```
 10 '********************************************************
 15 '*                                                      *
 20 '*          DATA GUIDE TO COMPUTER PROGRAMS             *
 30 '*                  Chapter 14                          *
 40 '*                                                      *
 45 '********************************************************
 46 '
 48 '
 50 '       Listed below is a condensed manual describing the
 60 ' necessary data for running some of the popular computer
 70 ' programs.  For each program the required input information
 80 ' and resulting computer output, along with a detailed
 90 ' description of input lines, are presented.
100 '
101 '
110 '
111 ' 1  CRAFT
112 '
120 '
130 ' 1.1   Input Data
140 '       The data required to run Craft are:
150 '       (1) An initial layout
160 '       (2) From-to chart that expresses the flow of materials
170 '       (3) The cost of moving materials between te different
180 '       (4) The number and location of any department(s) to be
190 '           fixed in its position as given in the initial
200 '           layout
210 '
220 ' 1.2   The Output
230 '       CRAFT will print an output that contains (1) the input
240 ' given to the program (flow data, cost, and the initial
250 ' layout); (2) a matrix that represents the product of move
260 ' cost and flow data; (3) the final layout.  In addition, the
270 ' change of location made, the resulting cost savings, and
280 ' the total cost associated with the final layout are also
290 ' printed.
300 '
310 ' 1.3   Input Lines
320 '       Four typees of lines are required to run CRAFT.  These
330 ' are explained below.
340 '
350 ' Line type 1:  one line is required.
360 ' Col.          Format                 Description
370 ' 1-2            I2            Number of departments in the layout,
380 '                             maximum number allowed is 40.
390 ' 3-4            I2            Number of rows in the scaled layout,
400 '                             maximum number is 30.
410 ' 5-6            I2            Number of columns in the scaled
420 '                             layout, maximum number is 30
430 ' 7-8            I2            A code to indicate the type of
440 '                             interchange to be attempted;
450 '                             00: 2 department interchange
460 '                             01: 3 department interchange
470 '                             02: 2 department followed by
480 '                                 3 department interchanges
490 '                             03: 3 department followed by 2
500 '                                 department interchanges
510 '                             04: select the best of 2 and 3
520 '                                 interchanges
```

```
530 ' 9-10            I2          A code to indicate whether to
540 '                             print a layout after each
550 '                             iteration or not:
560 '                             00: print only the initial and
570 '                                 the  final layouts
580 '                             01: printed a layout at the end of
590 '                                 each iteration
600 ' 11-12           I2          Debugging parameters:
610 '                             00: do not print error messages
620 '                             01: print the exchange failure
630 '                             02: print the exchange failure
640 '                                 and search for the best
650 '                                 interchanges
660 ' 13-14           I2          The number of departments to be
670 '                             fixed in its position.  Two columns
680 '                             initial layout
690 ' 15-79           I2          The number of the department to be
700 '                             fixed in its position.  Two columns
710 '                             for each department.
720 '
730 ' Line type 2
740 '       The number of lines required here equals the number of
750 ' departments.  Each line corresponds to a row in hte from-to
760 ' chart.  If the number of departments is more than 20, then
770 ' a second line is used to represent the same row in the
780 ' from-to chart.  Format (20 F 4.0)
790 '
800 ' Line type 3
810 '       The number of lines required here is also equal to the
820 ' number of departments.  Each line corrosponds to a row in
830 ' the cost matrix.  If the number of departments is greater
840 ' that 20, a second line can be used to represent the same
          row in the matrix.  The  diagonal of this matrix should be
          zeros
860 ' since the move cost between a department and itself is
870 ' zero.  Format ( 20 F 4.3)
880 '
890 ' Line type 4
900 '       This line is used to input the initial layout.  Each
910 ' line corresponds to a row in the scaled layout (which
920 ' represents, the initial layout).  The numbers 01, 02, 03,.
930 ' ..., 26 are punched on this line corrosponding to the
940 ' department letters A, B, C, . . . . ., Z on the scaled
950 ' layout.  Format (30 I2)
960 '
970 '
980 '
985 '
987 '
990 ' 2   CORELAP
995 '
996 '
1000 '2.1  Input Data
1010 '      The data required for running CORELAP are:
1020 '      (1)  Relationship chart.  The user has to assign
1030 '           arbitrary numerical values to the elements of
                 this chart. (e.g. A = 6, B = 5, etc.)
1050 '      (2)  The number of departments and their areas.
```

```
1060 '      (3)   Scale, required length to width ratio for the
1070 '            final layout, and the departments required to be
1080 '            fixed in a particular position.
1090 '
1100 '
1110 '2.2   The Output
1120 '      The output provided by the program contains the input
1130 'data, the total closeness rating of each department, the
1140 'final layout, and the distances between all the
     departments.
1160 '
1170 '
1180 '2.3   Input Lines
1190 '      Five types of lines are required to run CORELAP.   In
1200 'addition, more than one problem can be solved in the same
1210 'run.
1220 '
1230 'Line type 1:  one line is required.   The user specifies in
1240 'this line the numerical values he wishes to assign to the
1250 'ratings of the relationship chart.   The higher the rating,
1260 'the higher the numerical value assigned to it.   If a
1270 'department is to be fixed in a position, it should have the
1280 'highest numerical value.
1290 'Col.                   Format                  Description
1300 '1-5                     I5           The numerical value the user
1310 '                                     assigns to X rating.
1320 '6-10                    I5           Value of U rating.
1330 '11-15                   I5           Value of O rating.
1340 '16-20                   I5           Value of I rating.
1350 '21-25                   I5           Value of E rating.
1360 '26-30                   I5           Value of A rating.
1370 '31-35                   I5           Value for a preassigned
1380 '                                     department.
1390 '
1400 'Line type 2:  one line is need.   This line is used to
1410 'Col.                   Format                  Description
1420 '1-3                     I3           Problem identification number
1430 '5-80                   I9A4          Anything to identify the
                                          problem
1440 '
1450 'Line type 3:  one line is required
1460 'Col.                   Format                  Description
1470 '1-3                     I3           No. of departments, maximum
1480 '                                     number is 70.
1490 '4-6                     I3           Scale, i. e., the side length
1500 '                                     of a unit square in the
                                          output.
1510 '                                     This scale should be such that
1520 '                                     the number of unit squares in
1530 '                                     the output is less than 1520.
1540 '7-9                     I3           Max. length to width ratio
1550 '                                     (optional).
1560 '10-14                  F5.2          Desired length to width ratio
1570 '                                     which specifies the shape of
1580 '                                     the area available for the
1590 '                                     layout (optional).
1600 '15-17                  F3.2          Layout filling ratio indicates
1610 '                                     the proportion of the layout
1620 '                                     area to be filled by the
                                          actual departments.  Must be
                                          less than 1 (optional).
```

```
1650 '18                    I1              Type '1' if a punched deck of
1660 '                                      the final layout is to be
1670 '                                      produced (optional).
1680 '19                    I1              Put '1' if partial layouts
1690 '                                      are desired (optional).
1700 '        Note that all the fields specified as optional can be
1710 'left blank.
1720 '
1730 'Line type 4:  The number of lines required is equal to the
1740 'number of departments.
1750 'Col.                  Format                        Description
1760 '1-6                   I6              Area of the department.   The
1770 '                                      first department area is
1780 '                                      specified on the first line,
1790 '                                      the second department on the
1800 '                                      second line, etc.
1810 '7-10                  4I1             Indicates which side of the
1820 '                                      layout, the department should
1830 '                                      be placed (i.e., this field
1840 '                                      is used if a department is to
1850 '                                      be preassigned in a particular
1860 '                                      position
1870 '                                          Col. 7 represents North
1880 '                                          Col. 8 represents West
1890 '                                          Col. 9 represents South
1900 '                                          Col. 10 represents East
1910 '                                      Put "1" in one of the above
1920 '                                      columns to represent the side
1930 '                                      on which the department is to
1940 '                                      be placed.  If a "1" is placed
1950 '                                      in two adjacent columns, the
1960 '                                      department will be placed in
1970 '                                      the corner.  If a department
1980 '                                      is not preassigned then leave
1990 '                                      this field blank.
2000 '11-80                 70I1            The elements of the
2010 '                                      relationship chart are
2020 '                                      recorded here.   Enter
2030 '                                          1 for X
2040 '                                          2 for U
2050 '                                          3 for O
2060 '                                          4 for I
2070 '                                          5 for E
2080 '                                          6 for A
2090 '                                      This matrix is symmetrical.
2100 '     If more than one problem is to be solved in the same
2110 'run, the data for the second problem should be entered
2120 'after line type 4, and it starts with line type 2.
2130 '
2140 'Line type 5:  one line is needed.  This line signals the
2150 'end of the run.  Put "-1" in col. 1-3.
2160 '
2170 '
2180 '
2190 '3   ALDEP
2200 '
2210 '
2220 '3.1   Input Data
2230 '        The data required to run ALDEP are:
2240 '        (1) The area of each department.
```

```
2250 '      (2) Relationship Chart.  However, the user does not
2260 '          have to assign numerical values to the ratings in
2270 '          this chart since the program does this for him.
2280 '      (3) A random number seed, scale, the number of layouts
2290 '          to be generated, and the minmum score a layout
2300 '          should have in order to be printed.  That minimum
2310 '          score can be a zero.
2320 '      (4) The number of unit squares that constitutes the
2330 '          width of a department in the output.
2340 '
2350 '3.2  The Output
2360 '     The output provided by the program contains the
2370 'relationship chart with numerical values assigned to its
2380 'elements, the score of each layout generated, and the
2390 'layout generated.
2400 '
2410 '3.3  Input Lines
2420 '     Six types of lines are required to run the program.
2430 'These are explained below.
2440 '
2450 'Line type 1:  one line is required.
2460 'Col.                  Format                    Description
2470 '1-4                     I4            An odd number used as a
                                           random number seed.
2490 '5-6                     I2            Column number to begin
                                           placing departments for
                                           the top floor.
2510 '7-8                     I2            Department width for the
                                           top floor.  Leave it
                                           blank if there is no top
                                           floor.
2540 '9-12                    I4            Number of unit square to
                                           be available on the top
                                           floor.
2560 '                                     This number is the area
                                           of the floor divided by
                                           the scale used.  Leave
                                           it blank if there is no
                                           top floor.
2600 '13-16                   I4            Layout outline width in
                                           number of units squares
                                           on the top floor.  This
                                           is equal to the width of
                                           the floor divided by
2640 '                                     the side of the unit
                                           square used as a scale.
                                           Leave it blank if there
                                           is no top floor.
2680 '17-20                   I4            Layout outline length in
2690 '                                     number of unit squares on
                                           the top floor.
2710 '
2720 '     The above fields, except for the seed number, are for
2730 'the top floor.  Data for the main floor are punched on the
2740 'same line, in the same way as above, with the same format.
2750 'The columns used are:
2760 '     (21-22), (23-24), (25-28), (29-32), (33-36).
2770 'The basement data are punched in a similar way on the same
2780 'lines in the following columns:
2790 '     (37-38), (39-40), (41-44), (45-48), (49-52).
```

```
2800 '
2810 '53-57              F5.1                 Number of square feet in
                                              a unit square.  (i.e.,
                                              scale)
2830 '58-61              F4.2                 Rounding off factor for
                                              the program to use when
                                              it calculated the number
                                              of unit squares for each
                                              department.
2870 '62-65              I4                   Number of layouts to be
2880 '                                        generated.  Maximum
                                              number is 20.
2900 '66-69              I4                   Minumum allowable score a
2910 '                                        layout must have if it is
                                              to be printed.
2930 '70-72              I3                   Minimum rating a
                                              department must have with
                                              a selected department in
                                              order to be considered
                                              for placement beside the
                                              selected one.
2980 '                                        Enter:
2990 '
3000 '                                        64 if the relationship
3010 '                                        required is A.
3020 '                                        16 for E
3030 '                                        4 for I
3040 '                                        1 for O
3050 '
3060 'Line type 3:  one line is required.
3070 'Col.                Format                           Description
3080 '1-8                 2A4                  Width of the top floor in
3090 '                                         square units, plus 2,
                                               followed by I3 enclosed
                                               in parameters.
3110 '9-16                2A4                  Same as above but for
                                               main floor.
3130 '17-24               2A4                  Same as above but for
                                               basement.
3140 '
3150 'Line type 4:  the number of lines required is equal to the
3160 'number of departments.  This line is used to input
3170 'department number and area.  Department 1 should be input
3180 'as 111, 2 as 112, etc.  The maximum number of departments
3190 'is 64.
3200 'Col.                Format                           Description
3210 '1-4                 I4                   Department number.
3220 '7-13                F7.0                 Area of the department.
3230 'After these lines, a blank line is inserted.
3240 '
3250 'Line type 5:  the number of lines required is equal to the
3260 'number of departments.  These are used for the relationship
3270 'chart.  The chart is inserted as a triangular matrix.  Each
3280 'line corresponds to a row in the chart.  Enter the letter
3290 'ratings and put an "S" on the diagonal.
3300 'Col.                Format                           Description
3310 '1-3                 I3                   Department number, put in
                                               the same way as in card
                                               type 4.
```

```
3330 '4-67                 63A1                Elements of the
                                               relationship
3340 '                                         chart in a row.
3350 '
3360 'Line type 6:  the number of lines required here is
3370 'variable, depending on the number of restrictions.  If
3380 'there are no restrictions on the layout do not include this
3390 'line.
3400 'Col.                 Format                   Description
3410 '1                    A1                  Put A if an area is to be
3420 '                                         assigned for a particular
                                               use.
3430 '                                         Put F if a department is
                                               to be assigned to a
                                               specific floor.
3450 '2-3                  I2                  If the data of the first
                                               line are to be entered
                                               in this line only then
                                               put "1" in this field.
                                               If it is to be continued
                                               on subsequent
                                               line, then mark that
                                               continuation line by
                                               putting 2, 3, ...
                                               in this field.
3530 '4-7                  I4                  Type of area being
                                               restricted, or department
                                               number:
3550 '                                             1 = dock
3560 '                                             2 = elevator
3570 '                                             3 = stairwell
3580 '                                             4 = lobby
3590 '                                            19 = aisle
3600 '                                            88 = dummy department
3610 '8-9                  I2                  The floor number in which
                                               the restrictions apply:
3630 '                                             1 = top floor
3640 '                                             2 = main floor
3650 '                                             3 = basement
3660 '10-12                I3                  Total number of unit
                                               square allocated to the
                                               designated area.
3690 '13-72                30I2                Column and row members
                                               for each designated unit
                                               square.  Each number should
                                               be a two digit number.
                                               Use a second line if
                                               the number of unit squares
                                               is more than 15.
3750 '
3760 'Last line:  at the end of the data, put a line with the
3770 'letter "A" in col. 1.
3780 '
3790 '
3800 '
3810 '4   PLANET
3820 '
3830 '
3840 '4.1   Input Data
```

```
3850 '      The data required to run the program are:
3860 '      (1) Flow data
3870 '      (2) The number of departments and their areas
3880 '      (3) Scale
3890 '      (4) Placement priority, if not the same for all
3900 '          departments.
3910 '
3920 '4.2   The Output
3930 '      PLANET gives the output, a listing of departmental
3940 'areas and their priorities, normalized flow data and
3950 'normalized flow-between cost chart, and the resulting
3960 'layout with order of placement.
3970 '
3980 '4.3   Input Lines
3990 '      Three types of lines are required to run PLANET.
4000 '
4010 'Line type 1:  only one line is required.
4020 'Col.            Format              Description
4030 '1               I1          Identification code 1 must be
                                  entered in this column
4050 '2-3             I2          day of the month of running PLANET
4060 '4-6             A3          month, 3 letter abreviation
4070 '7-8             I2          year, last 2 digits
4080 '9-50            7A6         Headings to be printed on the
                                  output
4090 '51-52           I2          Number of departments
4100 '53-60           F8.0        scale, number of square feet in a
4110 '                           unit square
4120 '61-62           A2          blank
4130 '63-64           I2          Code identifying the type of
4140 '                           materials flow data:
4150 '                              01: part list
4160 '                              02: from-to chart
4170 '                              03: relationship chart
4180 '65-66           I2          This field is used if selection
4190 '                           method A is used:
4200 '                              00: do not utilize method A
4210 '                              01: use A, and print final
                                      layout
4220 '                              02: use A, and print layout
                                      after each department is
                                      placed
4240 '67-68           I2          The same as 65-66, but for B
4250 '69-70           I2          The same as 65-66, but for C
4260 '
4270 'Line type 2:  The number of lines required here is equal to
4280 'the number of departments.
4290 'Col.            Format              Description
4300 '1               I2          Identification code 2 must be
4310 '                           entered
4320 '2-3             A2          Department identifier in alphabetic
4330 '                           character
4340 '4-11            F8.0        The area of the department
                                  specified in Col. 2-3
4360 '12-13           I2          Placement priority
4370 '14-79           I1A6        Optional, any description  or
                                  user's comment.
4390 '
4400 'Line type 3:  This line is used to input the flow data.
```

```
4410 'Depending on the type of flow data used, only one of the
4420 'following methods should be used.
4430 'I.   Flow data is represented as parts list.  The number of
4440 '     lines is equal to the number of parts + 1.
4450 'Col.           Format                       Description
4460 '1               I1           Identification code 3
4470 '2-5             A4           Part identification code
4480 '6-7             I2           the number of departments the part
4490 '                             will flow through, maximum number
4500 '                             is 30.
4510 '8-10            I3           Frequency of movement of the part,
4520 '                             per unit of time
4530 '11-20           F10.0        Cost per move per 100 feet
4540 '21-80           30A2         Each 2 col. represents the cost of
4550 '                             a department through which the part
4560 '                             will flow.
4570 '
4580 'After the last line there should be a line with the number
4590 '3 in Col. 1, and 99 in Col. 6-7 to signal the end of the
4600 'parts.
4610 '
4620 'II.   From-to chart.  The number of lines required here is
4630 '      equal to the number of departments.  Each line
4640 '      corrosponds to a row in the from-to chart.  A second
4650 '      line can be used if needed.
4660 'Col.           Format                       Description
4670 '1               I1           Identification code 4
4680 '2-3             A2           Department identifier
4690 '4-5             I2           If each row in the from-to chart
4700 '                             occupies one line then put '01' in
4710 '                             this field.  If a second line is
4720 '                             needed then put 02 on Col. 4-5 of
4730 '                             the second line, if a third line is
4740 '                             needed, then put 03, ..., etc.
4750 '6-80            I5F5.0       Elements of the from-to chart.  If
4760 '                             a second line is needed, it should
4770 '                             start on Col. 6.
4780 '
4790 'III.  Relationship Chart.  The number of lines required
4800 '      here is equal to the number of departments.  The user
4810 '      should assign numbers to the closeness ratings and
4820 '      form a score matrix.  These numbers are the ones to
4830 '      be entered as data.  If needed, a second line can be
4840 '      used.
4850 'Col.           Format                       Description
4860 '1               I1           Identification code 5
4870 '2-3             A2           Department identifier
4880 '4               I1           If each row in the score matrix
4890 '                             occupies one line, put '1' in this
4900 '                             col.  If a second line is needed,
4910 '                             put '2' in col. 4 of the second
4920 '                             line.
4930 '5-80            38F.0        The elements of the score matrix
4940 '                             are inserted here.  If a second
4950 '                             line is needed, it should start at
4960 '                             Col. 5.
4970 '
```

```
1   '*************************************************************
2   '*                                                          *
5   '*                   FINANCIAL ANALYSIS                     *
7   '*                      Chapter 15                          *
8   '*                                                          *
9   '*************************************************************
10  '
15  ' *** This program performs the detailed financial analysis
for the income
20  ' statement ***
30     DIM A2(5),A3(5),A4(5),A9(5)
40     DIM A6(3,5),A7(8,5),A8(3,5),B$(8)
45     DIM C1(5),C2(5),C3(5)
50     DATA "Volume(units)        ","Selling Price($)     ","Total
Sales($)      "
60     DATA "* Raw Materials      ","* Purchased Parts    ","* Hourly
Employees    ",           "- Salaried Employees","- Insurance
","* Maintenance        ",           "* Utilities          ","-
Miscellaneous       "
70     DATA "Production Equipment",7,"Tool Room Equipment ",7,"Mech
Handling Equip ",7,          "Aux Handling Equip   ",12
80     DATA "Office Equipment     ",10,"Building             ",30
90     DATA "Loan Balance         ","Annual Payment       ","Principle
Paid        ",           "Interest On B.C.     ","Amount Of W.C.
","Interest On W.C.     "
100    DATA "Gross Tax Income     ","Loss Carry Forward   ","Net Tax
Income       ",           "Gross Tax            ","Investment Credit
","Net Tax              "
110    DATA "Sales Revenues       ","Operating Expenses   ","Gross
Profit       ",           "Depreciation         ","Interest
","Gross Tax Income     ",           "Net Tax              ","Net
Income       "
120    DATA "Net Income           ","Add Depreciation     ","Net Cash
Flow        "
130    CLS : LOCATE 11,11 : PRINT "financial analysis": PRINT :
PRINT
140    PRINT TAB(12) "by Shamik Bafna"
150    LOCATE 23,12 : PRINT "What is the name of your company? ";
160 LINE INPUT A$
170    Z = 40 - INT (LEN (A$) / 2)
180    CLS : LOCATE 7,15 : PRINT "important": PRINT : PRINT "1.turn
the epson printer off"
190    PRINT : PRINT "2.make sure the paper is torn at": PRINT "
the perforations"
200    PRINT : PRINT "3.rotate the roller to move the top": PRINT "
edge of the page in line with the": PRINT " upper most part of
the paper support": PRINT " wire rack"
210    PRINT : PRINT "4.turn the printer power switch on"
220    PRINT : PRINT "5.press return to continue";: INPUT X$
230    GOSUB 290: GOSUB 1130: GOSUB 2730: GOSUB 2030: GOSUB 3400:
GOSUB 3950
240    GOSUB 4560
250    END
260    ' *********************
270    ' *    Total Revenues    *
280    ' *********************
290    CLS
300    PRINT "how many products";
310    INPUT A1
313 A1=3
```

```
320    PRINT
330    FOR J = 1 TO A1
340    PRINT "plant capacity for product ";J;
350    INPUT A2(J)
360    PRINT
370    PRINT "first year selling price for ";J;
380    INPUT A3(J)
390    PRINT
400    PRINT
410    PRINT
420    NEXT J
430    GOSUB 4500
440    IF LEFT$ (Z1$,1) = "y" THEN 290
450    FOR I = 1 TO 3
460    READ B$(I)
470    NEXT I
480    CLS
490    FOR J = 1 TO 5
500    PRINT "percent of capacity for year ";J;
510    INPUT A4(J)
520    PRINT
530    A9(J) = 0
540    NEXT J
550    PRINT : PRINT
560    PRINT "average rate (percent) of price"
570    PRINT "increase per year";
580    INPUT A5
590    GOSUB 4500
600    IF LEFT$ (Z1$,1) = "y" THEN 480
610    FOR J = 1 TO A1
620    A7(J,1) = A3(J)
630    FOR I = 1 TO 5
640    A6(J,I) = A2(J) * (A4(I) / 100)
650    IF I = 1 THEN   GOTO 670
660    A7(J,I) = A7(J,I - 1) + (A7(J,I - 1) * (A5 / 100!))
670    A8(J,I) = A6(J,I) * A7(J,I)
680    A9(I) = A9(I) + A8(J,I)
690    NEXT I
700    NEXT J
710    GOSUB 4370
720    LPRINT  TAB( 33);"Sales Revenues"
730    LPRINT  TAB( 33);"--------------"
740    GOSUB 1540
750    LPRINT  TAB( 6);"Percent Of Capacity ";
760    FOR J = 1 TO 5
770    T9 = A4(J)
780    GOSUB 1800
790    NEXT J
800    LPRINT
810    LPRINT
820    FOR J = 1 TO A1
830    LPRINT  TAB( 6);"Product ";J;":"
840    LPRINT  TAB( 6);"----------"
850    FOR N = 1 TO 3
860    LPRINT  TAB( 6);B$(N);
870    FOR I = 1 TO 5
880    IF N = 1 THEN T9 =   INT (A6(J,I) + .5)
890    IF N = 2 THEN T9 =   INT (A7(J,I) * 100! + .5) / 100!
900    IF N = 3 THEN T9 =   INT (A8(J,I) + .5)
```

```
910   GOSUB 1800
920   NEXT I
930   LPRINT
940   LPRINT
950   NEXT N
960   LPRINT
970   NEXT J
980   LPRINT
990   LPRINT   TAB( 6);"Total Revenue($)      ";
1000  FOR I = 1 TO 5
1010  T9 =   INT (A9(I) + .5)
1020  GOSUB 1800
1030  NEXT I
1040  LPRINT
1050  LPRINT
1060  LPRINT
1070  LPRINT   TAB( 11);"Note:  1.An Annual Selling Price Increase
Of  ";A5;"% Is Assumed"
1080  FOR I = 2 TO A1 + 1
1090  LPRINT   TAB( 17);I;".Installed Capacity For Product ";I - 1;"
Equals ";A2(I - 1);" Units": LPRINT   TAB( 19);"Per Year"
1100  NEXT I
1110  GOSUB 4470
1120  RETURN
1130  ' ***********************
1140  ' * Operating Expenses   *
1150  ' ***********************
1160  FOR I = 1 TO 8
1170  READ B$(I)
1180  NEXT I
1190  CLS
1200  PRINT "give the following costs for": PRINT "the first year:"
1210  PRINT
1220  FOR I = 1 TO 8
1230  PRINT   RIGHT$ (B$(I), LEN (B$(I)) - 2);
1240  INPUT A7(I,1)
1250  PRINT
1260  NEXT I
1270  PRINT "average annual cost increase (%)";
1280  INPUT A5
1290  GOSUB 4500
1300  IF   LEFT$ (Z1$,1) = "y" THEN 1190
1310  FOR I = 1 TO 5
1320  A3(I) = 0
1330  NEXT I
1340  FOR J = 1 TO 8
1350  FOR I = 1 TO 5
1360  ON J GOTO 1370,1370,1370,1400,1400,1370,1370,1400
1370  A7(J,I) = (A7(J,1) / A4(1)) * A4(I) * (1 + (A5 / 100!)) ^ (I
- 1)
1380  A3(I) = A3(I) + A7(J,I)
1390  GOTO 1420
1400  A7(J,I) = A7(J,1) * (1 + (A5 / 100!)) ^ (I - 1)
1410  A3(I) = A3(I) + A7(J,I)
1420  NEXT I
1430  NEXT J
1440  FOR I = 1 TO 5
1450  A3(I) =   INT (A3(I) + .5)
1460  A2(I) = A9(I) - A3(I)
```

```
1470 NEXT I
1480 GOSUB 4370
1490 CLS
1500 LPRINT  TAB( 31);"Operating Expenses"
1510 LPRINT  TAB( 31);"------------------"
1520 GOSUB 1540
1530 GOTO 1640
1540 ' *******************
1550 ' *  table heading  *
1560 ' *******************
1570 LPRINT
1580 LPRINT
1590 LPRINT  TAB( 6);"Item                            Year 1     Year 2
Year 3     Year 4     Year 5"
1600 LPRINT  TAB( 6);"---------------------------------------------
------------------------"
1610 LPRINT "   ";
1620 LPRINT
1630 RETURN
1640 FOR J = 1 TO 8
1650 GOSUB 1710
1660 NEXT J
1670 GOTO 1890
1680 ' *******************
1690 ' *   prints table  *
1700 ' *******************
1710 LPRINT  TAB( 6);B$(J);
1720 FOR I = 1 TO 5
1730 T9 =  INT (A7(J,I) + .5)
1740 GOSUB 1800
1750 NEXT I
1760 LPRINT "   ";
1770 LPRINT
1780 LPRINT
1790 RETURN
1800 ' ***********************
1810 ' * right-justification *
1820 ' ***********************
1830 T1$ = "                    "
1840 T9$ =  STR$ (T9)
1850 T8 =  LEN (T9$)
1860 T7 = 10 - T8
1870 LPRINT  LEFT$ (T1$,T7);T9$;
1880 RETURN
1890 LPRINT  TAB( 28);"--------  --------  --------  --------   ---
-----"
1900 LPRINT
1910 LPRINT  TAB( 6);"Total Expenses      ";
1920 FOR I = 1 TO 5
1930 T9 = A3(I)
1940 GOSUB 1800
1950 NEXT I
1960 LPRINT
1970 LPRINT
1980 LPRINT
1990 LPRINT  TAB( 11);"Note: 1.An Annual Cost Increase Of ";A5;"%
Is Assumed"
2000 LPRINT  TAB( 17);"2.Asterisks (*) Indicate Directly Variable
Costs.": LPRINT  TAB( 19);"Dashes (-) Indicate Fixed Costs."
2010 GOSUB 4470
```

```
2020 RETURN
2030 ' ********************
2040 ' *    interest       *
2050 ' ********************
2060 CLS
2070 PRINT "what percent of capital is borrowed";
2080 INPUT B1
2090 PRINT
2100 PRINT "what is the interest rate(%) of": PRINT "borrowed
capital";
2110 INPUT B2
2120 PRINT
2130 PRINT "in how many years is the loan": PRINT "paid off";
2140 INPUT B3
2150 PRINT
2160 PRINT
2170 PRINT "what percent of annual revenues": PRINT "is the
working capital";
2180 INPUT W1
2190 PRINT
2200 PRINT "what is the interest rate on": PRINT "working
capital";
2210 INPUT W2
2220 GOSUB 4500
2230 IF  LEFT$ (Z1$,1) = "y" THEN 2060
2240 T = (1 + B2 / 100!) ^ B3
2250 A7(1,1) = Z5 * (B1 / 100!)
2260 FOR I = 1 TO 5
2270 A4(I) = 0
2280 A7(2,I) = A7(1,1) * ((B2 / 100! * T) / (T - 1))
2290 A7(4,I) = A7(1,I) * (B2 / 100!)
2300 A7(3,I) = A7(2,I) - A7(4,I)
2310 IF I = 5 THEN  GOTO 2330
2320 A7(1,I + 1) = A7(1,I) - A7(3,I)
2330 A7(5,I) = W1 / 100 * A9(I)
2340 A7(6,I) = A7(5,I) * (W2 / 100!)
2350 A4(I) = A7(6,I) + A7(4,I)
2360 NEXT I
2370 FOR I = 1 TO 6
2380 READ B$(I)
2390 NEXT I
2400 GOSUB 4370
2410 LPRINT  TAB( 30);"Interest Calculations"
2420 LPRINT  TAB( 30);"--------------------"
2430 GOSUB 1540
2440 LPRINT  TAB( 6);"Borrowed Capital"
2450 LPRINT  TAB( 6);"----------------"
2460 FOR J = 1 TO 4
2470 GOSUB 1710
2480 NEXT J
2490 LPRINT
2500 LPRINT  TAB( 6);"Working Capital"
2510 LPRINT  TAB( 6);"---------------"
2520 FOR J = 5 TO 6
2530 GOSUB 1710
2540 NEXT J
2550 LPRINT
2560 LPRINT
2570 LPRINT  TAB( 6);"Total Interest      ";
2580 FOR I = 1 TO 5
```

```
2590 T9 =  INT (A4(I) + .5)
2600 GOSUB 1800
2610 C1(I) = A2(I) - A5 - A4(I)
2620 NEXT I
2630 LPRINT
2640 LPRINT
2650 LPRINT
2660 LPRINT    TAB( 11);"Note: 1.Loan Is ";B1;"% Of Total Capital"
2670 LPRINT    TAB( 11);"      2.Capital Is Borrowed At ";B2;"%  Per
Year"
2680 LPRINT    TAB( 17);"3.Loan Is Paid Off In Equal Annual
Installments In ";B3;" Years"
2690 LPRINT    TAB( 11);"      4.Working Capital Is Assumed To Be
";W1;"% Of Annual Revenues"
2700 LPRINT    TAB( 11);"      5.Interest On Working Capital Is
Charge At ";W2;"% Per Year"
2710 GOSUB 4470
2720 RETURN
2730 ' *******************
2740 ' *   depreciation *
2750 ' *******************
2760 FOR I = 1 TO 6
2770 READ B$(I)
2780 READ A7(I,3)
2790 NEXT I
2800 CLS
2810 A7(7,1) = 0:A7(7,4) = 0:A7(7,5) = 0
2820 C5 = 0
2830 PRINT "give the following values:"
2840 PRINT
2850 FOR J = 1 TO 6
2860 PRINT B$(J);" cost";
2870 INPUT A7(J,1)
2880 PRINT "salvage value";
2890 INPUT A7(J,2)
2900 PRINT
2910 A7(7,1) = A7(7,1) + A7(J,1)
2920 A7(J,4) = (A7(J,1) - A7(J,2)) / A7(J,3)
2930 A7(J,5) = A7(J,1) - (5 * A7(J,4))
2940 A7(7,4) = A7(7,4) + A7(J,4)
2950 A7(7,5) = A7(7,5) + A7(J,5)
2960 IF (3 <  = A7(J,3)) AND (A7(J,3) < 5) THEN T = (1 / 30)
2970 IF (5 <  = A7(J,3)) AND (A7(J,3) < 7) THEN T = (1 / 15)
2980 IF 7 <  = A7(J,3) THEN T = .1
2990 C5=0
3000 NEXT J
3010 PRINT "land";
3020 INPUT P
3030 GOSUB 4500
3040 IF  LEFT$ (Z1$,1) = "y" THEN 2800
3050 A7(7,1) = A7(7,1) + P
3060 A7(7,5) = A7(7,5) + P
3070 Z5 = A7(7,1)
3080 Z6 = A7(7,5)
3090 GOSUB 4370
3100 LPRINT    TAB( 28);"Depreciation Calculations"
3110 LPRINT    TAB( 28);"-------------------------"
3120 LPRINT
3130 LPRINT
```

```
3140 LPRINT   TAB( 31);"First    Salvage     Tax-     Annual  Bk
Value"
3150 LPRINT   TAB( 6);"Item"; TAB( 31);"Cost     Value          Life
Deprec  Sixth Yr"
3160 LPRINT   TAB( 6);"---------------------------------------------
------------------------"
3170 LPRINT
3180 FOR J = 1 TO 6
3190 GOSUB 1710
3200 NEXT J
3210 LPRINT   TAB( 6);"Land                    ";
3220 T9 = P: GOSUB 1800
3230 LPRINT   TAB( 65);"0"
3240 LPRINT
3250 LPRINT
3260 LPRINT   TAB( 6);"Total                   ";
3270 T9 =  INT (A7(7,1) + .5): GOSUB 1800
3280 LPRINT   TAB( 56);
3290 FOR I = 4 TO 5
3300 T9 =  INT (A7(7,I) + .5)
3310 GOSUB 1800
3320 NEXT I
3330 LPRINT
3340 A5 = A7(7,4)
3350 LPRINT
3360 LPRINT
3370 LPRINT   TAB( 11);"Note: Straight Line Depreciation Is Assumed
In All Cases"
3380 GOSUB 4470
3390 RETURN
3400 ' ********************
3410 ' *      net tax      *
3420 ' ********************
3430 A7(2,1) = 0:A7(5,1) = C5
3440 FOR I = 1 TO 5
3450 A7(1,I) = C1(I)
3460 IF A7(2,1) < A7(1,I) THEN  GOTO 3550
3470 IF I = 5 THEN 3490
3480 A7(2,I + 1) = A7(2,I) - A7(1,I)
3490 A7(3,I) = 0
3500 A7(4,I) = 0
3510 A7(6,I) = 0
3520 IF I = 5 THEN 3540
3530 A7(5,I + 1) = A7(5,I)
3540 GOTO 3710
3550 A7(3,I) = A7(1,I) - A7(2,I)
3560 IF I = 5 THEN 3580
3570 A7(2,I + 1) = 0
3580 IF (0 <  = A7(3,I)) AND (A7(3,I) <  = 50000!)  THEN A7(4,I) =
3590 IF (50000! < A7(3,I)) AND (A7(3,I) <  = 75000!)  THEN A7(4,I)
= 7500 + .25 * (A7(3,I) - 50000!)
3600 IF (75000! < A7(3,I)) AND (A7(3,I) <  = 100000!)   THEN
A7(4,I) = 20000 + .34 * (A7(3,I) - 75000!)
3610 IF (100000! < A7(3,I)) THEN ABC=.05*(A7(3,I)-100000!)
3615 IF (ABC < 11750) THEN A7(4,I)=28500+.34*(A7(3,I)-100000!)
3620 IF (ABC > 11750) THEN A7(4,I)=.34*A7(3,I)
3630 IF A7(5,I) < A7(4,I)  THEN  GOTO 3680
3640 IF I = 5 THEN  GOTO 3660
3650 A7(5,I + 1) = A7(5,I) - A7(4,I)
```

```
3660 A7(6,I) = 0
3670 GOTO 3710
3680 IF I = 5 THEN 3700
3690 A7(5,I + 1) = 0
3700 A7(6,I) = A7(4,I) - A7(5,I)
3710 C2(I) = A7(6,I)
3720 C3(I) = C1(I) - C2(I)
3730 NEXT I
3740 FOR I = 1 TO 6
3750 READ B$(I)
3760 NEXT I
3770 GOSUB 4370
3780 LPRINT  TAB( 32);"Tax Computation"
3790 LPRINT  TAB( 32);"---------------"
3800 GOSUB 1540
3810 FOR J = 1 TO 2
3820 GOSUB 1710
3830 NEXT J
3840 LPRINT  TAB( 26);"---------------------------------------------
--------"
3850 LPRINT
3860 FOR J = 3 TO 4
3870 GOSUB 1710
3880 NEXT J
3890 LPRINT  TAB( 26);"---------------------------------------------
--------"
3900 LPRINT
3910 J = 6
3920 GOSUB 1710
3930 GOSUB 4470
3940 RETURN
3950 ' **********************
3960 ' *    income statement  *
3970 ' **********************
3980 FOR I = 1 TO 5
3990 A7(1,I) = A9(I)
4000 A7(2,I) = A3(I)
4010 A7(3,I) = A2(I)
4020 A7(4,I) = A5
4030 A7(5,I) = A4(I)
4040 A7(6,I) = C1(I)
4050 A7(7,I) = C2(I)
4060 A7(8,I) = C3(I)
4070 NEXT I
4080 FOR I = 1 TO 8
4090 READ B$(I)
4100 NEXT I
4110 GOSUB 4370
4120 LPRINT  TAB( 32);"Income Statement"
4130 LPRINT  TAB( 32);"----------------"
4140 GOSUB 1540
4150 FOR J = 1 TO 2
4160 GOSUB 1710
4170 NEXT J
4180 LPRINT  TAB( 26);"---------------------------------------------
--------"
4190 LPRINT
4200 FOR J = 3 TO 5
4210 GOSUB 1710
4220 NEXT J
```

```
4230 LPRINT  TAB( 26);"----------------------------------------------
--------"
4240 LPRINT
4250 FOR J = 6 TO 7
4260 GOSUB 1710
4270 NEXT J
4280 LPRINT  TAB( 26);"----------------------------------------------
--------"
4290 LPRINT
4300 J = 8
4310 GOSUB 1710
4320 GOSUB 4470
4330 RETURN
4340 ' ********************
4350 ' *    pr#1 & pr#0    *
4360 ' ********************
4370 '
4390 FOR K = 1 TO 7: LPRINT : NEXT K
4400 LPRINT  TAB( Z);A$
4410 LPRINT  TAB( Z);
4420 FOR I = 1 TO LEN (A$)
4430 LPRINT "-";
4440 NEXT I
4450 LPRINT : LPRINT : LPRINT : LPRINT
4460 RETURN
4470 LPRINT  CHR$ (12)
4480 '
4490 RETURN
4500 ' ********************
4510 ' *      abort input    *
4520 ' ********************
4530 LOCATE 23,11 : PRINT "do you want to change the above data";
4540 INPUT Z1$
4550 RETURN
4560 ' ********************
4570 ' *      cash flow    *
4580 ' ********************
4590 FOR I = 1 TO 3
4600 READ B$(I)
4610 NEXT I
4620 FOR I = 1 TO 5
4630 A7(1,I) = A7(8,I)
4640 A7(2,I) = A7(4,I)
4650 A7(3,I) = A7(1,I) + A7(2,I)
4660 NEXT I
4670 I = 1!
4680 P1 = 0!
4690 P2 = 0!
4700 T1 = 1 + I / 100!
4710 FOR J = 1 TO 4
4720 P1 = P1 + A7(3,J) / T1 ^ J
4730 NEXT J
4740 P1 = P1 + (A7(3,5) + Z6) / T1 ^ 5
4750 IF P1 < = Z5 THEN 4790
4760 P2 = P1
4770 I = I + 1
4780 GOTO 4680
4790 IF P1 = Z5 THEN 4810
4800 I = I - ((Z5 - P1) / (P2 - P1))
4810 I1 =  INT (I * 100! + .5) / 100!
```

```
4820 P1 = 0!
4830 P2 = 0!
4840 FOR I = 1 TO 5
4850 P1 = P1 + A7(3,I)
4860 IF P1 >  = Z5 THEN 4900
4870 IF I = 5 THEN 4940
4880 P2 = P1
4890 NEXT I
4900 IF P1 = Z5 THEN 4920
4910 I = I -((P1 - Z5) / (P1 - P2))
4920 I2 =  INT (I * 100! + .5) / 100!
4930 GOTO 4950
4940 I2 = 6
4950 GOSUB 4370
4960 LPRINT  TAB( 29);"Earnings On Investment"
4970 LPRINT  TAB( 29);"---------------------"
4980 GOSUB 1540
4990 FOR J = 1 TO 2
5000 GOSUB 1710
5010 NEXT J
5020 LPRINT  TAB( 26);"-----------------------------------------
--------"
5030 LPRINT
5040  J = 3
5050 GOSUB 1710
5060 LPRINT : LPRINT : LPRINT : LPRINT
5070 LPRINT  TAB( 6);"Return On Investment Over A 5-Year Period Is
";I1;"%"
5080 LPRINT : LPRINT
5090 IF I2 = 6 THEN 5120
5100 LPRINT  TAB( 6);"Payback On Investment = ";I2;" Years"
5110 GOTO 5130
5120 LPRINT  TAB( 6);"Payback On Investment Is More Than 5 Years"
5130 GOSUB 4470
5140 RETURN
```

Index